Optical Bistability:
Controlling Light with Light

39.50
phy
T

D1738310

Academic Press Rapid Manuscript Reproduction

QUANTUM ELECTRONICS — PRINCIPLES AND APPLICATIONS

A Series of Monographs

EDITED BY

PAUL F. LIAO
Bell Telephone Laboratories
Murray Hill, New Jersey

PAUL KELLEY
Lincoln Laboratory
Massachusetts Institute of Technology
Lexington, Massachusetts

A list of books in this series is available from the publisher on request.

Optical Bistability: Controlling Light with Light

Hyatt M. Gibbs

Optical Sciences Center
University of Arizona
Tucson, Arizona

1985

ACADEMIC PRESS, INC.
Harcourt Brace Jovanovich, Publishers
Orlando San Diego New York Austin
London Montreal Sydney Tokyo Toronto

COPYRIGHT © 1985 BY ACADEMIC PRESS, INC.
ALL RIGHTS RESERVED.
NO PART OF THIS PUBLICATION MAY BE REPRODUCED OR TRANSMITTED IN ANY FORM OR BY ANY MEANS, ELECTRONIC OR MECHANICAL, INCLUDING PHOTOCOPY, RECORDING, OR ANY INFORMATION STORAGE AND RETRIEVAL SYSTEM, WITHOUT PERMISSION IN WRITING FROM THE PUBLISHER.

ACADEMIC PRESS, INC.
Orlando, Florida 32887

United Kingdom Edition published by
ACADEMIC PRESS INC. (LONDON) LTD.
24-28 Oval Road, London NW1 7DX

LIBRARY OF CONGRESS CATALOGING-IN-PUBLICATION DATA

Gibbs, Hyatt M.
Optical bistability.

Bibliography: p.
Includes index.
1. Optical bistability. I. Title.
QC446.3.065G55 1985 535'.2 85-48069
ISBN 0-12-281940-3 (alk. paper)

PRINTED IN THE UNITED STATES OF AMERICA

85 86 87 88 9 8 7 6 5 4 3 2 1

To

Pip and Maa

Lethia, Alex, and Vanetta

and all of my bistability friends

CONTENTS

PREFACE

This is a research book intended for new entrants and active workers in the field of optical bistability. The treatment interprets optical bistability broadly to include all of the steady-state and transient characteristics of nonlinear optical systems which exhibit bistability under some operating conditions. It is restrictive in placing the emphasis on passive (non-laser) systems which exhibit reversible bistability with input intensity as the hysteresis variable. The book is motivated by the desire to summarize the beauty of the physics and to describe the potential applications of such systems for nonlinear optical signal processing.

This book may even be useful as a reference for advanced specialized courses. The first draft was written in the fall of 1981 as the author taught what may have been the only course on optical bistability. Parts of it were used again in the spring of 1984 as part of a course on the semiclassical description of coherent optical phenomena.

The history of optical bistability in lasers and passive systems is summarized in Chapter 1. Steady-state theories of optical bistability are presented in Chapter 2. Both intrinsic and hybrid experiments are described in some detail in Chapters 3 and 4, respectively. Light-by-light control is the focus of Chapter 5 which treats pulse reshaping and external switching. In contrast, the transient phenomena of Chapter 6 occur with a steady input, i.e., they are intrinsic instabilities. Considerations important for applications to optical signal processing and computing are discussed in Chapter 7.

This is a research book about a very rapidly expanding field. This fact has made it difficult to decide when to publish. Much of the underlying physics of the behavior of these nonlinear optical systems is now understood, even though intense studies on instabilities, transverse effects, and nonlinearity mechanisms will surely result in many new breakthroughs. Work on applications to very-high-speed switching and to massively parallel processing has just begun. Hopefully this book captures the beautiful and diverse behavior of nonlinear optical systems in a manner that will serve both as a basis for further physics research and as a ready reference for nonlinear optical signal processing.

The author has had many rewarding collaborations in twelve years of working on optical bistability. Thanks to Sam McCall for introducing the possibility of seeing bistability to me and convincing me that collaborating with him was essential to avoid a catastrophic explosion during one of his sodium-oven cleaning operations! Thanks to Dick Slusher for earlier laser technology transfer and fruitful collaborations on coherent optical phenomena. To Sam, Venky Venkatesan, George Churchill, and Al Passner for sharing the excitement of those first Bell Labs experiments in Na, ruby, and GaAs; somehow 2 a.m. data are special! And without the beautiful GaAs samples of Art Gossard and Bill Wiegmann, I would have never made the gas-to-solid transition. In 1980 the Justice Department frightened me into establishing "Bell Labs West" at the Optical Sciences Center, resulting in a greatly expanded effort with the help of a fantastic group of graduate students: Shin-Sheng Tarng, Jack Jewell, Ed Watson, David Kaplan, Doreen

Weinberger, Mike Rushford, Kuo-Chou Tai, Shlomo Ovadia, Lois Hoffer, Matt Derstine, Mial Warren, Yong Lee, Greg Olbright, Arturo Chavez, John Valley, Hans Kulcke, George Gigioli, Lon Wong, Ruxiang Jin. Thanks to Professors Fred Hopf, Jerry Moloney, Dror Sarid, and Rick Shoemaker for bistability collaborations and to them, George Stegeman, and many others for helping in the formation of the Optical Circuitry Cooperative. A special thanks to Assistant Professor Nasser Peyghambarian for his hard work and close friendship during the last three years.

Although Sam McCall never found time to coauthor this book as first planned, much of what I know about bistability has come from him and I hope I have captured much of his insight. Thanks to David Holm and Jerry Moloney for writing Appendices F and G, respectively. Thanks to Jeannette Gerl and Lisa DuBois for early versions of the book and to Kathy Seeley and Norma Laguna for the tedious and seemingly endless preparation of the camera-ready manuscript. Thanks for helpful comments on the manuscript from Professors Howard Carmichael, Luigi Lugiato, Pierre Meystre, Des Smith, and Doreen Weinberger and from Drs. Martine LeBerre, Elisabeth Ressayre, and Andrée Tallet.

CHAPTER 1

INTRODUCTION TO OPTICAL BISTABILITY

Optical bistability is a rapidly expanding field of current research because of its potential application to all-optical logic and because of the interesting phenomena it encompasses. Since the first observation of optical bistability in a passive, unexcited medium of sodium (Na) vapor in 1974 (McCall, Gibbs, Churchill, and Venkatesan, 1975), bistability has been observed in many different materials including tiny semiconductor etalons. Current applied research is focused on optimizing these devices by decreasing their size, switching times, and operating power, and operating them at room temperature. Both improved nonlinear materials and more efficient device configurations are being sought. Current fundamental research centers on the interesting physical behavior of simple bistable systems. Many bistable devices consist of a nonlinear medium within an optical resonator, just as do lasers, except the passive bistable devices are excited only by the incident coherent light. The counterparts of many of the phenomena studied in lasers, such as fluctuations, regenerative pulsations, and optical turbulence, can be observed in passive bistable systems, often under better controlled conditions. Optical bistability in lasers, which was seen prior to passive bistability, is treated briefly in Section 1.3 although it is not the main subject of this book.

1.1. DEFINITION AND TYPES OF OPTICAL BISTABILITY

A system is said to be optically bistable if it has two output states I_T for the same value of the input I_I over some range of input values. Thus a system having the transmission curve of Fig. 1.1-1 is said to be bistable between $I_\downarrow$ and $I_\uparrow$. Such a system is clearly nonlinear, i.e., I_T is not just a multiplicative constant times I_I. In fact, if I_I is between $I_\downarrow$ and $I_\uparrow$, knowing I_I does not reveal I_T. Nonlinearity alone is insufficient to assure bistability. It is feedback that permits the nonlinear transmission to be multivalued, i.e., bistable. It is this restricted definition of optical bistability defined by Fig. 1.1-1, with the nonlinear medium unexcited, that is adopted here. This definition implies that the bistable system can be cycled completely and repeatedly by varying the input intensity. Systems that exhibit hysteresis as a function of some other parameter but not light intensity are not of interest here. This restricted definition rules out "bistable" optical systems that cannot be reset merely by reducing the input intensity, such as a burglar alarm or a card in a laser beam powerful enough to burn through the card. Even an optical damage device that can be restored by irradiation with light of a different wavelength is not in the spirit here of an all-optical completely recyclable passive system.

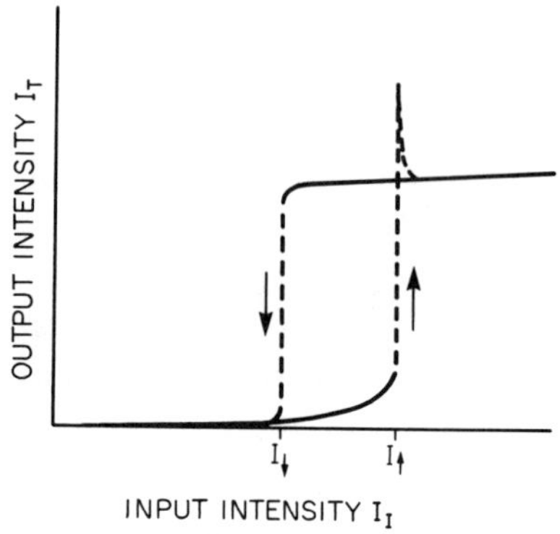

Fig. 1.1-1. Characteristic curve for an optical bistable system.

An example of a system exhibiting optical bistability is a Fabry-Perot interferometer containing a saturable absorber; see Fig. 1.1-2. A simple analysis of such a nonlinear etalon reveals the possibility of bistability. For weak input intensity, I_I, the intracavity absorption spoils the finesse of the cavity even though the laser frequency ν and cavity frequency ν_{FP} of peak transmission are coincident. Therefore the intracavity intensity I_C at $z = 0$ is simply I_I times the input mirror transmission T. At the cell exit

$$I_C(L) = e^{-\alpha L} T I_I \tag{1.1-1}$$

and the transmitted intensity is

$$I_T = e^{-\alpha L} T^2 I_I . \tag{1.1-2}$$

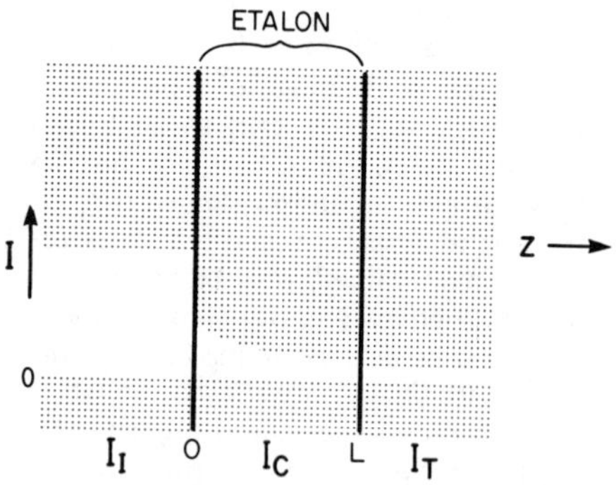

Fig. 1.1-2. Etalon intensities for an intracavity intensity much less than the saturation intensity.

Equation (1.1-2) holds as long as the saturation intensity I_s of the medium is large compared with the intracavity intensity, i.e., if

$$I_s > TI_I \qquad (1.1\text{-}3)$$

is satisfied sufficiently. Figure 1.1-3 depicts the case of strong input intensity in which the medium is bleached, the finesse is high, and the etalon transmits perfectly; i.e., $I_T \simeq I_I$ and $I_C \simeq I_T/T$. This clearly holds for $I_C \gg I_s$, i.e., if

$$I_I > TI_s \qquad (1.1\text{-}4)$$

is satisfied sufficiently. The possibility of bistability is suggested by noting that both Eqs. (1.1-3,4) can be satisfied by the same input intensity. For example, take $I_I = I_s$, then both inequalities require that T be less than 1 as it always is. This physical argument is substantiated by the more rigorous derivation in Section 2.1.

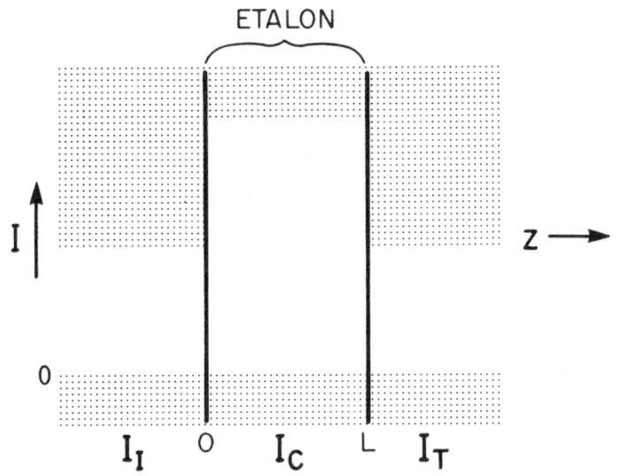

Fig. 1.1-3. Etalon intensities for an intracavity intensity much larger than the saturation intensity.

There are two useful classifications of bistable systems. A system may be absorptive or dispersive, and it may be intrinsic or hybrid. For example, the nonlinear Fabry-Perot interferometer just discussed is an absorptive intrinsic system. A system is absorptive or dispersive depending on whether the feedback occurs by way of an intensity-dependent absorption or refractive index. Clearly this distinction is not sharp, since both absorptive and refractive mechanisms may be significant simultaneously. The distinction between intrinsic (all-optical) and hybrid (mixed optics and electronics) is sharp. In an intrinsic system the intensity dependence arises from a direct interaction of the light with matter. In a hybrid system the intensity dependence arises from an electrical signal from a detector monitoring the transmitted intensity, usually applied to an intracavity phase shifter. Experimental embodiments of intrinsic and hybrid systems are described in Chapters 3 and 4, respectively.

For further reading: a simple introduction to optical bistability is Gibbs, McCall, and Venkatesan (1979) and recent collections of papers are: Bowden, Ciftan, and Robl (1981); Bowden, Gibbs, and McCall (1984); and A. Miller, Smith, and Wherrett (1984). Apart from this book, the most extensive review of optical bistability, both theory and experiment, is Abraham and Smith (1982a). Lugiato (1984) gives a more recent and thorough review of the theory of optical bistability. Goldstone (1984) is a good introduction, especially for dynamic effects.

1.2. OPTICAL LOGIC WITH BISTABLE DEVICES

The transmission of information as signals impressed on light beams traveling through optical fibers is replacing electrical transmission over wires. The low cost and inertness of the basic materials of fibers and the small size and low loss of the finished fibers are important factors in this evolution. Furthermore, for the very fast transmission systems, for example, for transmitting a multiplexed composite of many slow signals, optical pulses are best. This is because it is far easier to generate (Hochstrasser, Kaiser, and Shank, 1980; Shank, Ippen, and Shapiro, 1978) and propagate (Bloom, Mollenauer, Lin, Taylor, and DelGaudio, 1979) picosecond optical pulses than electrical pulses. With optical pulses and optical transmission a reality, the missing component of an all-optical signal processing system is an optical logic element in which one light beam or pulse controls another. The optical bistable systems described in this book have many desirable properties of an all-optical logic element. Hopefully they are the forerunners of tiny, low-energy, subpicosecond, room-temperature devices. The high frequencies of optical electromagnetic radiation give optical devices a potential for subpicosecond switching and room-temperature operation unavailable to Josephson junctions or electronics. The fact that electrical charges are not used or are used only in tiny beam-interacting regions makes an all-optical system much more immune to electromagnetic interference from electrically noisy industrial environments or the electromagnetic pulses from a nuclear explosion. If this book aids and accelerates the understanding and development of such all-optical systems, it will have served its purpose.

Bistable devices have already performed a host of logic functions. Both two-state (Fig. 1.2-1) and many-state (Fig. 1.2-2) optical memories have been demonstrated. The amount of transmitted light reveals the past history of the input light; i.e., the system "remembers" whether or not the input ever exceeded a particular threshold value. By modifying the operating conditions, an optical transistor or transphasor mode of operation is achieved as in Fig. 1.2-3. A small modulation on the input (or on a second signal beam) is amplified. An optical discriminator (Fig. 1.2-4) transmits pulses with intensities above the threshold and suppresses those below. An optical limiter (Fig. 1.2-5) shows little change in the transmitted power. This function could serve to limit the power reaching something or someone or to decrease the percentage noise level. Figure 1.2-6 illustrates optical discrimination in which a signal is faithfully transmitted whenever it exceeds the threshold level. Changes in pulse shape can be accomplished with nonlinear etalons. For example, in Fig. 1.2-7 the etalon initially transmits the light well, but the energy absorbed from the pulse heats the etalon, detunes it from the laser frequency, and terminates the transmission long before the input pulse turns off. Sometimes the transmitted signal oscillates when the input is perfectly constant, resulting in an optical oscillator (Fig. 1.2-8). Or an etalon placed in a continuous wave (cw) beam and initially detuned from the laser frequency can be swept using a pulse from a second laser to gate out a short pulse from the cw beam (Fig. 1.2-9). These examples illustrate that existing bistable optical devices have the desired characteristics of many all-optical operations. Nonetheless, smaller, faster, cheaper, room-temperature, efficient devices are needed.

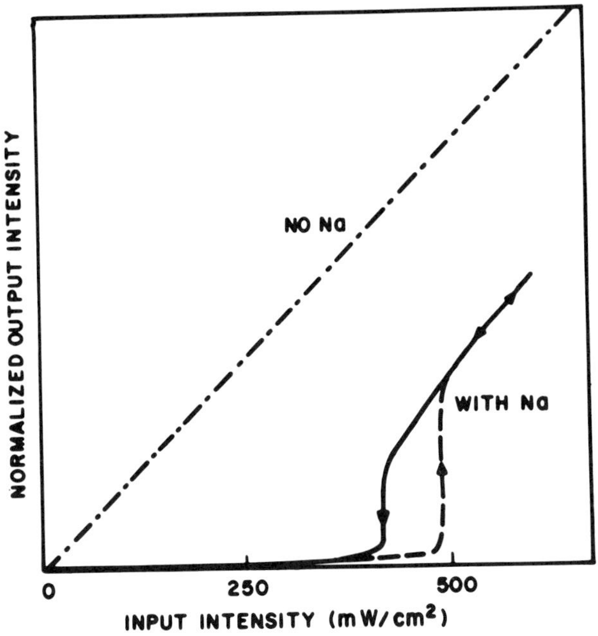

Fig. 1.2-1. Dispersive optical bistability in Na vapor showing stable upper and lower states that could serve as the 1 and 0 states of an optical memory. $\alpha_0 L \simeq 2.75$; ν about 150 MHz above the Na $^2S_{1/2}$, F = 2 to $^2P_{3/2}$ transitions. Gibbs, McCall, Venkatesan (1976).

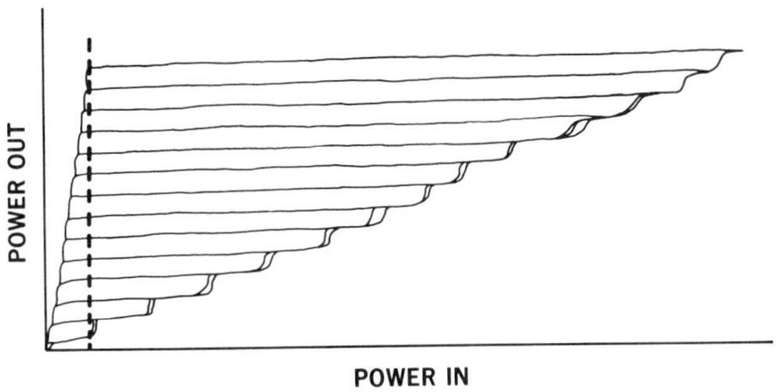

Fig. 1.2-2. Many-state optical memory or analog-to-digital converter in a hybrid bistable optical device. Smith, Turner, Maloney (1978). © 1978 IEEE

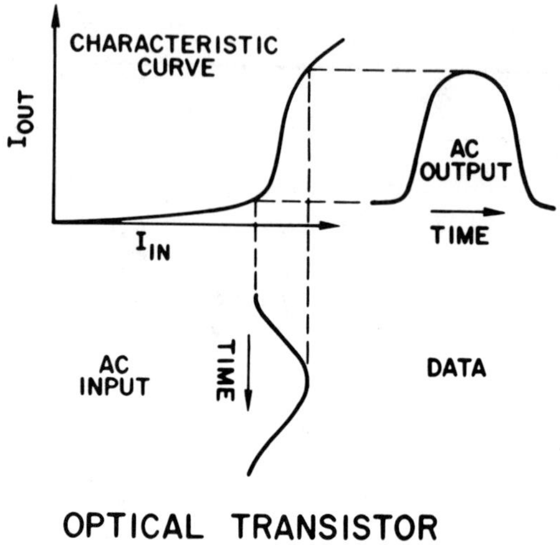

Fig. 1.2-3. Differential gain in Na vapor. Ac gains at 1 kHz exceeding 2 were observed, in spite of only 40% maximum Fabry-Perot transmission. Gibbs, McCall, Venkatesan (1976).

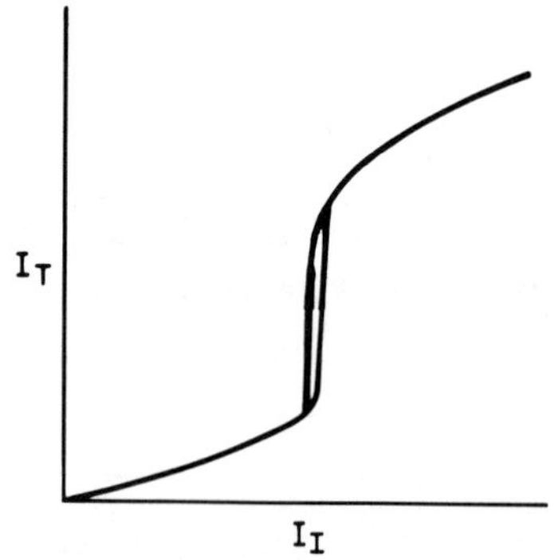

Fig. 1.2-4. Optical discriminator in room-temperature ruby. Venkatesan and McCall (1977a).

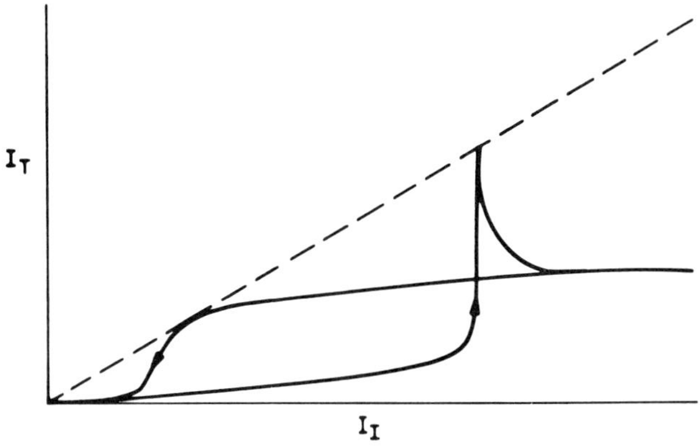

Fig. 1.2-5. Optical bistability and limiter action in a 2-mm-long etalon consisting of a polished Corning 3-142 filter with R = 0.8. In the on state, the output changes very little for a factor-of-4 change in the input. Note the spike in the transmission as the device turns on and the etalon's frequency of peak transmission is swept through the laser frequency. Similar results were obtained in a 60-μm etalon of Corning 4-74 with R = 0.9. McCall and Gibbs (1978).

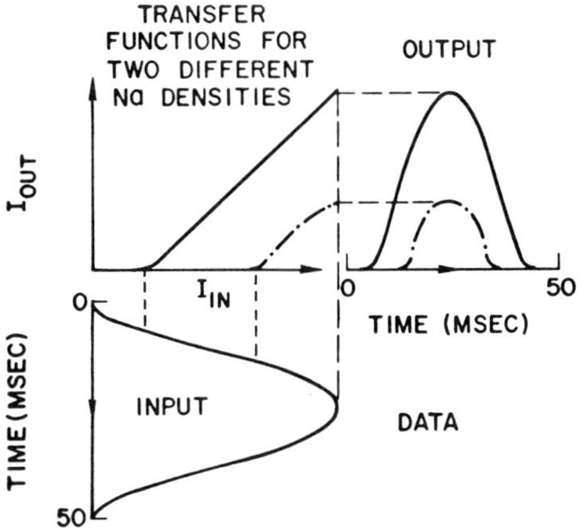

Fig. 1.2-6. Optical discriminator action in Na vapor. Gibbs, McCall, Venkatesan (1976).

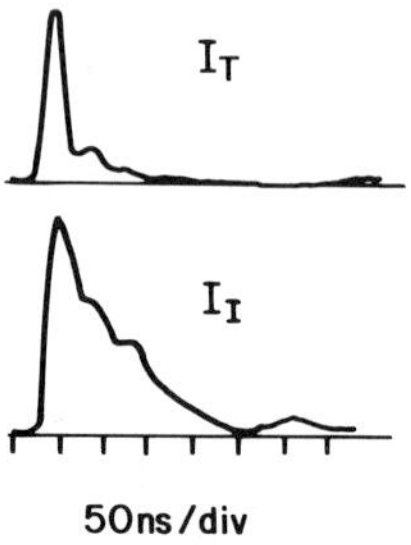

Fig. 1.2-7. Pulse compression by thermal self-sweeping of a gallium arsenide (GaAs) etalon.

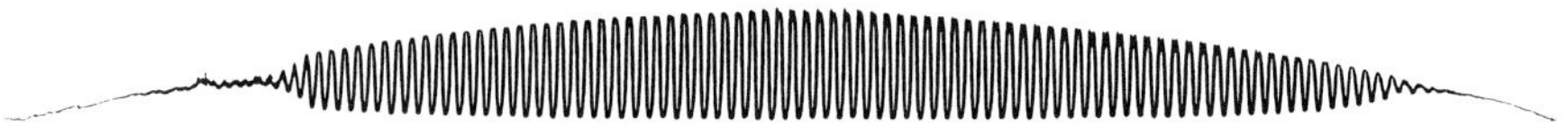

Fig. 1.2-8. Regenerative pulsations at 0.237 Hz in a hybrid bistable device with two feedback time constants of 1/3 and 2/3 s. McCall (1978).

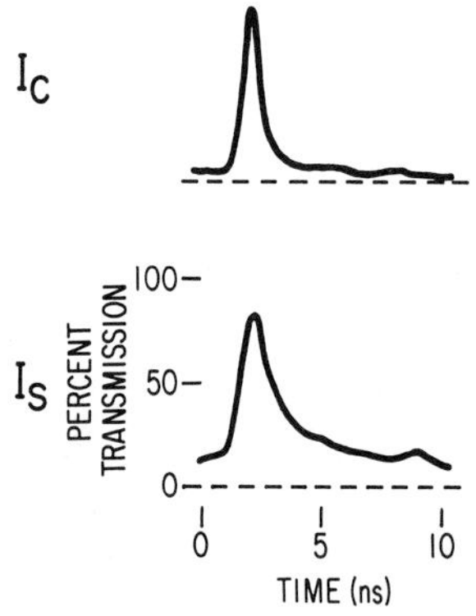

Fig. 1.2-9. Optical gating of a 10-mW cw dye laser signal beam I_S by a 2.5-W 514.5-nm 200-ps mode-locked argon (Ar) laser control pulse I_C. Gibbs, Venkatesan, McCall, Passner, Gossard, and Wiegmann (1979).

Figure 1.2-10 illustrates the use of a single nonlinear Fabry-Perot etalon to perform various optical logic gates, emphasizing that bistable devices need not be operated in a bistable mode to be useful for optical logic. In fact, the idea of Fig. 1.2-10 is to use two input logic pulses followed by a probe pulse. The etalon finesse should be high at the probe wavelength, but the inputs might be at a highly absorbing wavelength. Figure 1.2-10 shows how various logic operations can be performed by appropriate initial detuning between the probe and etalon peak wavelengths. Figure 1.2-11 displays NOR, NAND, XOR, OR, and AND gate operations using a room-temperature multiple-quantum-well etalon consisting of 63 periods of 76-Å GaAs and 81-Å aluminum gallium arsenide (AlGaAs) clad by AlGaAs layers and sandwiched between ≃ 97% reflecting mirrors. The observed responses are similar to the computer simulations. To increase the 82-MHz rate significantly would require a reduction of the carrier lifetime. Related measurements show that the Fabry-Perot peak shifts in ≃ 1 ps, so the logic decision is made that fast [Migus, Antonetti, Hulin, Mysyrowicz, Gibbs, Peyghambarian, and Jewell (1985)]. The device is probably not optimized, and the energy required per input pulse is only 3 pJ.

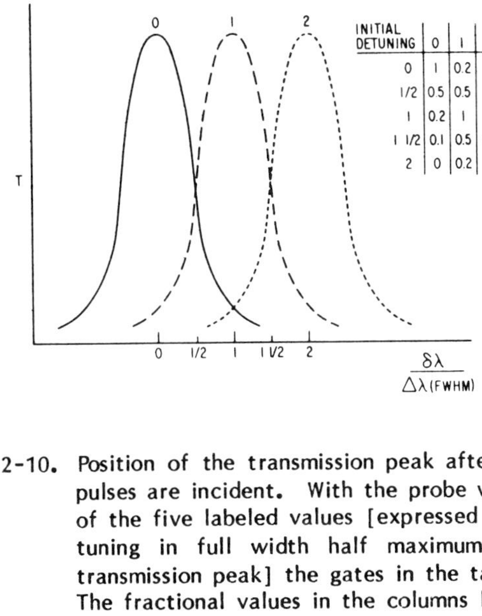

INITIAL DETUNING	0	1	2	T	R
0	1	0.2	0	NOR	OR
1/2	0.5	0.5	0	NAND	(AND)
1	0.2	1	.2	XOR	$\overline{\text{XOR}}$
1 1/2	0.1	0.5	0.5	OR	(NOR)
2	0	0.2	1	AND	NAND

Fig. 1.2-10. Position of the transmission peak after 0, 1, or 2 input pulses are incident. With the probe wavelength at one of the five labeled values [expressed by the initial detuning in full width half maximum (FWHM) of the transmission peak] the gates in the table are obtained. The fractional values in the columns below 0, 1, and 2 (of inputs) are the approximate transmissions when each input shifts the peak by 1 FWHM. In reflection the AND and NOR have poor contrast. Jewell, Rushford, and Gibbs (1984).

For further reading on passive optical bistability and all-optical signal processing, see Garmire (1979, 1981a,b), Marburger and Garmire (1979), Lugovoi (1979d), S. D. Smith and Miller (1980), S. D. Smith (1980, 1981a, 1982), Gibbs, McCall, and Venkatesan (1980), Collins (1980a), D. A. B. Miller, Seaton, and Smith (1981), P. W. Smith and Tomlinson (1981a), P. W. Smith (1981), Arecchi and Salieri (1982), Abraham and Smith (1982 a,b), D. A. B. Miller (1982a), Shen (1982), Gibbs (1982, 1983a,b), Peyghambarian (1983), S. D. Smith, and Wherrett (1983), S. D. Smith, and Walker (1984), and Laval (1984).

The use of lasers for performing optoelectronic logic operations is discussed by Kosonocky (1964), Basov (1965), and Basov, Culver, and Shah (1972). A GaInAsP 1.3-μm wavelength cleaved-coupled-cavity (C^3) crescent laser was

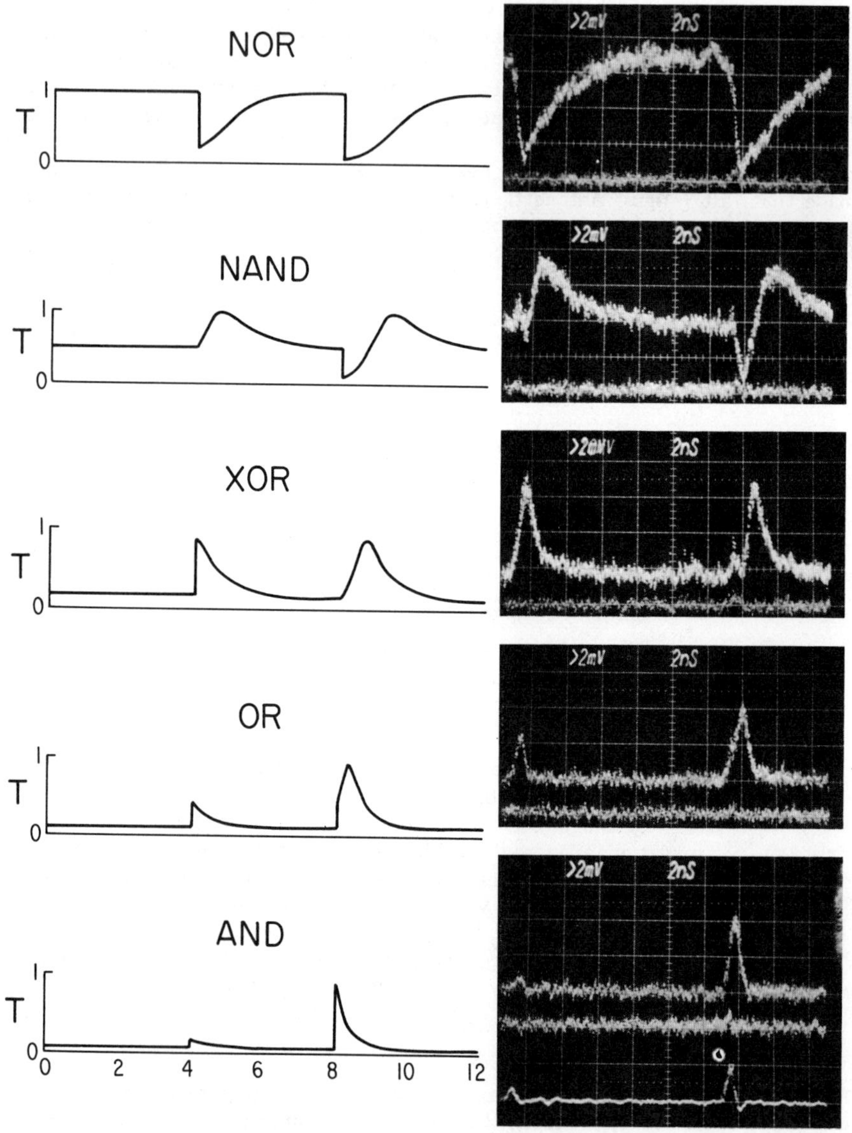

Fig. 1.2-11. Optical logic gate operation by means of (left) computer simulations and (right) a GaAs/AlGaAs multiple-quantum-well etalon. Jewell, Lee, Warren, Gibbs, Peyghambarian, Gossard, and Wiegmann (1984).

employed by Tsang, Olsson, and Logan (1983) [also see Tsang and Olsson (1983)] to demonstrate the complete set of basic logic operations: AND, OR, EXCLUSIVE OR, and INVERTER. The device operates on electrical inputs and produces optical outputs; the switching time was subnanosecond.

1.3. OPTICAL BISTABILITY IN LASERS

There are three reasons for making a strong distinction here between optical bistability in passive systems and optical bistability in lasers. First is the guess that passive systems are more likely to become practical because they are simpler (they do not require an inverted medium) and hence are likely to be smaller and require less power. Second, the two fields have evolved independently. Third, the author is less familiar with bistability in lasers and chooses not to treat it. Of course, from a unified theory perspective, it is clear that there are great similarities: both are described by coupled Maxwell-Bloch equations with boundary conditions, i.e., both involve the interaction of a nonlinear medium with light under feedback conditions. However, the active system is more complicated: the frequency of the laser may change or several modes may lase simultaneously, and the external excitation required to produce population inversion generally results in a complex environment. Since bistability was studied first in lasers, we mention some of that work even though it seems to have had nothing to do with initiating the field of passive optical bistability (see Section 1.4). In fact, the first cross reference between the two fields in either direction that we have found was Bonifacio, Gronchi, and Lugiato (1978).

In his classic paper on laser theory, Lamb (1964) states that the first theory of an oscillator capable of multi-frequency operation was given by Van der Pol (1922, 1934); see also page 46 of Sargent, Scully and Lamb (1974). The Van der Pol oscillator consists of two R-L-C circuits with resonance frequencies Ω_1 and Ω_2. The necessary nonlinearities arise from cubic terms in the current-voltage characteristic of a triode vacuum tube. Van der Pol found steady oscillations only near Ω_1 and Ω_2 but never on more than one frequency at a time. He found hysteresis: the oscillation frequency depended upon past history. Lamb notes the close connection between Van der Pol's problem and a multimode laser. The Fabry-Perot cavity modes correspond to the resonant constituents of Van der Pol's plate circuit. However, the nonlinear response of the laser atoms differs qualitatively from the tube characteristic, so the multimode laser behaves very differently. Lamb's analysis shows that under some conditions, a two-mode laser oscillates simultaneously at two frequencies. Under other conditions single-frequency oscillation occurs exhibiting hysteresis with tuning. Spencer and Lamb (1972a) predict that the intensity of a laser coupled to an empty resonant cavity will have hysteresis as a function of tuning of the laser cavity.

Lasher (1964) proposed a bistable laser consisting of a Fabry-Perot injection laser whose plated p contact is divided into two electrically isolated portions by a slit parallel to the light-reflecting sides of the crystal; see Fig. 1.3-1. An inversion is produced in one portion by an injection current, but the other portion is biased so that a negligible injection current flows through the junction, which then acts as a nonlinear absorber. Lasher defined "bistable" as having two states for the same current level and suggested that such a bistable laser could be switched by suitable current or light pulses. Nathan, Marinace, Rutz, Michel, and Lasher (1965) soon reported bistable operations in GaAs using the proposed structure and observed switching in a detector-limited time of less than 5 ns. Recently, Bogatov, Eliseev, Okhotnikov, and Pak (1978) have studied hysteresis of a cw AlGaAs heterolaser with planar stripe contacts as a function of injection current that they attribute to self-focusing arising from plasma and Burstein shift effects. Kawaguchi and Iwane (1981) use InP/InGaAsP/InP double heterostructure lasers with periodic excitation stripe geometry. They attribute the hysteresis of output versus current to the presence of saturable absorbers in the laser resonator. Switch-off times of 1 ns have been achieved [Kawaguchi (1981, 1982a,b)]. Harder, Lau, and Yariv (1981, 1982a,b,c) and Lau, Harder and Yariv (1982a,b) have seen current-controlled bistability and pulsations with a controlled amount of saturable absorption using a Lasher design. Lasher's double diode is the basic device behind the optical logic proposals of Kosonocky (1964), Basov, Culver, and Shah (1972), and Tsang, Olsson, and Logan (1983).

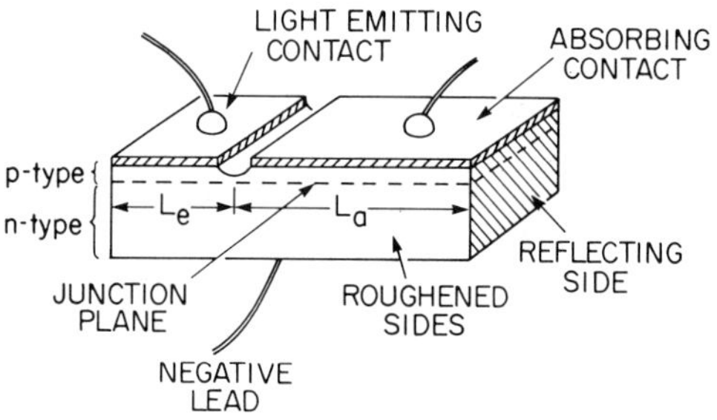

Fig. 1.3-1. GaAs bistable laser. Lasher (1964).

Fork, Tomlinson, and Heilos (1966) studied the competition in a helium-neon (He-Ne) laser between two modes of nearly the same frequency but with orthogonal polarizations. Over certain regions of tuning the polarization of the oscillating mode was determined by which mode was oscillating at the time the laser was tuned into the region. Similar effects were seen by Doyle and Gerber (1968). Lisitsyn and Chebotaev (1968) saw hysteresis with only one mode and with a large number of modes, ruling out mode interactions. They placed an absorbing medium inside the cavity of a He-Ne laser, much the same as Nathan et al. (1965) did for GaAs. They observed hystereses of three types: by changing the cavity loss, the cavity gain (Fig. 1.3-2) or absorption, and the cavity frequency. Ruschin and Bauer (1979, 1981) observed hysteresis of the output of a carbon dioxide (CO_2) laser containing sulfur hexafluoride (SF_6) as a saturable intracavity absorber as a function of gain, using the linear dependence of gain upon helium partial pressure; they also studied transient effects and self-oscillations. Scott, Sargent, and Cantrell (1975) show that a laser oscillator with a saturable absorber can be described as analogous to a first-order phase transition, explaining the results of Lisitsyn and Chebotaev (1968). Lugiato (1978), treating the on-resonance absorptive case, concludes that a laser with injected signal is not a bistable system and does not exhibit a first-order phase transition. Spencer and Lamb (1972b) find a hysteresis of laser output versus injected signal for an extremely limited range of parameters when the injected signal frequency differs from the normal cavity lasing frequency. They do not discuss passive bistability or reference any of the earlier work on passive bistability. However, their model is for a Fabry-Perot interferometer containing two-level atoms described by Bloch equations including off-resonance effects. Clearly their formulation could be applied to passive bistability and, in fact, their state equation [their Eq. (65)] is essentially Eq. (2.1-31) for mixed absorptive and dispersive bistability. Because the connection to passive bistability was not appreciated by others or pointed out by them, other workers independently formulated microscopic theories of passive bistability.

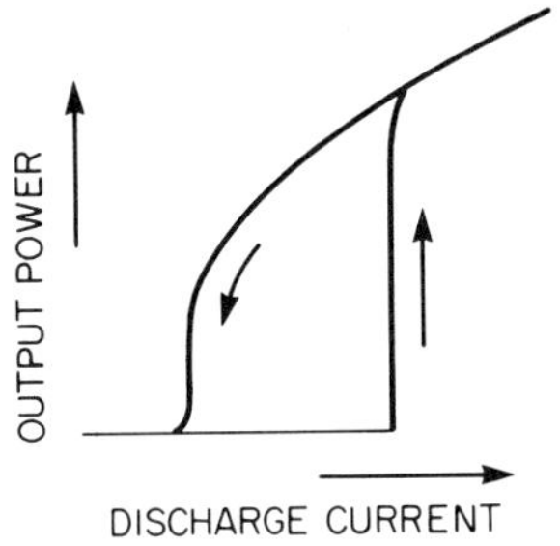

Fig. 1.3-2. Hysteresis in a He-Ne laser containing a nonlinear absorber. Lisitsyn and Chebotaev (1968).

The first observation of bistability in a laser in which the hysteresis variable was input light intensity is that of Levin and Tang (1979). Their system is a hybrid dispersive system consisting of an Ar-laser-pumped dye laser tuned by an electro-optical birefringent tuner. The tuner voltage is partly proportional to the laser intensity through the feedback circuit. They also observed multistability, i.e., multiple stable states, and differential gain of 10^4 where a 2-μW He-Ne beam controls the 20-mW dye laser output (with the help of electrical power in the feedback circuit).

Some of the previously described bistable lasers could be operated with the input intensity as the hysteresis variable in a similar manner. For example, the injection current of the GaAs laser could be made proportional to the incident intensity. This is clearly a hybrid mode of operation in the sense of mixed optics and electronics.

For other papers on optical bistability in lasers look for (1.3) after references.

1.4. EARLY HISTORY OF PASSIVE OPTICAL BISTABILITY

Just as for optical bistability in lasers, the early work was performed by several researchers studying independently. Seidel (1969) is apparently the first to officially record the idea of a passive bistable optical device. Figure 1.4-1 is a figure from his patent. He clearly appreciated its potential: "Use in an optical memory and optical data-processing system is contemplated." His predicted hysteresis loop, i.e., bistability curve, is shown in Fig. 1.4-2. His description of the operation of his device is a lucid account of intrinsic absorptive optical bistability:

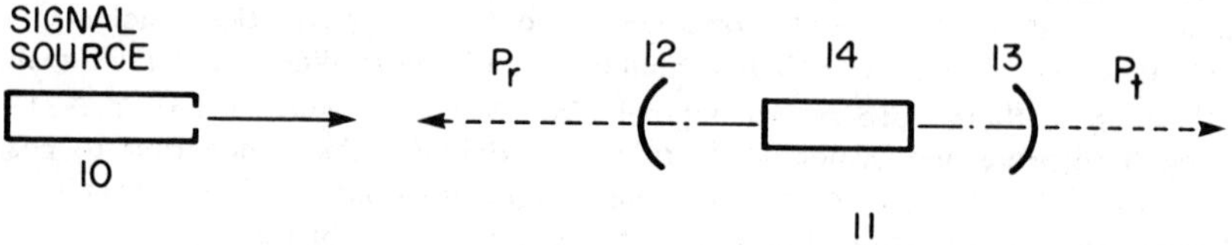

Fig. 1.4-1. Seidel's (1969) embodiment of the invention of a bistable optical circuit consisting of a signal source (10) of variable output intensity, a cavity (11), defined by reflectors (12) and (13), tuned to resonate at the signal frequency, and a saturable absorber (14) within the cavity.

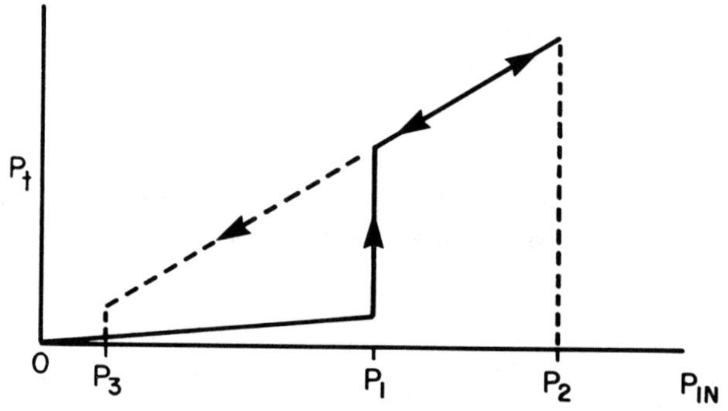

Fig. 1.4-2. Seidel's (1969) predicted bistability curve of transmitted power P_t versus input power P_{IN}.

"Saturable absorbers are characterized by power absorption characteristics that decrease with increasing applied power. The power range over which this transition occurs and the total change in attenuation depend upon the material. Many such materials are known, and thus the particular material that is used would depend upon the frequency and the available intensity range of the signal. Typically, however, the change in absorption is relatively gradual. The present invention is based upon the recognition that this change can be made to occur abruptly by locating the saturable absorber in a resonant cavity. This is the result of a regenerative process that occurs in the vicinity of the saturation curve. In accordance with this process, a slight increase in applied power causes a slight decrease in the cavity absorption, which, in turn, permits the cavity to accept more power. This added power further saturates the saturable absorber, which again permits more power to be coupled into the cavity. This process is cumulative, with the net result that the change in the power level within the cavity exceeds the original small change in the incident power. Thus, there is a threshold level at which the system is unstable and abruptly switches states.

As incident power is decreased from above P_1, the device responds in a somewhat different manner. Because of the high field intensity that builds up within the cavity, the saturable absorber 14 is maintained in its low attenuation state even though the incident power is decreased below the threshold level P_1. As a consequence, the level of reflected power remains low. Essentially, all the incident power continues to be transmitted. This state of affairs continues until the incident power reaches a lower threshold level P_3, at which the signal intensity within the cavity is incapable of sustaining the saturable absorber in its low loss state. The absorber then switches to its high loss state, accompanied by a detuning of the cavity. This results in an abrupt increase in the reflected power and a correspondingly sudden decrease in the transmitted power."

He further predicted limiting action for the reflected power in the "on" state, i.e., that the amount of reflected power remains essentially constant as the incident power is increased.

Equally pioneering and independent work on absorptive bistability was done by Szöke, Daneu, Goldhar, and Kurnit (1969). They analyzed a Fabry-Perot interferometer containing a saturable absorber and derived the condition for purely absorptive bistability $\alpha_0 L/T > 8$, where α_0 is the peak absorption coefficient, L is the etalon length, and T is the transmission of the etalon reflecting surfaces (see Section 2.2). They mentioned several of the problems still under study today: standing-wave effects, residual or unsaturable absorption, production of an infinite pulse train from a cw input, and crosstalk between nearby beams that could lead to adding and memory operations performed in parallel. They conducted experiments using a CO_2 laser operating in a single axial mode on the P(16) 10.6-μm transition and a 2-cm-long resonator filled to various pressures of SF_6. They saw some nonlinear transmission that suppressed the tail of a pulse, but "the pressure of the absorber was not high enough to bring us to the bistable region." The saturable resonator was also used as a passive cavity dumper by substituting it for one of the laser mirrors. A rotating mirror was used to initiate Q-switching; with sufficiently high intracavity intensity the resonator turned on, dumping out a 50-ns pulse (compared with a 35-ns laser round-trip time). Switch-on appeared to have occurred, but a high-gain characteristic with no bistability would dump the cavity similarly. Again bistability was not demonstrated. It is amazing how similar the Szöke et al. cavity dumper is to the "passive transmission switch" in a patent by Bjorkholm (1967) filed two years before. Although bistability is not spelled out in Bjorkholm's work, the concept of rapid switch-on of a Fabry-Perot cavity containing a saturable absorber certainly is.

Some of these ideas were elaborated upon further in a patent by Szöke (1972). In that patent he addresses the problem of switching off a bistable device without reducing the input intensity (See Section 5.5). He proposed the use of two coupled bistable devices to form an optical flip flop that could be switched off as well as on; see Fig. 1.4-3. If I_Z is close to but less than $I_\uparrow$, device 6 is initially off. A pulse I_{on} such that $I_Z + I_{on}$ exceeds $I_\uparrow$ can turn 6 on, and it can be made to stay on by adjusting the feedback I_R such that $I_R + I_Z$ exceeds $I_\downarrow$ (with device 22 off). Now pulse I_{OFF} can turn on device 22 long enough to interrupt I_R causing device 6 to turn off. If mirror 20 is made partially transmitting, then a portion of beam 7 is the desired flip-flop output. Although active switch-off of a single bistable etalon has been achieved in a special case (Section 5.5) and would be simpler, it is important that two-device schemes such as Szöke's permit fast optical switching without it.

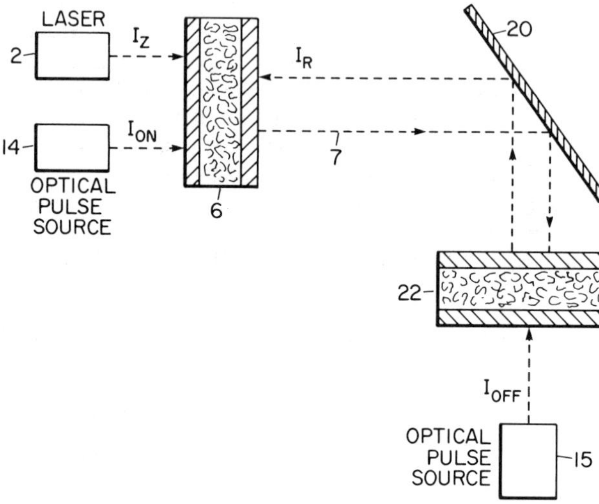

Fig. 1.4-3. Szöke's (1972) optical flip flop utilizing two bistable resonators.

Prompted by the work of Szöke et al. (1969), Austin and Deshazer (1971) and Spiller (1971, 1972) searched for absorptive optical bistability by studying the transmission of nonlinear Fabry-Perot cavities containing absorbing dyes; also, Austin and Deshazer studied the transmission of single-mode ruby laser pulses through a nonlinear Fabry-Perot cavity containing cryptocyanine in methanol. They observed pulse narrowing of up to 50% and "some pulse asymmetrizing, which is believed due to optical hysteresis, was noticed." The difference between their peak transmission of 1%, compared with 50% theoretically, was attributed to mode-matching problems. Spiller (1972) numerically calculated resonator transmission as a function of reflectivity and unsaturable absorption. He found that the power level for down-switching moves to higher power levels for increasing losses, but the power level for up-switching is practically unaffected. The overall effect of an increasing linear (unsaturable) loss is a drastic decrease in the area of the hysteresis curve and in the maximum transmission. He found that detuning the resonator has qualitatively the same effect as a linear loss; he did not treat dispersive effects. Spiller did observe highly nonlinear transmission, but the maximum transmission was only 2%; the saturable absorbers used did not bleach enough to fulfill the conditions for bistable behavior. Zhu and Garmire (1983a,b,c) have succeeded in observing bistability in dyes, but it arises from intensity-dependent phase shifts rather than saturation of absorption; see Section 3.5.

Unaware of the earlier work on absorptive bistability, McCall (1974) pointed out that a nonlinear Fabry-Perot cavity could be bistable. He analyzed in considerable detail the amplifying properties of a resonant medium in the presence of a strong light beam (Appendix A and Section 6.2).

The first observation of optical bistability (in a passive device, i.e., containing no gain medium) was reported by McCall, Gibbs, Churchill, and Venkatesan (1975); see Fig. 1.2-1 and also McCall, Gibbs, and Venkatesan (1975). They used Na vapor between the mirrors of a plane Fabry-Perot interferometer with 11-cm plate separation. This first observation of optical bistability was due to nonlinear refractive index effects, not the anticipated nonlinear absorptive effects.

A simple picture of intrinsic dispersive optical bistability is as follows. The optical cavity is tuned such that one of its transmission maxima lies close to the laser frequency, but is far enough away so that the cavity's transmission is low for low input intensity. Most of the input light is then reflected from the cavity. If the nonlinear index has the sign appropriate to compensate for the initial detuning, at sufficiently high input intensities the light penetrating the cavity will be strong enough to shift the cavity maximum back toward the laser frequency. This intrinsic feedback further increases the cavity's transmission which further increases the nonlinear phase shift, etc.; see Fig. 1.1-1. So the device turns "on" rapidly, the cavity's peak sweeps rapidly through the laser frequency, and comes to equilibrium with its peak detuned to the other side of the laser frequency where negative feedback provides stability. Further increases in input result in very little increase in output since larger phase shifts decrease the device's percentage transmission; the device functions as an optical limiter. (If the cavity has repeating orders, sufficiently high inputs result in turning "on" to successively higher orders, i.e., multistability results.) In the "on" state, the cavity has high Q with constructive interference giving rise to an intense internal field. As the input intensity is reduced, this intracavity field keeps the cavity shifted into resonance with the laser for lower input intensities than were required to turn on the device; see Fig. 1.1-1 again. This hysteresis cycle is a direct consequence of the optical feedback.

The experimental observation of optical bistability in Na sparked a rapidly expanding growth of the field; we will attempt to summarize the important developments in this book. Henceforth, we will use "optical bistability" to mean passive optical bistability and will make it clear if any bistable system involves a gain medium.

CHAPTER 2

STEADY-STATE MODEL OF OPTICAL BISTABILITY

The purpose of Section 2.1 is to summarize succinctly the Bonifacio-Lugiato model of optical bistability that is both from first principles and analytic. Those preferring an historical progression of models of increasing complexity should read Sections 2.2 through 2.4.2 and then return to Section 2.1. Those familiar with the boundary conditions of an optical cavity (Appendix B) and the coupled Maxwell-Bloch equations (Appendix C) should find the derivation in Section 2.1 straightforward and the solution amazingly general. It follows Bonifacio and Lugiato (1978d); Bonifacio, Gronchi, and Lugiato (1979a); Gronchi and Lugiato (1980). For an extensive and lucid review article on the theory of optical bistability see Lugiato (1984).

2.1. MEAN-FIELD MODEL OF MIXED ABSORPTIVE AND DISPERSIVE BISTABILITY INCLUDING INHOMOGENEOUS BROADENING

There are a number of ways of defining the complex electric field and polarization as discussed in Appendix C. Adopt the following convention for a single polarization component of a field of angular frequency $\omega = ck/n_0 = 2\pi c/\lambda$ propagating in the positive z direction:

$$E(z,t) = (E_r + iE_i)\, e^{i(\omega t - kz)} + \text{c.c.} \tag{2.1-1}$$

The polarization created in a system of two-level atoms of density N and electric dipole moments p can be written as

$$P(z,t) = Np[\bar{u}(z,t) - i\bar{v}(z,t)]\, e^{i(\omega t - kz)} + \text{c.c.} \tag{2.1-2}$$

The components of P in phase and out of phase with the electric field are labeled $\bar{u}$ and $\bar{v}$, respectively. The third component of a three-dimensional pseudo-polarization vector or Bloch vector is the inversion w. The Bloch equations [Eqs. (B-68 to B-70)] are:

$$\dot{\bar{u}} = \bar{v}\Delta\omega - \kappa E_i w - \gamma_T \bar{u}\,, \tag{2.1-3a}$$

$$\dot{\bar{v}} = -\bar{u}\Delta\omega - \kappa E_r w - \gamma_T \bar{v}\,, \tag{2.1-3b}$$

$$\dot{w} = \kappa E_r \bar{v} + \kappa E_i \bar{u} - \gamma_L [w+1]\,, \tag{2.1-3c}$$

with $\Delta\omega = \omega_a - \omega$, where ω_a is the atomic resonance angular frequency. Equations (2.1-3) can be written more compactly by defining

$$Q \equiv \overline{v} + i\overline{u} , \tag{2.1-4}$$

$$F \equiv \frac{\kappa(E_r + iE_i)}{(\gamma_T \gamma_L)^{1/2}} ; \tag{2.1-5}$$

i.e.,

$$\dot{Q} = -(\gamma_T \gamma_L)^{1/2} Fw - (\gamma_T - i\Delta\omega)Q , \tag{2.1-6a}$$

$$\dot{w} = (\gamma_T \gamma_L)^{1/2} \frac{FQ^* + F^*Q}{2} - \gamma_L(w+1) . \tag{2.1-6b}$$

Therefore, in steady state,

$$w = -\frac{1 + \Delta^2}{1 + \Delta^2 + |F|^2} , \tag{2.1-7a}$$

$$Q = \left[\frac{\gamma_L}{\gamma_T}\right]^{1/2} \frac{(1 + i\Delta)F}{1 + \Delta^2 + |F|^2} , \tag{2.1-7b}$$

with $\Delta \equiv \Delta\omega/\gamma_T$. Appendix C also shows that from Maxwell's equations and their second-order wave equation one can obtain, assuming slowly varying amplitude and phase, a first-order complex equation

$$\frac{\partial E}{\partial z} + \frac{n_0}{c}\frac{\partial E}{\partial t} = -\frac{2\pi\omega N p}{n_0 c}\int_{-\infty}^{\infty} g(\Delta\omega)Q(\Delta\omega)d(\Delta\omega) \tag{2.1-8}$$

from Eq. (B-25) by adding the integral over the inhomogeneous distribution of atomic line centers $g(\Delta\omega)$ normalized such that

$$\int_{-\infty}^{\infty} g(\Delta\omega)\, d(\Delta\omega) = 1 . \tag{2.1-9}$$

In equilibrium and using Eq. (2.1-5):

$$\frac{dF(z)}{dz} = -\frac{\alpha_0}{2} F\sigma(|F|^2) , \tag{2.1-10}$$

where

$$\sigma(|F|^2) = \int_{-\infty}^{\infty} g(\Delta\omega)\frac{1 + i\Delta}{1 + \Delta^2 + |F|^2}\, d(\Delta\omega) \tag{2.1-11}$$

and

$$\alpha_0 = \frac{8\pi\omega N p^2}{n_0 \hbar c \gamma_T} . \qquad (2.1\text{-}12)$$

So far we have described the nonlinear properties of a two-level atom subjected to a coherent field. In order to ensure the feedback essential for bistability, assume the atoms are within a ring cavity with input and output mirrors having reflectivity $R = 1 - T$ and fold mirrors having unity reflectivity; see Fig. 2.1-1.

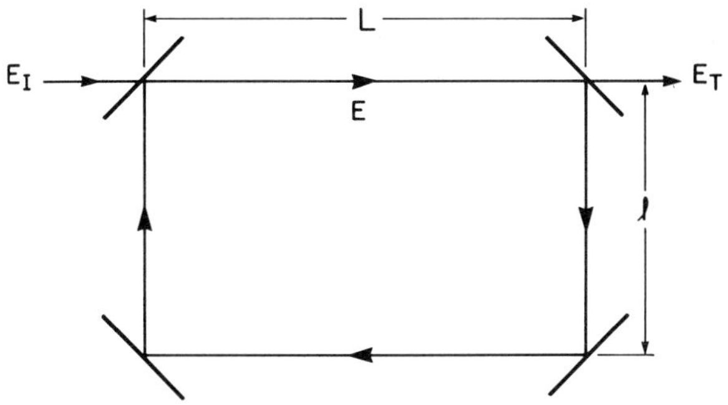

Fig. 2.1-1. Nonlinear ring cavity.

The boundary conditions are then

$$E_T(t) = \sqrt{T}\, E(L,t) \qquad (2.1\text{-}13)$$

$$E(0,t) = \sqrt{T}\, E_I(t) + Re^{i\beta_0} E(L,t-\Delta t) , \qquad (2.1\text{-}14)$$

where

$$\Delta t = \frac{L_T - L}{c} , \qquad (2.1\text{-}15)$$

L_T is the total length $(2L+2\ell)$ of the cavity, and

$$\beta_0 - 2\pi m = \beta_0' = 2\pi \frac{\nu_c - \nu}{c/L_T} \qquad (2.1\text{-}16)$$

is the cavity-laser phase detuning, where ν_c is the frequency of peak cavity transmission. In steady-state $E(L,t-\Delta t) = E(L,t)$. With the definitions

$$y \equiv \frac{\kappa E_I}{(\gamma_T \gamma_L T)^{1/2}} \qquad (2.1\text{-}17)$$

$$x \equiv \frac{\kappa E_T}{(\gamma_T \gamma_L T)^{1/2}} = \frac{\kappa E(L,t)}{(\gamma_T \gamma_L)^{1/2}} = F(L,t) \qquad (2.1\text{-}18)$$

the boundary condition Eq. (2.1-14) becomes

$$F(0,t) \simeq Ty + Rx\, e^{i\beta_0}. \tag{2.1-19}$$

In the mean-field theory one takes the limit

$$\alpha_0 L \to 0, \quad T \to 0, \quad \frac{\alpha_0 L}{T} = \text{constant}; \tag{2.1-20}$$

$\alpha_0 L \to 0$ ensures that the electric field changes little in traversing the medium, and $T \to 0$ enables the field to traverse the medium many times to give large nonlinearities. Integrating Eq. (2.1-10) in the mean-field limit,

$$\int_0^L F(z)\sigma(|F(z)|^2)dz \simeq LF(L)\sigma(|F(L)|^2)$$

$$= -\frac{2}{\alpha_0}[F(L) - F(0)] = -\frac{2}{\alpha_0}(x - Ty - Rx\, e^{i\beta_0}); \tag{2.1-21}$$

therefore

$$y = \frac{x(1 - Re^{i\beta_0})}{T} + 2Cx \int_{-\infty}^{\infty} g(\Delta\omega) \frac{1 + i\Delta}{1 + \Delta^2 + |x|^2} d(\Delta\omega), \tag{2.1-22}$$

where

$$C \equiv \frac{\alpha_0 L_{RC}}{4T} \tag{2.1-23}$$

and L_{RC} is the length of the nonlinear medium in the ring cavity. Close to a cavity resonance

$$e^{i\beta_0} = e^{i\beta_0'} \simeq 1 + i\beta_0'; \tag{2.1-24}$$

therefore

$$y \simeq x[1 + 2C\sigma_r(X)] + ix[2C\sigma_i(X) - \theta], \tag{2.1-25}$$

where

$$X \equiv |x|^2 \tag{2.1-26}$$

$$Y \equiv |y|^2 \tag{2.1-27}$$

$$\theta \equiv \frac{R\beta_0'}{T} = \frac{2\pi}{cT}(\nu_c - \nu)L_T R \tag{2.1-28}$$

and

$$\sigma(X) \equiv \sigma_r(X) + i\sigma_i(X) \qquad (2.1\text{-}29)$$

$$\sigma(X) = \int_{-\infty}^{\infty} g(\Delta)\,\frac{1 + i\Delta}{1 + \Delta^2 + X}\,d\Delta\ . \qquad (2.1\text{-}30)$$

Equation (2.1-25) is Eq. (21) of Bonifacio and Lugiato (1978d). Or, expressing the normalized input intensity Y as a function of the normalized output intensity X,

$$Y = X\{[1 + 2C\sigma_r(X)]^2 + [2C\sigma_i(X) - \theta]^2\}\ . \qquad (2.1\text{-}31)$$

This is the steady-state mean-field state equation for mixed absorption and dispersion and including inhomogeneous broadening.

Purely dispersive bistability is obtained for $\sigma_r = 0$; for example, for very large Δ, $\sigma_i >> \sigma_r$. Then

$$Y \simeq X\,\{1 + [2C\sigma_i(X) - \theta]^2\}\ . \qquad (2.1\text{-}32)$$

For small input Y, the output X is much smaller than Y because the atomic contribution to the intracavity phase shift is much larger than the laser-empty cavity phase shift β_0. For large enough Y, $2C\sigma_i = \theta$ and X = Y giving 100% transmission.

For purely absorptive bistability, $\sigma_i = 0$ and $\theta = 0$, yielding

$$Y = X[1 + 2C\sigma_r(X)]^2\ . \qquad (2.1\text{-}33)$$

Here for small Y, X << Y because $2C\sigma_r >> 1$. For large enough Y, $\sigma_r \to 0$ and X = Y. So saturation is essential for purely absorptive bistability. For purely dispersive bistability, any unsaturable index can be negated by initial detuning.

For a homogeneously broadened line and pure absorption, $\Delta = 0$ and

$$\sigma_r(X) = \frac{1}{1 + X} \qquad (2.1\text{-}34)$$

so

$$Y = X\left[1 + \frac{2C}{1 + X}\right]^2 . \qquad (2.1\text{-}35)$$

The condition for optical bistability is

$$\frac{dY}{dX} < 0\ . \qquad (2.1\text{-}36)$$

The extrema of dY/dX occur for $d^2Y/dX^2 = 0$, which occurs for $X_e = (2C+1)/(C-1)$

giving

$$\left.\frac{dY}{dX}\right|_{X=X_e} = \left[\frac{4 - C}{3C}\right]\left[\frac{2C + 1}{3}\right]^2 . \qquad (2.1\text{-}37)$$

The derivative dY/dX is negative if

$$C > 4 \qquad (2.1\text{-}38)$$

or

$$\frac{\alpha_0 L_{RC}}{4T} > 4 . \qquad (2.1\text{-}39)$$

It will be shown in Section 2.5.1 that for a detuning Δ_0 from a homogeneous resonance, the condition for bistability is

$$C > 2[1 + (1 + \Delta_0^2)^{1/2}] ;$$

thus dispersive bistability is harder than absorptive assuming no frequency-dependent background absorption (Section 2.5.3).

For a discussion of mixed absorptive and dispersive optical bistability, proceed to Section 2.4.3.

2.2. SZÖKE ET AL. MODEL OF ABSORPTIVE OPTICAL BISTABILITY

The first model of purely absorptive optical bistability was by Szöke, Daneu, Goldhar, and Kurnit (1969). From the boundary conditions for the light interacting with the etalon and from the saturation equations of a homogeneously broadened two-level system, they derived an inequality that must be satisfied for bistability to occur. They found a tradeoff between cavity finesse and absorption.

Figure 2.2-1 illustrates the fields associated with a plane Fabry-Perot interferometer with spacing L between mirrors of intensity reflectivity R. Assume that the laser frequency ν is coincident with the Fabry-Perot frequency ν_{FP}, that there are no intensity-dependent refractive index effects, and that standing-wave effects can be neglected. Then in equilibrium, the boundary conditions are (see Appendix B):

$$E_T = \sqrt{T}\, E_F(L) \qquad (2.2\text{-}1)$$

and

$$E_F(0) = \sqrt{T}\, E_I + Re^{i\beta}\, e^{-(\alpha/2)2L}\, E_F(0) , \qquad (2.2\text{-}2)$$

where β is the cavity-laser phase detuning, α is the intensity absorption coefficient, E_I, E_R, E_F, E_B, and E_T are the incident, reflected, forward, backward, and transmitted electric field slowly varying complex amplitudes. For purely absorptive bistability one selects $\nu = \nu_{FP}$; one can then choose the phase reference such that $e^{i\beta} = 1$. From Eq. (2.2-2)

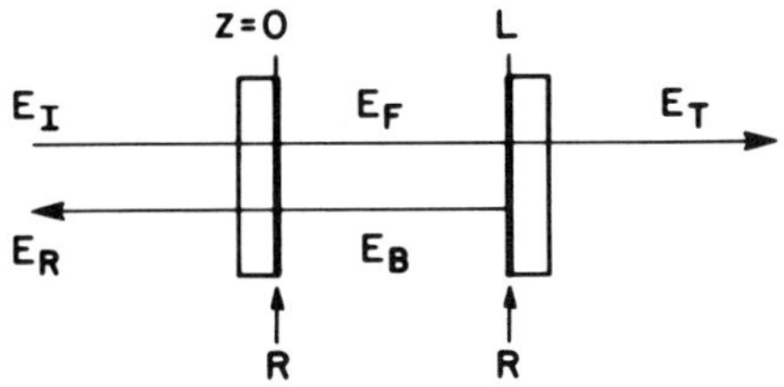

Fig. 2.2-1. Nonlinear optical device electric fields.

$$\frac{E_F(0)}{E_I} = \frac{\sqrt{T}}{1 - Re^{-\alpha L}} . \tag{2.2-3}$$

As a further simplification assume that the absorption is small, i.e., $\alpha L \ll 1$, but that the empty cavity finesse $\mathscr{F} = \pi\sqrt{R}/(1-R)$ is sufficiently large that

$$k \equiv \frac{R\alpha L}{1 - R} \tag{2.2-4}$$

is much larger than 1. Basically this implies $T \ll \alpha L$, which is consistent with the end result of the present calculation, namely $\alpha L/T > 8$. Expanding Eq. (2.2-3)

$$\frac{E_F(0)}{E_I} \simeq \frac{\sqrt{T}}{1 - R(1 - \alpha L)} = \frac{\sqrt{T}}{(1 - R)\left[1 + \dfrac{R\alpha L}{1 - R}\right]} = \frac{1}{\sqrt{T}\,(1 + k)} . \tag{2.2-5}$$

The field E_T is simply related to E_F:

$$E_T = \sqrt{T}\, E_F(L) = \sqrt{T}\, e^{-\alpha L/2}\, E_F(0) \simeq \sqrt{T}\, E_F(0) . \tag{2.2-6}$$

Therefore

$$\frac{I_T}{I_I} = \frac{|E_T|^2}{|E_I|^2} \simeq \frac{1}{(1 + k)^2} . \tag{2.2-7}$$

Although it is of no consequence in this ratio, we will generally set $I = |E|^2$ and not write out $I = n_0 c|E|^2/4\pi$.

I_T is very simply related to I_I for low intensities, but at high intensities α and k are intensity dependent. Assume that α arises from a two-level, lifetime-broadened transition between upper state a and lower state b. Then the rate equation for the upper state density N_a is, for no cavity,

$$\dot{N}_a = -\frac{N_a}{T_1} + \frac{\alpha I_F}{\hbar\omega} , \tag{2.2-8}$$

where $T_1 = \gamma_L^{-1}$ is the homogeneous relaxation time. The term $\alpha I_F/\hbar\omega$ is the number of atoms per unit volume excited per unit time:

$$\frac{\text{energy absorbed}}{(\text{area})(\text{time})} = -\frac{dI_F}{dz}L = \alpha I_F L = \frac{(\text{energy})L}{(\text{volume})(\text{time})}$$

$$\frac{\text{number}}{(\text{volume})(\text{time})} = \frac{\text{energy/[vol x time]}}{\text{energy per atom}} = \frac{\alpha I_F}{\hbar\omega}.$$

In steady state $\dot{N}_a = 0$:

$$N_a = \frac{\alpha I_F T_1}{\hbar\omega}. \tag{2.2-9}$$

The absorption coefficient α satisfies the equation of stimulated emission,

$$\alpha = \frac{(N_b - N_a)\,\alpha_0}{N}, \tag{2.2-10}$$

where α_0 is the low-light-level absorption coefficient, and N is the total density, i.e., $N = N_b + N_a$. Therefore,

$$\alpha = \frac{(N - 2N_a)\alpha_0}{N} = \frac{\alpha_0}{1 + I_F/I_s}, \tag{2.2-11}$$

where the saturation intensity I_s is given by

$$I_s = \frac{N\hbar\omega}{2T_1\alpha_0}. \tag{2.2-12}$$

For atoms in a Fabry-Perot cavity, α also depends on I_B and the standing wave between E_F and E_B (see Section 2.5.2), but those effects change only I_s, and not the functional form, to a good approximation. In Section 2.1, saturation of the form of Eq. (2.2-11) was derived directly from the Bloch equations. Then

$$k = \frac{R\alpha L}{1 - R} = \frac{k_0}{1 + I_F/I_s}, \tag{2.2-13}$$

where

$$k_0 = \frac{R\alpha_0 L}{1 - R}. \tag{2.2-14}$$

Equation (2.2-7) can be rewritten as

$$Y = X\left[1 + \frac{k_0}{1 + X}\right]^2, \tag{2.2-15}$$

where $Y \equiv I_I/TI_s$ and $X \equiv I_F/I_s = I_T/TI_s$ are the normalized input and output intensities, respectively. The choice of X and Y can be remembered by noting that, to be single-valued, I_I must be plotted against I_T and usually the vertical axis is labeled "y." With C defined as in Eq. (2.2-19), Eq. (2.2-15) is the same as Eq. (2.1-35) since $R \simeq 1$.

The state equation (2.2-15) gives the input as a single-valued function of the output intensity. Compare Fig. 2.2-2a, a plot of Eq. (2.2-15) for various k_0, with Fig. 2.2-2b, which is a Szöke et al. computer calculation without assuming $\alpha L \ll 1$ and $k \gg 1$; $\exp(-\alpha L)$ is always replaced by

$$\exp\left[-\int_0^L \alpha(z)dz\right].$$

Equation (2.2-15) is reasonably good even for $\alpha L \gtrsim 1$. The $\alpha_0 L = 2$ curve in Fig. 2.2-2b is seen to be bistable corresponding to $k_0 = 8$, which is the minimum value of k_0 for bistability according to Eq. (2.2-15). Bistability occurs if $dI_I/dI_T < 0$ implying $dY/dX < 0$. Differentiating Eq. (2.2-15)

$$\frac{dY}{dX} = \left[1 + \frac{k_0}{1+X}\right]^2 + 2X\left[1 + \frac{k_0}{1+X}\right]\left[-\frac{k_0}{(1+X)^2}\right]. \quad (2.2\text{-}16)$$

For $dY/dX = 0$

$$X^2 + (2 - k_0)X + 1 + k_0 = 0,$$

or

$$X = \frac{k_0 - 2 \pm (k_0(k_0-8))^{1/2}}{2}. \quad (2.2\text{-}17)$$

So for zero slope to occur for a physically meaningful X, i.e., mathematically real, k_0 must equal or exceed 8; therefore the condition for absorptive optical bistability is

$$k_0 \equiv \frac{\alpha_0 L R}{1 - R} > 8. \quad (2.2\text{-}18)$$

But since $R \simeq 1$, this is equivalent to

$$C \equiv \frac{\alpha_0 L_{FP}}{2T} > 4, \quad (2.2\text{-}19)$$

where L_{FP} is the length of nonlinear material in the Fabry-Perot device. Neglected effects, such as linear background absorption, standing waves, and transverse averaging increase the required α_0 (see Section 2.5).

One can use the Szöke et al. model to calculate the maximum differential gain

$$G = \left.\frac{dI_T}{dI_I}\right| = \left[\left.\frac{dY}{dX}\right|\right]^{-1}, \quad (2.2\text{-}20)$$

where $|$ represents evaluation at the inflection point defined by

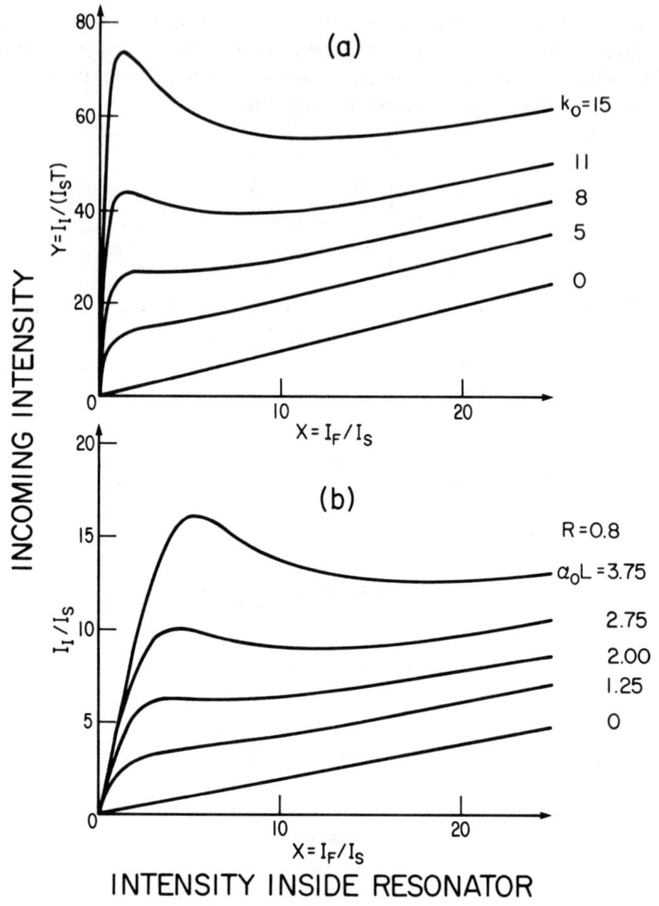

Fig. 2.2-2. Static characteristics of a Fabry-Perot resonator filled with a saturable absorber. (a) From Eq. (2.2-15) and (b) from computer calculation. Szöke et al. (1969).

$$\frac{d^2Y}{dX^2} = 0 , \qquad (2.2\text{-}21)$$

which occurs for

$$X_e = \frac{2(k_0 + 1)}{k_0 - 2} \qquad (2.2\text{-}22)$$

and yields

$$G = \frac{27k_0}{(k_0 + 1)^2(8 - k_0)} . \qquad (2.2\text{-}23)$$

At the transition between high gain and bistability at $k_0 = 8$, $G = \infty$. Gain can be seen in Fig. 2.2-2. Appendix A shows how a cavityless two-level system can exhibit differential gain without population inversion.

Note that the Szőke et al. model of absorptive bistability is analytic, so that later publications claiming to have the first model or the first analytic model were misguided.

2.3. SIMPLE MODEL OF DISPERSIVE OPTICAL BISTABILITY

Before the experimental observation of bistability in Na vapor by Gibbs, McCall, and Venkatesan (1976), all discussions of intrinsic optical bistability were for the purely absorptive case. The fact that most of the observations of the Na experiment (see Section 3.2) were inconsistent with absorptive bistability led to the discovery of dispersive bistability. Although the analysis of the Na experiment simultaneously treated absorptive and dispersive effects, the purely dispersive case is exceedingly simple and will be treated in this section (Gibbs, McCall, and Venkatesan (1977)).

Refer again to Fig. 2.2-1 and the boundary conditions Eqs. (2.2-1,2). This time, set $\alpha = 0$ and retain $e^{+i\beta}$, so that Eq. (2.2-3) becomes

$$\frac{E_F(0)}{E_I} = \frac{\sqrt{T}}{1 - Re^{+i\beta}} . \qquad (2.3\text{-}1)$$

Of course, for a system of two-level atoms there is no phase shift ($\sim\sigma_i$) if there is no absorption ($\sim\sigma_r$), since $\sigma_i/\sigma_r = \Delta$; for large enough Δ the absorption can be neglected. Also, as in Section 2.2, standing wave effects are neglected. Since

$$E_T = \sqrt{T}\, e^{+i\beta/2}\, E_F(0) , \qquad (2.3\text{-}2)$$

$$E_I = \frac{(1 - Re^{i\beta})E_T}{T\, e^{i\beta/2}} . \qquad (2.3\text{-}3)$$

Then

$$I_I = |1 - Re^{i\beta}|^2 \frac{I_T}{T^2} . \qquad (2.3\text{-}4)$$

If $\beta^2 << 12$, i.e., in the vicinity of a transmission peak,

$$I_I \simeq (1 + R\beta^2/T^2)I_T \quad \text{or} \quad Y \simeq X(1 + R\beta^2/T^2) \qquad (2.3\text{-}5)$$

for $T << 1$. Equation (2.3-5) relates the transmitted and incident intensities as a function of the plate reflectivity $R = 1 - T$ and the round-trip phase β, which may be intensity dependent. Equation (2.3-5) could have been obtained from the Fabry-Perot transmission formula, $I_T = I_I[1 + 4R\sin^2(\beta/2)/(1-R)^2]^{-1}$, by expanding for small β.

Suppose the phase shift is linearly dependent upon the intracavity intensity, which is proportional to I_T, i.e.,

$$\beta = \beta_0 + \beta_2 I_T , \tag{2.3-6}$$

where β_0 contains all intensity-independent phase shifts such as background index or plate detuning contributions. Phase shifts of the form of Eq. (2.3-6) arise when there is a nonlinear refractive index, $n_0 + n_2 I$, or for Bloch equations far enough off resonance.

For maximum transmission, β must vanish, i.e., $\beta_2 I_T = -\beta_0$. One might anticipate that in order for the transmission to be low at low intensities, the initial detuning must be larger than the cavity instrument width: $|\beta_0| \geq 2\pi/\mathscr{F} = 2T/\sqrt{R}$, where $\mathscr{F}$ is the finesse, i.e., free spectral range divided by instrument width. For $I_T = 0$, $\beta = \beta_0 \simeq 2T/\sqrt{R}$ so the $R\beta^2/T^2$ term in Eq. (2.3-5) is of order unity and is as important as the unity term. A value of unity for $R\beta^2/T^2$ implies $\beta^2 \simeq 10^{-2}$ if $R = 0.9$; thus β^2 is much smaller than 12 as needed in Eq. (2.3-5).

This simple model, Eqs. (2.3-5,6), yields both bistable (two equilibrium values of I_T for one value of I_I) and differential gain ($dI_T/dI_I > 1$) characteristic curves as shown in Fig. 2.3-1. For fixed reflectivity R, only the initial plate detuning β_0 is varied to go from conditions of no gain, to high gain, to infinite gain, and finally to bistability. The conditions on β_0 are easily derived. For differential gain one needs $0 < dI_I/dI_T < 1$. At the inflection point $d^2I_I/dI_T^2 = 0$, yielding $|\beta_2 I_T/\beta_0| = 2/3$ and

$$0 < |\beta_0| < \sqrt{3}\,\frac{T}{\sqrt{R}} = \sqrt{3}\,\frac{\pi}{\mathscr{F}} \qquad \text{(ac gain)} \tag{2.3-7}$$

as the condition for gain whose magnitude is

$$G = \frac{dI_T}{dI_I} = \frac{1}{1 - R\beta_0^2/3T^2} . \tag{2.3-8}$$

Transfer functions for gains of 4 and ∞ are shown in Fig. 2.3-1. For bistability one needs a region of $dI_I/dI_T < 0$, i.e., the phase shift term in Eq. (2.3-5) is large enough to produce a valley in the I_I versus I_T plot. This implies

$$|\beta_0| > \sqrt{3}\,\frac{T}{\sqrt{R}} = \sqrt{3}\,\frac{\pi}{\mathscr{F}} . \qquad \text{(bistability)} \tag{2.3-9}$$

Note that the phase shift required for bistability, $\sqrt{3}\,\pi/\mathscr{F}$, is close to the value $2\pi/\mathscr{F}$ argued physically above.

In Section 2.2 the condition for absorptive bistability was found to be $\alpha_0 L/T > 8$, where the intensity was assumed to be sufficient to saturate the absorption. Here a nonsaturable intensity-dependent refractive index has been assumed, so that the condition Eq. (2.3-9) states that the initial detuning must be sufficiently large. Implicit is the assumption that the intensity is large enough that $\beta_2 I_T$ can equal $-\beta_0$; clearly the smaller the nonlinearity (β_2 or n_2/λ), the larger I_I must be. See Sections 2.5.3 and Appendix D for background-absorption effects and Section 2.9 for finite-beam discussions.

The basic principles of a dispersive bistable system are described well by this simple model, which agrees with the simple physical picture. Note that this is a model of intrinsic dispersive bistability, that it is analytic, and that it is given in Gibbs, McCall, and Venkatesan (1976) (their Fig. 3a), so that the later claims to the first model or the first analytic model of dispersive bistability are misguided.

Clearly this simple model of purely dispersive bistability can be extended to include many more details such as standing-wave effects, nonlinear absorption, medium complexities such as many transitions on and off resonance, and inhomogeneous broadening, for example. Standing-wave and nonlinear-absorption effects are included in computer simulations presented by Gibbs, McCall, and Venkatesan (their Fig. 3b).

Kastal'skii (1973) suggested the use of a hybrid Fabry-Perot resonator with electrical feedback to an intracavity electro-optic phase shifter to produce electrical instabilities or to stabilize the intensity of a light beam. Neither Gibbs et al. nor P. W. Smith and Turner (1977b), who first observed hybrid dispersive bistability, were aware of his research. Had that paper been known, it may have expedited or stimulated dispersive bistability, but its interpretation in terms of optical bistability is not transparent even though in retrospect it is clear that his circuit would exhibit bistability.

Marburger and Felber (1978) studied the theory of the bistability of a nonlinear Fabry-Perot cavity containing a Kerr medium and suggested carbon disulfide (CS_2) and the nematic liquid crystal MBBA for experiments. They pointed out a useful graphical solution for determining the hysteresis loop (Section 2.6). They analyzed Gaussian-transverse-profile beams in interferometers with spherical mirrors in the limit that forward and backward powers are equal (high-finesse case). Because of the self-induced waveguiding, the intensity required for bistable operation is reduced substantially relative to that needed for a uniform-plane-wave plane-mirror analysis.

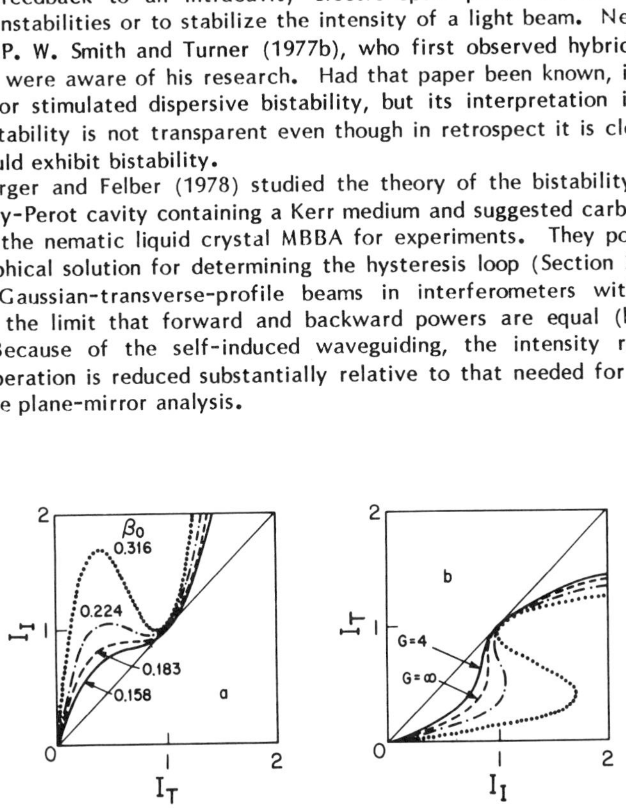

Fig. 2.3-1. Characteristic transfer functions calculated from the simple dispersive model of Eqs. (2.3-5,6). (a) I_I versus I_T. The equations are most easily plotted as I_I versus I_T. (b) I_T versus I_I. Their effect as a device is best shown in I_T versus I_I. R = 0.9. The β_0 = 0.183 curve in (a) and G = ∞ curve in (b) separate the ac gain and bistable regions. The signs of β_0 and β_2 are assumed to be opposite. I_I and I_T are in units of $|\beta_0/\beta_2|$. Gibbs, McCall, Venkatesan (1977).

2.4. BONIFACIO-LUGIATO MODELS

2.4.1. Mean-Field Theory of Absorptive Bistability

Probably the best known optical bistability model is that of Bonifacio and Lugiato (1976), which leads to the state equation

$$y = \frac{2Cx}{1 + x^2} + x \qquad (2.4\text{-}1)$$

and the bistability condition $C > 4$, where y is proportional to $\sqrt{I_I}$ and x to $\sqrt{I_T}$. We will follow the derivation in Bonifacio and Lugiato (1978a) and then show that Eq. (2.4-1) is the same as our Eq. (2.2-15) from the model of Szöke et al. (1969) and that $C > 4$ is identical to Eq. (2.2-19), namely $\alpha_0 L/T > 8$.

The coupled Maxwell-Bloch equations (after slowly varying envelope, rotating wave, forward-only, and plane-wave approximations) for an electric field with slowly varying amplitude ξ and phase ϕ (where ξ and ϕ are real quantities),

$$E(z,t) = \xi(z,t)\, e^{i(\omega t - kz - \phi(z,t))} + \text{c.c.}, \qquad (2.4\text{-}2)$$

and polarization with in-phase dispersive component u and out-of-phase absorptive component v,

$$P(z,t) = Np[u(z,t) - iv(z,t)]\, e^{i[\omega t - kz - \phi(z,t)]} + \text{c.c.}, \qquad (2.4\text{-}3)$$

consist of [McCall and Hahn (1969), Allen and Eberly (1975); see Appendix C]:

$$\frac{\partial \xi}{\partial z} + \frac{n_0}{c}\frac{\partial \xi}{\partial t} = -\frac{2\pi\omega}{n_0 c} Npv\,, \qquad (2.4\text{-}4a)$$

$$\xi\left[\frac{\partial \phi}{\partial z} + \frac{n_0}{c}\frac{\partial \phi}{\partial t}\right] = \frac{2\pi\omega}{n_0 c} Npu\,, \qquad (2.4\text{-}4b)$$

$$\dot{u} = (\omega_a - \omega)v - \gamma_T u\,, \qquad (2.4\text{-}5a)$$

$$\dot{v} = -(\omega_a - \omega)u - \gamma_T v - w\kappa\xi\,, \qquad (2.4\text{-}5b)$$

$$\dot{w} = -\gamma_L(w + 1) + v\xi\kappa\,. \qquad (2.4\text{-}5c)$$

Here n_0 is the background refractive index. The factor N is the total density of two-level systems having transitions at ω_a with electric dipole transition moment p, longitudinal or energy relaxation rate γ_L, and transverse or polarization relaxation rate γ_T; $\kappa \equiv 2p/\hbar$. The components of the pseudo-polarization vector (u,v,w) are defined, in terms of the density matrix $\underset{\sim}{\rho}$ of a two-level atom with upper state a and lower state b in a frame rotating at ω, as follows:

$$u \equiv \tilde{\rho}_{ab} + \tilde{\rho}_{ba} = \langle\sigma_x\rangle \quad (2.4\text{-}6a)$$

$$v \equiv i(\tilde{\rho}_{ba} - \tilde{\rho}_{ab}) = -\langle\sigma_y\rangle \quad (2.4\text{-}6b)$$

$$w \equiv \tilde{\rho}_{aa} - \tilde{\rho}_{bb} = \langle\sigma_z\rangle . \quad (2.4\text{-}6c)$$

In steady state and on resonance, the equilibrium values are

$$u_0 = 0 \quad (2.4\text{-}7a)$$

$$v_0 = \frac{\kappa\xi_0/\gamma_T}{1 + \kappa^2\xi_0^2/\gamma_L\gamma_T} = \left[\frac{F}{1 + F^2}\right]\left[\frac{\gamma_L}{\gamma_T}\right]^{1/2} \quad (2.4\text{-}7b)$$

$$w_0 = -\frac{1}{1 + \kappa^2\xi_0^2/\gamma_L\gamma_T} = -\frac{1}{1 + F^2} , \quad (2.4\text{-}7c)$$

where

$$F = \frac{\kappa\xi_0}{(\gamma_L\gamma_T)^{1/2}} . \quad (2.4\text{-}8)$$

Since

$$\frac{\alpha}{2} \equiv -\frac{\partial\xi/\partial z}{\xi}\Big|_{\dot{\xi}=0} , \quad (2.4\text{-}9)$$

the factor α is proportional to v_0/ξ_0, so that Eq. (2.4-7b) has the same form as Eq. (2.2-11) and

$$\alpha_0 = \frac{8\pi\omega N p^2}{n_0\hbar c\gamma_T} . \quad (2.4\text{-}10)$$

For simplification, normalized incident and transmitted field amplitudes are introduced:

$$y \equiv \frac{\kappa\xi_I}{(\gamma_T\gamma_L T)^{1/2}} , \quad (2.4\text{-}11)$$

$$x \equiv \frac{\kappa\xi_T}{(\gamma_T\gamma_L T)^{1/2}} = F(L), \quad (2.4\text{-}12)$$

$$F(z) \equiv \frac{\kappa\xi(z)}{(\gamma_T\gamma_L)^{1/2}} . \quad (2.4\text{-}13)$$

In equilibrium $\dot{v} = \dot{w} = \dot{\xi} = 0$, yielding $\alpha = \alpha_0 (1 + x^2)^{-1}$ as in Eq. (2.2-11). The boundary conditions Eq. (2.2-1,2) and mean-field approximation ($T \ll \alpha L \ll 1$) lead to Eqs. (2.2-5,6) which yield Eq. (2.4-1) with $C = \alpha_0 L/2T$ and α_0 given by Eq. (2.4-10). Equation (2.2-15) is the square of Eq. (2.4-1) with $y^2 = Y$, $x^2 = X$, and $2C = k_0$. Both were derived in the same spirit, namely for small absorption ($\alpha_0 L \ll 1$) and small transmission ($T \ll 1$), so a field independent of position could be assumed. The condition for bistability follows exactly as in Section 2.2, so $\alpha_0 L/T > 8$ or $C > 4$.

Bonifacio and Lugiato have used this basic model to discuss many interesting features of optical bistability: critical slowing down (Section 5.6), self-pulsing (Section 6.2), and resonance fluorescence spectra (Section 2.8), for example. They have also used it to discuss the role of atomic cooperation. Note that C is controlled by varying the density, length, and reflectivity. In a one-atom theory $E_T = E_I$ or $x = y$, which is the solution Eq. (2.4-1) with $C = 0$. The value of the upper state density N_a is then proportional to N as seen from the formulas for the equilibrium values:

$$v = \left[\frac{\gamma_L}{\gamma_T}\right]^{1/2} \frac{x}{1 + x^2}, \qquad (2.4\text{-}14)$$

$$w = \frac{-1}{1 + x^2} \qquad (2.4\text{-}15)$$

$$N_a = \frac{N}{2}(w + 1) = \frac{N}{2}\frac{x^2}{1 + x^2}. \qquad (2.4\text{-}16)$$

Note that these are equivalent to Eq. (2.4-7) since $x \simeq F$ in this case. The cooperative nature of the C term is also suggested by the fact that C is related to the superfluorescence time τ_R:

$$C = \frac{\alpha_0 L}{2T} = \frac{T_2}{\tau_R T} \qquad (2.4\text{-}17)$$

using $\alpha_0 L = 2T_2/\tau_R = 2/\gamma_T \tau_R$, i.e.,

$$\tau R = \frac{8\pi \tau_0 n_0}{3N\lambda^2 L}, \qquad (2.4\text{-}18)$$

where τ_0 is the spontaneous lifetime of the two-level system. Equation (2.4-17) can be interpreted as the ratio of the coherence relaxation time T_2 to the superfluorescence time τ_R shortened to $T\tau_R$ by feedback.

Clearly the C term in Eq. (2.4-1) has a typical saturable absorption structure. In the saturation region, $x \gg 1$, the cooperative term is negligible. Equation (2.4-1) is plotted in Fig. 2.4-1 for a small value of C, showing $x \simeq y$, and a large value of C, showing that x differs markedly from y for $y < C$. For $C > 4$, Eq. (2.4-1) has one maximum $y_\uparrow$ for $x = x_{\uparrow L}$ and one minimum $y_\downarrow$ for $x = x_{\downarrow U}$, where the upper and lower branches are distinguished by subscripts U and L, respectively:

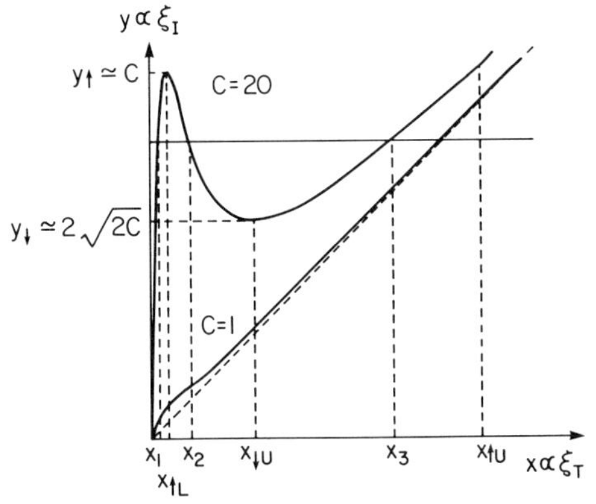

Fig. 2.4-1. Plot of the function $y = x + 2Cx/(1 + x^2)$ for C = 1 and C = 20. For C >> 1 one has $x_{\uparrow L} \simeq 1$, $y_\uparrow \simeq C$, $x_{\downarrow U} \simeq \sqrt{2C}$, and $y_\downarrow \simeq \sqrt{8C}$. Points x_1 and x_3 are stable; points x_2 are unstable. Bonifacio and Lugiato (1978a).

$$x_{\uparrow L} = \left[\frac{2C + 1}{C - 1 + (C^2 - 4C)^{1/2}}\right]^{1/2}, \tag{2.4-19}$$

$$x_{\downarrow U} = [C - 1 + (C^2 - 4C)^{1/2}]^{1/2} . \tag{2.4-20}$$

For C >> 1

$$x_{\uparrow L} \simeq 1, \quad y_\uparrow \simeq C, \tag{2.4-21}$$

$$x_{\downarrow U} \simeq \sqrt{2C}, \quad y_\downarrow \simeq 2\sqrt{2C} . \tag{2.4-22}$$

Using Eqs. (2.4-11) and (2.4-17), one finds for absorptive bistability that the switch-down intensity is independent of T whereas the switch-up intensity is proportional to 1/T.

To calculate the stable lower-state transmission, x_1, note that the x term in Eq. (2.4-1) contributes little to y_1, i.e.

$$y_1 \simeq \frac{2Cx_1}{1 + x_1^2}, \tag{2.4-23}$$

to which

$$x_1 = \frac{[C - (C^2 - y^2)^{1/2}]}{y} = \frac{y}{C + (C^2 - y^2)^{1/2}} \tag{2.4-24}$$

is a solution. To calculate the stable upper-state transmission, x_3, neglect 1 compared with x^2 in Eq. (2.4-1), obtaining

$$x_3 = \frac{y}{2}[1 + (1 - 8C/y^2)^{1/2}] . \tag{2.4-25}$$

Since for $y > C$, x_3 practically coincides with the one-atom solution, x_3 is called "one-atom stationary state." On the other hand, x_1 arises from atomic cooperation, so it is termed "cooperative stationary state." The region with negative slope dy/dx is unstable so that a switch-on (with $\Delta x \simeq x_{\uparrow U} - x_{\uparrow L} \simeq C$ from Fig. 2.4-1) occurs at $y = y_\uparrow$ and a switch-off at $y = y_\downarrow$.

Equations (2.4-14 to 16) reveal that a hysteresis cycle occurs in all of the relevant physical quantities; see Figs. 2.4-2,3 for hysteresis cycles for the fluorescence intensity (proportional to N_a) and atomic polarization (proportional to Nv). In the one-atom stationary state x_3, corresponding to practically complete saturation (i.e., $x_3 >> 1$), one has

$$v_3 \simeq \left[\frac{\gamma_L}{\gamma_T}\right]^{1/2} \frac{1}{x_3} = \left[\frac{\gamma_L}{\gamma_T}\right]^{1/2} \left[\frac{2/y}{1 + (1-8C/y^2)^{1/2}}\right] \tag{2.4-26}$$

$$N_{a3} \simeq \frac{N}{2} . \tag{2.4-27}$$

In this regime the fluorescence intensity is proportional to N, as for independent atoms. In the cooperative stationary state x_1

$$v_1 \simeq \left[\frac{\gamma_L}{\gamma_T}\right]^{1/2} x_1 = \left[\frac{\gamma_L}{\gamma_T}\right]^{1/2} \frac{y}{[C + (C^2 - y^2)^{1/2}]} \simeq \frac{y}{2C}\left[\frac{\gamma_L}{\gamma_T}\right]^{1/2} \tag{2.4-28}$$

for small y, and

$$N_{a1} \simeq \frac{Nx^2}{2} \simeq \frac{N}{2C^2} \frac{y^2}{[1 + (1 - y^2/C^2)^{1/2}]^2} . \tag{2.4-29}$$

In this regime the polarization is almost independent of N, and the total fluorescence intensity, which is proportional to N_a, is inversely proportional to the number of atoms. In this cooperative stationary state the incident field interacts with the atomic system as a whole, the atoms cooperatively create a reaction field that counteracts the incident field, and most of the light is reflected from the cavity. Note, however, that this so-called atomic cooperativity comes about by way of the intracavity field, i.e., it is a cavity effect. This can be seen in Eq. (2.4-28) in which v_1 depends on x_1; it is the cavity that makes $x_1 \simeq y/C$ so that $v_1 \propto N^{-1}$ and $Nv_1 \propto N^0$. In the one-atom stationary state the field overpowers the atoms by coherently saturating them, and the system is practically transparent.

For general discussions of their bistability theories see Bonifacio and Lugiato (1978b), Lugiato and Bonifacio (1978), and Bonifacio, Lugiato, and Gronchi (1979).

Most bistability experiments have been dominated by dispersive effects. Observations of purely absorptive bistability are discussed in Section 3.2.5. In the microwave range, good agreement has been found between the above theory and data taken using a spherical Fabry-Perot cavity containing ammonia gas by Arimondo, Gozzini, Lovitch, and Pistelli (1981) and Gozzini, Maccarrone, and Longo

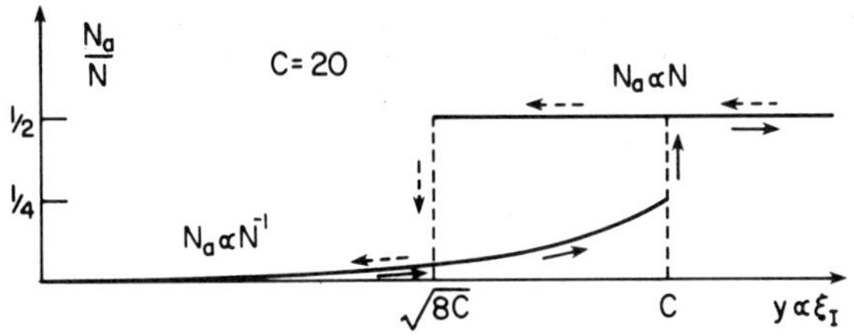

Fig. 2.4-2. Hysteresis cycle of the fluorescence intensity per atom (proportional to N_a/N). Full (dotted) line arrows indicate the variation obtained increasing (decreasing) the input field ξ_I. The upper part of the plot corresponds to the one-atom stationary state x_3. Bonifacio and Lugiato (1978a).

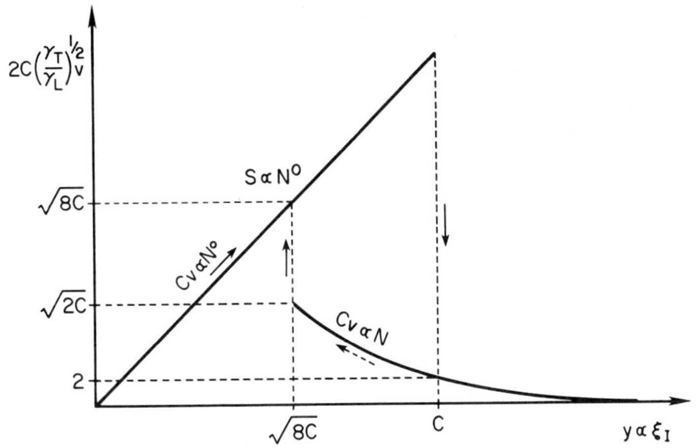

Fig. 2.4-3. Hysteresis cycle of the polarization (proportional to Nv or Cv). Arrow convention as in Figs. 2.4-1 and 2.4-2. The part of the plot beginning from the origin corresponds to the cooperative stationary state x_1, while the other branch corresponds to the one-atom stationary state. Bonifacio and Lugiato (1978a).

(1982a,b). Optical bistability achieved without a cavity by increasing absorption is treated in Section 2.10.

2.4.2. Analytical Theory with Spatial Variation for Absorptive Optical Bistability in a Ring Cavity

Two-mirror Fabry-Perot interferometers necessarily involve standing waves that make analytical solutions difficult. Sometimes rapid diffusion of excitation or inhomogeneous broadening effectively eliminates standing-wave effects.

However, a one-way ring cavity does not have standing waves at all and is experimentally achievable. Consequently the ring cavity has been a favorite configuration for theorists. Consider the ring cavity geometry of Fig. 2.1-1. The steady-state boundary conditions are:

$$\xi(0) = \sqrt{T}\,\xi_I + R\xi(L)\,, \tag{2.4-30}$$

$$\xi_T = \sqrt{T}\,\xi(L)\,, \tag{2.4-31}$$

$$\xi_R = \sqrt{RT}\,\xi(L) - \sqrt{R}\,\xi_I = \sqrt{R}\,(\xi_T - \xi_I)\,, \tag{2.4-32}$$

where the effects of intracavity absorption will appear as a difference between $\xi(0)$ and $\xi(L)$. The coupled Maxwell-Bloch equations are given by Eqs. (2.4-4,5) with $\omega_a = \omega$.

The stationary solutions of the coupled Maxwell-Bloch equations are then given by Eqs. (2.4-7) and

$$\frac{dF}{dz} = -\frac{\alpha_0}{2}\,\frac{F}{1 + F^2} \tag{2.4-33}$$

from Eq. (2.4-4) where F(z) is given by Eq. (2.4-13). The mean-field theory of Section 2.4.1 amounts to integrating Eq. (2.4-33) assuming that the z dependence of F(z) can be neglected:

$$\int_0^L dz\,\frac{F(z)}{1 + F^2(z)} \simeq L\,\frac{F(L)}{1 + F^2(L)} \tag{2.4-34a}$$

$$\simeq -\frac{2}{\alpha_0}\int_0^L \frac{dF}{dz}\,dz = -\frac{2}{\alpha_0}\,[F(L) - F(0)]\,. \tag{2.4-34b}$$

With the definitions Eq. (2.4-11, 12) for y and x, Eqs. (2.4-30, 31, 34) yield

$$F(0) = Rx + Ty \tag{2.4-35}$$

with $x \equiv F(L)$. Then

$$y = \frac{2Cx}{1 + x^2} + x \tag{2.4-1}$$

where for the ring cavity

$$C = \frac{\alpha_0 L_{RC}}{4T} \tag{2.4-36}$$

where L_{RC} is the length of the absorber in the ring cavity. Bonifacio and Lugiato (1978a) derive Eq. (2.4-1) with $C = \alpha_0 L/2T$ by defining α_0 as the field absorption coefficient; recall that $C = \alpha_0 L/2T$ for a Fabry-Perot cavity where α_0 is the intensity absorption coefficient. It is clear why C is $\alpha_0 L_{RC}/4T$ and $\alpha_0 L_{FP}/2T$: in one round trip a field encounters a transmission T twice in both cavities but experiences absorption over a length L only once in the ring cavity and twice in the Fabry-Perot cavity. So it is reasonable that a ring cavity needs twice the length of the absorber as a Fabry-Perot cavity for the same effect. Or, equivalently, if $T_{RC} = T_{FP}/2$ and $L_{RC} = L_{FP}$, the bistability effects are the same, again neglecting standing-wave effects.

Bonifacio and Lugiato (1978c) have integrated Eq. (2.4-33) directly:

$$\frac{\alpha_0}{2}\,dz = -\frac{dF}{F} - F\,dF \tag{2.4-37}$$

$$\frac{\alpha_0 L}{2} = \ln\frac{F(0)}{x} + \frac{1}{2}\left[F^2(0) - x^2\right] \tag{2.4-38}$$

where $x = F(L)$. Equations (2.4-38, 35) lead to

$$\boxed{\frac{\alpha_0 L}{2} = \ln\left[1 + T\left[\frac{y}{x} - 1\right]\right] + \frac{x^2}{2}\left[\left[1 + T\left[\frac{y}{x} - 1\right]\right]^2 - 1\right],} \tag{2.4-39}$$

which gives an exact relation between the normalized transmitted field x and the normalized incident field y for absorptive bistability of a ring cavity containing homogeneously broadened two-level atoms. This equation depends on $\alpha_0 L$ and T separately, not just on the ratio C as does the approximate Eq. (2.4-1). In the limit $T \to 0$, $\alpha_0 L \to 0$, and $\alpha_0 L/4T = C$ (constant)

$$\frac{\alpha_0 L}{2} \simeq T\left[\frac{y}{x} - 1\right] + \frac{x^2}{2}\left[1 + 2T\left[\frac{y}{x} - 1\right] - 1\right],$$

which is exactly Eq. (2.4-1), which was derived in the limit $T \to 0$, $\alpha_0 L \to 0$, and $C \gg 1$. Hence the mean-field approximation is valid for $T \to 0$ and $\alpha_0 L \to 0$, but for an arbitrary value of C. Figure 2.4-4 compares the mean-field state equation (Eq. (2.4-1)) for C = 10 with the exact equation (2.4-39), which can be plotted by assuming a value for $y/x \equiv \delta$, calculating x from Eq. (2.4-39), and then $y = x\delta$. The mean field equation is seen to be a close approximation even for $\alpha_0 L = 4$, which is far above the $\alpha_0 L \ll 1$ assumed in its derivation. Note that as $\alpha_0 L$ and T become large, the bistability disappears even though $C = \alpha_0 L/4T = 10 > 4$.

The exact equation can be used to study $T \to 1$. In the limit T = 1, i.e., no mirrors, Eq. (2.4-39) becomes

$$\frac{\alpha_0 L}{2} = \ln\left[\frac{y}{x}\right] + \frac{y^2}{2} - \frac{x^2}{2}. \tag{2.4-40}$$

Then

$$\frac{dy}{dx} = \left[\frac{1 + x^2}{x}\right]\left[\frac{y}{1 + y^2}\right], \tag{2.4-41}$$

which is never negative, so bistability never occurs. Thus a homogeneously broadened saturable absorber subjected to a plane-wave incident beam does not exhibit bistability without optical feedback, i.e., R = 1 - T must exceed zero.

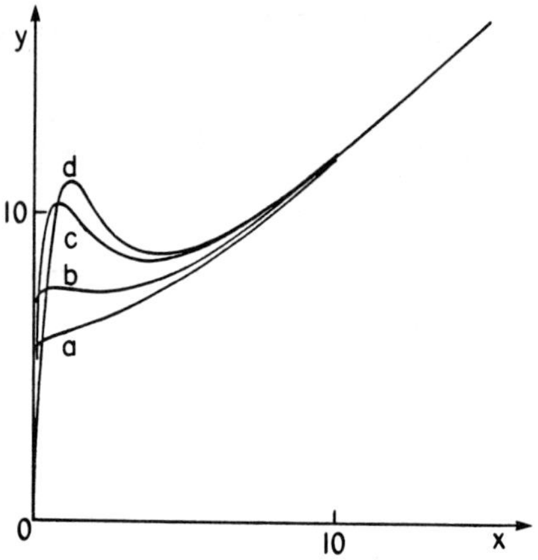

Fig. 2.4-4. Plot of incident light vs transmitted light for $C = \alpha_0 L/4T$ fixed equal to 10 and different values of $\alpha_0 L$ and T. For $\alpha_0 L \to 0$, $T \to 0$ one approaches the behavior predicted by the mean field theory. (a) $\alpha_0 L = 40$, $T = 1$; (b) $\alpha_0 L = 20$, $T = 0.5$; (c) $\alpha_0 L = 4$; $T = 0.1$; (d) mean field. Bonifacio and Lugiato (1978c)

2.4.3. Mixed Absorptive and Dispersive Optical Bistability

Absorptive bistability can in principle be seen with no dispersive effects and vice versa, but in practice both usually contribute. Mixed absorptive and dispersive bistability including inhomogeneous broadening was treated in Section 2.1. Readers who skipped Section 2.1 to follow the historical and progressively more difficult treatments in Section 2.2 through 2.4.2 should now return to Section 2.1 before continuing this section, which discusses the solutions presented in Section 2.1.

For analytic solutions it is convenient to assume a Lorentzian inhomogeneous atomic line shape centered at $\omega_a = \omega_a{}^0$ with $\gamma_2{}^* = 1/T_2{}^*$:

$$g_L(\Delta\omega) = \frac{\gamma_2{}^*/\pi}{(\Delta\omega - \Delta\omega^0)^2 + (\gamma_2{}^*)^2}, \qquad (2.4\text{-}42)$$

where $\Delta\omega^0 = \omega_a{}^0 - \omega \equiv \gamma_T \Delta_0$. Then Eq. (2.1-11) becomes [Gronchi and Lugiato (1980)]:

$$\sigma_L(|F|^2) = \frac{1}{\Delta_0^2 + (\gamma_2^*/\gamma_T + (1+|F|^2)^{1/2})^2}\left[\frac{\gamma_2^*/\gamma_T + (1+|F|^2)^{1/2}}{(1+|F|^2)^{1/2}} + i\Delta_0\right], \tag{2.4-43}$$

and Eq. (2.1-31) can be written in closed form.

Equation (2.1-25) can be compared with previous results. For a homogeneously broadened transition the inhomogeneous distribution function becomes a delta function

$$g(\Delta\omega) = \delta(\Delta\omega - \Delta\omega^0), \tag{2.4-44}$$

so that Eq. (2.1-25) becomes

$$y \simeq x\left[1 + \frac{2C}{1+|x|^2+\Delta_0^2}\right] + ix\left[\frac{2C\Delta_0}{1+|x|^2+\Delta_0^2} - \frac{R\beta_0'}{T}\right]. \tag{2.4-45}$$

If the laser, cavity, and atomic frequencies are coincident, $\Delta_0 = 0$ and $\beta_0' = 0$, and Eq. (2.4-45) reduces to the Szöke et al. result of Eq. (2.2-15) and the Bonifacio and Lugiato Eq. (2.4-1). In the case of a transition far off resonance from the laser frequency such that $\Delta_0^2 \gg 1 + |x|^2$, then

$$y \simeq x\left[1 - \frac{iR\beta_0'}{T} + 2C(1+i\Delta_0)\Delta_0^{-2}\left[1 - \frac{1+|x|^2}{\Delta_0^2}\right]\right]. \tag{2.4-46}$$

For sufficiently large Δ_0 the absorption term is negligible, resulting in an expression of the form

$$y \simeq x[1 + i(\beta'_2|x|^2 - \theta)], \tag{2.4-47}$$

where θ and β'_2 are independent of x and y. Equation (2.4-47) has the same form as the purely dispersive model of Section 2.3: substitute Eq. (2.3-6) into Eq. (2.3-5).

Bonifacio, Gronchi, and Lugiato (1979a) also conclude that for a homogeneously broadened line, C must exceed 4 for bistability even when dispersive effects are included in a mean-field approximation (see Section 2.5.1). Thus dispersive bistability is not easier than absorptive bistability in a purely homogeneous saturable system. They also find that for a given C, the largest hysteresis loops are obtained on resonance with no cavity detuning.

For an inhomogeneously broadened system the conclusions are quite different. For given values of the off-resonance parameter Δ_0, initial detuning β_0', and inhomogeneous broadening parameter γ_2^*/γ_T, there is bistability provided C is larger than a minimum value C_{min}, which depends upon Δ_0, β_0' and γ_2^*/γ_T. C_{min} increases rapidly with γ_2^*/γ_T. For $\gamma_2^*/\gamma_T \gg 1$ there are values of C such that the system is bistable for Δ_0 and β_0' large enough but not for $\Delta_0 = \beta_0' = 0$. Thus with inhomogeneous broadening, dispersive bistability may be possible when absorptive is not. This is shown in Fig. 2.4-5 for both absorptive and dispersive effects with inhomogeneous broadening described by a Lorentzian lineshape.

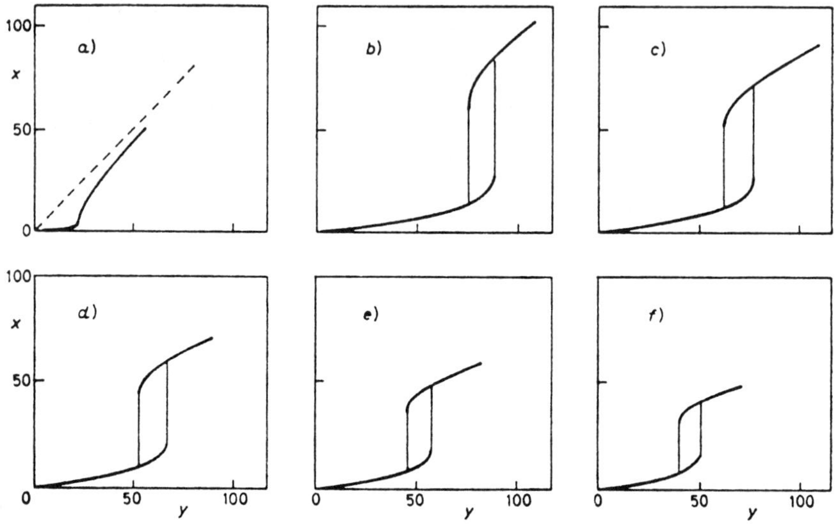

Fig. 2.4-5. Hysteresis cycles for an inhomogeneously broadened ring-cavity system with C = 180 and $\gamma_2^*/\gamma_T = 15$ as obtained from Eqs. (2.1-31) and (2.4-43); $\theta = R\beta_0'/T$. (a) $\theta = 0$, $\Delta_0 = 0$: the system does not exhibit absorptive bistability. The other plots are drawn for $\Delta_0 = 30$ and (b) $\theta = 1.0$, (c) $\theta = 1.5$, (d) $\theta = 2.0$, (e) $\theta = 2.5$, and (f) $\theta = 3.0$. Bonifacio, Gronchi, and Lugiato (1979a).

Gronchi and Lugiato (1980) have integrated Eq. (2.1-10) exactly using Eqs. (2.4-43) and (2.1-19), thus eliminating the mean-field approximation. The solution is complicated and not very revealing upon inspection, so it is not reproduced here. However, Figs. 2.4-6 and 7 compare the mean-field and exact bistability curves for the purely dispersive and purely absorptive inhomogeneously broadened cases, respectively. For the conditions of Fig. 2.4-6, bistability does not occur on resonance, i.e., absorptive bistability fails. Of course, for $\Delta\omega^0 \equiv \gamma_T\Delta_0 \equiv 2\gamma_2^*$ the inhomogeneous broadening is less important than the detuning, i.e., an inhomogeneous line appears homogeneous for sufficient detuning.

Equation (2.1-31) was also obtained by Hassan, Drummond, and Walls (1978) [see also Drummond (1979)] and, in the homogeneous broadening limit, by Agrawal and Carmichael (1979). Roy and Zubairy (1980a) have also derived exact analytic solutions for absorptive and dispersive ring-cavity bistability in the presence of inhomogeneous broadening and reduced them to Eq. (2.1-31) in the mean-field limit. Hermann and Carmichael (1982) show that Stark or light-shift terms associated with nonresonant transitions can affect dispersive bistability by changing the frequency detuning between the laser frequency and the resonant transition.

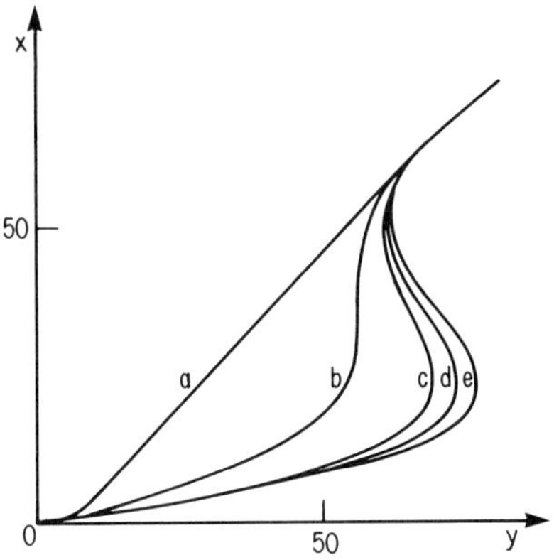

Fig. 2.4-6. Comparison of Lorentzian-atomic-line-shape exact solutions with mean-field solution for purely dispersive ring-cavity case. Plot of the transmitted amplitude x versus the incident amplitude y for C = 180, $\gamma_2{}^*/\gamma_T = 15$, $\Delta_0 = 30$: $\theta = 1.5$; and (a), $\alpha_0 L = 720$, T = 1; (b) $\alpha_0 L = 360$, T = 0.5; (c) $\alpha_0 L = 144$, T = 0.2; (d) $\alpha_0 L = 72$, T = 0.1; (e) $\alpha_0 L = 7.2$, T = 0.01 coincident with the mean-field result. Gronchi and Lugiato (1980).

2.5. CONDITIONS FOR OPTICAL BISTABILITY

2.5.1 Homogeneous Nonlinear Absorption and Nonlinear Refractive Index within a Ring Cavity

From Eqs. (2.1-29-31) one has, for a saturable susceptibility of the form

$$\sigma = \sigma_r + i\sigma_i = \frac{1 + i\Delta}{1 + \Delta^2 + X} \tag{2.5-1}$$

a mean-field state equation

$$Y = X\Big[[1 + 2C\sigma_r(X)]^2 + [2C\sigma_i(X) - \theta]^2\Big]. \tag{2.5-2}$$

It will now be shown [McCall and Gibbs (1980)] that the requirement for bistability of Eq. (2.5-2) is that

$$\frac{1}{2C} < \mathrm{Sup}\left[-\sigma_r - X\frac{d\sigma_r}{dX} + X\left[\left[\frac{d\sigma_r}{dX}\right]^2 + \left[\frac{d\sigma_i}{dX}\right]^2\right]^{1/2}\right]. \tag{2.5-3}$$

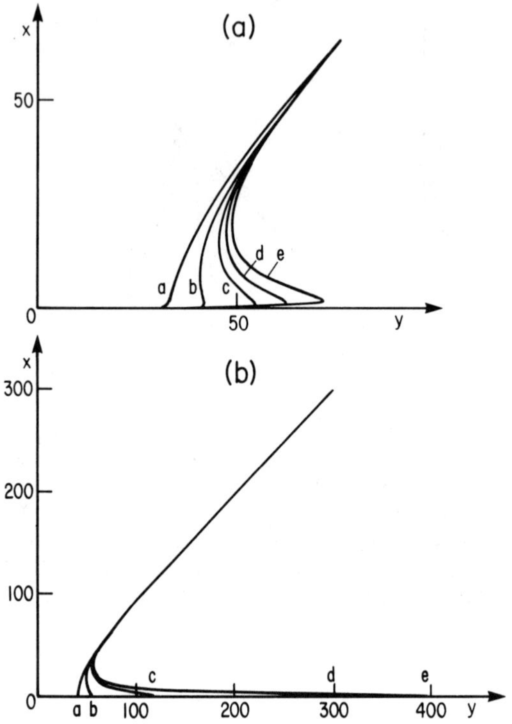

Fig. 2.4-7. Effect of inhomogeneous broadening in absorptive ring-cavity bistability C = 400; $\Delta_0 = \theta = 0$, and curve (e) coincides with the mean-field result. (a) $\gamma_2^*/\gamma_T = 8$, a) $\alpha_0 L = 1600$, T = 1; b) $\alpha_0 L = 800$, T = 0.5; c) $\alpha_0 L = 320$, T = 0.2; d) $\alpha_0 L = 160$, T = 0.1; e) $\alpha_0 L = 16$, T = 0.01. (b) $\gamma_2^*/\gamma_T = 0$; a) $\alpha_0 L = 1600$, T = 1; b) $\alpha_0 L = 800$, T = 0.5; c) $\alpha_0 L = 160$, T = 0.1; d) $\alpha_0 L = 16$, T = 0.01; e) $\alpha_0 L = 1.6$, T = 0.001. Gronchi and Lugiato (1980).

By Sup is meant the largest value of the subsequent quantity when X is allowed to vary over all positive values. For bistability, the requirement is that $dY/dX < 0$ for some $X > 0$

$$\frac{dY}{dX} = (1 + 2C\sigma_r)^2 + (2C\sigma_i - \theta)^2$$

$$+ 2X\left[(1 + 2C\sigma_r)\, 2C\,\frac{d\sigma_r}{dX} + (2C\sigma_i - \theta)2C\,\frac{d\sigma_i}{dX}\right]. \tag{2.5-4}$$

Now choose θ to minimize dY/dX:

$$\frac{d(dY/dX)}{d\theta} = -2(2C\sigma_i - \theta) - 4XC\frac{d\sigma_i}{dX} = 0$$

$$= -2\left[-\theta + 2C\sigma_i + 2CX\frac{d\sigma_i}{dX}\right]. \qquad (2.5\text{-}5)$$

But dY/dX can be written as

$$\frac{dY}{dX} = \left[1 + 2C\sigma_r + 2C\frac{d\sigma_r}{dX}X\right]^2 + \left[-\theta + 2C\sigma_i + 2CX\frac{d\sigma_i}{dX}\right]^2$$

$$- 4C^2X^2\left[\left[\frac{d\sigma_r}{dX}\right]^2 + \left[\frac{d\sigma_i}{dX}\right]^2\right]. \qquad (2.5\text{-}6)$$

Note that the center term on the right side of Eq. (2.5-6) is zero according to Eq. (2.5-5) if θ is selected to minimize dY/dX. Then for dY/dX to be less than zero, Eq. (2.5-6) implies

$$1 + 2C\sigma_r + 2C\frac{d\sigma_r}{dX}X < 2CX\left[\left[\frac{d\sigma_r}{dX}\right]^2 + \left[\frac{d\sigma_i}{dX}\right]^2\right]^{1/2},$$

which is the same as Eq. (2.5-3).

For homogeneously broadened two-level saturable systems,

$$\sigma = \frac{1 + i\Delta_0}{1 + \Delta_0^2 + X} = \frac{1 + i\Delta_0}{D},$$

where $D \equiv 1 + \Delta_0^2 + X$,

$$\frac{d\sigma_r}{dX} = -D^{-2},$$

$$\frac{d\sigma_i}{dX} = -\Delta_0 D^{-2}.$$

Then Eq. (2.5-3) becomes

$$\frac{1}{2C} < \mathrm{Sup}[-D^{-1} + XD^{-2} + X(D^{-4} + \Delta_0^2 D^{-4})^{1/2}],$$

$$\frac{1}{2C} < \mathrm{Sup}\left[[-1 - \Delta_0^2 + X(1 + \Delta_0^2)^{1/2}]D^{-2}\right].$$

Taking the derivative with respect to X of the right-hand side and setting it equal to zero, gives for the value of X for the supremum,

$$X = (1 + \Delta_0^2)^{1/2} [2 + (1 + \Delta_0^2)^{1/2}] , \quad (2.5\text{-}7)$$

so that

$$\frac{1}{2C} < \frac{(1 + \Delta_0^2)[1 + (1 + \Delta_0^2)^{1/2}]}{\left[2(1 + \Delta_0^2)^{1/2}[1 + (1 + \Delta_0^2)^{1/2}]\right]^2}$$

$$\boxed{C > 2[1 + (1 + \Delta_0^2)^{1/2}] , \quad (2.5\text{-}8)}$$

where $C = \alpha_0 L_{RC}/4T$. For purely absorptive bistability $\Delta_0 = 0$, so the condition is $C > 4$. For $\Delta_0 > 0$, C must exceed 4, i.e., dispersive bistability is more difficult than absorptive bistability for a completely saturable homogeneously broadened transition (the same conclusion was reached in Section 2.4.3). Evaluating σ_i and $d\sigma_i/dX$ for X given by Eq. (2.5-7), the optimum θ of Eq. (2.5-5) becomes

$$\theta_{min} = \frac{\Delta_0}{1 + (1 + \Delta_0^2)^{1/2}} \quad (2.5\text{-}9)$$

using the minimum value for C given in Eq. (2.5-8). θ is related to the cavity-laser detuning $\Delta\nu_C$ by Eq. (2.1-28). Compared with the empty-cavity FWHM instrument width $\Delta\nu_{RC} = c/L_T\mathscr{F} = cT/L_T\pi\sqrt{R}$,

$$\frac{\Delta\nu_C}{\Delta\nu_{RC}} \simeq \frac{\theta}{2\sqrt{R}} . \quad (2.5\text{-}10)$$

For $R \simeq 1$, $\Delta\nu_C/\Delta\nu_{RC} \simeq \theta/2$. For $\Delta_0 >> 1$, $\theta_{min} \simeq 1$ and the best cavity detuning is half an instrument width. For $\Delta_0 << 1$, $\Delta\nu_C/\Delta\nu_{RC} \simeq \Delta_0/4$.

Equation (2.5-8) for a ring cavity or for a Fabry-Perot cavity, neglecting standing wave effects, is compared with the condition including standing waves in Section 2.5.2.

Tewari and Tewari (1979) and Tewari, Tewari, and Das (1980) do not allow for cavity-laser detuning, so they arrive at an equality different from Eq. (2.5-8); their result still shows C must always exceed 4. Conditions for bistability in homogeneously broadened media are discussed in many of the papers cited in Section 2.1.

2.5.2. Standing-Wave Effects

Most theoretical treatments of optical bistability have been for a ring cavity to avoid the effects of the standing waves in a two-mirror Fabry-Perot cavity. Most experiments are performed with a Fabry-Perot cavity either for convenience or because of the micrometer dimensions of semiconductor devices. Even in a Fabry-Perot configuration the effect of standing waves may be reduced or eliminated by diffusion of the excitation from strong-field regions to weak-field regions in a time much less than other characteristic times. In any case it is interesting to determine how important standing-wave effects can be.

The electric field inside a Fabry-Perot cavity may be written as the sum of a forward field and backward field

$$\mathcal{E} = E_F e^{i(\omega t - kz)} + E_B e^{i(\omega t + kz)} + c.c. \tag{2.5-11}$$

Because of the interference between the forward and backward waves, the superposition field may vary markedly in a wavelength λ, contrary to the usual slowly varying envelope approximation (SVEA): $\lambda \partial E/\partial z \ll E$. However, if $\alpha\lambda \ll 1$ then E_F and E_B still satisfy the SVEA; only their sum shows rapid variations that can be accounted for separately. The traveling Maxwell equation is

$$e^{-ikz}\left[\frac{\partial E_F}{\partial t} + c\frac{\partial E_F}{\partial z}\right] + e^{+ikz}\left[\frac{\partial E_B}{\partial t} - c\frac{\partial E_B}{\partial z}\right] = -\frac{c\gamma_T\alpha_0}{2\kappa} Q, \tag{2.5-12}$$

where the kz exponentials remain in Q. The Bloch equations are

$$\dot{Q} = -\kappa E w - (\gamma_T - i\Delta\omega)Q \tag{2.5-13a}$$

and

$$\dot{w} = \frac{+\kappa(EQ^* + E^*Q)}{2} - \gamma_L(w + 1), \tag{2.5-13b}$$

where $E = E_F e^{-ikz} + E_B e^{ikz}$. Meystre (1978), following Fleck (1968), expanded Q and w:

$$Q = Q_F e^{-ikz} + Q_B e^{ikz} + Q_{-3} e^{-i3kz} + Q_3 e^{i3kz} + \dots \tag{2.5-14a}$$

$$w = w_0 + w_2 e^{2ikz} + w_{-2} e^{-2ikz} + \dots \tag{2.5-14b}$$

so that Eqs. (2.5-13) become an infinite number of coupled equations. Meystre kept w_2 and w_{-2} but not Q_3 and Q_{-3}. At the end of this section his results will be compared with the more accurate procedure of McCall (1974). McCall expanded the complex optical susceptibility σ, e.g., Eq. (2.5-1) for a homogeneous two-level system:

$$\sigma = \sigma(I) = \sigma_0 + \sigma_{2k} e^{2ikz} + \sigma_{-2k} e^{-2ikz} + \dots, \tag{2.5-15}$$

where

$$I = |E_F e^{-ikz} + E_B e^{+ikz}|^2 . \tag{2.5-16}$$

Succeeding terms in σ will be reduced in their effect by at least order $(\alpha\lambda)^2$. Hence the series is truncated after σ_{-2k}. The equations used by Meystre (1978) do not include all contributions to σ_0 and $\sigma_{\pm 2k}$. Following McCall and Gibbs (1980), one finds in steady state that Eq. (2.5-12) becomes

$$e^{-ikz}\frac{\partial E_F}{\partial z} - e^{ikz}\frac{\partial E_B}{\partial z}$$

$$= -\frac{\alpha_0}{2}(E_F\, e^{-ikz} + E_B\, e^{ikz})(\sigma_0 + \sigma_{2k}\, e^{2ikz} + \sigma_{-2k}\, e^{-2ikz}) \quad (2.5\text{-}17)$$

yielding, for the e^{-ikz} coefficients,

$$\frac{\partial E_F}{\partial z} = -\frac{\alpha_0}{2}(\sigma_0 E_F + \sigma_{-2k}E_B) \quad (2.5\text{-}18a)$$

and, for the e^{+ikz} coefficients,

$$\frac{\partial E_B}{\partial z} = +\frac{\alpha_0}{2}(\sigma_0 E_B + \sigma_{2k}E_F)\,. \quad (2.5\text{-}18b)$$

In the limit $\alpha_0 L$, $T \to 0$,

$$E_F(0) \simeq E_F(L) + \frac{\alpha_0 L}{2}[\sigma_0 E_F(L) + \sigma_{-2k}\, E_B(L)]\,, \quad (2.5\text{-}19a)$$

$$E_B(0) \simeq E_B(L) - \frac{\alpha_0 L}{2}[\sigma_0 E_B(L) + \sigma_{+2k}E_F(L)]\,. \quad (2.5\text{-}19b)$$

The Fabry-Perot boundary conditions are (see Appendix B):

$$E_F(L) = \frac{E_T}{\sqrt{T}} \quad (2.5\text{-}20a)$$

$$E_B(L) = \frac{\sqrt{R}\, E_T}{\sqrt{T}} \quad (2.5\text{-}20b)$$

$$E_F(0) = \sqrt{T}\, E_I + \sqrt{R}\, e^{i\beta_0}\, E_B(0)\,. \quad (2.5\text{-}20c)$$

Substituting Eqs. (2.5-19) into Eq. (2.5-20c) and using Eqs. (2.5-20a, 20b) gives

$$E_I = \left[\frac{1 - Re^{i\beta_0}}{T} + \frac{\alpha_0 L}{2T}\left[\sigma_0(1 + Re^{i\beta_0}) + \sqrt{R}\,\Sigma\,(1 + e^{i\beta_0})\right]\right] E_T\,, \quad (2.5\text{-}21)$$

where $\Sigma = \sigma_{+2k}(z{=}L) = \sigma_{-2k}(z{=}L)$ as will be shown below. Using $\alpha_0 L$, T, $\beta_0 \to 0$,

$$E_I = E_T\left[1 + 2C\sigma_r + i\left[2C\sigma_i - \frac{\beta_0}{T}\right]\right]; \quad (2.5\text{-}22)$$

$\sigma_r + i\sigma_i = \sigma_0 + \Sigma$. The square of Eq. (2.5-22) is the same as Eq. (2.5-2) for $R \to 1$. Since the phases of E_F and E_B do not enter in the calculations of σ_r and σ_i, as will be shown below, σ_r and σ_i are functions of the fields only through the intensity given by Eq. (2.5-16). Consequently, the inequality Eq. (2.5-3) still applies as the condition for bistability.

With this motivation let us return to the calculation of σ_0 and $\sigma_{\pm 2k}$. From the definition Eq. (2.5-15)

$$\sigma_0 = \frac{1}{2\pi}\int_0^{2\pi} d\phi\,\sigma(|E_F|^2 + |E_B|^2 + E_F E_B{}^* e^{-i\phi} + E_F{}^* E_B e^{i\phi}) \tag{2.5-23}$$

with $\phi = 2kz$, and

$$\sigma_{\pm 2k} = \frac{1}{2\pi}\int_0^{2\pi} d\phi\, e^{\mp i\phi}\,\sigma(|E_F|^2 + |E_B|^2 + E_F E_B{}^* e^{-i\phi} + E_F{}^* E_B e^{i\phi})\,. \tag{2.5-24}$$

Let $E_F E_B{}^* = e^{-i\theta}|E_F E_B|$, then

$$\sigma_{+2k} = \frac{1}{2\pi}\int_0^{2\pi} d\phi\, e^{-i\phi}$$

$$\times\, \sigma(|E_F|^2 + |E_B|^2 + 2|E_F E_B|\cos(\theta + \phi))$$

$$= \frac{e^{+i\theta}}{2\pi}\int_0^{2\pi} d(\theta + \phi)\, e^{-i(\theta + \phi)}\,\sigma(|E_F|^2 + |E_B|^2 + 2|E_F E_\beta|\cos(\theta + \phi))$$

$$= \frac{E_F{}^* E_B}{|E_F E_B|}\,\Sigma \tag{2.5-25}$$

with

$$\Sigma = \frac{1}{2\pi}\int_0^{2\pi} d\phi\,\sigma(|E_F|^2 + |E_B|^2 + 2|E_F E_B|\cos\phi)\cos\phi\,, \tag{2.5-26}$$

where one has used the fact that $\int d\phi\sigma \sin\phi$ is zero, because σ is an even function of ϕ. Note that $\sigma_{+2k}(z=L) = \Sigma = \sigma_{-2k}(z=L)$.

As a specific example consider the important case of a saturable absorber driven off-resonance by a frequency difference $\Delta\omega$. Then the steady-state solution of Eqs. (2.5-13) yields

$$\sigma = \frac{1 + i\Delta}{1 + \Delta^2 + I}\,, \tag{2.5-27}$$

where I is Eq. (2.5-16) evaluated at $z = L$ and multiplied by $\kappa^2/\gamma_L\gamma_T$. Define

$$A \equiv 1 + \Delta^2 + \frac{(|E_F|^2 + |E_B|^2)\kappa^2}{\gamma_L\gamma_T} \tag{2.5-28}$$

and

$$b \equiv \frac{2\kappa^2|E_FE_B|}{A\gamma_L\gamma_T}. \tag{2.5-29}$$

Then one has

$$\sigma_0 = \frac{1}{2\pi}\int_0^{2\pi} \frac{1 + i\Delta}{A + (E_FE_B^* e^{-i\phi} + E_F^*E_B e^{i\phi})\kappa^2/\gamma_L\gamma_T}\, d\phi$$

$$= \frac{1 + i\Delta}{2\pi A}\int_0^{2\pi} \frac{1}{1 + b\cos(\phi + \theta)}\, d\phi$$

$$= \frac{1 + i\Delta}{A(1 - b^2)^{1/2}}, \tag{2.5-30}$$

using integral tables for the definite integral. Similarly Eq. (2.5-26) for Σ:

$$\Sigma = \frac{1 + i\Delta}{2\pi A}\int_0^{2\pi} \frac{\cos\phi}{1 + b\cos\phi}\, d\phi$$

$$= \frac{1 + i\Delta}{2\pi Ab}\int_0^{2\pi} \frac{(1 + b\cos\phi) - 1}{1 + b\cos\phi}\, d\phi$$

$$= \frac{1 + i\Delta}{2\pi Ab}\left[2\pi - \frac{2\pi}{(1 - b^2)^{1/2}}\right]$$

$$= \frac{(1 + i\Delta)[1 - (1 - b^2)^{-1/2}]}{Ab}. \tag{2.5-31}$$

Comparisons with no-standing-wave models can be made by assuming that the irradiated atoms diffuse a distance $\lambda/4$ in times short compared to the energy relaxation time. Then the high-frequency spatial modulation of the light is not transferred to the medium, so $\Sigma = 0$ and $\sigma_0 = \sigma(|E_F|^2 + |E_B|^2)$; for $\Delta = 0$ and $z = L$

$$\sigma = \frac{1}{1 + \kappa^2(|E_F|^2 + |E_B|^2)/\gamma_L\gamma_T} = \frac{1}{1 + 2X}, \tag{2.5-32}$$

where $E_F = E_B = E_T/\sqrt{T}$ and $X = \kappa^2|E_F|^2/\gamma_L\gamma_T$. Then $\sigma_0 = \sigma_r = \sigma$ and by Eq. (2.5-3), $1/2C < \text{Sup}[-\sigma_r - 2X\,(d\sigma_r/dX)]$ since $d\sigma_r/dX < 0$; therefore

$$\frac{1}{2C} < \text{Sup}\left[\frac{2X - 1}{(1 + 2X)^2}\right] = \frac{1}{8} \tag{2.5-33}$$

for $X = 1.5$, i.e., $C > 4$ as found by Szöke et al. (Section 2.2) and Bonifacio and Lugiato (Section 2.4.1) for purely absorptive bistability neglecting standing waves.

Now assume Eq. (2.5-27) holds and standing-wave effects are fully present ($\Delta = 0$ still); then by Eq. (2.5-30)

$$\sigma_0 = (1 + 4X)^{-1/2}; \tag{2.5-34}$$

and by Eq. (2.5-31):

$$\Sigma = \frac{\left[1 - \dfrac{1 + 2X}{(1 + 4X)^{1/2}}\right]}{2X}. \tag{2.5-35}$$

Then $\sigma_i = 0$ and

$$\sigma_r = \sigma_0 + \Sigma = \frac{\left[1 - \dfrac{1}{(1 + 4X)^{1/2}}\right]}{2X} \tag{2.5-36}$$

$$\frac{1}{2C} < \text{Sup}\left[\frac{1 - 2(1 + 4X)^{-1/2} + (1 + 4X)^{-3/2}}{2X}\right]. \tag{2.5-37}$$

The right-hand side has a maximum at $X = 1.268$, so that $2C > 9.937$, about 24% larger than 8. One concludes that standing waves increase the $\alpha_0 L/T$ needed for bistability.

Now return to Eq. (2.5-27) and retain the dispersive Δ terms. The no-standing-wave case was worked out in Section 2.5.1, yielding

$$2C > 4[1 + (1 + \Delta^2)^{1/2}]. \tag{2.5-8}$$

In the absence of diffusion, standing-wave effects are fully effective and $\sigma_0 + \Sigma$ is given by the sum of Eqs. (2.5-30, 31):

$$\sigma_r + i\sigma_i = \frac{(1 + i\Delta)\left[1 - \left[\dfrac{1 + \Delta^2}{1 + \Delta^2 + 4X}\right]^{1/2}\right]}{2X}; \tag{2.5-38}$$

and Eq. (2.5-3) becomes with $A_0 \equiv 1 + \Delta^2$

$$\frac{1}{2C} < \mathrm{Sup}\left[\frac{\left[A_0^{1/2} - \dfrac{1 + 3A_0^{1/2}}{2(1 + 4X/A_0)^{1/2}} + \dfrac{1 + A_0^{1/2}}{2(1 + 4X/A_0)^{3/2}}\right]}{2X}\right], \tag{2.5-39}$$

and the standing-wave state equation is

$$Y = X\left[1 + \frac{C}{X}\left[1 - \left[\frac{1 + \Delta^2}{1 + \Delta^2 + 4X}\right]^{1/2}\right]\right]^2$$

$$+ X\left[\frac{C\Delta}{X}\left[1 - \left[\frac{1 + \Delta^2}{1 + \Delta^2 + 4X}\right]^{1/2}\right] - \frac{R\beta_0'}{T}\right]^2 . \tag{2.5-40}$$

The results of Eq. (2.5-8) for no standing waves and Eq. (2.5-39) for standing waves are shown in Fig. 2.5-1; as detuning increases there is a decrease in the percentage increase in C as a result of standing wave effects.

Standing waves can decrease the input power required to switch-up the bistable system as shown in Fig. 2.5-2. For values of 2C greater than those given by Eqs. (2.5-8) and (2.5-39), one may find incident and transmitted switch-up and switch-down powers with state equations, given a value of the detuning parameter β_0. For Fig. 2.5-2

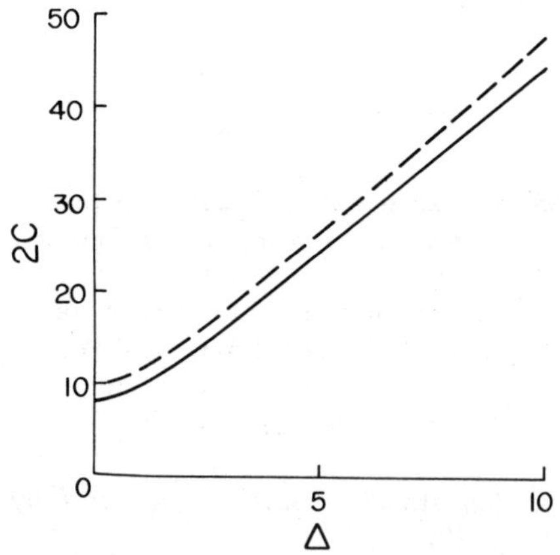

Fig. 2.5-1. Values of $2C = \alpha_0 L/T$ required for optical bistability as a function of Δ for a nonlinear Fabry-Perot cavity. The lower solid curve is for the no-standing-wave case of Eq. (2.5-8) while the dashed curve is for the standing-wave model of Eq. (2.5-39). McCall and Gibbs (1980).

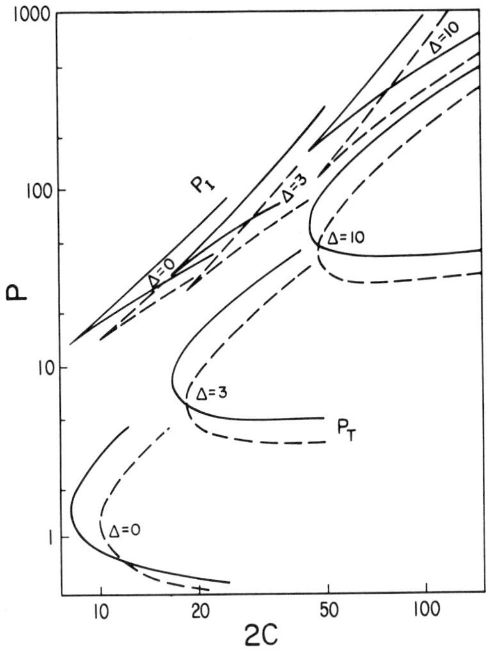

Fig. 2.5-2. Switching powers for three values of Δ as a function of $2C = \alpha_0 L/T$. Dashed curves are for standing-wave models and solid curves are for no-standing-wave models. The upper curves are input switching powers, and the lower curves are output switching powers. Recalling that powers are defined by $P_I = (E_I)^2/T$ and $P_T = (E_T)^2/T$, the variation $2C = \alpha_0 L/T$ should be regarded as due to $\alpha_0 L$ changing and T constant. McCall and Gibbs (1980).

$$\beta_0 = \frac{2C_{min}T}{R} \frac{d}{dX} (X\sigma_i)$$

satisfying Eq. (2.5-5), where C_{min} is determined by Eq. (2.5-3). Under these assumptions the standing-wave effects lower switching powers by typically 35%.

Comparison with other work will now be made. Most other treatments do not use the freedom to choose the initial laser-cavity detuning, i.e., β_0 or θ, in such a way as to minimize the power required. Some others do arrive at the same state equation (2.5-2) with the same nonlinear optical susceptibility. For example, Eq. (65) of Carmichael and Hermann (1980) is the same as Eq. (2.5-38) and was found by linearizing exact analytic solutions. In the mean-field limit standing-wave state equations have been found by averaging the standing mode function in the denominator of the steady-state polarization [Agrawal and Carmichael (1979)] and by adopting a harmonic expansion for both polarization and inversion and truncating the resultant Bloch hierarchy to include standing waves to first order [Bonifacio and Lugiato (1978a); Meystre (1978); Abraham, Bullough, and Hassan (1979); Roy and Zubairy (1980b)]. Within the truncated Bloch hierarchy approximation, Roy and Zubairy (1980b) give analytic results for both Fabry-Perot and ring-cavity geometries. The truncation approximation seems unavoidable in transient studies, so it is instructive to compare it with untruncated results in steady-state. Quantitatively, they often differ by 10% to 30%, e.g., for purely absorptive bistability Eq. (2.5-37) gives $2C > 9.937$, whereas Meystre (1978) finds $2C > 8$ in the mean-field limit. Carmichael (1980) has made a nice comparison by showing

$$y_{\text{truncated}} = x + \frac{2Cx}{1 + 3x^2}, \tag{2.5-41}$$

$$y_{\text{full}} = x + \frac{C}{x}\left[1 - \frac{1}{(1 + 4x^2)^{1/2}}\right], \tag{2.5-42}$$

$$y_{\text{average}} = x + \frac{2Cx}{1 + 2x^2}. \tag{2.5-43}$$

With $Y = y^2$, $X = x^2$, $\beta_0 = 0$, and $\Delta = 0$, Eq. (2.5-40) gives Eq. (2.5-42). Maintaining the same intensity normalization, the Fleck-Meystre truncation yields Eq. (2.5-41), and an average of the interference term over a wavelength yields Eq. (2.5-43). These three state equations are compared with exact numerical results in Fig. 2.5-3. The exact results for $R = 0.99$, $\alpha_0 L = 0.4$ are indistinguishable from the plot of Eq. (2.5-42). Hermann (1980) reaches the same conclusions.

Carmichael and Agrawal (1980, 1981) have extended the full hierarchy treatment to inhomogeneous broadening and to a Doppler-broadened medium in particular. In Fig. 2.5-4 are shown the effects of Doppler broadening on absorptive bistability in the mean-field limit. In this figure C is held constant while γ_2^*/γ_T is varied. But as γ_2^* is increased, the peak absorption decreases for fixed C. So the effective C, $\alpha_{\text{peak}}L/2T$, might fall below 8, even when $\alpha_0 L/2T$ is well above. By keeping the weak-field absorption constant as γ_2^*/γ_T changes, as in Fig. 2.5-5, one finds that the $\gamma_2^*/\gamma_T = 10$ curve in Fig. 2.5-4 is far from bistable whereas in Fig. 2.5-5 it is strongly bistable. Also the reduction in Fig. 2.5-4 of the switch-up field with increased inhomogeneous broadening is seen to result from reduced absorption.

2.5.3. Unsaturable Background Absorption

In the first three attempts to observe optical bistability, nonlinear transmission and pulse reshaping were seen but bistability was not [Szöke, Daneu, Goldhar, and Kurnit (1969); Spiller (1971, 1972); Austin and Deshazer (1971) and Austin (1972)]. It is believed that bistability was prevented by the detrimental effects of an unsaturable background absorption coefficient α_B. An α_B can arise from off-resonance absorption by other transitions with higher effective saturation intensities or from excited-state absorption. That an α_B makes bistability more difficult can be argued by treating $\alpha_B L$ as a mirror transmission since both are loss mechanisms for a cavity. The condition Eq. (2.2-19), derived assuming small $\alpha_0 L$, then becomes

$$\frac{\alpha_0 L}{T + \alpha_B L} > 8. \tag{2.5-44}$$

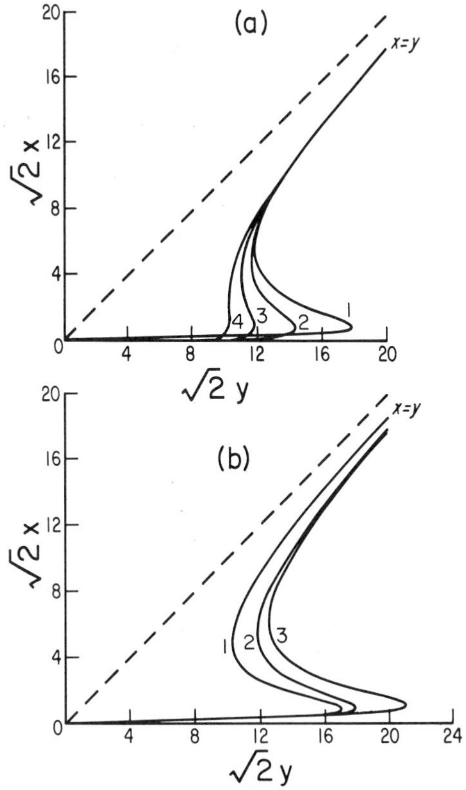

Fig. 2.5-3. Comparison of Fabry-Perot standing-wave approximations. (a) Bistability curves for $C = \alpha_0 L/2T = 20$ calculated numerically: (1) $R = 0.99$, $\alpha_0 L = 0.4$; (2) $R = 0.8$, $\alpha_0 L = 8$; (3) $R = 0.6$, $\alpha_0 L = 16$; (4) $R = 0.4$, $\alpha_0 L = 24$. (b) Bistability curves from the three state equations for $C = 20$: (1) Eq. (2.5-41); (2) Eq. (2.5-42); (3) Eq. (2.5-43). Note that curve (1) in (a) is indistinguishable from curve (2) in (b). Carmichael (1980).

In the limit $T \to 0$, Eq. (2.5-44) states that $\alpha_0/\alpha_B > 8$, i.e., the ratio of saturable to unsaturable loss must exceed 8. Since mirror transmission and losses are never zero, in practice $\alpha_0/\alpha_B \simeq 20$ is a good target. Many transitions do not saturate that completely. For the off-resonance case, α saturates as

$$\alpha = \frac{\alpha_0}{1 + \Delta_0^2 + I/I_s} + \alpha_B = \alpha_0\left[\sigma_r + \frac{\alpha_B}{\alpha_0}\right]; \qquad (2.5\text{-}45)$$

it is of no help if α_B does saturate but only for I much higher than I_s. The addition of α_B/α_0 to σ_r simply adds α_B/α_0 to 1/2C [see Eq. (2.5-3) in the derivation of Eq. (2.5-8)]. Therefore

$$\frac{\alpha_0 L}{T + \alpha_B(\Delta_0)L} > 4[1 + (1 + \Delta_0^2)^{1/2}] ; \qquad (2.5\text{-}46)$$

thus α_0/α_B must be even larger than 8 for dispersive bistability. However, in Eq. (2.5-46) α_B has been written as $\alpha_B(\Delta_0)$ to emphasize that it may be frequency dependent. As will be seen in Section 3.6, the decrease of α_B with frequency below the GaAs exciton resonance was used to satisfy Eq. (2.5-46) off-resonance, even though it could not be satisfied on-resonance.

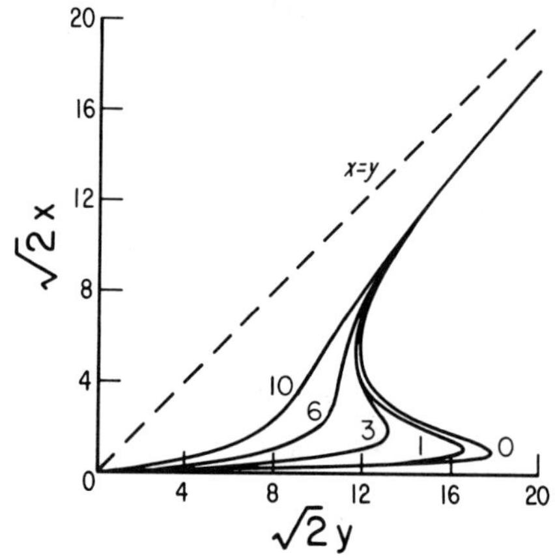

Fig. 2.5-4. Fabry-Perot full-Bloch-hierarchy numerical calculations: the effects of Doppler broadening on absorptive bistability in the mean-field limit. Steady-state transmission characteristics for C = 20, $\Delta_0 = \theta = 0$, $\gamma_T/\gamma_L = 0.5$, and γ_2^*/γ_T values given in the figure. Carmichael and Agrawal (1981).

A small linear loss can have a large effect on finesse and an even more drastic effect on peak cavity transmission. One can show that

$$I_T = \frac{I_0 T^2 \exp(-\alpha_B L/\cos\theta)}{(1 - R_\alpha)^2 \left[1 + \frac{4R_\alpha}{(1 - R_\alpha)^2} \sin^2 \left[\frac{2\pi L}{\lambda} n_0 \cos\theta\right]\right]} \qquad (2.5\text{-}47)$$

for a Fabry-Perot cavity of length L, index n_0, loss A = 1 - R - T in each coating, and unsaturable absorption coefficient (intensity) of α_B; $R_\alpha \equiv R \exp(-\alpha_B L/\cos\theta)$. The light inside the Fabry-Perot cavity makes an angle θ with respect to the normal. With $\theta = 0$, the finesse is

$$\mathcal{F} = \frac{\pi (R_\alpha)^{1/2}}{1 - R_\alpha} \qquad (2.5\text{-}48)$$

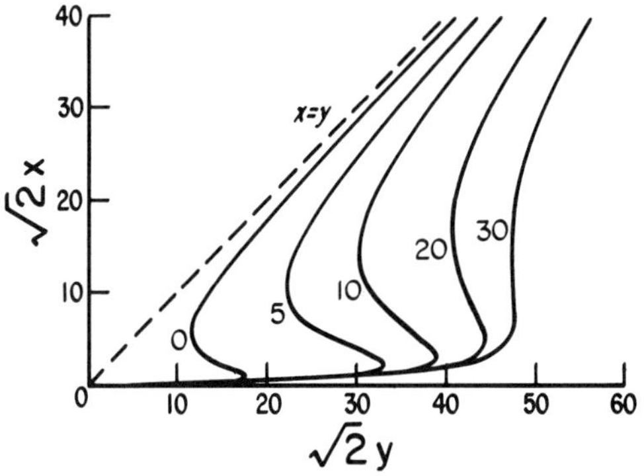

Fig. 2.5-5. Fabry-Perot full-Bloch-hierarchy numerical calculations: the effects of Doppler broadening on absorptive bistability in the MFL for systems showing equal weak-field absorption. Steady-state transmission characteristics for $\Delta_0 = \theta = 0$, $\gamma_T/\gamma_L = 0.5$, γ_2^*/γ_T as labeled in the figure, and C adjusted to keep the equal weak-field absorption: $\gamma_2^*/\gamma_T = 0$ (C = 20), 5 (C = 69.74), 10 (C = 125.9), 20 (C = 238.6), and 30 (C = 251.3). Carmichael and Agrawal (1981).

where $\mathscr{F}$ is defined as the free spectral range divided by the instrument width. Also,

$$\frac{I_T(\min)}{I_T(\max)} = \left[1 + \frac{4R_\alpha}{(1 - R_\alpha)^2}\right]^{-1} = [1 + (2\mathscr{F}/\pi)^2]^{-1} . \tag{2.5-49}$$

A relatively small amount of absorption can require a large increase in required input intensity to achieve the same I_C.

Spiller (1972) has calculated the effect of $\alpha_B L = 0.02$ on a saturable Fabry-Perot cavity with $\alpha_0 L = 0.5$; see Fig. 2.5-6. Clearly the peak transmission and the width of the bistable loop are greatly reduced. Since his formula did not account for an intensity-dependent index of refraction, he concluded that detuning of the resonator has qualitatively the same effect as a linear loss. Austin (1972) and Okuda and Onaka (1977a,b) also show the ill effect of α_B, the latter for proposed corrugated waveguide bistable structures. See Appendix D for a discussion by Miller (1981a) of the optimization of dispersive bistability etalons containing a linear absorption characterized by an α_B.

2.6. GRAPHICAL SOLUTIONS

It is not always easy to obtain a state equation relating I_I and I_T. Felber and Marburger (1976) are apparently the first to use a graphical solution. If one knows the transmission of the device as a function of some parameter and also knows the intensity dependence of that parameter, then a graphical solution is straightforward. Suppose

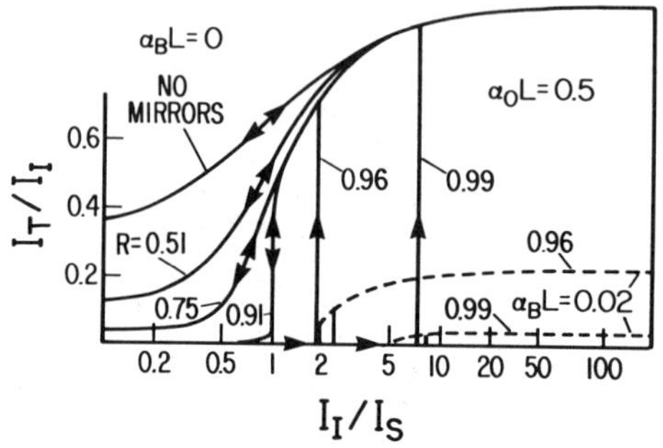

Fig. 2.5-6. Transmission of a saturable resonator vs the incident power densities for different mirror reflectivities R. The characteristics for R = 0.91 are also the down-switching parts of all characteristics with higher mirror reflectivity. All curves have $\alpha_0 L = 0.5$; $\alpha_B L$ is 0 for the solid curves and 0.02 for the two dashed hysteresis loops. Spiller (1972).

$$\mathcal{T} = \frac{I_T}{I_I} = \frac{1}{\left[1 + \frac{4R}{(1 - R)^2} \sin^2(\beta/2)\right]} \tag{2.6-1}$$

and that the round-trip phase is given by $\beta = \beta_0 + \beta_2 I_T$. Then the transmission can also be written as

$$\mathcal{T} = \frac{I_T}{I_I} = \frac{\beta - \beta_0}{\beta_2 I_I} . \tag{2.6-2}$$

The steady-state solutions of I_I and I_T must satisfy both Eqs. (2.6-1,2). Both are plotted in Fig. 2.6-1a for an initial detuning of -0.75π and R = 0.7. Equation (2.6-1) is then a single Fabry-Perot transmission curve, but Eq. (2.6-2) is a family of straight lines with slopes inversely proportional to I_I. For input intensities below that of curve A, there is a single intersection of the line with the curve; the system is in the lower state. For inputs between those of A and C there are three intersections giving the three values of I_T for a given I_I in the usual S-shaped bistable curve. The negative-slope portion of the S curve is unstable (Appendix E and Section 6.2). $\mathcal{T}$, not I_T, is plotted in Fig. 2.6-1b, so the usual bistability loop looks quite different. This plot emphasizes that the percentage transmission in the upper state may be quite low at the switch-up input intensity because of overshoot. $\mathcal{T}$ approaches unity just before switch-down occurs.

Switch-up occurs for inputs exceeding that of situation C because there is no longer a solution for small $(\beta - \beta_0)/\pi$. Similarly, if I_I is reduced after having exceeded that of C, switch-down occurs at A where there are no longer any intersections for large $(\beta - \beta_0)/\pi$. For bistability to occur there must be a region of three intersections; this requires that the slope of the curve exceeds the slope of the straight line. If that condition is met, then, with the appropriate detuning, bistability can occur. In the first InSb bistability experiment (Section 3.7), this condition could not be met in the first three orders of the Fabry-Perot etalon but was satisfied for the fifth order.

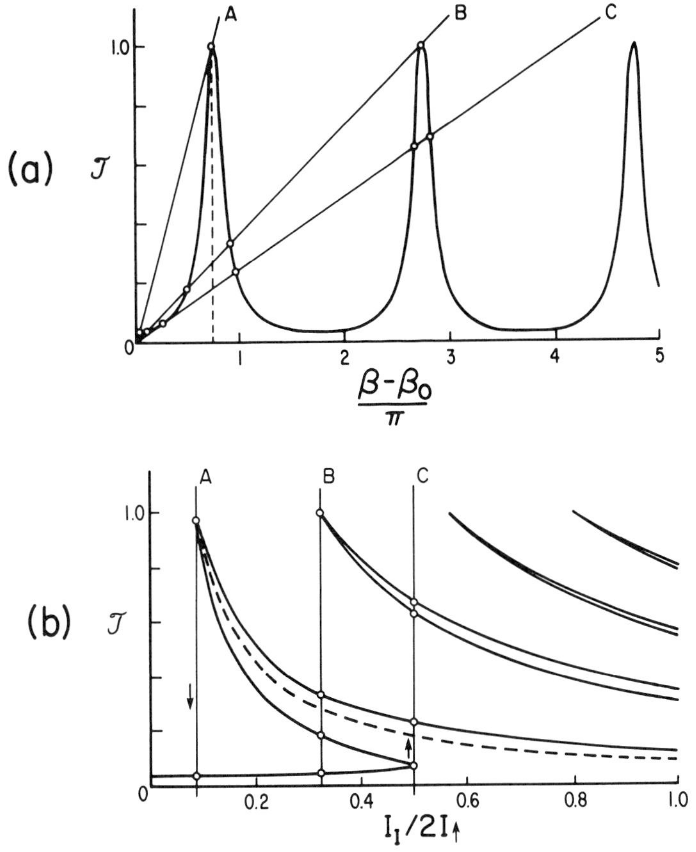

Fig. 2.6-1. (a) The curve is a plot of the Fabry-Perot transmission, Eq. (2.6-1), as a function of the phase shift. The straight lines show the relation between phase shift, transmission, and input intensity via Eq. (2.6-2). $R = 0.7$, $\beta_0 = -0.75\pi$. (b) Transmission versus input intensity graphically constructed from (a). Felber and Marburger (1976).

The possibility of multistability (see also Section 4.1), i.e., more than one switch-up with increased I_I, is clear from Fig. 2.6-1. Each time the straight line becomes tangent to the trough between two peaks, switch-up to the other side of the next peak occurs. Multistability of more than a dozen steps is shown in Fig. 1.2-2.

The discussion here of graphical solutions has been restricted to the simple case of purely dispersive bistability. Graphical solutions in the absorptive case are less straightforward since the cavity transmission function depends upon both α and β, which both depend upon intensity. If an absorption peak or edge shifts without changing shape as I_I is increased, a graphical construction is again simple and can be used to discuss increasing absorption optical bistability; see Section 2.10.

2.7. POTENTIAL WELL DESCRIPTION

Bistability can be viewed from the point of view of a potential well. For example, consider a circular rubber diaphragm supported around its circumference. Its own weight will cause the center spot to be the lowest. Place a marble in that potential minimum. Now place a hook in the diaphragm some distance from the minimum. By adding weights to that hook, one pulls the diaphragm down further and further until the marble rolls from the center to the new absolute minimum. Clearly a hysteresis occurs, i.e., removing an infinitesimal part of the weight does not cause the marble to return. One must remove weights until it is all "down hill" from the new minimum to the original one. If noise is introduced by shaking the diaphragm or using a jumpy marble, state switching back and forth can occur. This fluctuation switching works best if the two minima have the same depth and the barrier between them is low relative to the noise; see Section 6.5.

Lugiato and Bonifacio (1978) have given a nice potential-well description of bistability. Let x be a quantity observed as parameter y is externally varied, resulting in the hysteresis cycle shown in Fig. 2.7-1a. For any given value of y there is a suitable stationary probability distribution function $P_{st}(x)$, typically the solution of a master or Fokker-Planck equation. The quantity

$$V(x) \equiv -\ln[P_{st}(x)/P_{st}(0)] \tag{2.7-1}$$

plays the role of a generalized free energy. Figure 2.7-1b shows qualitatively the shape of $P_{st}(x)$ and $V(x)$ for some values of y. For $y = y_1$, $P_{st}(x)$ has only one peak centered at $x = x_1$; correspondingly, $V(x)$ has a single minimum at $x = x_1$. In the bistable regime ($y = y_2$, y_c, or y_3), $P_{st}(x)$ has two peaks and $V(x)$ two minima. A minimum in $V(x)$ above the absolute minimum can be thought of as a metastable state. In Fig. 2.7-1a the upper branch between $y_\downarrow$ and y_c and the lower branch between y_c and $y_\uparrow$ are metastable states. A large fluctuation can make the "particle" tunnel from a metastable state to the ground state (and vice versa).

See also Abraham and Smith (1982) and Lugiato (1983). For a nice discussion of optical bistability as a first-order phase transition and how it relates to a laser at threshold, a laser with saturable absorber, and superfluorescence see Arecchi (1981).

2.8. SPECTRA

Periodic cycling at the Rabi frequency of the atomic probability between states a and b is sporadically interrupted by spontaneous decay from a to b. In the upper bistable state there are many absorptions and emissions of laser photons between each spontaneous fluorescent emission of a non-laser photon. Clearly a nonperturbative approach is needed for such a multiphoton process. A complete

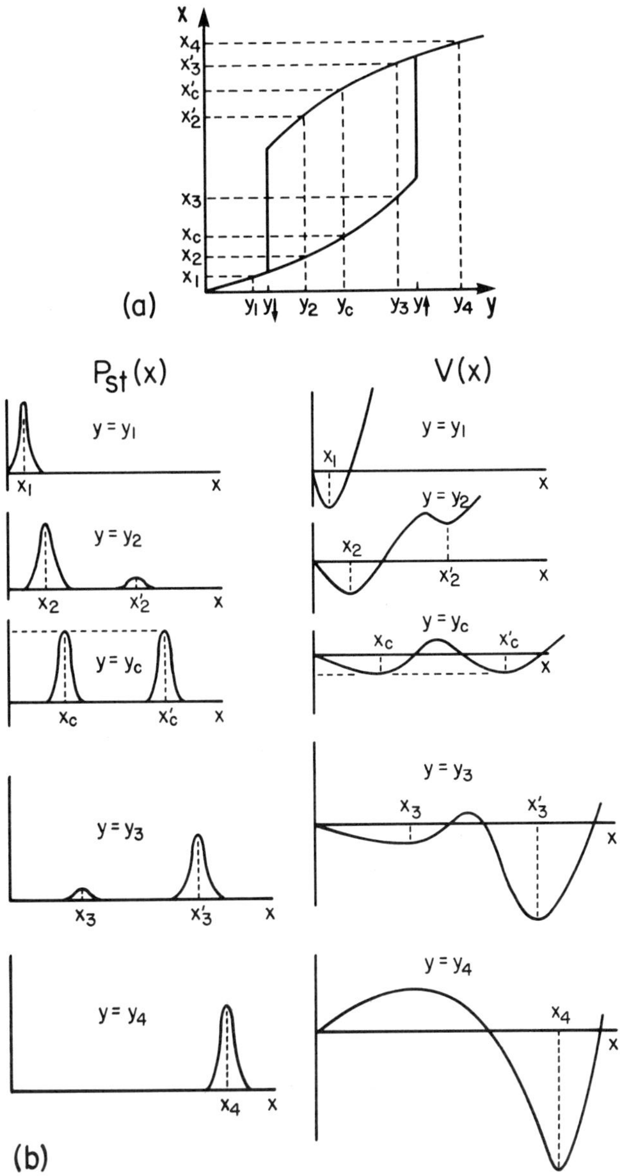

Fig. 2.7-1. (a) Hysteresis cycle of the observed quantity x versus the external parameter y. (b) Qualitative shape of the stationary probability distribution $P_{st}(x)$ and of the generalized free energy V(x) for the values of y indicated in (a). Lugiato and Bonifacio (1978).

treatment would require quantization of the electromagnetic field; however, we can obtain much insight from the semiclassical Maxwell-Bloch equations with the aid of the quantum regression theorem.

Let us consider the on-resonance case ($\Delta\omega^0 = 0$) of homogeneously broadened two-level atoms in a ring cavity. Equations (2.1-3) become

$$\dot{\bar{u}} = -\kappa E_i w - \gamma_T \bar{u} \qquad (2.8\text{-}1a)$$

$$\dot{\bar{v}} = -\kappa E_r w - \gamma_T \bar{v} \qquad (2.8\text{-}1b)$$

$$\dot{w} = +\kappa E_r \bar{v} + \kappa E_i \bar{u} - \gamma_L [w + 1] \, , \qquad (2.8\text{-}1c)$$

and Eq. (2.1-8) reduces to

$$\frac{\partial E}{\partial z} + \frac{n_0}{c}\frac{\partial E}{\partial t} = -\frac{2\pi\omega Np(\bar{v} + i\bar{u})}{n_0 c} = -\frac{\gamma_T \alpha_0}{2\kappa}(\bar{v} + i\bar{u}) . \qquad (2.8\text{-}2)$$

The equation for $\bar{u}$ is retained because $\bar{u}$ is not always zero when fluctuations occur. Now assume that the relaxation rates γ_L and γ_T are much smaller than the cavity decay rate, $\gamma_c = cT/(2Ln_0)$. Then the field can be eliminated adiabatically ($\dot{E} \simeq 0$):

$$\int_0^L \frac{\partial E}{\partial z} dz \simeq -\frac{\gamma_T \alpha_0}{2\kappa}[\bar{v}(L) + i\bar{u}(L)]\, L = E(L) - E(0) \, , \qquad (2.8\text{-}3)$$

in the mean-field case. The boundary conditions Eqs. (2.1-13,14) are

$$E_T = \sqrt{T}\, E(L) \, , \qquad (2.8\text{-}4)$$

$$E(0) = \sqrt{T}\, E_I + RE(L) \, . \qquad (2.8\text{-}5)$$

Then, with Eqs. (2.1-17,18), $L = L_{RC}$, and $C = \alpha_0 L_{RC}/4T$,

$$E(L) = [\gamma_L \gamma_T]^{1/2}\frac{y}{\kappa} - \frac{2C\gamma_T}{\kappa}[\bar{v}(L) + i\bar{u}(L)] = [\gamma_L \gamma_T]^{1/2}\frac{x}{\kappa} \, . \qquad (2.8\text{-}6)$$

The time dependence can now be followed using Eqs. (2.8-1):

$$\dot{\bar{u}} + i\dot{\bar{v}} = -\gamma_T(\bar{u} + i\bar{v}) - i\kappa w E^* \qquad (2.8\text{-}7)$$

$$\dot{\bar{u}} - i\dot{\bar{v}} = -\gamma_T(\bar{u} - i\bar{v}) + i\kappa w E \qquad (2.8\text{-}8)$$

$$\dot{w} = -\gamma_L(w + 1) - \frac{i\kappa E}{2}(\bar{u} + i\bar{v}) + \frac{i\kappa E^*}{2}(\bar{u} - i\bar{v}) \, . \qquad (2.8\text{-}9)$$

With $S^{\pm} = \bar{u} \pm i\bar{v}$ and Eq. (2.8-6) in Eqs. (2.8-7 to 9)

$$\dot{S}^{+} = -\gamma_T S^{+} - i[\gamma_L \gamma_T]^{1/2} yw + 2C\gamma_T S^{+}w , \qquad (2.8\text{-}10a)$$

$$\dot{S}^{-} = -\gamma_T S^{-} + i[\gamma_L \gamma_T]^{1/2} yw + 2C\gamma_T S^{-}w , \qquad (2.8\text{-}10b)$$

$$\dot{w} = -\gamma_L(w + 1) - i[\gamma_L \gamma_T]^{1/2} \frac{y(S^{+} - S^{-})}{2} - 2C\gamma_T S^{+}S^{-} , \qquad (2.8\text{-}10c)$$

with y taken to be real.

The spectrum of the transmitted light is proportional to the Fourier transform of $\langle x^{*}(t'+t)x(t')\rangle$ which involves the atoms only through the atomic correlation function, $\langle S^{+}(t'+t)S^{-}(t')\rangle$; see Eq. (2.8-6). In particular, the steady-state value is $\lim_{t'\to\infty} \langle S^{+}(t'+t)S^{-}(t')\rangle \equiv \langle S^{+}(t)S^{-}\rangle$.

To solve for this correlation, define the set of correlation functions χ:

$$\chi \equiv \begin{pmatrix} \langle (S^{+}(t) - S_0^{+})(S^{-}(0) - S_0^{-})\rangle_0 \\ \langle (S^{-}(t) - S_0^{-})(S^{-}(0) - S_0^{-})\rangle_0 \\ \langle (w(t) - w_0)(S^{-}(0) - S_0^{-})\rangle_0 \end{pmatrix} = \begin{pmatrix} \chi_1 \\ \chi_2 \\ \chi_3 \end{pmatrix} , \qquad (2.8\text{-}11)$$

where the subscript zero denotes steady state.

The quantum regression theorem [Lax (1963)] states that the correlations obey exactly the same dynamical law of evolution as do the Bloch variables themselves. In the case of a single atom in an intense field, $C \to 0$, and the nonlinear terms such as $S^{+}w$ in Eq. (2.8-10) drop out. The feedback in the bistability case introduces the nonlinear terms. An approximate evolution matrix is obtained by linearizing Eq. (2.8-10), i.e., $\delta(S^{+}w) \simeq S_0^{+}\delta w + w_0\delta S^{+}$.

The steady-state solutions to Eq. (2.8-10) were given in Eqs. (2.1-7):

$$S_0^{\pm} = \pm iv_0 = \pm i \left[\frac{\gamma_L}{\gamma_T}\right]^{1/2} \frac{x}{1 + x^2} , \qquad (2.8\text{-}12)$$

$$w_0 = -\frac{1}{1 + x^2} , \qquad (2.8\text{-}13)$$

$$y = x + \frac{2Cx}{1 + x^2} . \qquad (2.8\text{-}14)$$

Then

$$V(t) \equiv \begin{pmatrix} S^{+}(t) - S_0^{+} \\ S^{-}(t) - S_0^{-} \\ w(t) - w_0 \end{pmatrix} = \begin{pmatrix} \delta S^{+} \\ \delta S^{-} \\ \delta w \end{pmatrix} \qquad (2.8\text{-}15)$$

evolves according to

$$\dot{V} = \gamma_T M V , \qquad (2.8\text{-}16)$$

where

$$M = \begin{pmatrix} -1 + 2Cw_0 & 0 & -i\left[\frac{\gamma_L}{\gamma_T}\right]^{1/2} y + 2CS_0^+ \\ 0 & -1 + 2Cw_0 & i\left[\frac{\gamma_L}{\gamma_T}\right]^{1/2} y + 2CS_0^- \\ \frac{-iy}{2}\left[\frac{\gamma_L}{\gamma_T}\right]^{1/2} - 2CS_0^- & \frac{+iy}{2}\left[\frac{\gamma_L}{\gamma_T}\right]^{1/2} - 2CS_0^+ & -\frac{\gamma_L}{\gamma_T} \end{pmatrix}, \tag{2.8-17}$$

which can be read by inspection from Eqs. (2.8-10). One can then use Eqs. (2.8-12 to 14) to write

$$M = \begin{pmatrix} -\frac{y}{x} & 0 & -i\left[\frac{\gamma_L}{\gamma_T}\right]^{1/2} x \\ 0 & -\frac{y}{x} & +i\left[\frac{\gamma_L}{\gamma_T}\right]^{1/2} x \\ -i\left[\frac{\gamma_L}{\gamma_T}\right]^{1/2} \frac{2x - y}{2} & i\left[\frac{\gamma_L}{\gamma_T}\right]^{1/2} \frac{2x - y}{2} & -\frac{\gamma_L}{\gamma_T} \end{pmatrix}. \tag{2.8-18}$$

By the quantum regression theorem then

$$\dot{\chi} = \gamma_T M \chi . \tag{2.8-19}$$

This, of course, could be shown directly.

The desired correlation function is χ_1, which has a frequency spectrum

$$\chi_1(\Delta\nu) = 2\mathrm{Re}\int_0^\infty d\tau\; e^{-i\Delta\nu\tau}\, \chi_1(\tau) . \tag{2.8-20}$$

Laplace transforms reduce first-order time differential equations to algebraic equations for their transforms:

$$x(t) = x_0 + \int_0^t dt'\, \dot{x}(t') . \tag{2.8-21}$$

The Laplace transform of $x(t)$ is $\tilde{x}(\lambda)$:

$$\tilde{x}(\lambda) = \int_0^\infty dt\, e^{-\lambda t}\, x(t) \tag{2.8-22}$$

$$= \frac{x_0}{\lambda} + \int_0^\infty dt\, e^{-\lambda t} \int_0^t dt'\, \dot{x}(t')\ .$$

Integrating by parts, i.e., $(uv)\,\big|_0^\infty = \int_0^\infty v\,du + \int_0^\infty u\,dv$, with $u = \int_0^t dt'\dot{x}(t')$ and $v = -\lambda^{-1}\exp(-\lambda t)$, one finds

$$\tilde{x}(\lambda) = \frac{x_0}{\lambda} + \frac{1}{\lambda}\int_0^\infty dt\, e^{-\lambda t}\, \dot{x}(t)$$

$$= \frac{x_0}{\lambda} + \frac{1}{\lambda}\tilde{\dot{x}}(\lambda)$$

i.e.,

$$\tilde{\dot{x}}(\lambda) = \lambda\tilde{x}(\lambda) - x_0\ . \tag{2.8-23}$$

Applying Eq. (2.8-23) to Eq. (2.8-19), one has

$$\tilde{\dot{\chi}} = \lambda\tilde{\chi}(\lambda) - \chi_0 = \gamma_T M \tilde{\chi}$$

$$\left[\frac{\lambda}{\gamma_T} - M\right]\tilde{\chi} = \frac{\chi_0}{\gamma_T}\ . \tag{2.8-24}$$

So the steady-state values of χ_1, χ_2, χ_3 are needed to solve Eq. (2.8-24). $\tilde{\chi}_1(\lambda)$, for example, is found by the methods of linear algebra:

$$\tilde{\chi}_1(\lambda) = \begin{vmatrix} \dfrac{\chi_{10}}{\gamma_T} & 0 & +i\left[\dfrac{\gamma_L}{\gamma_T}\right]^{1/2} x \\[2ex] \dfrac{\chi_{20}}{\gamma_T} & \dfrac{\lambda}{\gamma_T} + \dfrac{y}{x} & -i\left[\dfrac{\gamma_L}{\gamma_T}\right]^{1/2} x \\[2ex] \dfrac{\chi_{30}}{\gamma_T} & -i\left[\dfrac{\gamma_L}{\gamma_T}\right]^{1/2} & \dfrac{2x - y}{2} & \dfrac{\lambda}{\gamma_T} + \dfrac{\gamma_L}{\gamma_T} \end{vmatrix} \cdot \left[\det\left[\frac{\lambda}{\gamma_T} - M\right]\right]^{-1}\ . \tag{2.8-25}$$

It is then the $\det[(\lambda/\gamma_T) - M]$ that determines the resonances in the denominator: thus the eigenvalues of M determine the width and frequency shift of the Lorentzian spectral components. χ_0 is needed only to determine the relative weightings of those components. Now

$$\det\left[\frac{\lambda}{\gamma_T} - M\right] = \left[\frac{\lambda}{\gamma_T} + \frac{y}{x}\right]\left[\left[\frac{\lambda}{\gamma_T}\right]^2 + \left[\frac{y}{x} + \frac{\gamma_L}{\gamma_T}\right]\frac{\lambda}{\gamma_T} + \frac{\gamma_L}{\gamma_T}\left[2x^2 - xy + \frac{y}{x}\right]\right], \tag{2.8-26}$$

so the eigenvalues are

$$\lambda_\phi = -\gamma_T \frac{y}{x}, \tag{2.8-27}$$

$$\lambda_\pm = -\gamma_T \frac{\left[\frac{y}{x} + \frac{\gamma_L}{\gamma_T} \pm \left[\left[\frac{y}{x} - \frac{\gamma_L}{\gamma_T}\right]^2 - 4\frac{\gamma_L}{\gamma_T} x(2x - y)\right]^{1/2}\right]}{2}. \tag{2.8-28}$$

For spontaneous emission without a cavity, $\gamma_L = 2\gamma_T = 1/\tau_0$, $C = 0$, and $y = x$; so $\lambda_\phi = -1/2\tau_0$ and $\lambda_\pm = -3/4\tau_0 \mp i\kappa E$ for $y \to \infty$. Then one can use the eigenvalues to construct a partial fraction's expansion of $\tilde{\chi}_1(\lambda)$:

$$\tilde{\chi}_1(\lambda) = \sum_i \frac{c_i}{\lambda - \lambda_i} \tag{2.8-29}$$

$$\tilde{\chi}_1(\lambda)\big|_{C=0} = \frac{c_\phi}{\lambda + \frac{1}{2\tau_0}} + \frac{c_+}{\lambda + \frac{3}{4\tau_0} + i\kappa E} + \frac{c_-}{\lambda + \frac{3}{4\tau_0} - i\kappa E} \tag{2.8-30}$$

$$\chi_1(t)\big|_{C=0} = c_\phi e^{-t/2\tau_0} + c_+ \exp\left[-\left[\frac{3}{4\tau_0} + i\kappa E\right]t\right] + c_- \exp\left[-\left[\frac{3}{4\tau_0} - i\kappa E\right]t\right]. \tag{2.8-31}$$

From Eqs. (2.8-20 and 22), one has

$$\chi_1(\Delta\nu)\big|_{C=0} = 2\,\mathrm{Re}[\tilde{\chi}_1(\lambda = i\Delta\nu)\big|_{C=0}]$$

$$= \frac{\frac{c_\phi}{\tau_0}}{(\Delta\nu)^2 + \left[\frac{1}{2\tau_0}\right]^2} + \frac{\frac{3c_+}{2\tau_0}}{\left[(\Delta\nu + \kappa E)^2 + \left[\frac{3}{4\tau_0}\right]^2\right]} + \frac{\frac{3c_-}{2\tau_0}}{\left[(\Delta\nu - \kappa E)^2 + \left[\frac{3}{4\tau_0}\right]^2\right]}, \tag{2.8-32}$$

which is the familiar three-peak spectrum of strong-field isolated-atom resonance fluorescence [Mollow (1969)]. One physical interpretation of the three-peak spectrum is that in the presence of the strong electric field the "dressed atom" has both its lower and upper states Stark-split by the Rabi frequency κE. There are four possible transitions, but the two intermediate frequencies are equal [Cohen-Tannoudji (1976)].

For sufficiently large y, y = x even if $C \neq 0$, so that the three-peak spectrum of Eq. (2.8-32) applies then as well, where E is the field amplitude inside the ring cavity. Near the switch-up from the lower branch, $x_{\uparrow L} \simeq 1$ and $y_{\uparrow} \simeq C$ for large C [see Eq. (2.4-21)]. Then $\lambda_- \to 0$ and $\chi_1(\Delta\nu)$ is a single line since all the roots are real. Of course, when an eigenvalue goes to zero, the corresponding time constant, e.g., Eq. (2.8-31), goes to infinity and one has a sluggishness of response known as critical slowing down (see Section 5.6).

Lugiato (1979) has evaluated the initial conditions of Eq. (2.8-11) using a linearized Fokker-Planck equation and determined the c_i coefficients in Eq. (2.8-29). He is then able to determine the incoherent part of the spectrum $\chi_1(\Delta\nu)$ of the transmitted light as shown in Fig. 2.8-1 for spontaneous emission relaxation for which $\gamma_L = 2\gamma_T$. Figure 2.8-2 shows the corresponding points on the hysteresis curve. Figure 2.8-3 shows the case $\gamma_L/\gamma_T = 0.002$, i.e., for transverse relaxation much faster than the longitudinal.

An analysis of the incoherent part of the spectrum of the transmitted light similar to the one here was outlined in Bonifacio and Lugiato (1976) and in more detail in Bonifacio and Lugiato (1978a,b,e). Particularly lucid are Lugiato (1979); Narducci, Gilmore, Feng, and Agarwal (1979); Agarwal, Narducci, Gilmore, and Feng (1978). Agarwal and Tewari (1980) extend the analysis to include laser-atom detuning, but no low-intensity laser-cavity detuning is included, so dispersive bistability does not occur; the index contributions of the nonlinear susceptibility are therefore always harmful in their analysis. Drummond and Walls (1980) have analyzed the spectrum of the transmitted light for dispersive bistability by assuming a cubic nonlinearity in the polarization as did Marburger and Felber (1978). They find a delta-function peak corresponding to the radiation transmitted at the input frequency plus two quasi-Lorentzian peaks located symmetrically about the input frequency. To obtain a physical picture, let $\omega_{c'}$ be the effective cavity frequency including changes in the nonlinear refractive index. Consider a multiphoton scattering in which two laser photons at ω are "absorbed" and photons with frequencies $\omega_{c'}$ and $2\omega - \omega_{c'}$ are emitted. Then both peaks have the same height since one photon of each energy arises for each scattering event. Each peak is displaced by $\omega - \omega_{c'}$ from ω. The only treatment of the spectrum for mixed absorptive and dispersive bistability is Carmichael, Walls, Drummond, and Hassan (1983).

The spectrum $I(\omega)$ of the fluorescent light emitted at 90° from the incident light is the Fourier transform of the time correlation function

$$\sum_{i=1}^{N} \langle r_i^+(t) r_i^- \rangle_0 ,$$

where $r_i^{\pm}$ are the raising and lowering operators of the i-th two-level atom:

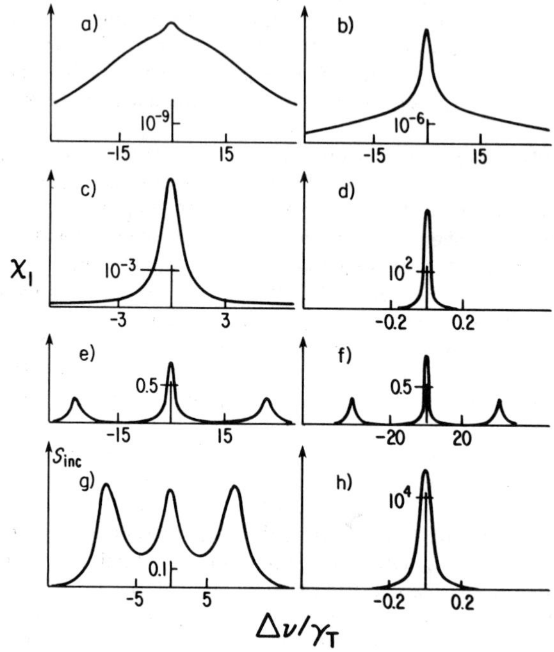

Fig. 2.8-1. Hysteresis cycle of the incoherent part of the spectrum χ_1 of the transmitted light for $\gamma_L = 2\gamma_T \ll \gamma_c$ and C = 20. χ_1 is given in units $C/2\pi\gamma_c$. Here and in the following figures one has: (i) $\Delta\nu = \nu_a - \nu$ where ν is the frequency of the injected field. (ii) The scale varies from diagram to diagram as indicated. (iii) The points of the (x,y)-plane corresponding to (a)-(h) are indicated in Fig. 2.8-2; the point corresponding to (f) does not appear because y is too large. (a) y = 4.06, (b) y = 14.193, (c) y = 20.589, (d) y = 21.026, (e) y = 21.995, (f) y = 31.33, (g) y = 13.39, (h) y = 12.48. Lugiato (1979).

$$I(\omega) \propto \frac{1}{\pi} \mathrm{Re} \int_0^\infty dt \exp[-i(\omega_a - \omega)t] \sum_{i=1}^{N} \langle r_i^+(t) r_i^-(0) \rangle_0 .$$

[See Fig. 3.2-19a for the observation of optical bistability by means of the fluorescence.] Carmichael (1981), Lugiato (1980a), and Agarwal (1982a) find no line narrowing in the resonance fluorescence at the boundaries of the hysteresis cycle. Earlier work had calculated narrowing to be accompanying critical slowing down in the spectra of the transmission <u>and</u> the fluorescence. The consensus now is that the fluorescence spectrum is given (apart from collective terms of order 1/N) by

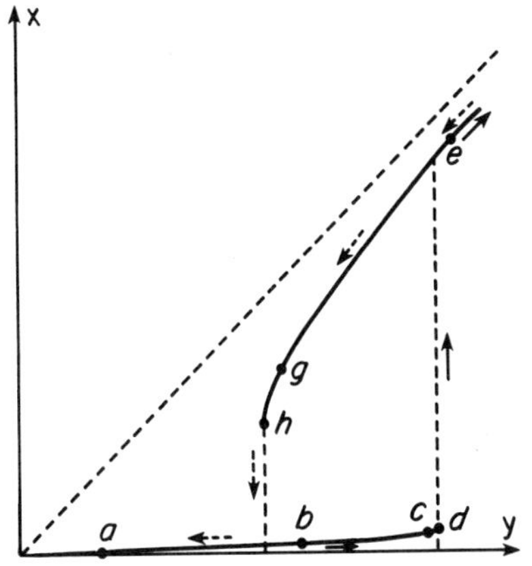

Fig. 2.8-2. Hysteresis cycle of the transmitted field vs the incident field. The spectra of the transmitted light in correspondence to points (a)-(h) are plotted in Fig. 2.8-1. Lugiato (1979).

the single-atom theory of resonance fluorescence [Mollow (1969)]. The earlier problem was the failure to distinguish like-atom and unlike-atom correlations when implementing the quantum regression theorem. The previous results are obtained if like-atom and unlike-atom correlations are assumed equal; at steady-state, unlike-atom correlations are smaller than like-atom correlations by a factor 1/N. Experimental observation of collective features by means of fluorescence will consequently require a system of a few atoms, which would, of course, exhibit fluctuations. One interpretation is that critical slowing down, typical of the relaxation of the collective variables, does not apply to single-atom fluctuations, thereby resulting in major differences in the linewidths of the radiation in the forward direction and in other directions [Agarwal (1982)]. Therefore, except for small N, observation of the spectrum of fluorescence from atoms undergoing optical bistability is of little interest; one could immediately deduce the same results by plotting known single-atom spectra versus intracavity intensity. Furthermore, experimental realization is complicated by the necessity of using a ring cavity under uniform-plane-wave conditions, since standing waves and nonuniform transverse profiles average out the Stark splitting.

Since the fluorescence exhibits no unique features in absorptive or dispersive large-N bistable systems, interesting spectra must be taken in transmission. This raises two questions. Are the interesting features of the incoherent spectrum detectable in the presence of the transmitted incident light, which has the same center frequency? Is it interesting to measure those features for an atomic system, since the spectrum of a hybrid device must also exhibit narrowing corresponding to the observed critical slowing down? Granted, the transmission of a hybrid will not exhibit the three-peak Stark splitting, but that is a single-atom feature. It should exhibit the narrowing, a cooperative feature arising from the feedback. Perhaps the time dynamics, i.e., critical slowing down (Section 5.6),

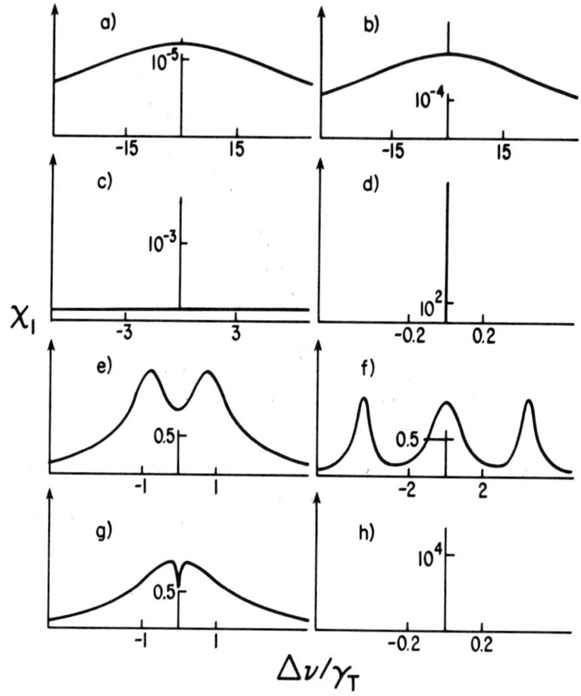

Fig. 2.8-3. Hysteresis cycle of the incoherent part of the spectrum χ_1 of the transmitted light for $\gamma_C >> \gamma_T$, $\gamma_L/\gamma_T = 0.002$ and $C = 20$. χ_1 is given in units $C/2\pi\gamma_C$. See also caption of Fig. 2.8-1. (a) y = 4.06, (b) y = 14.193, (c) y = 20.589, (d) y = 21.026, (e) y = 21.995, (f) y = 100.4, (g) y = 13.39, (h) y = 12.48. Lugiato (1979).

and fluctuations (Section 6.5) are better monitors than spectra of the cooperative features of bistable systems.

If the incoherent part of the spectrum of the transmitted light χ_1 can be measured, it would exhibit several interesting cooperative features. In a "bad" cavity (cavity linewidth much broader than the atomic fluorescence spectrum) and on the lower branch of the bistability loop, χ_1 is broad with a linewidth proportional to C; on the upper branch in the range of the hysteresis, the Rabi splittings and ratio of peak heights deviate significantly from those for a single atom even for an input intensity 10% or more above the switch-down value; see Fig. 2.8-1g. The linewidth narrows as the switching points are approached either from below [Fig. 2.8-1a to 1c] or from above [Fig. 2.8-1f to 1h]. Many predicted spectra for both absorptive and mixed bistability are displayed in Carmichael, Walls, Drummond, and Hassan (1983).

For other discussions of the spectra see: Lugiato (1984); Agarwal, Narducci, Feng, and Gilmore (1978); Carmichael and Walls (1977); Drummond (1979);

Drummond and Walls (1980); Narducci, Gilmore, Feng, and Agarwal (1979); Tewari (1980); Walls, Drummond, Hassan, and Carmichael (1978); Walls, Drummond, and McNeil (1981). Sargent, Zubairy, and DeMartini (1982) show that a straightforward quantum population-pulsation formalism, which avoids the quantum regression theorem, simultaneously describes all the strong-mode/weak-probe interactions: saturation spectroscopy, phase conjugation, resonance fluorescence, and laser/optical bistability; see Appendix F.

2.9. TRANSVERSE EFFECTS

Optical bistability experiments are generally performed with lasers of finite transverse dimensions, usually with a Gaussian radial intensity profile. As noted in Section 4.2, this has no effect on hybrid experiments performed so far, because the feedback phase shift is applied equally to all parts of the beam. But sometimes the nonlinearity in an intrinsic bistability experiment is a local effect; it depends on the intensity at the point in question. In that case the beam profile and mechanisms that change the profile, namely diffraction and self-focusing or defocusing, can substantially alter the bistability. Clearly, transverse effects do not eliminate bistability in general, since intrinsic bistability is observed using Gaussian beams. However, in some cases the bistability observed with a Gaussian beam is close to uniform-plane-wave predictions; this may be because diffusion of the excitation within the medium results in far less radial dependence in the phase shift than in the input beam, much like the hybrid case.

To study changes in the profile of the laser beam arising from diffraction and focusing effects, one must include the transverse part of the Laplacian ∇^2 in the wave equation:

$$\left[\nabla^2 - \frac{n_0{}^2}{c^2}\frac{\partial^2}{\partial t^2}\right]\mathbf{E} = \frac{4\pi\ddot{\mathbf{P}}}{c^2}\,. \tag{C-9}$$

In the plane-wave limit $\nabla^2 \to \partial^2/\partial z^2$; to include transverse effects $\nabla^2 = \partial^2/\partial z^2 + \nabla_T{}^2$ where $\nabla_T{}^2 = \partial^2/\partial x^2 + \partial^2/\partial y^2$ in general, or $\nabla_T{}^2 = \partial^2/\partial\rho^2 + \rho^{-1}\partial/\partial\rho$ if the problem has cylindrical symmetry. Various approximations have been made to obtain transverse behavior from Eq. (C-9). The usual slowly varying envelope approximation (SVEA) leads to

$$\left[-2ik\frac{\partial}{\partial Z} + \nabla_T{}^2\right]E(Z,\tau) = -\frac{4\pi\omega^2}{c^2}P(Z,\tau)\,, \tag{2.9-1}$$

where $Z = z$ and $\tau = t - z/c$ are the new retarded variables, E is the complex electric field envelope, and P is the polarization envelope.

Solutions of Eq. (2.9-1) require either numerical methods (Section 2.9.2) or a mode expansion (Section 2.9.1); the latter have limited applicability because of discontinuities in the transverse profile arising from the bistable states.

2.9.1. Analytical Approaches

Marburger and Felber (1978) assume a Kerr nonlinearity of the form $n_0 + n_2 I$ and work in the high-finesse limit so that forward and backward fields are equal. So Eq. (2.9-1) has the same form as the slowly varying envelope equation for a self-focusing beam traveling to the right, except for a factor of 3 (because the counterpropagating field causes twice the nonlinearity for the forward field as caused by the forward field itself; see Appendix D). They use a variational method to obtain analytical expressions for the phases of the forward and backward amplitudes, allowing the transmissivity to be expressed analytically in parametric form. The phases obtained this way are in almost perfect agreement with the numerical solutions of the SVEA self-focusing equations after the critical power P_C is renormalized to 1.07 P_C. They find that, because of the effective index change associated with self-induced waveguiding in the spherical mirror geometry, the powers required for bistable operation are reduced substantially relative to the plane-mirror geometry.

Drummond (1979, 1980, 1981a) and Ballagh, Cooper, Hamilton, Sandle, and Warrington (1981) [and Sandle, Ballagh, and Gallagher (1981)] have shown that when the lowest-order mode of a cavity with spherical mirrors is excited and higher modes are lossy, the state equation can be approximated by

$$Y \simeq X \left|1 - i\theta + 2C\langle\sigma\rangle\right|^2 \qquad (2.9\text{-}2)$$

in the mean-field ($\alpha_0 L \ll 1$, $T \ll 1$) and weak reshaping regime ($\alpha_0 L \ll 1$, $L \ll$ Rayleigh length). Small $\alpha_0 L$, where L is the length of the nonlinear medium, ensures that self-focusing and defocusing are small and that the field amplitude changes very little across the medium. $T \ll 1$ ensures that the backward wave approximately equals the forward wave. The average of the saturable susceptibility σ is defined as

$$\langle\sigma(X)\rangle \equiv \int_0^\infty \frac{4}{w_0^2} \exp\left[-\frac{2\rho^2}{w_0^2}\right] \sigma\left[X \frac{4}{w_0^2} \exp\left[-\frac{2\rho^2}{w_0^2}\right]\right] \rho\, d\rho \qquad (2.9\text{-}3a)$$

$$= \int_0^\infty e^{-\eta}\, \sigma(X_0\, e^{-\eta})\, d\eta\ , \qquad (2.9\text{-}3b)$$

where $X_0 \equiv 4X/w_0^2$ and $\eta = 2\rho^2/w_0^2$. Formulas for σ are given in Eq. (2.5-1) for a unidirectional spherical-mirror ring cavity:

$$\sigma_{RC} = \frac{1 + i\Delta}{1 + \Delta^2 + X}\ , \qquad (2.9\text{-}4)$$

and in Eq. (2.5-38) for a spherical-mirror Fabry-Perot:

$$\sigma_{FP} = \frac{1 + i\Delta}{2X}\left[1 - \left[\frac{1 + \Delta^2}{1 + \Delta^2 + 4X}\right]^{1/2}\right]. \qquad (2.9\text{-}5)$$

The above authors make a mode expansion and show that the lowest-order TEM_{00} mode decouples from other modes under the above assumptions. The nonlinear response of the medium couples the various modes, but since this coupling is only of order $\alpha_0 L$, the higher-order mode amplitudes generated in a single pass are small. The higher-order amplitudes from many passes are suppressed because these modes are not degenerate with the TEM_{00} mode and are not resonant at the laser frequency. The average given in Eq. (2.9-3) results, as one might have guessed, and the details of the many-mode proof are not worth the added encumbrance here. Performing the integral of Eq. (2.9-3),

$$\langle\sigma_{RC}\rangle = \frac{1 + i\Delta}{X_0} \ln\left[1 + \frac{X_0}{1 + \Delta^2}\right], \tag{2.9-6}$$

$$\frac{2\langle\sigma_{FP}\rangle}{1 + i\Delta} = \int_0^\infty \frac{e^{-\eta}}{X_0\, e^{-\eta}}\left[1 - \frac{1}{(1 + a\, e^{-\eta})^{1/2}}\right] d\eta\,, \tag{2.9-7}$$

where $a = 4X_0/(1 + \Delta^2)$. Let $z \equiv (1 + a\, e^{-\eta})^{1/2}$, then $d\eta = 2z\,dz/(1 - z^2)$ and

$$\frac{2\langle\sigma_{FP}\rangle}{1 + i\Delta} = \frac{1}{X_0}\int_{\sqrt{1+a}}^{1} \frac{z - 1}{z}\,\frac{2z\,dz}{1 - z^2} = \frac{-2}{X_0}\int_{\sqrt{1+a}}^{1} \frac{d(z + 1)}{z + 1}$$

$$= \frac{2}{X_0} \ln \frac{1}{2}\left[\sqrt{1 + a} + 1\right]$$

$$\langle\sigma_{FP}\rangle = \frac{1 + i\Delta}{X_0} \ln \frac{1}{2}\left[1 + \left[1 + \frac{4X_0}{1 + \Delta^2}\right]^{1/2}\right]. \tag{2.9-8}$$

The state equations then become

$$Y_0^{RC} \simeq X_0^{RC}\left[\left[1 + \frac{2C_{RC}}{X_0^{RC}} \ln\left[1 + \frac{X_0^{RC}}{1 + \Delta^2}\right]\right]^2 + \left[\theta - \frac{2C_{RC}\Delta}{X_0^{RC}} \ln\left[1 + \frac{X_0^{RC}}{1 + \Delta^2}\right]\right]^2\right] \tag{2.9-9}$$

and

$$Y_0^{FP} = X_0^{FP}\left[\left[1 + \frac{2C_{FP}}{X_0^{FP}} \ln Z\right]^2 + \left[\theta - \frac{2C_{FP}\Delta}{X_0^{FP}} \ln Z\right]^2\right] \tag{2.9-10}$$

where

$$Z = \frac{1}{2}\left[1 + \left[1 + \frac{4X_0 FP}{1+\Delta^2}\right]^{1/2}\right]. \qquad (2.9\text{-}11)$$

Equation (2.9-10) is compared with an experiment in Na vapor in Section 3.2.5a and explains the data much better than the plane-wave theory. Averaging over the Gaussian mode softens the saturation of σ and sometimes destroys plane-wave bistability. It has a much larger effect upon absorptive bistability than upon dispersive bistability as shown in Table 2.1, where C_{min} is the minimum value of C for absorptive bistability and $Y_\uparrow$ is the normalized minimum switch-on intensity.

Firth and Wright (1982a) expand the field envelopes for a plane-mirror cavity in a set of orthonormal modes:

$$\sum_j \left[\frac{2}{\pi w^2}\right]^{1/2} L_j\left(\frac{2r^2}{w^2}\right) e^{-r^2/w^2}, \qquad (2.9\text{-}12)$$

where L_j is a Laguerre polynomial ($L_0 = 1$, $L_1(x) = 1 - x$, etc.). Numerical results show some indication that including each higher-order Laguerre-Gaussian mode has a smaller effect on the calculated output power and bistability thresholds. The

Table 2.1
Effect of Gaussian Average Upon Absorptive and Dispersive Bistability Thresholds. Ballagh, Cooper, Hamilton, Sandle, and Warrington (1981).

Type of bistability	Absorptive bistability (Δ=0, θ=0)		Dispersive bistability (Δ>>1, C>2Δ, Δ^2>>X^2)
Quantity	C_{min}	$Y_\uparrow$	$Y_\uparrow$
Plane-wave ring cavity	4	5.20	$0.88\left[\frac{\Delta^3}{C}\right]^{1/2}$
Plane-wave Fabry-Perot	4.97	5.30	$0.72\left[\frac{\Delta^3}{C}\right]^{1/2}$
Gaussian ring cavity	8.31	15.78	$1.24\left[\frac{\Delta^3}{C}\right]^{1/2}$
Gaussian Fabry-Perot	10.04	16.21	$1.01\left[\frac{\Delta^3}{C}\right]^{1/2}$

convergence is not fast, however, and they estimate that the computer time increases roughly as the fourth power of the mode number, making it impractical to proceed much beyond j = 6.

Hall and Dziura (1984) have studied the effects of Gaussian mode and gain profiles on bistable operation of lasers containing saturable absorbers.

2.9.2. Numerical Solutions

Perhaps the most significant results, from the point of view of determining beam-profile changes during bistability switching, have been obtained by direct numerical integration of Eq. (2.9-1); see Rozanov (1980a,b, 1981b); Rozanov and Semenov (1980, 1981); Watson (1981); Moloney, Belic, and Gibbs (1982); Moloney, Gibbs, Jewell, Tai, Watson, Gossard, Wiegmann (1982); Moloney and Gibbs (1982a,b); Mattar (1981); Moloney (1982b); Rozanov, Semenov, and Khodova (1982a,b); Mattar, Moretti, and Franceour (1981). The full numerical integration is much better than an expansion in a large number of transverse modes because it permits an accurate and, if desired, time-dependent description of the transverse profile even if sharp discontinuities are present.

Much of the work has been done in the limit that the medium response time is much faster than both the cavity round-trip time and changes in the input field amplitude or phase. In this limit the polarization in Eq. (2.9-1) is the steady-state polarization for the instantaneous field. For a unidirectional ring cavity containing saturable two-level atoms, the saturable susceptibility is given by Eq. (2.9-4), so the dynamical Eq. (2.9-1) becomes

$$\frac{\partial E(Z,\tau)}{\partial(Z/2L)} = -\alpha_0 L \left[\frac{1+i\Delta}{1+\Delta^2+|E|^2/E_s^2} + \frac{i\lambda}{2\pi\alpha_0 n_0}\nabla_T^2\right] E(Z,\tau)\,, \tag{2.9-13a}$$

where Z,τ are the retarded-time variables. If the transverse gradient is replaced by $\partial^2/\partial x^2$, Eq. (2.9-13a) becomes

$$\frac{\partial E(Z,\tau)}{\partial(Z/2L)} = -\alpha_0 L \left[\frac{1+i\Delta}{1+\Delta^2+|E|^2/E_s^2} + \frac{i(\ln 2)}{4\pi\alpha_0 LF}\frac{\partial^2}{\partial(x/w_0)^2}\right] E(Z,\tau) \tag{2.9-13b}$$

for an input profile

$$E(x) = E(0)\exp\left[-\frac{x^2}{w_0^2}\right], \tag{2.9-14}$$

where the Fresnel number F for a Gaussian intensity profile of half width half maximum (HWHM) radius a, length L, and refractive index n_0 has been defined as $n_0 a^2/\lambda L$. This definition for F is 9.06 times smaller than $n_0 \pi w_0^2/\lambda L$, another often-used definition. Clearly $\alpha_0 LF$ is a measure of the relative importance of the optical nonlinearity (first term on right-hand side of Eq. (2.9-13b)) compared with diffraction (second term). The presence of the cavity implies optical feedback, which can continue to change the output for much longer than a cavity build-up time. The use of fast Fourier transforms can reduce substantially the required computer time; see Appendix G. An assumption of only one transverse Cartesian coordinate can further reduce the costs and lead to exceedingly instructive preliminary studies, directly applicable to specialized planar waveguide structures.

(a) Dispersive Bistability in a Plane-Mirror Ring Cavity

These one-transverse-dimension simulations have led to several interesting results. For large Fresnel number, and primarily dispersive bistability, switch-on occurs quickly out to the radius for which the input intensity equals the plane-wave switch-on intensity; it then switches on more gradually out to the radius having an input equal to the plane-wave switch-off intensity. For Fresnel numbers of order unity or less and dispersive bistability, whole-beam switching occurs with minimal loss for $F \simeq 1$. The dispersive bistability hysteresis loop looks plane-wave-like for small Fresnel number but much less sharp for large F. For fast, low-power, whole-beam switching, a Fresnel number of order unity seems best. Finally, two beams can be brought within three beam diameters on the same device and still operate independently.

The spatial response of a bistable ring cavity to the sudden application of a one-dimensional Gaussian input beam (Fig. 2.9-1a) of peak intensity 2.7 times the plane-wave switch-up intensity $I_{\uparrow}$ is shown in Fig. 2.9-1 for a self-defocusing nonlinearity. In 40 to 50 round trips, i.e., in a cavity build-up time, the output switches on out to a transverse coordinate having an input equal to the plane-wave switch-up intensity $I_{\uparrow} \simeq 200\ I_s$; see Fig. 2.9-1b. Equilibrium has not yet been reached, however. Light diffracted from the sharp edges of the "on" inner part enables the adjacent "off" region to turn on, too. This expansion continues (Fig. 2.9-1c) until the "on" region extends out to the transverse coordinate having an input equal to the plane-wave switch-off intensity $I_{\downarrow} \simeq 60\ I_s$ (Fig. 2.9-1d). This slow expansion of the initially turned on region of a large-F cavity was first shown by Rozanov and Semenov (1980, 1981) using a Kerr nonlinearity and spatial filtering to remove self-focusing instabilities.

A self-focusing nonlinearity and large Fresnel number lead to very different results as shown in Fig. 2.9-2. The initial cavity buildup is similar, resulting in a flat profile with a sharp discontinuity between "on" and "off" regions. This discontinuity apparently works its way back in by diffraction and feedback, reminiscent of small-scale self-focusing instabilities. Figure 2.9-2d predicts a series of concentric bright and dark lines within the bistable one-transverse-dimension ring cavity. It seems bizarre indeed that there should be dark lines within an "on" cavity. Further studies by McLaughlin, Moloney, and Newell (1983) have given a transverse solitary wave interpretation to these lines, finding good agreement between the lowest-order analytic transverse profile and the numerical one. The Laplacian term in the bistability wave equation looks much like the kinetic energy term in Schroedinger's equation, so that transverse eigenmodes should be expected. The characteristic dimension of a transverse solitary wave is governed by a balancing of diffraction and self-focusing as in self-trapping. The line structure of Fig. 2.9-2 is then an example of solitary waves that are, when stationary, the fixed points of an infinite-dimensional map. The number of transverse solitary waves scales as the square root of the Fresnel number and as the incident laser

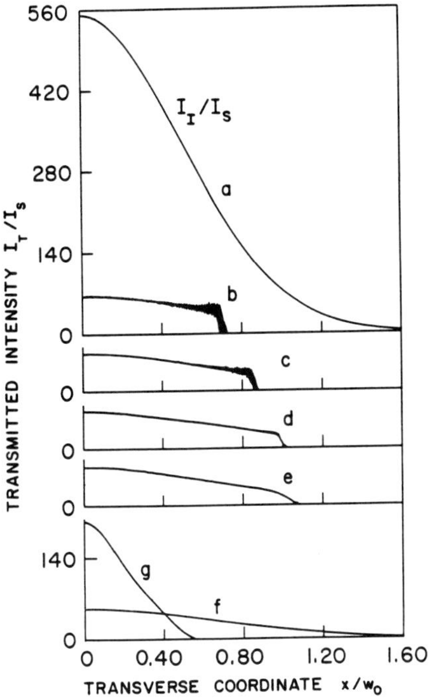

Fig. 2.9-1. Optical bistability switch-on in response to a step-function input as a function of Fresnel number. (a) Input profile I_I/I_s. (b)-(f) are for defocusing (Δ = +5) and (g) is for focusing (Δ = -5). (b)-(d); F = 2200 and round trips (b) 40 to 50, (c) 100 to 110, and (d) 230 to 240. (e) F = 220, round trips 98 to 100. (f)-(g): F = 2.2; round trips 40 - 50. For comparison the plane-wave $I_\uparrow$ is approximately 200 I_s and $I_\downarrow$ is 60 I_s, so the beam-center intensity here is approximately 2.7 $I_\uparrow$. Not shown: I_T/I_s for F = 22 and Δ = +5 falls off gradually, vanishing by $x/w_0 \simeq 1.2$; for F = 0.22 diffraction overcomes self-focusing, and whole-beam switching occurs. Other parameters are: mirror reflectivity R = 0.9; 0.4 radian detuning of laser from empty Fabry-Perot cavity (in same direction as Δ) or 2.48 radian detuning from the full Fabry-Perot peak at low intensity; $\alpha_0 L_{NL}$ = 30, where L_{NL} = 0.3L is the nonlinear medium length compared with the total length 2L of the ring cavity. L is used to calculate F. Moloney and Gibbs (1982a).

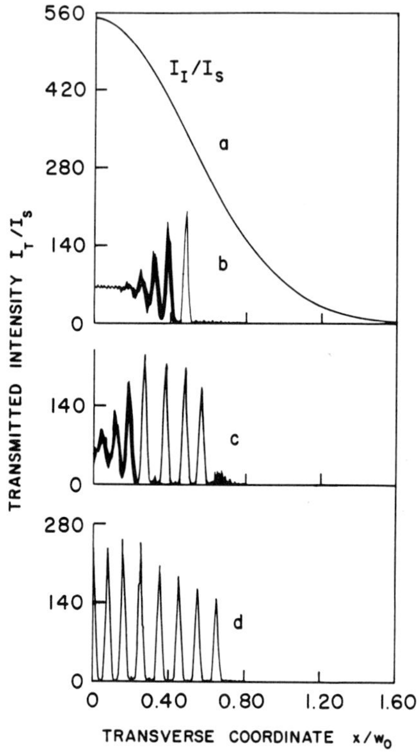

Fig. 2.9-2. Optical bistability switch-on in response to a step-function input as a function of round trips under high-Fresnel-number ($F = 870$), self-focusing ($\Delta = -5$) conditions. (a) Input. (b) 40 to 50 round trips. (c) 100 to 110. (d) 360 to 380. Even (d) may not have reached steady state. Other parameters the same as for Fig. 2.9-1. Moloney and Gibbs (1982a).

amplitude. Preliminary numerical evidence suggests that these solitary wave structures can period double while maintaining spatial coherence [Moloney, Newell, and McLaughlin (1984)].

Moloney has extended these calculations to two-transverse dimensions and has shown that the line structure of the one-transverse-dimension saturable-medium calculation goes over to a ring structure. For both saturable and Kerr media the initial switching on of the beam is identical, with a central bright spot developing around the twentieth resonator pass. Subsequent dynamical evolution, however, is significantly different. For a saturable medium, circular rings first appear on the outer edges of the transmitted profile; these are two-dimensional solitary waves of Eq. (2.9-13a). Sharp gradients on the outer edge of the profile cause the two-dimensional Laplacian in Eq. (2.9-13a) to be locally important. The solitary waves

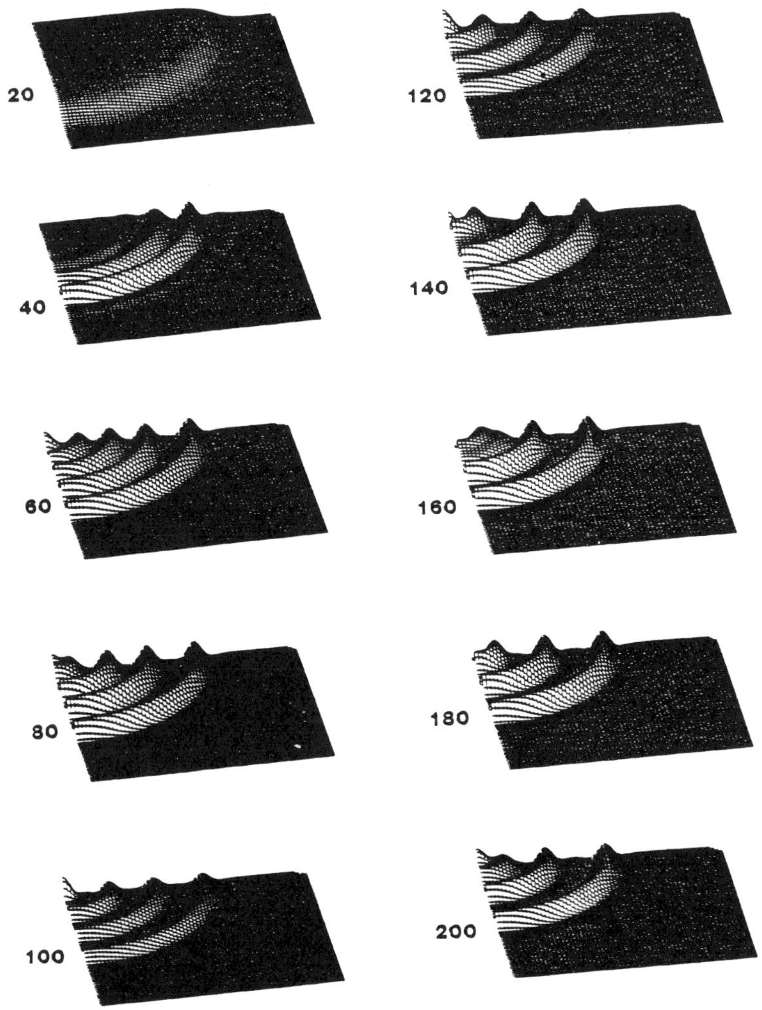

Fig. 2.9-3. Time evolution of the two-dimensional intracavity field amplitude (one quadrant is shown) as the beam switches to the high-transmission state, assuming a saturable nonlinearity in a ring cavity. The numbers refer to the ring-resonator pass. The two outer rings (solitary waves) have stabilized by the 160th pass, while the center is still undergoing slow oscillations. Moloney (1984a).

begin to evolve slowly inwardly toward the center of the beam in a wavelike fashion (see Fig. 2.9-3). Whether these structures eventually reach an asymptotically stable state or undergo complicated periodic oscillations will depend on the total energy trapped within the switched-on beam. If the total energy is such that an integral number of solitary waves can be accommodated then the structures will rapidly reach equilibrium. The situation for Kerr media is different. Circular rings initially appear uniformly over the full switched-on beam; their evolution is much slower. These rings then begin to break up into filament-like two-dimensional solitons. The two-dimensional solitons arrange themselves spatially so as to minimize their interaction energy.

The small-F defocusing case of Fig. 2.9-1f is much better behaved from the device viewpoint than the focusing case of Fig. 2.9-2: the transmission is still high, but the input power is much lower than that required for large F; and whole-beam switching occurs, reaching equilibrium in a few cavity build-up times. The strong diffraction and defocusing are responsible for the faster equilibration and whole-beam switching. Similar switching is shown for the self-focusing case in Fig. 2.9-1g, but other simulations show that self-focusing is more prone to develop rings or other instabilities.

The hysteresis loops of the output, integrated over the transverse coordinate, exhibit plane-wave-like bistability for $F \leq 1$. This results from the whole-beam switching; all parts of the beam switch simultaneously. The radial cutoff to the switch-on in the large-F case smears out the hysteresis loop. In fact, one might question whether the device is bistable at all, although there is a clear sharp-switching bistability loop at any given radius within the region that turns on. As a result, the switching time measured with the entire beam is dependent on the time dependence of the input. Again from the device viewpoint $F \simeq 1$ seems best in simplifying the hysteresis loop and minimizing the input power.

(b) Absorptive Bistability in a Plane-Mirror Ring Cavity

For purely absorptive bistability ($\Delta = 0$), the on-axis intensity for a large Fresnel number has a hysteresis loop that is very similar to the plane-wave loop (Fig. 2.9-4a). But for $F = 8.8$ the loop (Fig. 2.9-4c) has narrowed dramatically because of the strong absorption of energy from the low-intensity wings of the strongly diffracted beam profile. Figure 2.9-5 displays the corresponding whole-beam loops. At high F values the switch-on above $I_{\uparrow}$ is gradual, reflecting the radial dependence of the switching process. As a result, the up-sweep and down-sweep portions of the high-transmission branch no longer coincide. The strong radial dependence is explicitly shown in Fig. 2.9-6.

The whole-beam hysteresis loop for $F = 8.8$, Fig. 2.9-5b, again shows marginal bistability, but the switching is sharp, which is indicative of whole-beam switching. The corresponding profiles are shown in Fig. 2.9-7. At lower F values the hysteresis disappears, which may be why absorptive bistability has been difficult to observe experimentally. It is interesting that $F = 8.8$ corresponds to the first Na experiment (Section 3.2.3) in which absorptive bistability was reported with the laser tuned to a frequency having no nonlinear index and the intensity increased to saturate the medium. Figure 2.9-5b shows a very narrow bistable loop consistent with observations.

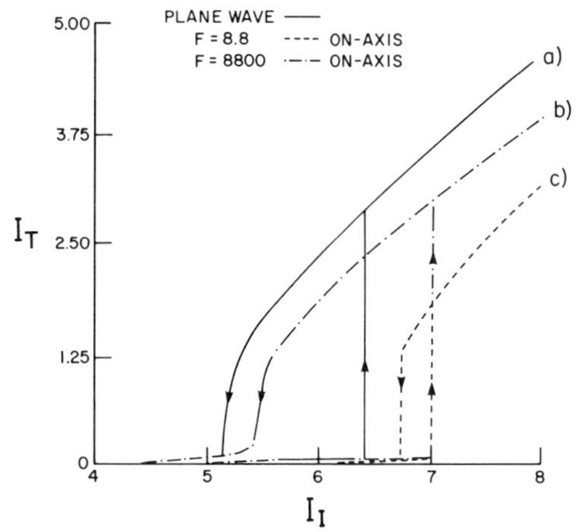

Fig. 2.9-4. Comparison of plane-wave and Gaussian-beam (on-axis) hysteresis loops. Parameters common to all calculations are $\alpha_0 L = 1.5$, $T = 0.9$, (a) plane-wave hysteresis loop (b) on-axis Gaussian-beam hysteresis with Fresnel number F = 8800 and (c) on-axis Gaussian-beam hysteresis with F = 8.8. Note that both low- and high-Fresnel-number beams switch-on at approximately the same input intensity. Moloney, Sargent, and Gibbs (1982).

(c) Crosstalk

An interesting question associated with the use of semiconductor etalons is how close can one bring two light beams on the same etalon and have independent bistable operation? This question has been studied numerically by Tai, Moloney, and Gibbs (1982a) in the limit that the only coupling is light diffraction in one transverse dimension. Particle or excitation diffusion might lead to the need for a larger beam separation. Figure 2.9-8 shows that beam separations of 2.94 and 2.35 FWHM beam diameters are sufficient for independent operation whereas 2.12 and 2 are not. Of course, a more stringent test would determine the separation needed to keep off a single device in a full two-dimensional array suddenly turned on simultaneously. The separation required to keep "off" a device surrounded by four identical devices is not much larger than that shown in Fig. 2.9-8 according to two-transverse-dimension simulations of Tai, Gibbs, Moloney, Weinberger, Tarng, Jewell, Gossard, and Wiegmann (1984).

One might expect that if the two nearby beams have opposite phase that they can be brought closer. Such is not the case; almost four FWHM are required if the phase difference is π. This is because the field inside the cavity is almost in phase with the input field in the "on" state but nearly out of phase in the "off" state; see Section 5.7.

These crosstalk calculations suggest that parallel operation of many devices on a single etalon is not prevented by diffractive coupling of light (Section 7.4). For thermal crosstalk on a dye etalon see Rushford et al. (1984).

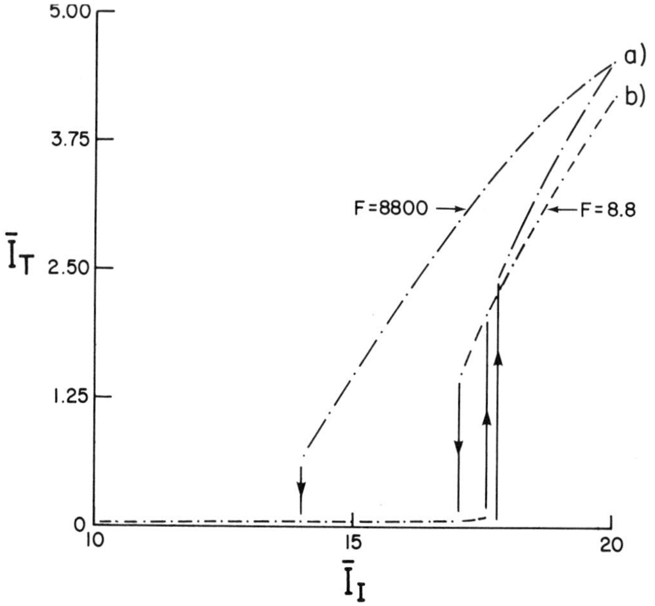

Fig. 2.9-5. Integrated whole-beam hysteresis loops for Fresnel number (a) F = 8800 and (b) F = 8.8. The bar over each intensity signifies an integration over the transverse dimension x. Moloney, Sargent, and Gibbs (1982).

2.9.3. Relation to Other Work

The simulations in Fig. 2.9-2 for large Fresnel numbers appear at first in conflict with the careful agreement found by Bischofberger and Shen (1978a,b; 1979a,b) between uniform-plane-wave numerical simulations and their Gaussian-beam experiment using CS_2 (Section 3.4). They detected the on-axis output with $F \simeq 480$ and $F = 13$, and the device was "on" for 75 to 200 round trips. Perhaps the line structure of Fig. 2.9-2 has insufficient time to evolve or perhaps other processes removed it. Note also that Fig. 2.9-2 is calculated for an initial detuning of 2.48 radians, i.e., almost midway between Fabry-Perot peaks, whereas Bischofberger and Shen used 0.63 radian. A larger change in index is then required for Fig. 2.9-2, enhancing self-focusing effects. Bischofberger and Shen did not report a radial dependence to the switching as observed in GaAs (Section 3.6).

The simulations of Section 2.9.2 were carried out in the same limit required for Ikeda instabilities (Section 6.3), namely, a medium response time much faster than a cavity round-trip time. Ikeda instabilities have been simulated for a plane-mirror ring cavity with the same program, but Δ and β_0 were selected to avoid those instabilities for the spatial dynamics described in Section 2.9.2.

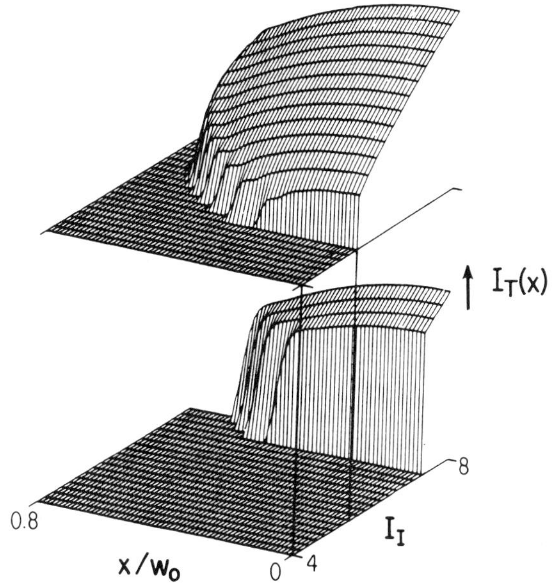

Fig. 2.9-6. The strong radial dependence of the switching to the high transmission state at high Fresnel number (F = 8800) for absorptive bistability. The input intensity was increased (decreased) in the lower (upper) plot. For symmetry reasons only half of the transverse spatial profile is displayed. Moloney, Sargent, and Gibbs (1982).

However, transverse effects do make major alterations in the bifurcation structure within the unstable regions as discussed in Section 6.3. Lugiato and Milani (1984) have shown the importance of transverse effects on self-pulsing and other instabilities using a mode-expansion approach for a ring cavity with spherical mirrors; see Section 6.2.

Associated with dispersive bistability is a change in spatial profile due to nonlinear wavefront encoding (Section 7.3). Hysteresis in the spatial profile has been calculated and observed and is at the heart of self-focusing optical bistability (Section 3.9). See Firth, Seaton, Wright, and Smith (1982); Khoo (1982); Gibbs, Jewell, Moloney, Rushford, Tai, Tarng, Weinberger, Gossard, McCall, Passner, and Wiegmann (1982); Gibbs, Jewell, Moloney, Tarng, Tai, Watson, Gossard, McCall, Passner, Venkatesan, and Wiegmann (1982); Gibbs, Derstine, Hopf, Jewell, Kaplan, Moloney, Shoemaker, Tai, Tarng, Watson, Gossard, McCall, Passner, Venkatesan,

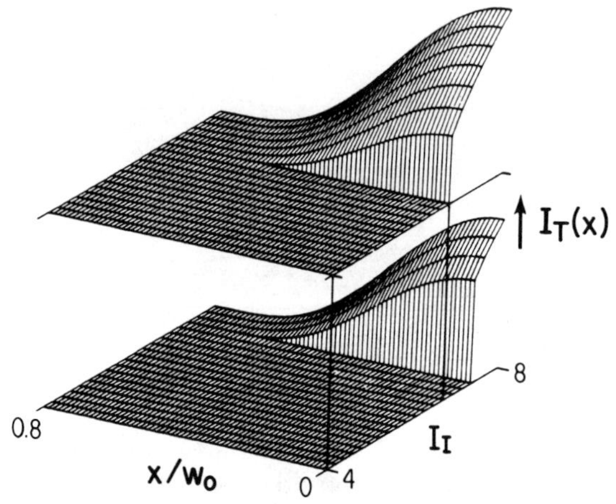

Fig. 2.9-7. Absorptive bistability whole-beam switching at low Fresnel number (F = 8.8). Note the significant narrowing of the hysteresis loop relative to Fig. 2.9-6. Moloney, Sargent, and Gibbs (1982).

and Wiegmann (1981); Tai, Gibbs, Moloney, Weinberger, Tarng, Jewell, Gossard, and Wiegmann (1984); Khoo, Yan, Liu, Shepard, and Hou (1984a,b); LeBerre, Ressayre, Tallet, Tai, Hopf, Gibbs, and Moloney (1984).

This section has concentrated on transverse effects arising from diffraction and self-lensing of the light beam. Clearly, diffusion of the excitation within the medium can be even more important under some conditions. For example, if the Fresnel number is large and the diffusion length ℓ_D is much larger than the beam diameter, whole-beam uniform-plane-wave-like hysteresis loops may result. Firth, Abraham, Wright, Galbraith, and Wherrett (1984) have studied diffusion effects. Similar to Eq. (2.9-13a) they write

$$\frac{\partial E_F(r)}{\partial(Z/L)} = \left[-\frac{\alpha L}{2} - \frac{i\lambda L \nabla_T^2}{4\pi n_0} + id(r)\right] E_F(r) \qquad (2.9\text{-}15)$$

$$-\frac{\partial E_B(r)}{\partial(Z/L)} = \left[-\frac{\alpha L}{2} - \frac{i\lambda L \nabla_T^2}{4\pi n_0} + id(r)\right] E_B(r) \qquad (2.9\text{-}16)$$

$$(-\ell_D^2 w_0^2 \nabla_T^2 + 1)d(r) = -\frac{2\lambda L n_2}{\pi w_0^2 n_0}\left(|E_F(r)|^2 + |E_B(r)|^2\right). \qquad (2.9\text{-}17)$$

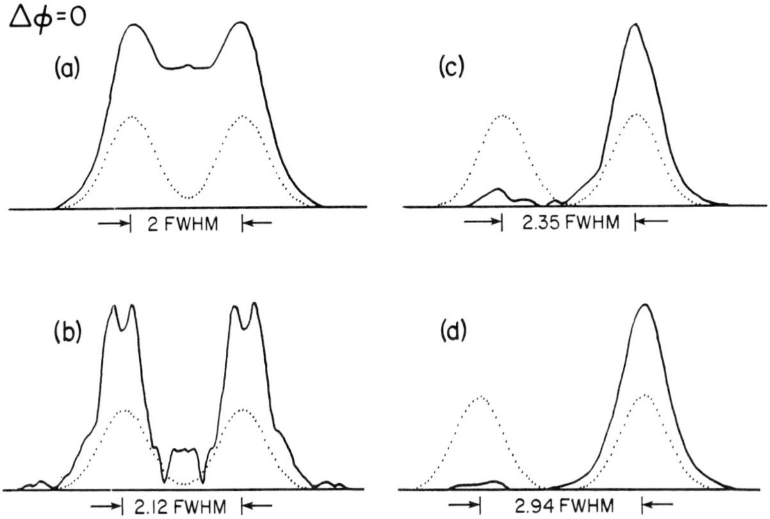

Fig. 2.9-8. The effects of diffraction coupling when both input beams have the same phase and the same input intensity--below the switch-on and above the switch-off values. Initially, both outputs are in the lower branch. (a)-(d) show the effects of coupling after switching the right beam to the upper branch. For (a) and (b), the left beam is turned on by diffraction coupling. For (c) and (d), the left one remains in the lower branch. The ring cavity parameters are as follows: $F = 0.55$, $\Delta = +5$ (defocusing side), $I_{I1}(x=0) = I_{I2}(x=0) = 200\ I_s$; other parameters: see Fig. 2.9-1. Note that the input profile, i.e., two equal-height Gaussians, is also superimposed on the figures. Tai, Moloney, and Gibbs (1982a).

The excitation density, d(r), is the nonlinear phase shift per unit propagation distance and is assumed to diffuse transversely with a diffusion length ℓ_D. They have studied these equations numerically and concluded that diffraction and diffusion give rise to qualitatively similar effects. Whereas diffraction establishes correlation over regions of area $\simeq \lambda L$, diffusion will be effective over areas of order $\ell_D{}^2$. Either effect may lead to whole-beam switching where a purely local theory would predict part-beam switching. There is evidence for whole-beam switching via diffusion in both GaAs (nice whole-beam loops and $P_\uparrow$ independent of w_0 for small w_0) and indium antimonide (InSb) (nice whole-beam loops and multistability with large Fresnel number).

For interesting effects of longitudinal diffusion upon increasing absorption optical bistability see Section 2.10.

In summary, transverse effects are inescapably significant in optical bistability. Under some conditions transverse communication by means of diffraction or diffusion can result in plane-wave-like behavior (at the expense of higher input intensity) much the same as for hybrid bistability. Under other conditions transverse effects can result in markedly different whole-beam behavior and complicated transverse profiles.

2.10. OPTICAL BISTABILITY WITHOUT EXTERNAL FEEDBACK: INCREASING ABSORPTION OPTICAL BISTABILITY

So far in this book, optical bistability has always involved external feedback by means of external mirrors or the naturally occurring reflectivity at the surfaces of a solid etalon. In Chapter 4, hybrid bistable systems are described in which the external feedback is electrical and the microscopic optical nonlinearity is replaced by a voltage-dependent phase shifter within a Fabry-Perot cavity or between crossed polarizers. Recently, a new class of bistability has been recognized in which external feedback is absent. In most of the cases demonstrated so far, the bistability arises from increased absorption with increasing input intensity, so it is called increasing absorption optical bistability and its hysteresis loop appears inverted.

We analyze increasing absorption bistability following D. A. B. Miller, Gossard, and Wiegmann (1984), although very similar derivations are given by Toyozawa (1978, 1979) and by Averbukh, Kovarsky, and Perelman (1979). Assume that a medium's optical absorption A depends on some parameter N of the medium, such as carrier density or temperature, that depends upon the degree of optical excitation:

$$A = A(N). \qquad (2.10\text{-}1)$$

If light of power P_I is incident on the medium, the absorbed power will be AP_I. If we assume for simplicity that, in the steady state, N is directly proportional to the power then

$$N = \eta A P_I, \qquad (2.10\text{-}2)$$

where η is a constant. That bistability can result can be seen by the graphical solution of Fig. 2.10-1. The dashed curve is a hypothetical curve, but it approximates well the disappearance of the transparency of a semiconductor due to the thermal shift of the band edge by means of absorption of just-below-bandgap light. The graphical construction, described in the figure caption and much like that in Section 2.6, shows the inverted hysteresis loop. The graphical construction makes it clear that the condition for multiple intersections of the straight line, Eq. (2.10-2), and the curve, Eq. (2.10-1), is that the curve be steeper than the straight line, i.e., the condition for optical bistability is

$$\frac{dA}{dN} > \frac{A}{N}\,; \qquad (2.10\text{-}3)$$

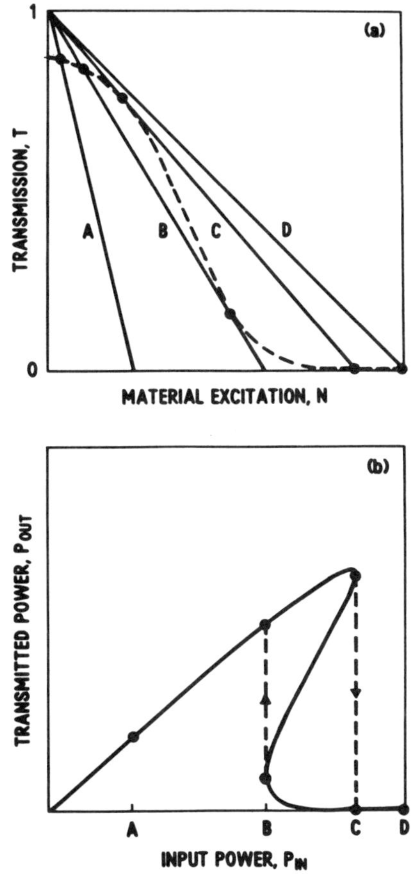

Fig. 2.10-1. Graphical solution of optical bistability due to a hypothetical absorption increasing with increasing excitation. (a) The dashed line is $T = 1 - A(N)$ [Eq. (2.10-1)]. Lines (A) to (D) correspond to increasing power in Eq. (2.10-2). Lines (A) and (D) intersect only once with the curve, indicating only one solution for these powers. Lines (B) and (C), each showing two intersections, represent the critical powers for switch-to-low-transmission and switch-to-high-transmission. Any lines between (B) and (C) would show three intersections as required for optical bistability. (b) The transmitted power, is plotted against the incident power using the solution method from (a). D. A. B. Miller, Gossard, and Wiegmann (1984).

A must be more than linearly proportional to N over some region of N. For a similar analysis of increasing absorption bistability including propagation and diffusion effects see Henneberger and Rossmann (1984).

There are several independently analyzed examples of increasing absorption optical bistability. Épshtein (1978) predicts a temperature hysteresis in the optical thermal breakdown of a semiconductor due to temperature-dependent absorption and attenuation of the light in the sample; in retrospect it is clear that the transmitted light exhibits increasing absorption bistability [see also Toyozawa (1978, 1979) and Rozanov (1981b)]. Bowden and Sung (1979, 1981) and Bowden (1981) proposed intrinsic mirrorless bistability based on atomic correlation in a small volume. Kaplan (1982b) predicted bistable behavior for a relativistic electron gas under optical illumination. Thermal increasing absorption optical bistability due to shrinkage of the optical bandgap with increasing temperature has been reported by Hajtó and Janossy (1983) and Forgács (1982) in amorphous $GeSe_2$ [≃ 15 mW focused to 207-μm diameter onto a 6-μm self-supporting film at 298 K, ≃ 1-s switching times], by D. A. B. Miller, Gossard, and Wiegmann (1984) [hybrid, ≃ 3 mW, ≃ 1.1 mm × ≃ 0.6-mm piece of multiple-quantum-well consisting of 375 periods of 8.7-nm-thick $Al_{0.3}Ga_{0.7}As$ and 8.5-nm-thick GaAs layers thermally isolated at 300 K, few-second modulation of input, 40 to 400-ms switch-back-on time], and by Dagenais, Sharfin, and Winful (1984) [≃ 3 mW, 10 to 20-μm cadmium sulfide (CdS) platelet held electrostatically to a thin glass slide (≃ 150 μm) at 2 to 50 K, < 1-μs switching times, >20 on/off contrast; see also Dagenais and Sharfin (1984)] in CdS using the bound exciton. In Rossmann, Henneberger, and Voigt (1983) [≃ 1 MW/cm^2, few-micrometer-thick uncoated etalons at 77 K, 4-μs pulses, no switching-back-on seen] and Henneberger and Rossmann (1984) it is broadening of the exciton that increases the CdS absorption whereas in Bohnert, Kalt, and Klingshirn (1983) [≃ 1 MW/cm^2, 1- to 5-μm-thick uncoated etalons at 1.8 K, 2.5-ns pulses, no switching-back-on seen with 0.8-ns resolution] and Schmidt, Haug, and Koch (1984) it is bandgap renormalization by means of first two-photon and then one-photon absorption. Shen (1984) reports the effect using the change of state of liquid crystals. Lavallard, Duong, and Itoh (1983) see reduced transmission through a 4-μm-thick cadmium selenide (CdSe) sample when the input intensity is increased from 0.5 to 5 kW/cm^2 and attribute it to the scattering of polaritons on bound excitons.

Hopf, Bowden, and Louisell (1984), considering the effect of local field corrections on the behavior of a dense collection of two-level atoms, predict mirrorless bistability. Inclusion of the local field gives a Stark shift of the optical transition to lower photon energy; reduction of the local field by means of optical excitation can increase the absorption for light on the high-energy side.

In the self-electro-optic-effect device (SEED) of D. A. B. Miller, Chemla, Damen, Gossard, Wiegmann, Wood, and Burrus (1984), a quantum-well electroabsorptive modulator is connected in series with a large resistor to a reverse bias supply. The laser wavelength is chosen so that decreasing voltage on the modulator increases the absorption. Over a large range of voltages the modulator also works as a photodetector of constant internal quantum efficiency, so that a photocurrent (N above) is generated that is directly proportional to absorbed power P_I. Increasing P_I increases N, which increases the voltage drop across the series resistor and decreases that across the modulator, thereby increasing the absorption.

Pepper and Klein (1979) observed bistability using the intensity dependence of the transmission of a 10.784-μm $C^{13}O_2^{16}$ laser beam through ammonia (NH_3) gas subjected to a Stark field proportional to the transmitted intensity. Their case may have been the first example of increasing absorption optical bistability. They explained their mirrorless bistability by a graphical solution similar to that of Section 2.6. Figure 2.10-2 shows the transmission of the ammonia-filled Stark cell as a function of the applied voltage. The multiple absorption peaks result from various transitions whose degeneracy is lifted by the electric field; only one transition is needed for bistability, so consider only the lowest voltage dip in transmission. The voltage V across the Stark electrodes is obtained from the detected output power P_T and can be written as

$$V = V_B - gP_T , \qquad (2.10\text{-}4)$$

where V_B is the bias voltage and g is a gain factor. Therefore, the transmission can also be written as

$$\mathscr{T} = \frac{P_T}{P_I} = \frac{V - V_B}{-gP_I} ; \qquad (2.10\text{-}5)$$

the straight lines in Fig. 2.10-2 are Eq. (2.10-5) plotted for various V_B and gP_I values. As gP_I is increased from straight line 2 to 1, the absorption increases at first and then switches to low absorption as the straight line passes the tangent point. Upon decreasing gP_I, switching to high absorption occurs at a lower gP_I than for the previous switching. Pepper and Klein's bistability, which uses no cavity or polarizers, might then be called "decreasing absorption optical bistability." Their setup could be used to demonstrate increasing absorption bistability by making the Stark voltage proportional to the absorbed power, i.e.,

$$V' = V_B + g(P_I - P_T); \qquad (2.10\text{-}6)$$

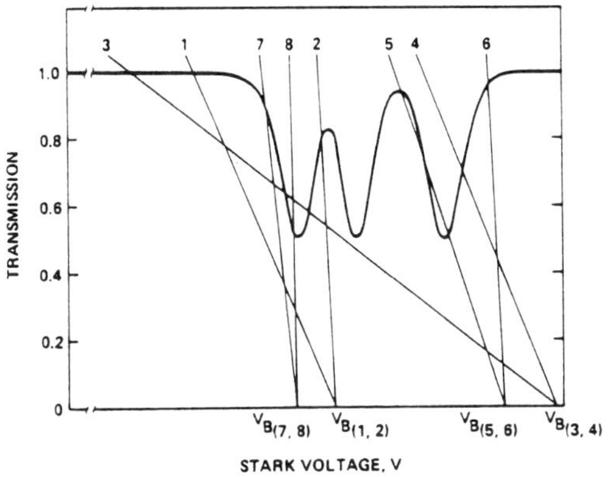

Fig. 2.10-2. Graphical solution for the mirrorless Stark bistable optical device in the absence of saturation. Pepper and Klein (1979).

then

$$\mathscr{T}' = \frac{P_T}{P_I} = 1 - \frac{V' - V_B}{gP_I} . \qquad (2.10\text{-}7)$$

A graphical solution shows that with increasing P_I switch-down occurs, and with decreasing P_I switch-up occurs, and that $P_\downarrow > P_\uparrow$, unlike the usual bistability. Even the decreasing absorption bistability of Pepper and Klein satisfies Miller's conditions if their V is the N of Miller: $A = 1 - \mathscr{T}(V) = A(V)$ and $V = V_B - gP_I + gAP_I$, so A depends on V and not merely on P_I.

Because the bistability is a local effect rather than a property of the entire device, there may be discontinuities in medium variables along the light beam somewhat similar to the transverse discontinuities discussed in Section 2.9. Koch, Schmidt, and Haug (1984) have solved the transport equations for the light intensity and the excitation density. Under appropriate pulse operation a kink in the excitation density is found, which moves discontinuously in the beam propagation direction and causes a sawtooth variation in the transmitted intensity. Because of absorption, the light intensity decreases as it propagates through the sample. Therefore, the front part of the sample reaches the condition (temperature, carrier density, or whatever) for switch-off before the remainder of the sample. Once the first portion switches off, the intensity downstream is even lower. However, if the transmission of the first portion is not too low and if the input intensity continues to increase, eventually the second portion will turn off, etc. If too thin a sheet switches off all at once, then the effect of the kink on I_T may be too small. The length of the sheet that switches off can be adjusted by the risetime of the input pulse: while the leading edge of a sheet whose I_I has exceeded $I_\downarrow$ is undergoing the positive-feedback switching, some of the remainder of the medium also exceeds $I_\downarrow$. Toyozawa (1978) and Henneberger and Rossman (1984) also recognized the possibility of domains in different states, but they did not describe a series of jumps leading to the dramatic sawtooth variation in I_T presented by Koch, Schmidt, and Haug (1984).

The transport equations of the light intensity I(z,t) and the free-carrier density $N_c(z,t)$ are:

$$\frac{\partial I}{\partial t} = -\frac{c}{n_0}\left[\alpha I + \frac{\partial I}{\partial z}\right] \qquad (2.10\text{-}8)$$

and

$$\frac{\partial N_c}{\partial t} = \frac{\alpha I}{\hbar\omega} - \frac{N_c}{\tau_M} + \frac{\partial}{\partial z} D \frac{\partial}{\partial z} N_c , \qquad (2.10\text{-}9)$$

where τ_M is the free-carrier lifetime and D is the diffusion coefficient. Koch, Schmidt, and Haug (1984) display beautiful kinks modeling bandgap renormalization in CdS. Gibbs, Olbright, Peyghambarian, and Haug (1985) have observed kinks in increasing absorption optical bistability in color filters. They selected filters that have sharp cutoffs that shift with temperature but whose absorption in the off state is not too high for sample lengths of a few millimeters. Input pulse lengths and diameters can then be chosen for which longitudinal conduction [D term in Eq. (2-10.9)] is much less important than transverse conduction [modeled by τ_M in Eq. (2.10-9)]. Figure 2.10-3 shows several switch-offs of subsequent portions of a color filter and corresponding switch-ons as the input is reduced. The

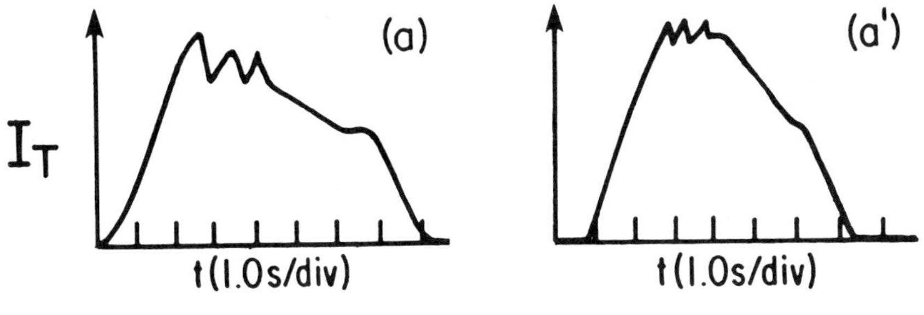

Fig. 2.10-3. Kinks (a) observed in a Corning 3-70 sharp cut color filter and (a') calculated. Gibbs, Olbright, Peyghambarian, Schmidt, Koch, and Haug (1985).

presence of kinks is an amusing manifestation of the local nature of increasing absorption optical bistability; the whole device need not switch simultaneously even in the longitudinal direction!

D. A. B. Miller, Gossard, and Wiegmann (1984), D. A. B. Miller (1984b), and Goldstone and Garmire (1984a,b) seem to be the first to recognize the relatedness of the various cases of increasing absorption bistability; see Miller (1984b) for a more detailed comparison of each case to his general model. Goldstone and Garmire (1984a,b) stress the difference between two classes of bistability. In the usual class, the subject of this book except for this section, the polarization P of the medium is expressible as a single-valued function of the optical field E. It is the optical feedback that leads to bistability. Increasing absorption optical bistability is a special case of the second class in which the internal optical field E is a single-valued function of the state of the medium as expressed by P (or the absorption A or degree of excitation N in Miller's model). The converse is not true, i.e., the state of the medium is many-valued as a function of E if bistability without feedback is to occur. This is equivalent to the impossibility of expanding P as a power series in E with constant coefficients. One can still argue that internal feedback is present: treat the bistable device as a black box with input I_I and output I_T, since there is no way to know I_T from I_I, but I_T does uniquely determine I_I, there must be feedback. Nonetheless, the distinction between whether external feedback or the medium itself introduces the multi-valuedness is operationally useful.

Increasing absorption optical bistability has been discovered and studied only recently, so it may take some time to evaluate its potential for practical devices. It is being promoted as particularly interesting from a practical standpoint because of the absence of a resonator and because the output signal is automatically inverted. So far, the demonstrated cases have mostly been thermal and hence rather slow. The fact that there is no resonator means that a large shift of the absorption is required for a high on-to-off contrast ratio; most of the observed ratios have been low. From the point of view of minimizing the heat dissipation in each device, it remains to be seen if devices based on absorption without a cavity can win out over devices based on nonlinear refraction within a cavity. Perhaps their simplicity and the fact that absorption is never avoided entirely in any device give some hope for practical increasing absorption bistable devices.

In some respects increasing absorption optical bistability is more like dispersive bistability than absorptive. In dispersive bistability the Fabry-Perot peak is shifted relative to fixed absorber and laser frequencies. In increasing absorption bistability, the absorption is shifted relative to the fixed laser frequency. Multistability, of course, cannot occur for increasing absorption bistability, although kinks involve multiple spatial states.

Increasing absorption optical bistability might also be called photodarkening bistability or self-induced darkness bistability or self-induced opaqueness bistability. Perhaps the author's bistability falls in the last category!

CHAPTER 3

INTRINSIC OPTICAL BISTABILITY EXPERIMENTS

3.1. EARLY SEARCHES FOR ABSORPTIVE OPTICAL BISTABILITY

All of the early attempts to see optical bistability were perceived as searches for absorptive bistability; dispersive bistability was not conceived until it was observed in Na (Section 3.2). In Section 1.4 the early history of optical bistability, in systems containing no gain medium, was reviewed. The conception of passive optical bistability by Seidel (1969) and Szöke, Daneu, Goldhar, and Kurnit (1969) is described there, as are experiments conducted by Szöke et al. (1969), Austin and DeShazer (1971), Austin (1972), and Spiller (1971, 1972). Optical bistability was not seen, but nonlinear transmission and the corresponding pulse reshaping (Section 5.2) were. It is likely that in all three cases too much unsaturable background absorption was present so that $\alpha_0 L/(T + \alpha_B L) > 8$ could not be satisfied [Eq. (2.5-46)].

3.2. SODIUM VAPOR: FIRST OBSERVATION OF PASSIVE OPTICAL BISTABILITY AND DISCOVERY OF NONLINEAR INDEX MECHANISM

The first observation of passive optical bistability was made in a Fabry-Perot resonator containing Na vapor; see Section 1.3 for earlier observations of optical bistability in lasers. The results of the Na experiment have appeared in McCall, Gibbs, Churchill, and Venkatesan (1975); McCall, Gibbs, and Venkatesan (1975); Gibbs, McCall, and Venkatesan (1975, 1976); and Venkatesan (1977). The principal results of the Na experiment were: the first observation of optical bistability, both absorptive and dispersive; the discovery and understanding (Section 2.3) of dispersive bistability; demonstration that neither inhomogeneous broadening nor a nonuniform transverse spatial profile prevents bistability in general; demonstration of other optical logic functions such as transistor (ac gain), clipping, and limiting action.

3.2.1. Experimental Details

At the conception of the Na experiment, the experimenters were unaware of the previous work on absorptive bistability by Seidel (1969); Szöke et al. (1969), Spiller (1971, 1972), and Austin and DeShazer (1971). They were guided only by McCall (1974). In any case, dispersive bistability was unknown, and inhomogeneous broadening appeared fatal for absorptive bistability. Consequently, a system was designed to provide a foreign gas to homogeneously broaden the Na. As shown in Fig. 3.2-1 the gas served the additional function of preventing Na from reaching the Fabry-Perot mirrors. Because of the interaction of Na with most glasses at high temperatures and the insertion loss of any window, a windowless system seemed preferable. As it turned out, argon gas was usually not used because the laser power was insufficient to saturate the total homogeneously broadened line and because dispersive bistability was seen without it. The long narrow water-cooled tubes and bellows connecting the hot Na chamber to the Fabry-Perot end mirrors reduced the solid angle for Na atoms to strike the end mirrors. Every week or so the mirrors had to be cleaned.

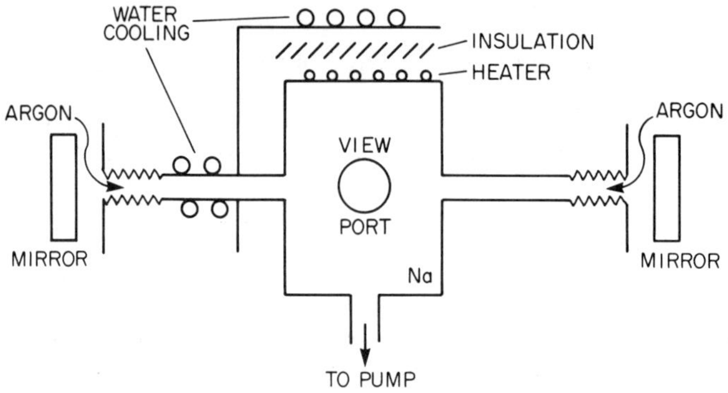

Fig. 3.2-1. Na optical bistability apparatus. Gibbs, McCall, and Venkatesan (1975).

The experimental setup is shown in Fig. 3.2-2. A 3-W argon laser pumped a home-built folded-cavity jet-stream dye laser which was stabilized to ≃ 2 MHz by locking it to an external tunable empty Fabry-Perot interferometer. Frequency calibration was achieved by saturation spectroscopy in an evacuated cell of Na. Even though ≃ 50 mW of single-mode power was emitted by the dye laser, a maximum of 13 mW reached the Fabry-Perot device containing Na vapor. The optical beam diameter there was 1.65 mm, large enough to make diffractive losses negligible for the multiple passes within the Fabry-Perot device. The mirrors were flat to $\lambda/100$ and had 90% reflectivity dielectric coatings. The Fabry-Perot structure, a forerunner of the present Burleigh interferometers, contained piezoelectric crystal mounts for up to kilohertz electronic scanning. Water cooling of the Na oven eliminated any serious problems from thermal changes in the cavity length. With all of these precautions, the overall system was marginally stable for observation of features, such as high ac gain, which were extremely sensitive to the cavity frequency, laser frequency, and Na density (typically 10^{-4} to 10^{-5} Torr over a 2.5-cm region midway between the mirrors spaced 11 cm apart).

An acousto-optic modulator was used to modulate the input intensity in a triangular wave for bistability hysteresis studies or in square steps for fast-switching measurements. The 40-MHz Doppler shift per pass through the modulator and the use of a Glans prism and $\lambda/4$ plate between the dye laser and Fabry-Perot device provided adequate isolation between the highly-reflecting (at times) Fabry-Perot device and the laser. Thus the observed effects did not arise from feedback interactions between the external cavity and the laser. The incident and transmitted intensities were detected by phototubes and monitored by an x-y oscilloscope display. The optical transmission of the properly tuned empty Fabry-Perot cavity was about 45% because of losses from mirror nonflatness and Na deposits on the mirrors. The vertical gain was adjusted so that the empty cavity input-output curve was a 45° line as in Fig. 3.2-3.

3.2.2. Observations of Nonlinear Transmission and Bistability

Measurements were made first on the nonlinear transmission properties of the Na vapor without any Fabry-Perot mirrors. Such a curve is shown in Fig. 3.2-3 for no argon; essentially complete saturation was seen, i.e., I_T versus I_I approached 45°. As argon was added the saturation became less complete. This is now known to be because the no-argon saturation is by optical pumping, which has a relaxation time much longer than the spontaneous lifetime characteristic of homogeneously broadened two-level saturation. With the low intensities available inside the plane Fabry-Perot cavity using the weak laser, it was easy to optically pump but difficult to saturate the entire homogeneously broadened line (the natural width is only 10 MHz compared with a Doppler width of about 2 GHz). The reduction of the switch-up power $P_\uparrow$ by means of optical pumping is discussed further in Ballagh, Cooper, and Sandle (1981) and Arecchi, Giusfredi, Petriella and Salieri (1982). With the addition of Fabry-Perot mirrors, still with no argon, optical bistability was observed as shown in Fig. 1.2-1. By observing the traces in time and by z-axis dashing of the rising portion of each triangular wave, it was established that the switch-up intensity $I_\uparrow$ was higher than the switch-down intensity $I_\downarrow$. The intensity modulation could be stopped and the input intensity adjusted to a holding intensity I_H between $I_\downarrow$ and $I_\uparrow$. If I_H was reached from above $I_\uparrow$, then the transmission state remained high; interruption of the beam resulted in stable low transmission again. The system was truly bistable in a cw sense, in contrast to some later devices which can turn off by means of thermal effects if they are not heat sunk.

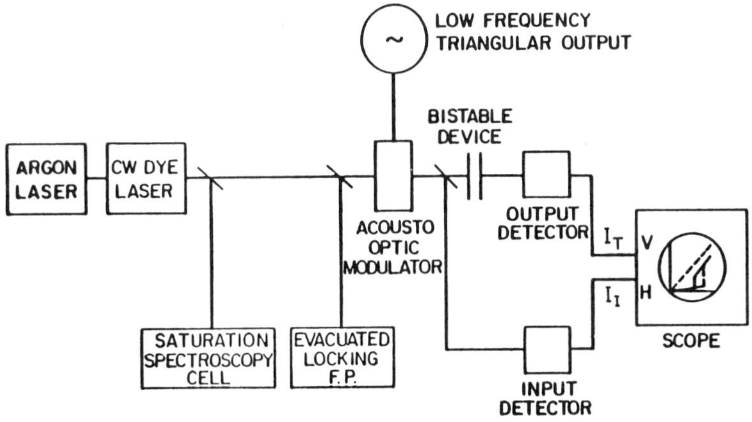

Fig. 3.2-2. Experimental setup for Na optical bistability. Gibbs, McCall, Venkatesan (1977).

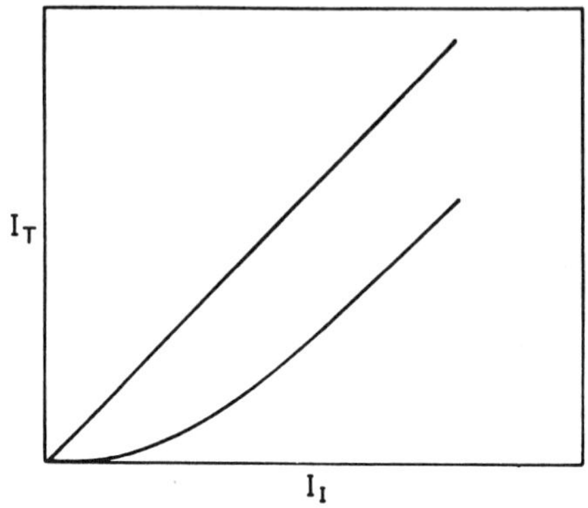

Fig. 3.2-3. Saturation of Na vapor with no Fabry-Perot cavity and no buffer gas. Venkatesan (1977).

3.2.3. Nonlinear Refractive Index and Asymmetric Fabry-Perot Scans

In Section 2.3 it was shown that for purely dispersive bistability, one can go from no bistability, to high gain, through better and better stability loops, to no bistability again merely by changing the laser-cavity detuning. That was shown first experimentally, as shown in Fig. 3.2-4. This was a puzzle when first observed as discussed below. A high-gain characteristic is not shown in Fig. 3.2-4 but is in Fig. 1.2-3 where the distortion in the amplification results from an input modulation signal exceeding the linear portion of the characteristic. Although this first device was not small like a transistor, it did exhibit optical ac gain with only optical inputs in much the same way that electrical ac gain with electrical inputs is observed with a transistor. By adjusting the Na density, operating frequency, and laser-cavity detuning in various ways, other characteristics such as thresholding (Fig. 1.2-6) and limiting were observed.

Let us return to Fig. 3.2-4, which was the key to the discovery of dispersive bistability. In Fig. 3.2-4 the parameter that is changed from curve to curve is the laser-cavity detuning in megahertz where 0 was arbitrarily assigned to the largest bistable loop. Then for a tiny shift in one direction, the bistability disappears altogether. For much larger shifts in the other direction the loop becomes narrower and $I_\uparrow$ decreases. With still larger shifts the bistability disappears, high gain comes and goes, and finally low transmission occurs on that side as well. As pointed out before, this is expected for dispersive bistability. At the time Fig. 3.2-4 was taken, only absorptive bistability was known. In purely absorptive bistability the laser-cavity detuning is not intensity dependent; bistability occurs by saturating the nonlinear absorption. Consequently, the bistability would be expected to degrade symmetrically with detuning. The clear asymmetry in Fig. 3.2-4 pointed toward dispersive effects. This was made even clearer by storage oscilloscope traces of I_T versus Fabry-Perot cavity tuning as shown in Fig. 3.2-5 for plots of the peak I_T versus tuning. For the empty cavity (solid traces) the Fabry-Perot transmission peak is symmetrical, and when the laser is shifted to ±360 MHz the peaks are still symmetrical and displaced equally. But when Na is added (dashed), the transmission peak mapped out with a linear voltage scan can

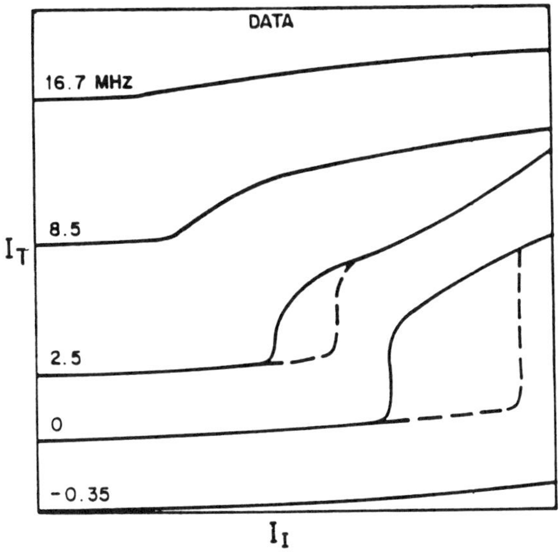

Fig. 3.2-4. Na transmission for various laser-cavity detunings in megahertz, where zero detuning has been arbitrarily (and erroneously) assigned to the largest bistability loop. Gibbs, McCall, and Venkatesan (1976).

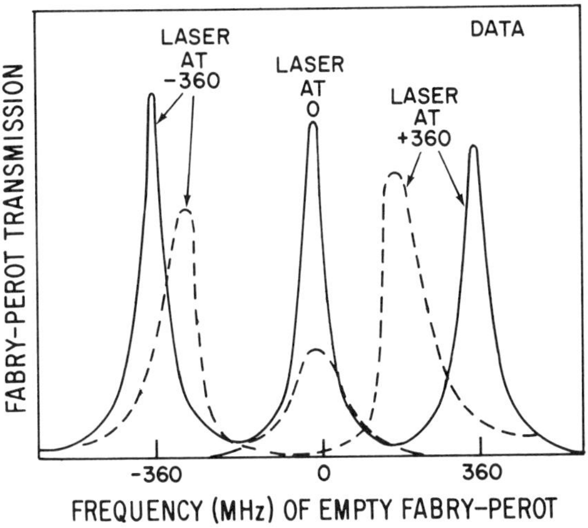

Fig. 3.2-5. Asymmetries in Fabry-Perot scans arising from intensity-dependent phase shifts in intracavity Na vapor. The change in peak heights of the empty cavity (solid line) results from laser intensity changes. With Na (dashed line) the left peak is lower because it is on the F = 2 → $^2P_{3/2}$ transition side, which is more absorbing than F = 1 → $^2P_{3/2}$. Venkatesan (1977).

be highly asymmetric and shifted from the voltage required to give maximum transmission with no Na. These effects are now understood to arise from cavity tuning by the intensity-dependent refractive index of Na vapor. Gardiner and Vaughan (1983) have reported similar changes in mode spacings in a cavity containing calcium.

The frequency dependence of these Fabry-Perot asymmetries is summarized in Fig. 3.2-6b where the scans are plotted under the Na D_2 hyperfine transitions (determined from the energy level diagram of Fig. 3.2-7 and the one-electron-atom calculations of relative transition probabilities). The Fabry-Perot scans were symmetrical when the laser frequency ν was at the crossover frequency ν_1, at ν_2 about 400 MHz lower than the $^2S_{1/2}$, F = 2 to $^2P_{3/2}$,F' transitions, or at ν_3 about 300 MHz above the $^2S_{1/2}$, F = 1 to $^2P_{3/2}$,F' transitions. Except for those three frequencies the scans were asymmetric with the asymmetry changing signs as ν was swept through ν_1, ν_2, or ν_3. This behavior and the bistability can be understood on the basis of hyperfine optical pumping and the accompanying changes in refractive index.

Hyperfine optical pumping can be described using Fig. 3.2-8. If ν is adjusted to resonance with the $^2S_{1/2}$, F = 1 to $^2P_{3/2}$,F' transitions, then the $^2P_{3/2}$ state can be populated. The lifetime of the $^2P_{3/2}$ state is only 16 ns, so radiative decay occurs with some branching ratio to both the F = 1 and F = 2 hyperfine levels of the $^2S_{1/2}$ ground state. Atoms which make a transition to the F = 2 level remain there for several microseconds; in fact, they finally drift out of the laser beam and unpumped atoms drift in. Atoms which end up in the F = 1 level can absorb light again and again until they finally branch to the F = 2 level, even if the branching to the F = 2 level is very small. In Na it is 5/8, so that pumping is quite efficient from either hyperfine level to the other. At low light intensities the laser beam interacts strongly only with those atoms with resonance frequencies (including Doppler shifts from motion) within a narrow range determined by a convolution of the laser and natural (10-MHz) linewidths. Therefore, a narrow notch is removed from the F = 1 level and appears as a bump in the F = 2 distribution. Clearly, pumping can occur from F = 2 to F = 1 in a similar manner.

Assuming that the Fabry-Perot asymmetry results from intensity-dependent refractive index changes, then at the frequencies ν_1, ν_2, and ν_3 no nonlinear index effects occur. At ν_1 they do not occur because at the crossover no optical pumping occurs. This is because in a Fabry-Perot cavity there are two running waves. An atom with a velocity giving it a $\nu_{F=1} - \Delta\nu_{hfs}/2$ resonance frequency for the right-going wave will have $\nu_{F=2} + \Delta\nu_{hfs}/2$ resonance frequency for a left-going wave, where $\Delta\nu_{hfs}$ is the 1772-MHz ground-state hyperfine splitting. So when the laser frequency ν satisfies $\nu = \nu_{F=1} - \Delta\nu_{hfs}/2 = \nu_{F=2} + \Delta\nu_{hfs}/2$, the two running waves compete. No nonlinear index occurs at ν_2 and ν_3 because at those two frequencies no change in index occurs when an atom is pumped from one hyperfine level to the other. In Fig. 3.2-6, the dashed curves are sketches of the contribution to the refractive index from an atom at rest in the F = 1 or F = 2 hyperfine level. At ν_2 and ν_3 the contribution is the same whether the atom is in the F = 1 or F = 2 level. Even though optical pumping occurs, and the absorption may saturate, the refractive index is not dependent upon the degree of pumping; hence, there is no nonlinear index. So dispersive optical bistability was not seen at ν_1, ν_2, and ν_3. Absorptive bistability was seen at ν_2, but inhomogeneous broadening makes the saturation less abrupt [McCall (1974)]: as the input intensity is increased so that atoms in exact resonance are completely pumped, then off-resonance atoms begin to be pumped. Absorptive bistability was seen only with the

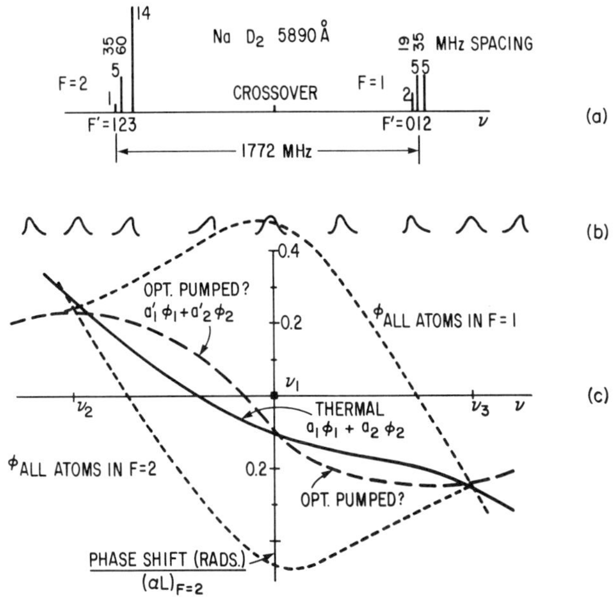

Fig. 3.2-6. (a) D_2 5890 Å Na transitions. (b) Fabry-Perot asymmetries as a function of laser frequency. (c) Phase shifts under various conditions. The two short-dash curves are the phase shifts if all the atoms are in F = 1 or in F = 2. The solid curve is the resultant phase shift for thermal equilibrium (a_1 = 3/8, a_2 = 5/8). The long-dash curve is a possible phase shift for partial optical pumping. Venkatesan (1977).

highest input intensity, probably by means of two-level saturation where power broadening rather than foreign-gas broadening made the transition approximate the homogeneous case.

Dispersive optical bistability was seen before absorptive and much more easily. Except in the vicinity of ν_1, ν_2, and ν_3, there is a change in refractive index when an atom is pumped from one hyperfine level to the other. The solid curve in the lower part of Fig. 3.2-6 depicts the frequency dependence of the round-trip phase shifts $\beta = (3/8)\phi_{F=1} + (5/8)\phi_{F=2}$ for thermal equilibrium where $\phi_{F=1}(\phi_{F=2})$ is the phase shift if all the atoms are in the F = 1 (F = 2) level. The curve with the long dashes is a possible curve for β for partial optical pumping. Notice that the index change Δn for an increase in pumping intensity is negative for $\nu < \nu_2$ or $\nu_1 < \nu < \nu_3$ and positive for $\nu_2 < \nu < \nu_1$ or $\nu > \nu_3$. Also notice that the calculated [see Gibbs, Churchill, and Salamo (1974)] $\Delta\beta_{cal}$, i.e., the difference between the long-dashed and solid curves, can easily be 0.1 $\alpha_0(F = 2)L$. In

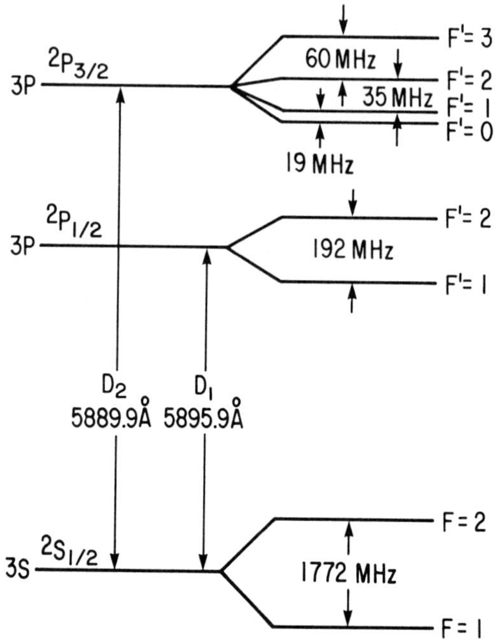

Fig. 3.2-7. Hyperfine energy level structure for the Na D lines.

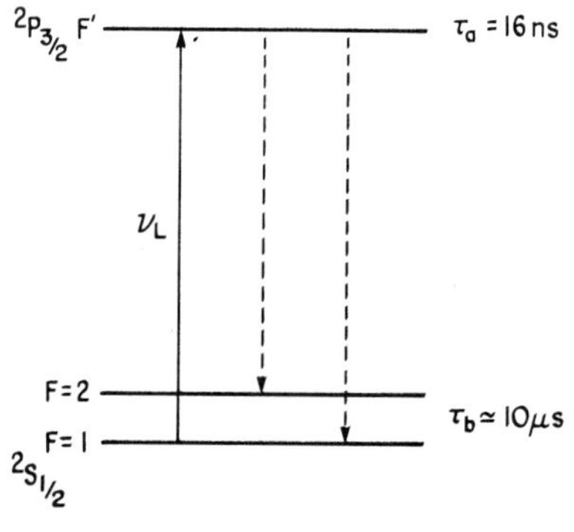

Fig. 3.2-8. Hyperfine optical pumping in Na.

the experiment, $\alpha_0(F = 2)L$ was about π so that $\Delta\nu_{cal} \simeq 70$ MHz. The measured instrument width was also 70 MHz. Thus the calculated frequency shift of the Fabry-Perot cavity can easily exceed an instrument width as required for dispersive bistability (Section 2.3).

The nonlinear index explanation is also consistent with the observed Fabry-Perot shifts and asymmetries. The formula for normal incidence constructive interference in a Fabry-Perot cavity is

$$2nL = m\lambda . \qquad (3.2\text{-}1)$$

Then if one keeps ν (and hence λ) fixed and follows a given order m by changing L to compensate for $\Delta n(I_C)$, one has

$$\Delta L = -\frac{L}{n}\Delta n . \qquad (3.2\text{-}2)$$

Consider Fig. 3.2-5 where the frequency "0" is the crossover frequency. Think of the horizontal axis as the voltage across the piezoelectric crystals or the length of the cavity (L decreases as ν increases). For $\nu = -360$, ν_{FP} occurs at $\simeq -360$ if the cavity is empty but occurs at $\simeq -300$ with Na at some density. For Fig. 3.2-6 we concluded that $\Delta n > 0$ at that frequency; so by Eq. (3.2-2) $\Delta L < 0$; i.e., the peak is shifted to the right as observed. For $\nu > \nu_1$, $\Delta n < 0$ so $\Delta L > 0$ and the peak is shifted left. The long tail side of each asymmetric curve is the optical bistability side for which Δn compensates for the initial detuning. The peak occurs where turn-on just barely occurs, i.e., the turned-on $\Delta\beta$ just equals the initial empty-cavity detuning β_0. Increased β_0 results in low transmission, and the sharp side of the curve results. This is easily seen in Fig. 3.2-4. The correspondence between Fabry-Perot transmission and the hysteresis loop is depicted in Fig. 3.2-9.

The dependence upon detuning computed for a two-level, homogeneously broadened medium and including standing-wave effects shows the same behavior as the data in Fig. 3.2-4. Similar agreement is obtained in Fig. 3.2-10 where $\alpha_0 L$ is insufficient for bistability. Figure 3.2-11 shows the computed effect of detuning in the purely absorptive case; detuning is always detrimental. Notice the similarity to the Na data taken at one of the frequencies, ν_1, ν_2, ν_3 for which the Fabry-Perot scans were symmetric. MacGillivray (1983) and Schulz, MacGillivray, and Standage (1983) have studied the differences in laser-frequency-scan bistability curves in flat-mirror Fabry-Perot and ring cavities.

3.2.4. Transient, Transverse, and Foreign-Gas Effects

Switching transients were examined using input changes that were slow compared with the switching times (Fig. 3.2-12). The switch-on time $\tau_\uparrow$ of a few microseconds was presumably determined by the hyperfine optical pumping time, i.e., it depends upon light intensity. The switch-down time $\tau_\downarrow$ varied from about 5 to 50 μs depending upon operating conditions, such as ν, $\nu-\nu_C$, density, etc. It is believed that $\tau_\downarrow$ is determined by the time required for pumped atoms to drift out of the light beam and for unpumped atoms to drift in (a thermal atom crosses the 1.65-mm diameter in 5 μs). Clearly, if the input intensity is suddenly decreased below $I_\downarrow$, the Fabry-Perot transmission will remain high until the medium has had time to become unpumped. If I_I is increased above $I_\downarrow$ before sufficient relaxation of the pumping occurs, the device will stay on even though $I_I < I_\uparrow$. In some of the switch-on curves a slight overshoot was seen; this characteristic feature of purely dispersive bistability was much more pronounced in ruby (Section 3.3), hybrid (Section 4), Kerr media (Section 3.4), color filter (Section 3.5), and GaAs multiple-quantum-well (Section 3.6) devices.

Under certain conditions, especially at high $\alpha_0 L$ and at particular frequencies, the transmitted intensity I_T became quite complicated exhibiting multiple switchings as in Fig. 3.2-13. By observing the output through a small aperture, it was found that these switchings originated from transversely independent regions.

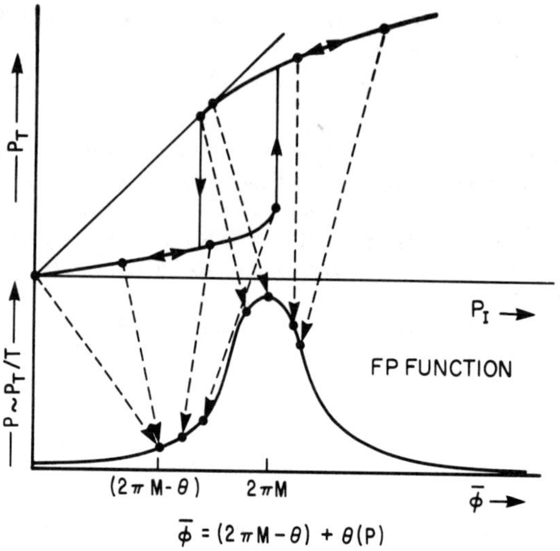

Fig. 3.2-9. Hysteresis in a nonlinear refractive index device. Venkatesan (1977).

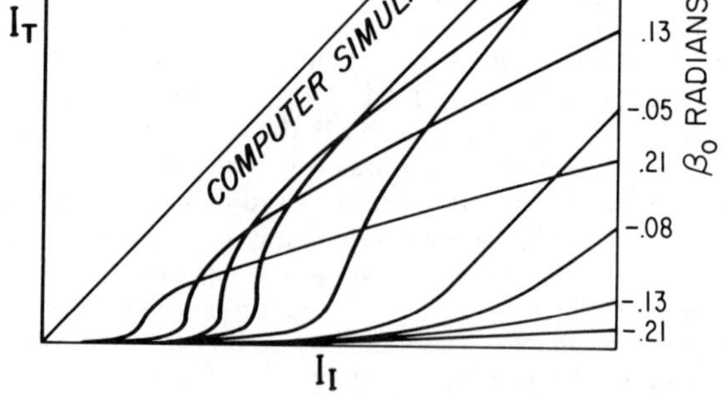

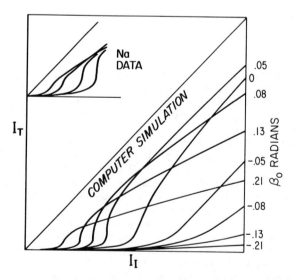

Fig. 3.2-10. Comparison of Na data with computer simulations: effect of Fabry-Perot detuning β_0 for an off-resonance case ($\Delta = 3$).

Filamentation of the laser beam inside the cavity was seen through a 90° window, as reported by Bjorkholm and Ashkin (1974). Presumably, each filament underwent its own hysteresis loop. Such observations raised interesting questions. How close can two beams be and undergo independent bistability? Out to what radius is a device "on" if the input is Gaussian? Can a finite-beam bistable system be turned off or on by a nearby beam? See Section 2.9.

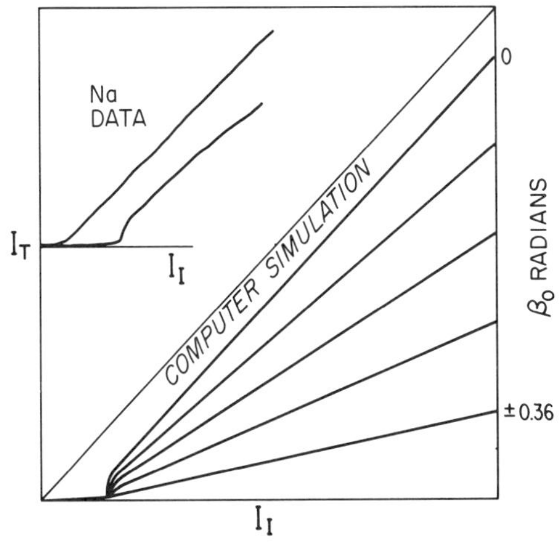

Fig. 3.2-11. Comparison of Na data with computer simulations: effect of Fabry-Perot detuning β_0 for the on-resonance purely absorptive case ($\Delta = 0$).

Most data were taken with no foreign gas, but bistability was seen with a low pressure of Ar (Fig. 3.2-14). The foreign-gas collisions increased the homogeneous width making it harder to saturate and increasing the bistability threshold.

3.2.5. Other Na Optical Bistability Experiments

Three experiments [Sandle and Gallagher (1981); Weyer, Wiedenmann, Rateike, MacGillivray, Meystre, and Walther (1981); and Grant and Kimble (1982, 1983)] contribute further insight into bistability in Na. Sandle and Gallagher use pressure broadening and the high intensities within a spherical Fabry-Perot cavity to compare data and theories of absorptive bistability in the homogeneous limit. Quantitative agreement necessitates a treatment of transverse effects. Weyer et al. and Grant and Kimble achieve purely absorptive bistability by using multiple atomic beams inside the cavity to greatly reduce inhomogeneous (Doppler) broadening. The system of Weyer et al. was found to exhibit some transient behavior even when investigated on a time scale about 100 times longer than the characteristic time of the cavity. Grant and Kimble (1982) achieve steady-state behavior and study critical slowing down (Section 5.6).

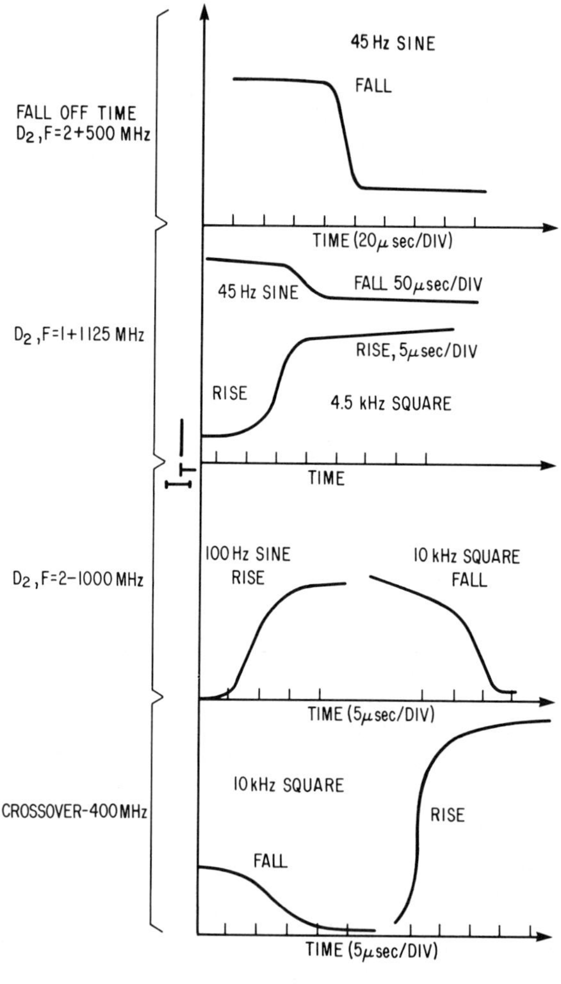

Fig. 3.2-12. Switching transients for Na optical bistability. Venkatesan (1977).

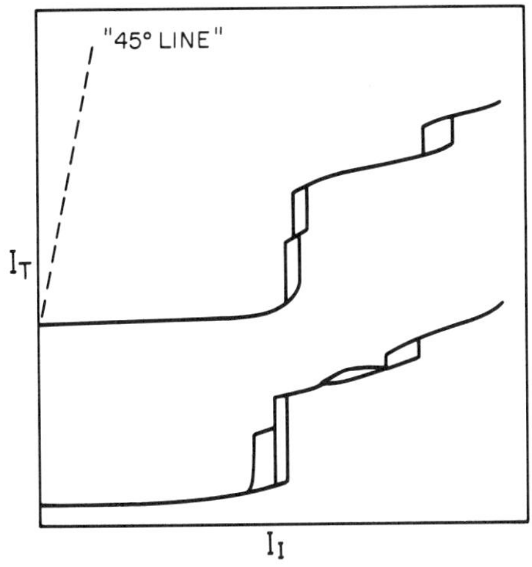

Fig. 3.2-13. Effect of self-focusing filamentation upon Na optical bistability. Venkatesan (1977).

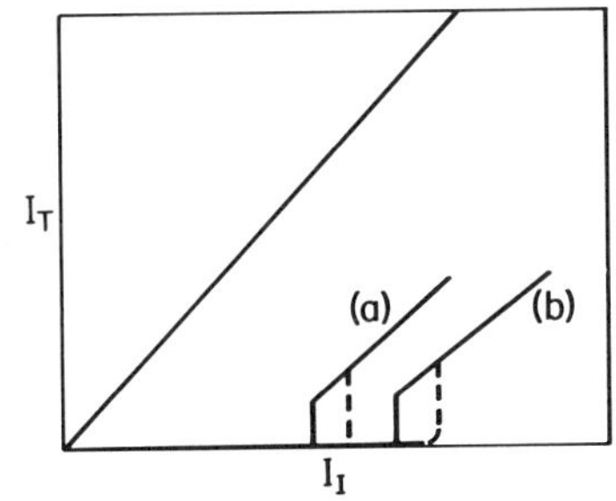

Fig. 3.2-14. Optical bistability in Na with foreign gas. (a) 0.4 Torr Ar; ν 500 MHz below D_2, F = 2. (b) 0.5 Torr Ar; ν 400 MHz below D_2, F = 2. Venkatesan (1977).

(a) Mean-Field Theories and Transverse Effects of a Focusing Fabry-Perot Cavity

In the Sandle and Gallagher experiment a cw dye laser was mode-matched to an optical cavity [Sandle (1980)]. The cavity consisted of two mirrors of 30-cm radius of curvature separated by about 59.2 cm (near-concentric arrangement), yielding a waist size of 80 μm with a Rayleigh length of 3.4 cm and a cavity free spectral range of 250 MHz. The mirror reflectivities were R_F = 0.986 and R_B = 0.992 for the input and output mirrors, respectively. Intracavity nonresonant losses of $e^{-\alpha_B L} \simeq 0.954$ reduced the empty cavity finesse of $\pi(R_\alpha)^{1/2}/(1 - R_\alpha)$ to 55 with $R_\alpha = e^{-\alpha_B L}(R_F R_B)^{1/2}$. The Na was contained within approximately a 10-cm length at the waist.

The laser frequency was scanned slowly at constant power across many widths of the Na transition. With such a long cavity, there were many narrow cavity resonances within the width of the Na D_1 transition. Dispersive bistability was seen off-resonance and absorptive bistability on-resonance. For absorptive bistability, the transition was made to approximate a homogeneously broadened two-level transition by using 20 to 45 Torr of Ar and by the fact that in the "on" state the power broadening is $\simeq$ 3 GHz. The collisional broadening FWHM width is about 0.4 GHz at 20 Torr, about one fourth the hyperfine splitting or the Doppler width. That a single two-level transition is approached [see Ballagh, Cooper, and Sandle (1981)] is indicated by the occurrence of a symmetrical Fabry-Perot profile at only one frequency, in spite of the two hyperfine transitions.

Dispersive bistability is seen in Fig. 3.2-15 under conditions similar to those of Fig. 3.2-14, namely low Ar pressure and large laser-atom detuning. Sandle, Ballagh, and Gallagher (1981) have made quantitative comparisons between their data and the plane-wave steady-state theory Eq. (2.1-31) in the homogeneous limit $g(\Delta\omega) = \delta(\Delta\omega - \Delta\omega^0)$, i.e.,

$$Y = X\left[\left[1 + \frac{2C}{1 + \Delta^2 + X}\right]^2 + \left[\theta - \frac{2C\Delta}{1 + \Delta^2 + X}\right]^2\right]. \qquad (3.2\text{-}3)$$

For the Fig. 3.2-15c dispersive case, they used the expected values C = 4,500 and $\Delta = -165$ and found a best fit value of 4225 for Y compared with $\simeq$ 36,100 deduced from measurements. In the absorptive case of Fig. 3.2-16 with $\Delta = 0$ they used C = 5 and Y = 37.62 and 37.61 just above and below threshold. C was selected for best fit of the ratio of pedestal-to-peak-height ratio; C $\simeq$ 110 was expected from the Na density and Ar pressure measurements. The value Y $\simeq$ 37 used in the computations should be compared with the experimentally deduced effective value of 4900 at the beam waist. So qualitatively the observations are well explained, but quantitatively the experimentally required input intensities are considerably higher than those predicted.

To further test the theory, they increased the pressure to 80 Torr, for which pressure optical bistability no longer occurred with their available power, and concentrated on the peak transmissions as a function of laser frequency; see Fig. 3.2-17. The solid curves are theoretical fits from Eq. (3.2-3) where the maximum X_P^{RC} of X for each input Y_P^{RC} to the ring cavity occurs for $\theta = 2C\Delta/(1 + \Delta^2 + X_P^{RC})$ assuming $d\Delta/d\theta \ll 1$:

$$Y_P^{RC} = X_P^{RC}\left[1 + \frac{2C_{RC}}{1 + \Delta^2 + X_P^{RC}}\right]^2. \qquad (3.2\text{-}4)$$

The dashed curves are X_P^{FP} for a Fabry-Perot cavity with standing-waves effects maximally included. From Eq. (2.5-40)

$$Y_P^{FP} = X_P^{FP}\left[1 + \frac{C_{FP}}{X_P^{FP}}\left[1 - \left[\frac{1+\Delta^2}{1+\Delta^2+4X_P^{FP}}\right]^{1/2}\right]\right]^2, \qquad (3.2\text{-}5)$$

where $C_{FP} = 2C_{RC} = \alpha_0 L/2T$. To compare a ring cavity with a Fabry-Perot cavity in which there are no standing-wave effects, for example by rapid diffusion, it is reasonable for a given $C_{FP} \equiv \alpha_0 L/2T$ to make C_{RC} numerically equal to C_{FP} by adjusting the parameters in the definition $\alpha_0 L/4T$ of C_{RC}. This can be done, for example, by doubling α_0 for the ring compared with the Fabry-Perot cavity since there is a double (single) traversal of the medium per pass in the Fabry-Perot (ring) cavity. Likewise in the Fabry-Perot cavity the saturating intensity ($\propto |E_F|^2 + |E_B|^2$) is twice that ($\propto |E_F|^2$) of the ring, so let $\overline{Y}_P^{FP} = 2Y_P^{FP}$ and $\overline{X}_P^{FP} = 2X_P^{FP}$ in Eq. (3.2-5) to obtain

$$\overline{Y}_P^{FP} = \overline{X}_P^{FP}\left[1 + \frac{2C_{FP}}{\overline{X}_P^{FP}}\left[1 - \left[\frac{1+\Delta^2}{1+\Delta^2+2\overline{X}_P^{FP}}\right]^{1/2}\right]\right]^2. \qquad (3.2\text{-}6)$$

The dashed curve in Fig. 3.2-17a is given by Eq. (3.2-6) with $C_{FP} = 10$ and the solid curve by Eq. (3.2-4) with $C_{RC} = 10$. It is clear that neither theoretical curve is in good agreement; in fact, for small detuning bistability is predicted but not observed.

Ballagh, Cooper, Hamilton, Sandle, and Warrington (1981) have extended the ring-cavity and standing-wave plane-wave models of bistability to account for the Gaussian envelope of the cavity mode in the limit that self-focusing and defocusing are small. See Section 2.9 for an outline of the theory and assumptions. Equation (3.2-6) is replaced by Eq. (2.9-10) with θ chosen to cancel the Δ term and using the above definition, with $\overline{X}_P^{FP} = 2(X_0^{FP})_P = 8X_p^{FP}/w_0^2$, etc.:

$$\overline{Y}_P^{FP} = \overline{X}_P^{FP}\left[1 + \frac{4C_{FP}}{\overline{X}_P^{FP}} \ln Z\right]^2, \qquad (3.2\text{-}7)$$

where

$$Z = \frac{1}{2}\left[\left[\frac{1+\Delta^2+2\overline{X}_P^{FP}}{1+\Delta^2}\right]^{1/2} + 1\right]. \qquad (3.2\text{-}8)$$

The effect of the transverse averaging is to soften the onset of saturation and to eliminate the bistability. Figure 3.2-17b shows good agreement between the data and Eq. (3.2-7). The $\overline{Y}_P^{FP}$ values are fitted and are about three times lower than the measured values. Ballagh et al. suggest that the discrepancy is experimentally significant and might arise from laser-fluctuation or radiation-trapping effects.

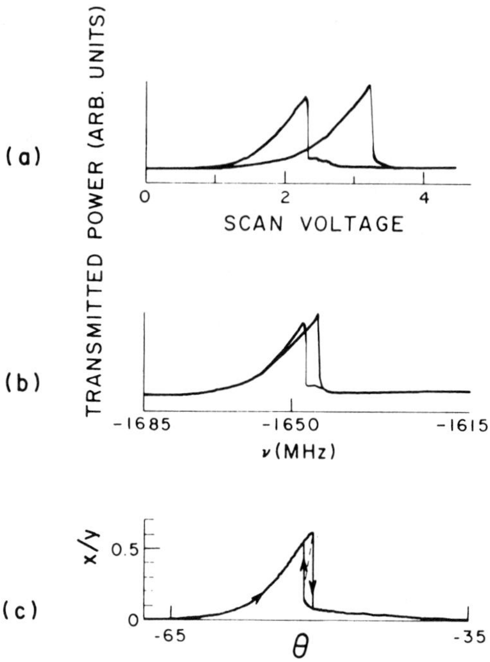

Fig. 3.2-15. Dispersive optical bistability of the transmission peak at laser detuning of -1.65 GHz, observed with a 100-MHz up-down scan. Experimental parameters are: Na density 9.1 x 10^{11} cm^{-3}, constant 60-mW laser power, 0.5-Torr argon pressure (2γ = 20 MHz, 10 MHz coming from natural linewidth). (a) Oscilloscope trace of transmitted cavity power as a function of laser frequency drive voltage. (b) Trace corrected for laser frequency-drive hysteresis. (c) Theoretical curve [from Eq. (3.2-3)] of X/Y as a function of laser detuning θ from the empty cavity peak transmission. Parameters are C = 4.5 x 10^3, Y = 4225, and Δ = -165 at the profile maximum. The dashed portion of the profile gives the unstable part of the analytic curve of X/Y versus θ. Sandle and Gallagher (1981).

(b) Absorptive Optical Bistability Using Multiple Atomic Beams

Weyer, Wiedenmann, Rateike, MacGillivray, Meystre, and Walther (1981) have eliminated Doppler inhomogeneous broadening and dispersive effects by using highly collimated Na atomic beams inside a confocal Fabry-Perot resonator. Na was evaporated in an oven and, by means of apertures, collimated to form 50 atomic beams with a width of about 1 mm each. The absorber linewidth was

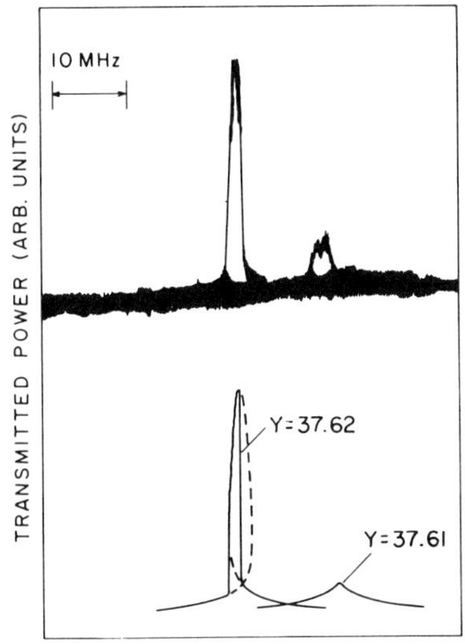

Fig. 3.2-16. Absorptive optical bistability of the profile at the center of the line at an Ar pressure of 20 Torr. The Na density is 4.5×10^{11} cm^{-3}. At the top is an oscilloscope trace of transmitted cavity power as a function of laser frequency for constant incident power (≃ 135 mW). For the profile on the left, the scan is right to left and the laser power is marginally above threshold, while for the profile on the right the scan is left to right and the laser power is marginally below threshold. Below are theoretical profiles from Eq. (3.2-3) with C = 5. The profile on the left is drawn for Y = 37.62; the one on the right has Y = 37.61, and Δ is zero at profile maximum. The peak value of X/Y is 0.32 for the left and 0.04 for the right profiles. The dotted portion of the profile gives the unstable portion of the analytic curve of X/Y versus θ, while the solid line corresponds to a right-to-left scan. Sandle and Gallagher (1981).

reduced to ≃ 14 MHz, close to the ≃ 10-MHz natural width. Circularly polarized light resonant with the D_2 $3^2S_{1/2}$, F = 2 to $3^2P_{3/2}$, F' = 3 transition optically pumped the Zeeman substates into a two-level system. The maximum $\alpha_0 L$ was about 0.8, so $C = \alpha_0 L/2T > 4$ was achieved using a high-finesse cavity (R = 0.99; 300-MHz free spectral range).

The 6-mW input power was modulated with an acousto-optic modulator with a pulse length of 6 µs to avoid intensity fluctuations at frequencies up to 100 kHz from laser frequency jitter. The cavity was swept piezoelectrically with a

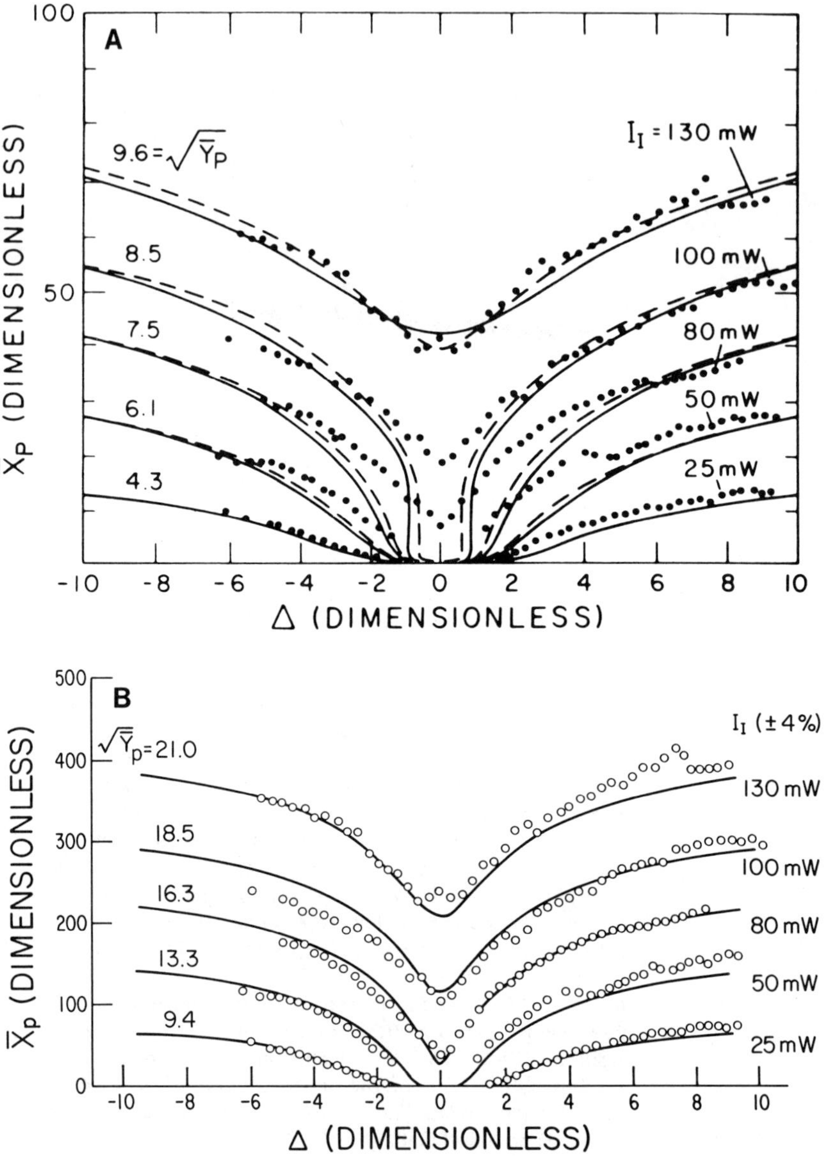

Fig. 3.2-17. Experimental peak Fabry-Perot profile heights (dots) compared with predictions from: (a) the plane-wave envelope equations; solid curves for Eq. (3.2-4) and dashed curves for Eq. (3.2-6) and (b) the Gaussian-profile, standing-wave envelope equation (3.2-7). Na density is 1.8×10^{11} cm^{-3} and Ar pressure is 80 Torr. C = 10; Δ = 10 corresponds to approximately 8-GHz detuning. Sandle and Gallagher (1981).

repetition rate of 20 Hz, and the modulator was triggered when resonance was reached. It was found that 6 μs was not sufficiently long for steady-state to be established. Even an empty Fabry-Perot cavity with resonator bandwidth $\gamma_c = cT/2L$ introduces a delay γ_c^{-1} between the output and input, resulting in a hysteresis for fast enough modulation. In order to separate the pure cavity effects from those of optical bistability, a delay was introduced in the input intensity detection channel to yield a straight line of slope 1 for the empty cavity.

The mean-field theory of absorptive bistability is based on Eqs. (2.4-4a and 5), which were solved numerically since steady-state was not reached. They assumed the experimental input intensity to be approximated by

$$Y(\tau) = Y_0 \exp[-(\tau - \tau_0)^2/\bar{\tau}^2] \qquad (3.2\text{-}9)$$

and inserted a delay to obtain a straight line for Y versus X for an empty cavity as in the experiments. Data and simulations are shown in Fig. 3.2-18. Clearly no bistability was seen in the usual sense, but good qualitative agreement between the measured and computed reshaped outputs is quite good. The first quantitative study of a transient nonlinear Fabry-Perot cavity was by Bischofberger and Shen (1978a,b; 1979a,b); see Sections 3.4 and 5.1; in their case the nonlinearity was purely refractive, arising from a Kerr medium. The present study is the first quantitative comparison between theory and experiment for a purely absorptive nonlinearity in a transient Fabry-Perot resonator.

Grant and Kimble (1982) have achieved steady-state absorptive optical bistability using five atomic beams of Na in a 0.25-m near-confocal cavity with a finesse of 210 ± 15. Before entering the intracavity field, the atoms in the atomic beams are optically pumped with circularly polarized light so that they behave as two-level atoms in the bistability experiments. The optical pumping and collimation of atom and light beams result in a single symmetric absorption line of 30-MHz FWHM. The output of a single-mode dye laser stabilized in frequency to 0.25 MHz rms provides a few milliwatts of power. Grant and Kimble (1982) observed the first unquestioned steady-state hysteresis loop of absorptive optical bistability using an $\alpha_0 L$ of 1.0 and with $P_\downarrow = 0.63$ mW and $P_\uparrow = 1.0$ mW. Figure 3.2-19 shows a later but similar hysteresis loop. Grant, Drummond, and Kimble (1982) report that a comparison with a Gaussian-average model of absorptive bistability with Gaussian inhomogeneous broadening produces reasonable overall agreement for the evolution of the switching characteristics ($P_\uparrow/P_\downarrow$ vs C) with discrepancies noted at the highest values of atomic cooperativity C explored ($\simeq 50$) and for the overall scaling of the intensities.

Similar discrepancies are found by Rosenberger, Orozco, and Kimble (1983a,b) using a ring cavity containing 10 atomic beams (0.5×0.5 mm^2 with ± 1 mrad divergence) with 13-MHz FWHM close to the 10-MHz natural width. This is the first observation of steady-state hysteresis in purely absorptive bistability for a nearly Doppler-free medium of two-level atoms within a ring resonator [see Sections 2.4.1 and 2.4.2]. Very good agreement is found with Eq. (3.2-7), averaged over the residual inhomogeneous broadening, for absorptive bistability (and also for dispersive bistability when Eq. (3.2-7) is appropriately extended) by Rosenberger, Orozco, and Kimble (1984b) and Orozco, Rosenberger, and Kimble (1984). See also Kimble, Grant, Rosenberger, and Drummond (1983); Orozco, Rosenberger, and Kimble (1983); Kimble, Rosenberger, and Drummond (1984).

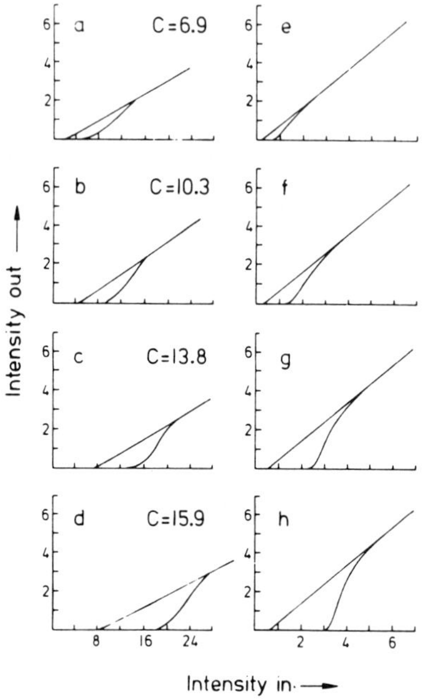

Fig. 3.2-18. Output intensity of the bistable system versus input intensity. Experimental and theoretical results for γ_C = 120 ns and different values of C parameters. Intensities are given in corresponding dimensionless arbitrary units. (a) to (d) Experimental results for values of C = 6.9, 10.3, 13.8, 15.9. (e) to (h) Theoretical results for a Gaussian input pulse of the form of Eq. (3.2-9) with $\overline{\tau}$ = 15 γ_C, τ_0 = 40 γ_C, and corresponding C values. The difference in intensity scales is due to the non-ideal response of the Fabry-Perot cavity. Weyer et al. (1981).

In related work Harrison, Firth, Emshary, and Al-Saidi (1984) and Emshary, Al-Saidi, Harrison, and Firth (1983) report optical bistability in an all-optical ring cavity containing a molecular gas (NH_3).

3.3. RUBY: FIRST SOLID; ROOM TEMPERATURE; USE OF UNDRIVEN STATES

Venkatesan and McCall (1977a,b) reported optical bistability, differential gain, discriminator, clipper, and limiter action using a Fabry-Perot resonator containing ruby between 85 and 296 K. An earlier report was submitted in the application for U.S. Patent 4,121,167 [Gibbs, McCall, and Venkatesan (1975)] and many more details are contained in the thesis of Venkatesan (1977). Ruby was the second system in which passive optical bistability was observed. It was important for

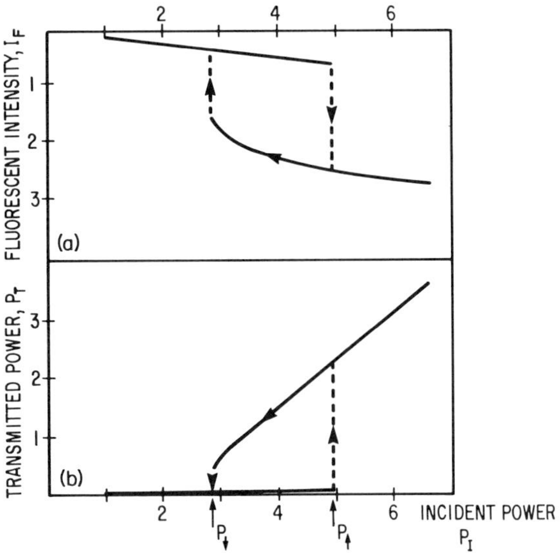

Fig. 3.2-19. Absorptive optical bistability. (a) Fluorescence intensity I_F and (b) transmitted power P_T as functions of incident laser power P_I for near-zero atomic and cavity detunings. The $\alpha_0 L$ was 1.08 and $P_\downarrow$ = 1.3 mW and $P_\uparrow$ = 2.4 mW. The 10 ms taken to sweep through a hysteresis cycle is much longer than either the cavity-decay time ($\tau_C \simeq$ 50 ns) or the atomic relaxation time ($\tau_M \simeq$ 16 ns). Grant and Kimble (1983).

several reasons. It was the first solid, and the first room-temperature solid. Overshoot was seen and analyzed for the first time. A plano-concave Fabry-Perot cavity was employed to counteract diffraction. And strong far-away undriven resonances were shown to be very effective as a mechanism for nonlinear refraction.

The design of the plano-concave Fabry-Perot cavity is shown in Fig. 3.3-1. The input was a TEM_{00} single-mode 20-mW beam from a cw ruby laser operating at 65 or 77 K; an external Fabry-Perot interferometer filtered the output of the ruby laser.

The cw ruby laser emitted 6934 Å radiation, nearly resonant with the narrow R_1 line at 77 K, about 0.2 Å wide. It was anticipated that a shift of the round-trip phase due to the resonant R_1 dispersion would be nonlinear because the R_1 line was strongly driven. The room-temperature R_1 and R_2 lines are about 5 Å wide and are centered at 6943 Å and 6929 Å, respectively. Consequently, the laser frequency lies between the lines and is weakly absorbed. The line strengths (E field perpendicular to crystal C axis; referred to as 0° ruby) are such that the R_1 and R_2 contributions to the nonlinear refractive index almost cancel. Therefore, bistable operation was expected only when the device was near 77 K. However, bistable operation was observed from 85 to 296 K (Fig. 3.3-2).

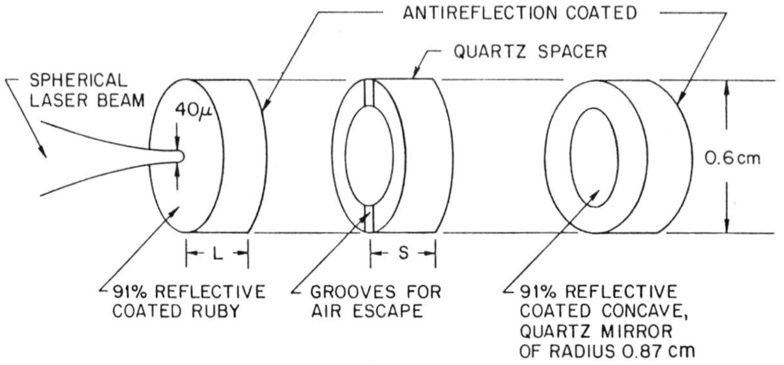

Fig. 3.3-1. Fabry-Perot cavity containing ruby. The TEM_{00} mode has a 40-μm beam waist at the ruby entry face. By using quartz spacers of different thicknesses, the 40-μm dimension was maintained when different thicknesses of ruby were used. The 0.6-cm diameter elements were enclosed in a spring-loaded brass holder thermally stabilized to within 0.1 K of the desired temperature. A number of effects (e.g., the nonspherical intracavity surface of the ruby) allowed only 40% (6328 Å) maximum transmission. Mode matching was achieved by adjusting two lenses in front of the dewar entry windows. The low-intensity room-temperature finesse was about 18. Mirror reflectivities R = 0.91, and transmissivities T = 0.09. Venkatesan (1977).

Experimentally observed switch-on and switch-off times varied from 3 to 20 ms comparable with the R-line radiative lifetime, which varies from 3 to 5 ms in this temperature range. The switching times did not depend upon the ruby temperature to within a factor of two, whereas calculated (Section 3.5) thermal diffusion times decreased (for a 40-μm diameter heat source) from 8 μs to 70 ns as the ruby temperature varied from 296 to 85 K. Thus a thermally induced nonlinear phase shift was ruled out.

Even though the Cr^{3+} concentration was only 0.03% or 0.05%, the pump and charge-transfer bands contributed to the ruby nonlinear refractive index. This contribution depends upon population changes induced by the weakly resonant 6934 Å laser. Calculations show that at room temperature, the refractive index of 0.05% ruby changes in value by about 2.2×10^{-6} upon applying an intensity that depopulates the ground state by 32%. This value is approximately that needed to explain the bistability observed with a ruby length of 1 cm (Fig. 3.3-3).

One can calculate the contribution to the nonlinear index from the faraway undriven states shown in Fig. 3.3-4. From Eq. (2.1-30) the steady-state nonlinear complex absorption coefficient for a two-level transition can be generalized to

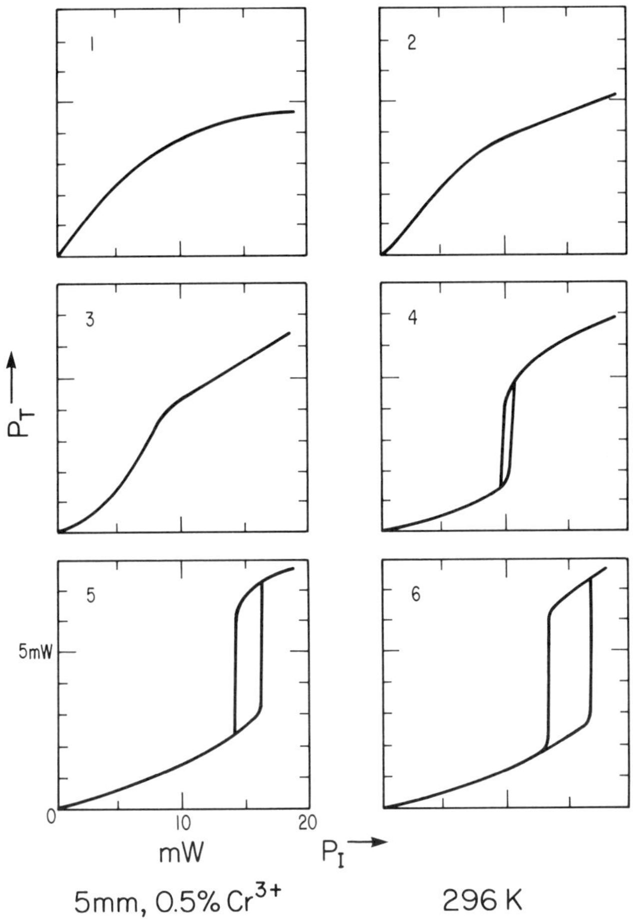

Fig. 3.3-2. Input-output characteristics for a 5-mm, 0.05% Cr^{3+}, and 0° ruby device at 296 K. The different curves are obtained by varying the low-intensity detuning of the cavity. The cavity detuning parameter β was controlled by regulating the pressure of helium in the quartz spacer region. The curves illustrate limiter, differential gain, discriminator (i.e., narrow bistability), and bistability operations. The laser is at 77 K. The horizontal scale runs from 0 to 20 mW; vertical from 0 to 7 mW [Venkatesan (1977)].

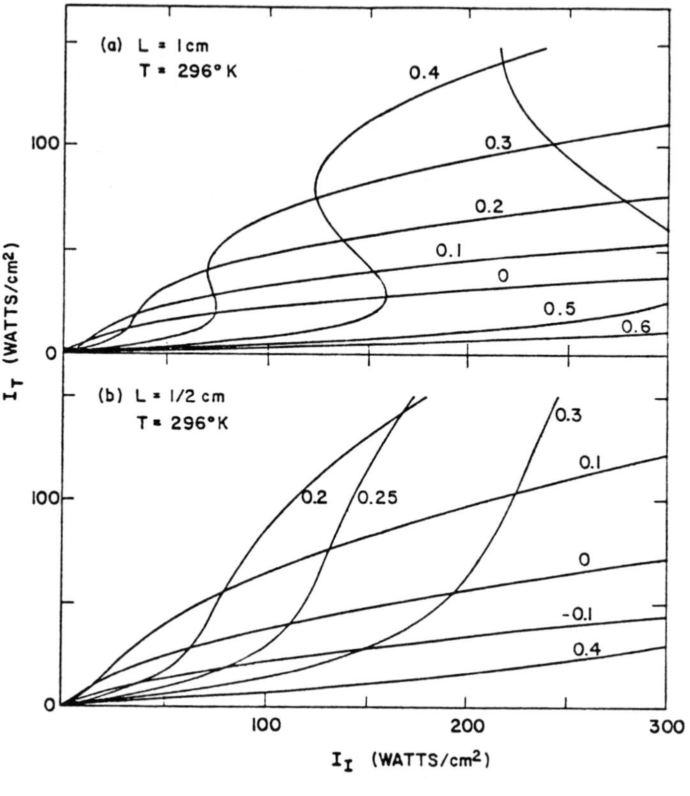

Fig. 3.3-3. Calculated input-output characteristic curves for 0°, 0.05% Cr^{3+} ruby at various detuning parameters. Numbers refer to the low-intensity round-trip phase shift detuning (modulo 2π) in radians. R = 0.91 and T = 0.09. Both nonlinear dispersion and absorption are taken into account, as well as the Boltzmann equilibration of excited states. Not taken into account is the effect of standing waves inside the cavity, which more detailed calculations indicate is an effective increase of roughly a factor of two in the nonlinearity. Venkatesan (1977).

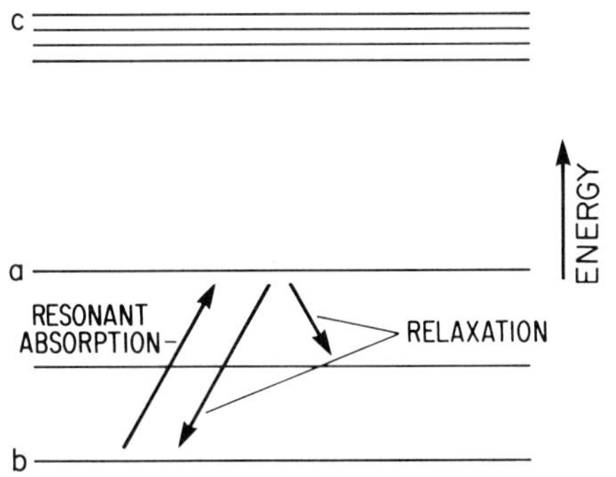

Fig. 3.3-4. Greatly simplified energy level diagram for ruby. Gibbs, McCall, and Venkatesan (1975).

$$\frac{1}{2}\int_{-\infty}^{+\infty}\left[\frac{1+i\Delta}{1+\Delta^2+X}\right]\alpha(\Delta)\gamma_T d\Delta \tag{3.3-1}$$

where $\alpha(\Delta) = \sum_m \alpha_0^m g_m(\Delta\omega)$ can represent many transitions. Far off-resonance ($\Delta^2 >> 1, X$), Eq. (3.3-1) is just a phase shift

$$\frac{i}{2\Delta}\int_{-\infty}^{+\infty}\alpha(\Delta)\gamma_T d\Delta. \tag{3.3-2}$$

The field round-trip phase shift β_b for a ground-state atom is then

$$\beta_b = \frac{L}{\Delta_b}\int_{-\infty}^{+\infty}\alpha_b(\Delta)\gamma_T d\Delta\ ; \tag{3.3-3}$$

and it is

$$\beta_a = \frac{L}{\Delta_a}\int_{-\infty}^{+\infty}\alpha_a(\Delta)\gamma_T d\Delta \tag{3.3-4}$$

for an excited-state atom. Note that in ruby the excited-state absorption $\alpha_a(\Delta)$ is quite different from the ground-state absorption $\alpha_b(\Delta)$. If w is the difference between the upper- and lower-state populations, i.e., $\rho_{aa} - \rho_{bb}$ in Fig. 3.3-4, the nonresonant contribution to the nonlinear phase shift, adjusted to be zero at zero light intensity, is

$$\beta = \beta_a\rho_{aa} + \beta_b\rho_{bb} - \beta_b = (\beta_a - \beta_b)\frac{1+w}{2}, \tag{3.3-5}$$

where all of the intensity dependence of ϕ is in w. Venkatesan found that the charge-transfer bands at 54,400 $\tilde{\text{cm}}^{-1}$ (1830 Å) are the major contributors to β_a and β_b; therefore $\gamma_T\Delta_b = \tilde{\nu}_c - \tilde{\nu}_b - \tilde{\nu} = 39983$ cm^{-1} (2501 Å) and $\gamma_T\Delta_a = \tilde{\nu}_c - \tilde{\nu}_a - \tilde{\nu} = 25600$ cm^{-1} (3906 Å), where γ_T is determined by the short-lived c bands. These bands can dominate in spite of the large detunings because of their huge $\alpha_0 \simeq 73.5$ cm^{-1} compared to < 4 cm^{-1} for closer transitions. Using the details of the ground- and excited-state absorption, Venkatesan computed β_a and β_b and used Eq. (3.3-5) in his simulations (for examples see Fig. 3.3-3). For room-temperature samples this charge-transfer mechanism was dominant because the fixed laser frequency ν was between the R_1 and R_2 lines, so their nonlinear index effects approximately cancelled. For samples at low temperatures ($\simeq$ 77 K) the R_1 absorption was much closer to ν, so its nonlinear refraction dominated.

Venkatesan (1977) also analyzed the dynamics of a nonlinear Fabry-Perot cavity, predicting and observing overshoot (Fig. 3.3-5). Overshoot occurs when the input is changed quickly, so that effectively the input is constant at its new value while the medium responds. For example, if the input is suddenly changed from below $I_\uparrow$ to well above it, the transmission peak of the device will sweep through its maximum to a self-consistent value on the other side (see Section 2.6). The output, which is of course the product of the input (constant at its new value) and the device transmission, will go through a maximum, i.e., overshoot, before settling into a lower steady-state value. Abrupt reduction of the input $I_\downarrow$ results in overshoot in switch-off, as the transmission peak sweeps back through the laser frequency.

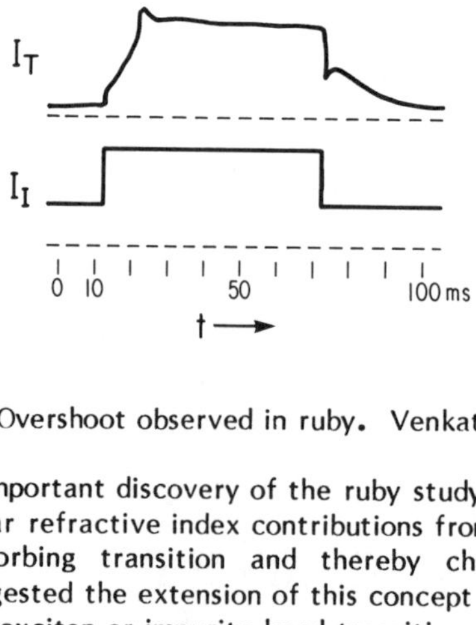

Fig. 3.3-5. Overshoot observed in ruby. Venkatesan (1977).

Perhaps the most important discovery of the ruby study was that it is possible to obtain large nonlinear refractive index contributions from nonresonant levels by driving a weakly absorbing transition and thereby changing the population distributions. This suggested the extension of this concept to other materials such as semiconductors with exciton or impurity level transitions (see Section 3.6).

3.4. KERR MEDIA: CS_2; NITROBENZENE; LIQUID CRYSTALS; Rb

Bischofberger and Shen (1978a,b; 1979a,b) have obtained excellent agreement between a careful experimental study and theoretical analysis of the dynamical transmission of a nonlinear Fabry-Perot cavity. These important quantitative results will be discussed in Section 5.1. It is appropriate in this chapter to point out that, following Na and ruby, the Kerr liquids CS_2 and nitrobenzene were the next successful intrinsic bistability systems. Transient studies were also made in the liquid-crystalline material n-p methoxybenzylidene-p-butylanile (MBBA) in the isotopic phase. The Fabry-Perot cavity consisted of $\lambda/200$, R = 0.98 plane mirrors separated by L = 1 cm giving a round-trip time $t_R \simeq 0.1$ ns. Losses in the nonlinear materials reduced the reflective finesse of 155 to an effective finesse of 13. The maximum intensity of the ruby laser pulses was 25 MW/cm^2, much less than the threshold for self-focusing.

Optical bistability hysteresis loops in nitrobenzene and CS_2 are shown in Figs. 3.4-1 and 3.4-2, respectively. The agreement between experimental data (dots) and simulations (curves) is impressive indeed. In both of these Kerr liquids the nonlinearity is a refractive index change resulting from molecular reorientation in the intense optical field, leading to purely dispersive bistability. Nonresonant nonlinearities have the advantage of being widely tunable, but they are much smaller than the resonant nonlinearities of Na, ruby, GaAs, InSb, etc. Consequently, relatively long interaction lengths are required to obtain an intensity-dependent frequency shift equal to the cavity width. For L = 1 cm and $n_0 = 1.5$, the cavity free spectral range is 10 GHz. With a finesse $\mathscr{F} = 10$, the instrument width is 1 GHz. The laser linewidth and laser-cavity stability need to be much narrower than 1 GHz to stabilize the nonlinear characteristics. So a nonresonant n_2 allows ν and ν_{FP} to be tuned over a wide range, but it imposes more stringent requirements on the laser stability.

Liquid crystals exhibit an extremely large effective nonlinear refractive index due to optical-field-induced reorientation of molecular alignment. In a typical case, an index change $\Delta n \simeq 0.1$ in a nematic layer 100 μm thick can be induced by a laser intensity of 100 W/cm^2 [Durbin, Arakelian, Cheung, and Shen (1982)], much lower (and slower) than for the isotopic phase. A Fabry-Perot interferometer filled with an 83-μm-thick nematic liquid crystal film has exhibited five or more orders of multistability and regenerative pulsations (Section 6.1); see Cheung, Durbin, and Shen (1983). Khoo (1982); Khoo and Hou (1982); Khoo, Hou, Normandin, and So (1983); Khoo, Normandin, and So (1982); Khoo and Shepard (1982); and Khoo, Shepard, Nadar, and Zhuang (1982) have studied nematic film bistability in Fabry-Perot, self-focusing (Section 3.9), and total-internal-reflection configurations using a few hundred milliwatts of cw power. Winful (1982a,b) points out that cholesteric liquid crystals are particularly interesting because they possess an intrinsic periodic structure that enables them selectively to reflect light whose wavelength matches that periodicity; he predicts bistability in transmission and reflection for sufficiently high input intensities. Liquid-crystal light-valve devices permit two-dimensional arrays of bistable devices (Section 7.4).

Grischkowsky (1978) has seen transient bistability in a nonlinear Fabry-Perot cavity using the electronic nonlinearity of rubidium (Rb) vapor, which is a Kerr nonlinearity described well by the adiabatic following model when the pumping light frequency is well out on the Lorentzian wing of the line. He observed sharp switching on the leading edge of the pulse, but never on the trailing edge. The switching caused the observed risetime of the output pulse to be limited by the detector risetime of 1 ns, compared to the risetime of over 2 ns for the input pulse. An optical limiting action was also observed for which the output pulses became flat-topped. The analysis was complicated by self-phase modulation and four-wave mixing.

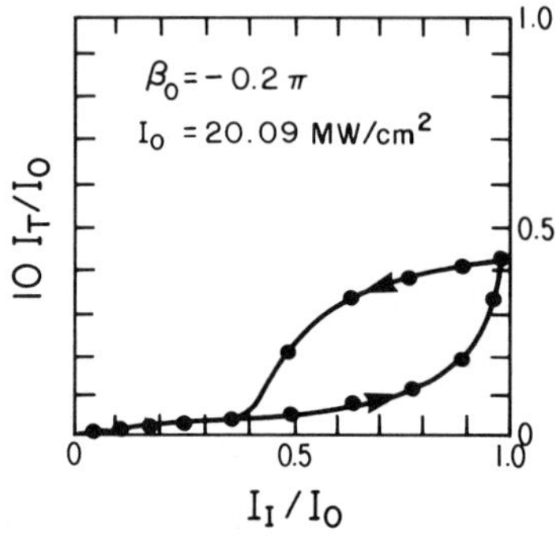

Fig. 3.4-1. Optical bistability in nitrobenzene. Bischofberger and Shen (1979b).

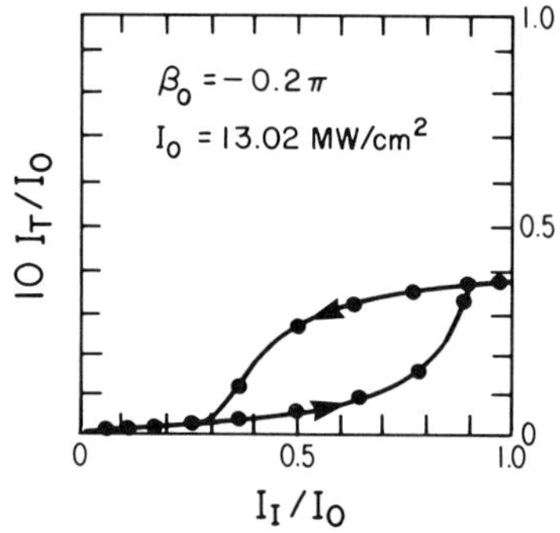

Fig. 3.4-2. Optical bistability in CS_2. Bischofberger and Shen (1979b).

3.5. THERMAL BISTABILITY: ZnS, ZnSe, COLOR FILTERS, GaAs, Si, DYES

The thermal devices contributed by demonstrating the ease of optical scanning of a Fabry-Perot cavity (see also Sections 5.2 and 5.3), reducing the length of devices to a few micrometers, and illustrating the ease of observing intrinsic bistability in reflection. The GaAs thermal device reduced the active length of GaAs to only 0.42 μm, and only 0.22 μm of zinc sulfide (ZnS) was the nonlinear material in an interference filter device. It is interesting that the origin of the absorption is a semiconductor in all of the thermal devices discussed here except the liquid dyes.

The basic idea of thermal optical bistability is to achieve purely dispersive bistability where the intensity dependence of the refractive index originates from heating [Karpushko and Sinitsyn (1978); McCall and Gibbs (1978)]. In Section 2.3 it was shown that for dispersive bistability one needs an intensity-dependent cavity shift of about one instrument width. From the Fabry-Perot formula $2nL = m\lambda$, this implies an optical path length change of

$$\delta(nl) = n\delta L + L\delta n = \frac{nL}{\lambda}(\delta\lambda) \qquad (3.5\text{-}1)$$

where $\delta\lambda$ is the cavity linewidth (FWHM). If some of the absorbed light energy is converted into heat, there will be both an index change δn and a length change δL. Clearly this thermal optical path-length mechanism is not an n_2 process in the rigorous sense. That is, the index does not depend upon the instantaneous value of the intensity according to $n_0 + n_2 I$. Instead the path length change due to the light depends upon the total temperature change, i.e., it depends upon competition between absorption and heat conduction. Roughly speaking it depends upon the energy absorbed during the last thermal conduction time.

Since the thermal conduction plays an important role in the dynamics of thermal optical bistable devices, a formula for the thermal conduction time is useful. If a laser beam suddenly heats a cylinder of length much greater than the beam radius r_0, solution of the heat diffusion equation leads to a thermal conduction time of

$$T_C = \frac{c_V \rho r_0^2}{(2.4)^2 K}, \qquad (3.5\text{-}2)$$

where c_V is the specific heat, ρ is the density, and K is the thermal conductivity. K and c_V both depend upon temperature, so that T_C can change by orders of magnitude. Also T_C is proportional to the square of the smallest conduction length; small dimensions can result in short conduction times. In fact, for GaAs at 5 K and a smallest dimension of 1 μm, T_C can be subnanosecond; i.e., a thermal process is not necessarily slow.

If the sample is not heatsunk, i.e., in close contact with a constant-temperature infinite drain for heat, the surroundings can gradually warm up and change the operating conditions. Heating will be greater in the "on" state because of the larger intracavity intensity, so that thermal effects may compete with other mechanisms and regenerative oscillations sometimes result (Section 6.1). The actual dynamics can be quite complicated and geometry dependent.

Optically-induced thermal shifts of absorption peaks or band edges can also lead to a different class of optical bistability for which external feedback is unnecessary; see Section 2.10 on increasing absorption optical bistability.

3.5.1. ZnS and ZnSe Interference Filters

Apparently the first passive optical bistability in a semiconductor was by Karpushko and Sinitsyn (1978, 1979) using nonlinearities of the ZnS intermediate layer of a dielectric interference filter. Later workers describing thermal bistability [McCall and Gibbs (1978)] and claiming the first observation of bistability in a semiconductor [Gibbs, McCall, Venkatesan, Gossard, Passner, and Wiegmann (1979)] were unaware of the earlier ZnS work. Similarly, Karpushko et al. (1977) did not reference earlier work on dispersive bistability [Gibbs, McCall, and Venkatesan (1975, 1976)] when they analyzed a nonlinear interferometer and suggested it as a logic element.

Upon hearing (in December 1981) about the 1977-1979 work of Karpushko and Sinitsyn, we immediately saw bistability in a 514.5-nm narrowband filter purchased in 1967 long before our work on Na, ruby, and GaAs. During all of those years of research, an optical bistable device lay unsuspected in our interference-filter drawer!

Classifying their interference filter bistable devices as thermal devices is somewhat controversial. Karpushko and Sinitsyn (1978) call the mechanism "two-photon photorefraction," which does not rule out a thermal mechanism via two-photon absorption and heating. A later paper [Karpushko and Sinitsyn (1982)] is no clearer:

> "It is worth noting, that for materials such as Na_3AlF_6, LuF_3, YbF_3 displaying no essential nonlinearity as crystals, the anomalous nonlinear response is observed on passing from crystals to films. This fact enables the mechanism of multilayer thin-film structure nonlinearity to be considered as unambiguously independent of the initial materials characteristics. It is to a great extent due to specific properties of thin films and the structural effects in them. As a result of the existence of a large number of boundaries, dislocations, and other structural heterogeneities in the thin-film systems, there arise additional energy states playing a part of the traps. Both the concentration of the traps and their properties such as lifetime, excitation energy, trapping time, etc. are determined largely by the production of thin films."

That paper compares the above materials with ZnS and zinc selenide (ZnSe) at 600 nm, finding ZnSe to have the largest nonlinearity: 25.9 nm/(J of transmitted energy) for the shift in the peak of a filter with 2.5-nm half width and 65% transmittance. But that paper does not clarify the mechanisms for ZnS and ZnSe, and it gives no switching times.

Weinberger, Gibbs, Li, and Rushford (1982) also studied bistability in a commercial filter and found the sign of the nonlinearity and the switching times (few milliseconds) to be consistent with a thermal mechanism. They observed a damage effect that shifted the peak of the etalon's transmission; the shift recovered in a few months [see also Rushford, Weinberger, Gibbs, Li, and Peyghambarian (1983)]. The damage in the commercial filter may have been caused by the glue holding on a protective cover. Olbright, Gibbs, Macleod, Peyghambarian, and Tai (1984) and Olbright, Peyghambarian, Gibbs, Macleod, and Van Milligen (1984) have grown filters and obtained damage-free bistability with 5-mW holding power: 514.5-nm ZnS filter with 5-nm bandwidth and 60% peak transmission, $\tau_{\uparrow} \simeq 10$ µs, $\tau_{\downarrow} \simeq 20$ µs, $I_{\uparrow} \simeq 70$ kW/cm^2; 632.8-nm ZnSe filter with 3.7-nm bandwidth and 23% peak transmission, $\tau_{\uparrow} \simeq 50$ µs, $\tau_{\downarrow} \simeq 75$ µs, $I_{\uparrow} \simeq 76$ kW/cm^2. They conclude that the bistability arises from one-photon absorption and a thermal nonlinearity. S. D. Smith, Mathew, Taghizadeh, Walker, Wherrett, and Hendry (1984) use an interference filter with a 2λ-thick ZnSe spacer peaked at 514-nm when tilted to 30° and having a 4.5-nm bandwidth and 20% peak transmission: $\tau_{\uparrow} \simeq \tau_{\downarrow} \simeq 5$ ms, $P_{\uparrow} \simeq 30$mW, $I_{\uparrow} \simeq 265$ W/cm^2. The spot diameters of 7.7 µm and 120 µm for the Arizona and Heriot-Watt experiments, respectively, and the different operating wavelengths account for the different results.

These interference filter devices are especially interesting because of their simplicity of construction, the low power ($\simeq$ 5 mW) they require, and their potential for parallel processing (Section 7.4). The shortest conceivable optical resonator would have a length of $\lambda/2n_0$; some of Karpushko and Sinitsyn's devices were only twice as long, namely λ/n_0, which was 0.22 μm for the ZnS intermediate layer; see Fig. 3.5-1.

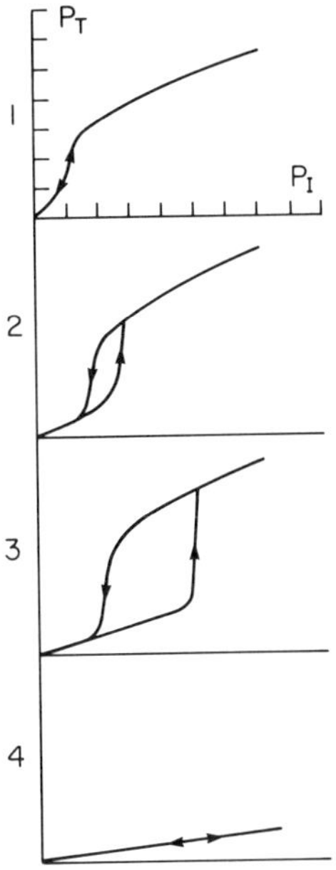

Fig. 3.5-1. Interference filter optical bistability. Experimental functions of P_T versus P_I averaged over a few cycles. Horizontal scale: 50 mW/division; vertical: 25 mW/division. Laser-etalon detuning in units of etalon halfwidths: (1) 0.6, (2) 1.2, (3) 1.4, (4) 1.8. The nonlinear interferometer was made by thin-film vacuum deposition: 0.22-μm-thick ZnS intermediate layer; 517-nm peak wavelength, 50% transmission, 1.1-nm halfwidth; dielectric eight-layer mirrors with thickness of 0.625 μm and R ≃ 0.98. Karpushko and Sinitsyn (1978).

3.5.2. Color Filters, GaAs, Si, Dyes

If one is satisfied with quasi steady-state, in which the input is turned on and off with triangular pulses with a duration time long compared with T_C but shorter than the time to reach complete thermal equilibrium with the surroundings, much can be learned about the thermal path-length changes. This has been done for an antireflection coated 2-mm-thick Corning 3-142 color filter placed inside a Fabry-Perot cavity [McCall and Gibbs (1978)], a Corning 4-74 color filter polished down to ≃ 57 μm and coated to give R = 0.9 [Gibbs, McCall, Gossard, Passner,

Wiegmann, Venkatesan (1980)], and a GaAs ≃ 5-μm etalon with R = 0.9 coatings [Gibbs, McCall, Venkatesan, Gossard, Passner, and Wiegmann (1979)]. None of these early devices was heatsunk, so only quasi steady-state was reached: the illuminated cylinder heated and cooled, but the surrounding medium never reached equilibrium. If instead of cycling, the input was held at a fixed value, the transmission would change for a time much longer than the normal switching or cycling time.

Two heatsunk samples were investigated: the 57-μm etalon was thermally contacted to a brass heatsink using transistor paste; a 3-142 color filter was polished into a 1-mm-diameter 3-mm-long etalon with spherical end faces of 17-mm radius of curvature and 80% coatings and epoxied into a hole in a brass heatsink for good thermal conduction. Both of these heatsunk devices had little long-term drift.

Before summarizing the experiments and the results, it is worth emphasizing the simplicity of the interpretation. An unsaturable background absorption of 5% to 20% is purposefully introduced to provide substantial absorption without reducing the finesse too much. Since the absorption is unsaturable, the finesse of the cavity is fixed, i.e., the transmission profile does not change with intensity; it only shifts. The graphical construction of Section 2.6 can then be used to analyze the bistability. With sufficient initial detuning and high enough intensity to sweep the cavity back into resonance, bistability will occur if the finesse is high enough. Section 5.3 contains a simple calculation of expected thermal shifts for color filters.

Studies of the 2-mm-thick 3-142 Corning filter were made with the 600-mW single-mode output of a krypton (Kr) ion laser. The operating wavelength was 647.1 nm for which the filter absorbed 20%. The filter was placed inside a cavity defined by R = 0.93, 15-cm-radius mirrors spaced 6 or 1.5 cm apart. Lenses mode-matched the input beam into the device. Figure 3.5-2 shows asymmetric Fabry-Perot profiles characteristic of dispersive bistability. Switching times were about 1 s in this device, which was not heatsunk.

Figures 1.2-5 and 3.5-3 show bistability in the 2-mm-long, 1-mm-diameter heatsunk 3-142 device; the overshoot in the switch-on, another characteristic of dispersive bistability, is quite pronounced. The input is reduced too slowly to see overshoot in switching off for the cooling rates operative in those figures. The heatsinking resulted in stable operation in either a high- or low-transmission state for a holding input intensity midway between $I_{\downarrow}$ and $I_{\uparrow}$.

The 57-μm etalon had an $\alpha_B L$ of about 10% at 647.1 nm. Bistability and limiting action are shown in Fig. 3.5-4. The short length and resultant large free spectral range made the single-mode operation of the Kr laser unnecessary. Consequently, up to 6 W was available; usually 1 to 2 W of power was used unfocused (≃ 1-mm diameter). The Fabry-Perot etalon could be scanned by tilting since the finesse deteriorated very little because of the short length and large beam diameter. The etalon could also be scanned thermally allowing normal incidence operation for tightly focused inputs. By optimum focusing of the ≃ 1-mm-diameter input beam by a 2.5-cm focal length lens, bistability was observed with as little as 18 mW.

Bistability in the 57-μm sample pasted to the copper heatsink was truly cw and the switching times were ≃ 1 ms. The heatsink prevented the observation of the bistability in transmission; instead, the reflected signal had to be used. Hybrid bistability had been observed in reflection earlier by P. W. Smith, Turner, and Mumford (1978); see Section 4.5. The reflected and transmitted signals

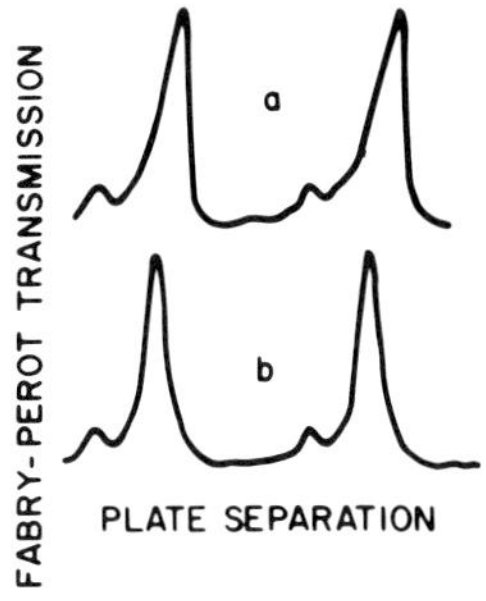

Fig. 3.5-2. (a) Asymmetry of a Fabry-Perot cavity scanned piezo-electrically under bistability conditions using an intracavity antireflection coated 2-mm-thick Corning 3-142 filter. (b) Symmetric scan in low-intensity linear regime. Small peak arises from improper mode matching. Gibbs, McCall, Gossard, Passner, Wiegmann, Venkatesan (1980).

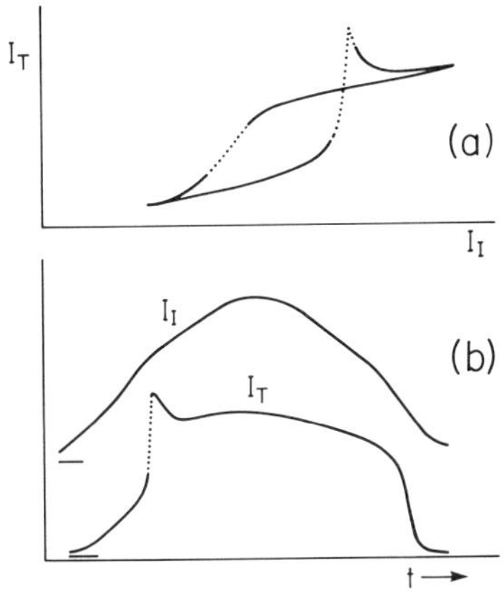

Fig. 3.5-3. Thermal optical bistability in the 3-142 Corning filter: 2 mm long, 1 mm diameter, R = 0.8, and heatsunk. The value of $\tau_{\uparrow}$ was typically several to 100 ms; $\tau_{\downarrow}$, about 100 ms. McCall, Gibbs, Greene, Passner (1978).

complement each other; in the absence of absorption they add up to the input signal. Experimentally there is often part of the input which is reflected unintentionally into the input detector. However, by displaying $I_I - aI_R$ where "a" is a constant gain factor, one sees a signal which differs from the transmitted signal I_T only because of absorption effects. These two signals are displayed in Fig. 3.5-5 for another device. The minimum intensity required for switch-up could be

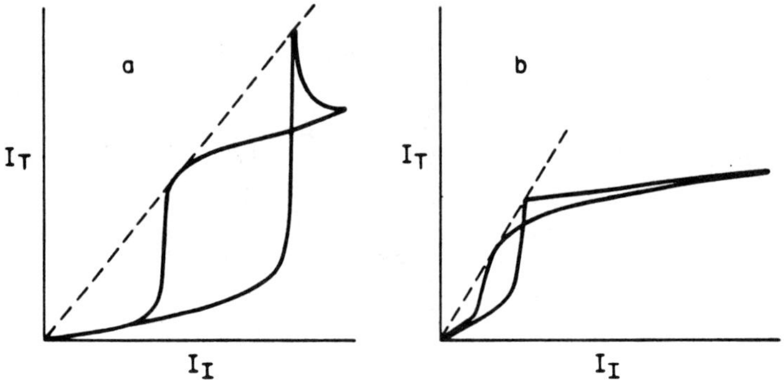

Fig. 3.5-4. (a) Optical bistability and (b) limiter action in a 57-μm Corning 4-74 color filter with R = 0.9. Input is 0.1-Hz triangular wave. Gibbs, McCall, Gossard, Passner, Wiegmann, Venkatesan (1980).

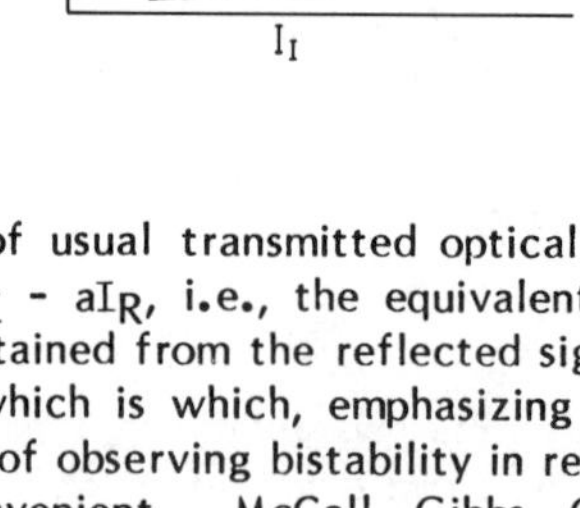

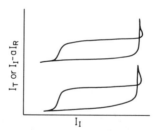

Fig. 3.5-5. Display of usual transmitted optical bistability signal I_T and of I_I - aI_R, i.e., the equivalent of the transmitted signal obtained from the reflected signal. It is no longer certain which is which, emphasizing their similarity and the ease of observing bistability in reflection when that is more convenient. McCall, Gibbs, Greene, and Passner (1978).

reduced by a factor of up to two by making the output reflectivity 100% when the transmitted beam is not needed (Appendix D). Wherrett (1983, 1984) discusses other possible advantages of taking the reflected signal as the output of interest.

Shortly after its observation in color filters, thermal optical bistability was seen in GaAs as shown in Fig. 3.5-6 [Gibbs, McCall, Passner, Gossard, Wiegmann, Venkatesan (1979)]. The input pulse was about 100-ms long; $\tau_{\uparrow}$ was a few

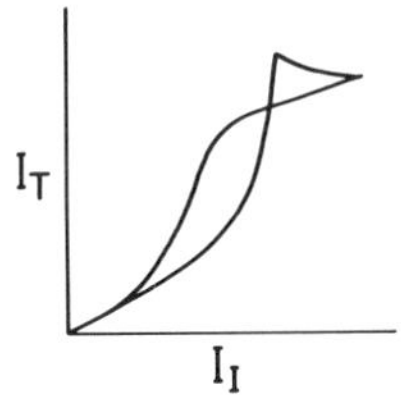

Fig. 3.5-6. Thermal optical bistability in GaAs at 220 K.

milliseconds and $\tau_{\downarrow}$ a few tens of milliseconds, typically. It was observed in an uncoated (R ≃ 0.3 from n_0 ≃ 3.4) and a coated (R ≃ 0.9) heterostructure of 0.42 μm of GaAs between 2.38- and 3.33-μm-thick AlGaAs layers. (Bistability has also been seen without coatings in InSb etalons--see Section 3.7.) The thermal bistability was observed from room temperature down to ≃ 100 K with the ≃ 100 mW available. Self-thermal scanning of a 100-mW beam through two peaks of a GaAs etalon is shown in Section 5.2 (Fig. 5.2-3). The GaAs thermal nonlinear index is positive and thus opposite in sign to that from the GaAs free-exciton resonance (for ν less than the exciton resonance), permitting regenerative oscillations to occur under certain conditions (Section 6.1).

Optical bistability has been seen in a 500-μm Si etalon (uncoated: R = 0.32, T = 0.65) using a 0.5-W cw beam from a neodymium: yttrium aluminum garnet (Nd:YAG) laser focused to a spot radius of 150 μm [Eichler, Massmann, Zaki, and Heritage (1982)]. With 3 W, four orders were switched into resonance successively [Eichler (1983)].

One of the simplest intrinsic optical bistable devices consists of a dye (and solvent) between two mirrors [Rushford et al. (1983)]. Thermal bistability has been seen with as little as 5 mW from a 632.8-nm He-Ne laser using cresyl violet perchlorate in ethylene glycol with 10-μm dye thickness and ≃ 10- to 20-μm-diameter focused spot. This very simple and inexpensive device should be a good device for student laboratories; see also Section 4.3. The low powers, wide wavelength range (certainly 510 to 633 nm), and simple adaptation of a plane-mirror Fabry-Perot interferometer make it a good candidate for two-dimensional studies at intermediate speeds; see Section 2.9 and 7.4. Optical crosstalk (Section 2.9) and external switch-off using a second nonlinear etalon (Section 5.3) have been demonstrated with dye etalons.

Zhu and Garmire (1983a,b,c) have observed nonthermal dispersive optical bistability in a four-level nonradiative dye BDN. Dyes generally do not bleach sufficiently to permit purely absorptive optical bistability (Section 2.5.3). The unsaturable absorption often arises from an excited state transition connected nonradiatively to the initial transition. Very rapid nonradiative relaxation keeps the upper levels of both transitions very nearly empty. Since the contributions to the refractive index are different for the two transitions, there is an intensity dependence to the index, similar to the ruby case (Section 3.3). The conditions for the observed bistability were: BDN dye dissolved in dichloroethane; Fabry-Perot mirrors with R = 0.9 and 0.95 separated by 1 mm; 20-ns, 5-MW, 1.06-μm pulses.

Khoo, Hou, Shepard, and Yan (1983) have observed low-power (approximately 0.1 W) thermal optical bistability using a novel cavity requiring no mirrors. They sandwich a thin-film ($\simeq$ 25 μm) nonlinear material (liquid crystal or a liquid with a trace of absorbing dye) between two higher-index dielectric media. Within a range of incident angles, the reflectivity of the two interfaces and the overlap of the multiply-reflected beam (both depend crucially on the incident angle) are such that high-finesse Fabry-Perot behavior can be obtained.

For a comparison of numerical simulations with data on the dynamics of bistability in semiconductor doped glasses see Thibault and Denariez-Roberge (1985). Golik, Grigor'yants, Elinson, and Balkarei (1983) present simulations and data (CO_2 laser and 2-mm-thick Ge) on thermo-optic waves that propagate outward as switch-on occurs. Thermal bistability without a cavity is described in Section 2.10; even an uncoated Corning 3-70 color filter exhibits increasing absorption optical bistability in a 1-W Ar laser beam!

3.6. GaAs

The first observation of passive nonthermal optical bistability using a resonantly enhanced nonlinearity in a semiconductor was made in approximately a 5-μm-long etalon of GaAs. See Section 1.3 for an earlier observation of bistability in a GaAs laser containing a saturable absorber. See Section 3.5.1 for the first observation of passive optical bistability in a semiconductor. The first observations of the GaAs bistability appeared in: Gibbs, McCall, Passner, Gossard, Wiegmann, and Venkatesan (1979; received Jan. 5); Gibbs, Gossard, McCall, Passner, Wiegmann, and Venkatesan (1979); Gibbs, McCall, Gossard, Passner, Wiegmann, and Venkatesan (1980); with the most complete being Gibbs, McCall, Venkatesan, Gossard, Passner, and Wiegmann (1979). The use of GaAs was suggested in Gibbs, McCall, and Venkatesan (1975).

The principal results of the GaAs experiments were: first observation of resonantly enhanced passive bistability in a semiconductor; use of a tiny device ($L \simeq 5$ μm; $\simeq$ 10-μm diameter; i.e., volume $< 10^{-9}$ cm^3); use of the nonlinear refraction of the free-exciton resonance for nanosecond bistability from 5 to 120 K; use of thermal index changes for millisecond bistability from 100 K to room temperature (see Section 3.5); achievement of room-temperature bistability in GaAs-AlGaAs superlattices and bulk etalons as well.

It is natural to turn to semiconductors in the search for practical bistable optical materials. Semiconductors have the attraction of providing adequate absorption ($\alpha_0 L \simeq 1$) in very short ($\simeq$ 1 μm, in favorable cases) lengths. A short length means a short round-trip time so that cavity-lifetime limitations (Sections 3.2.5b, 3.4, 5.1) are of little concern: the lifetime of a 90% reflectivity, 5-μm-long GaAs ($n_0 \simeq 3.4$) cavity is less than 1.2 ps. Switching times can then be very short if fast nonlinear mechanisms are used. A short length permits tighter focusing before beam-walkoff losses become significant, so that input powers and switching energies are reduced. Guided-wave designs (Section 3.10) may negate this advantage. The narrow exciton resonances [see Reynolds and Collins (1981)] and sharp band edges of some semiconductors suggest large nonlinearities. Two such giant nonlinearities in GaAs and InSb have given rise to optical bistability and will be discussed in this and the next section.

3.6.1. Bulk GaAs

The experimental setup and sample are shown in Figs. 3.6-1 to 3. The GaAs sample was grown by molecular beam epitaxy with 4.1 μm of GaAs between 0.21-μm $Al_{0.42}Ga_{0.58}As$ layers supported by a 150-μm GaAs substrate containing a 1- to 2-mm-diameter etched hole for optical access. For a discussion of etching techniques and the construction of sandwich devices consisting of substrate-free samples between coated cover slips see Jewell, Gibbs, Gossard, and Wiegmann (1982) and Jewell, Gibbs, Gossard, Passner, and Wiegmann (1983). Reflective coatings with R = 0.9 were added. The input laser beam was provided by a Coherent Radiation 590 dye laser containing oxazine 750 perchlorate dye pumped by the 647- and 676-nm, 7.5-W output of a Spectra Physics 171 Kr laser. The 0.6-W dye laser output, about 1 Å wide, easily tunable from 770 to 870 nm, was converted into 1-μs triangular pulses by an acousto-optic modulator. The laser beam was focused to 10-μm diameter on the sample. A silicon photodiode with a 0.6-ns full-width half-maximum pulse response and a fast oscilloscope were used to observe the input I_I and output I_T.

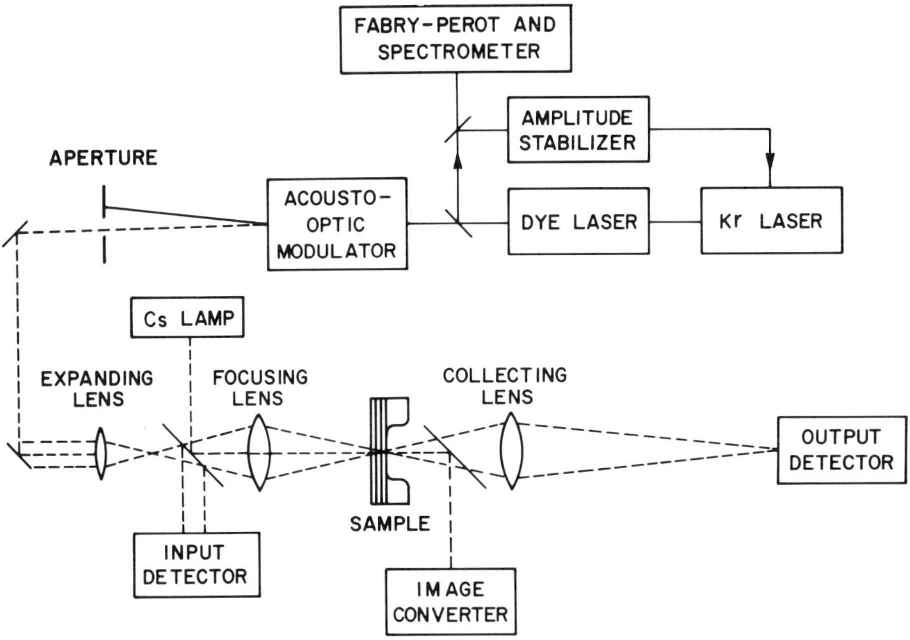

Fig. 3.6-1. Experimental arrangement for GaAs nonlinearity and bistability studies. Gibbs, Gossard, McCall, Passner, Wiegmann, and Venkatesan (1979).

Using this configuration, optical bistability was observed with a laser wavelength 10 to 20 Å longer than the wavelength of the free-exciton peak (see Fig. 3.6-4). A holding power (i.e., less than switch-up but greater than switch-down

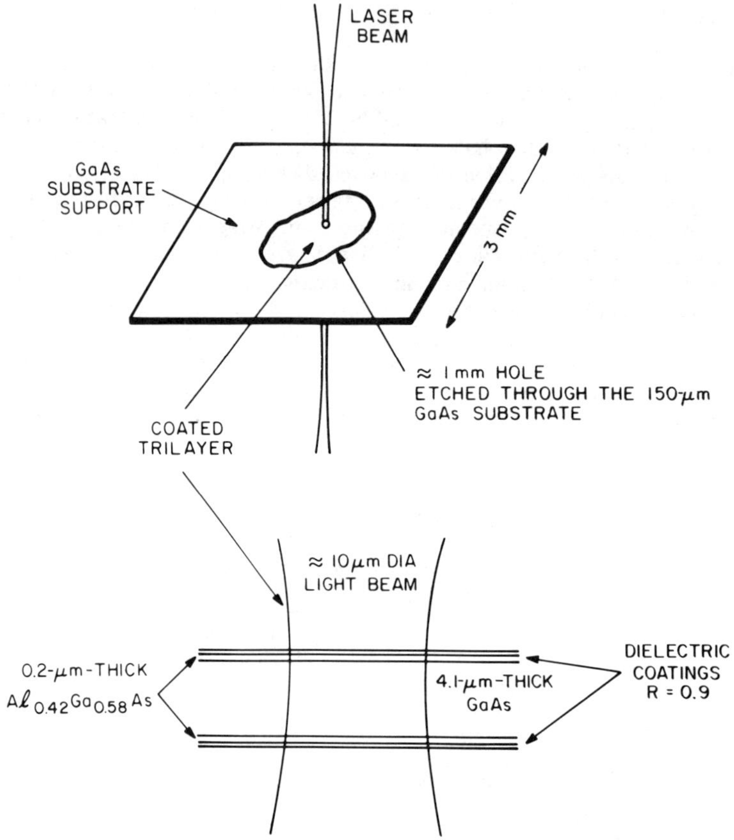

Fig. 3.6-2. GaAs sample construction: grow on substrate, etch hole through substrate, coat to desired reflectivity, cool, and illuminate. Gibbs, McCall, and Venkatesan (1980).

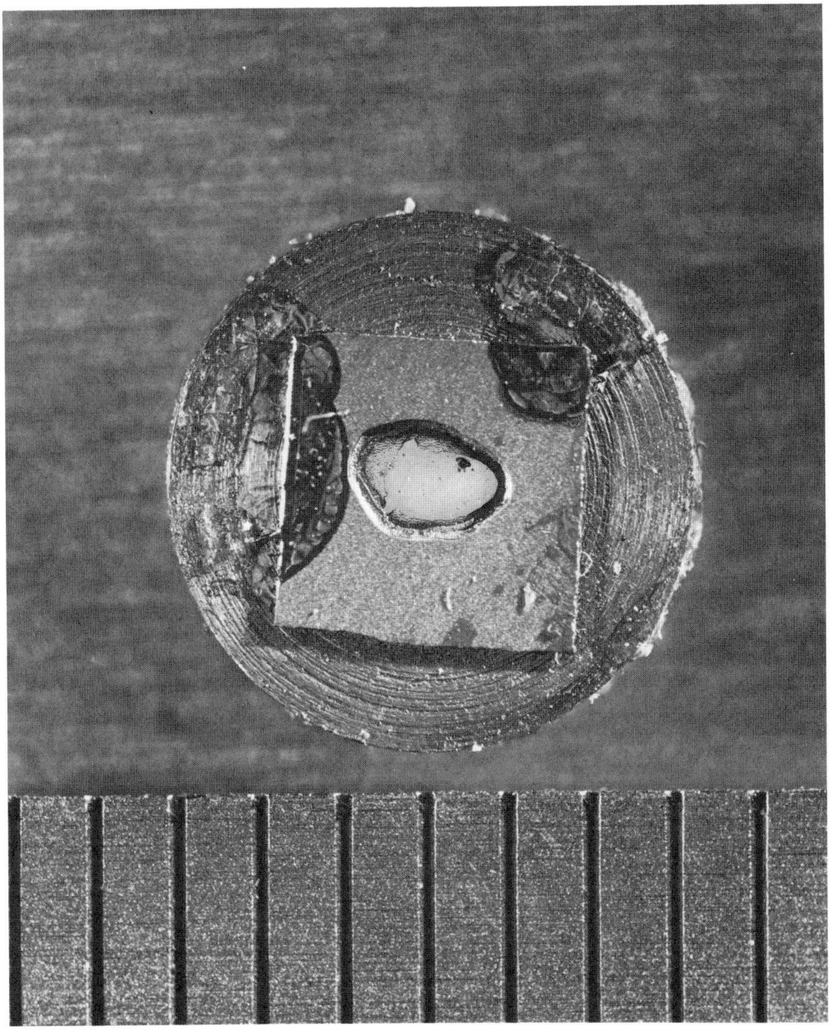

Fig. 3.6-3. GaAs etalon as seen through hole in substrate which is rubber cemented to a 6-mm-diameter washer. Scale at bottom is 1 mm/division. Gibbs, McCall, and Venkatesan (1980).

powers) of about 200 mW was required for the 5-μm-thick sample over a diameter of 10 μm. A switching energy of 8 nJ is the product of the 200-mW power and 40-ns switching time (Fig. 3.6-4), but only a fraction of this 820-nm energy was actually absorbed. By working at the holding power and using a 200-ps 590-nm pulse of 0.6-nJ energy, the device was turned on in a detector-limited time less than 1 ns (see Section 5.5). The 590-nm pulse illuminated a circle about 50 μm in diameter and was totally absorbed, so the 0.25 pJ/μm^2 energy/unit area corresponds to a carrier density of about 1.6 x 10^{17}/cm^3, or one per (185 Å)^3; the exciton Bohr radius is 140 Å. A 10-ps, 590-nm, 1-nJ pulse has turned on such a device in less than the 200-ps response time of the detector (see Section 5.5). By operating with I_I very close to $I_\uparrow$, switch-on has been achieved with as little as 10 pJ in a 10-ps, 800-nm pulse [Jewell, Tarng, Gibbs, Tai, Weinberger, and Ovadia (1984)]. The limiting turn-on time should be less than 1 ps for 1-μm-thick devices, based on measurements of Shank et al. (1979, 1981); this is confirmed by observations showing that the etalon peak is shifted in 1 to 2 ps by a 0.1-ps pulse [Migus, Antonetti, Hulin, Mysyrowicz, Gibbs, Peyghambarian, and Jewell (1985)].

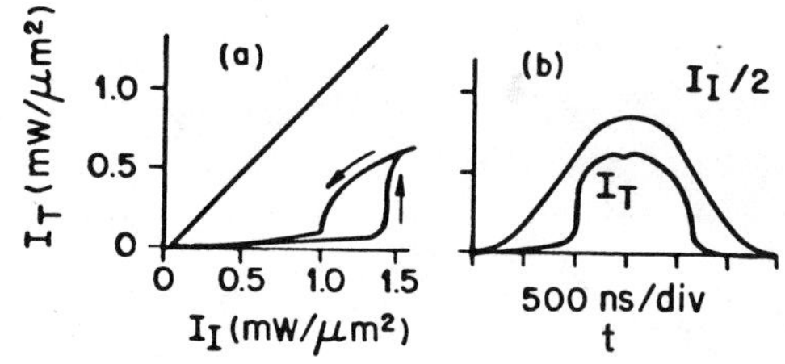

Fig. 3.6-4. Excitonic optical bistability in GaAs at 15 K and 819.9-nm laser wavelength. Bistability as seen in (a) x-y display and (b) time display. The 45° line in (a) shows etalon transmission at the next etalon peak (≃ 830 nm). Gibbs, McCall, Venkatesan, Gossard, Passner, and Wiegmann (1979).

Fig. 3.6-5 shows a hysteresis loop at 107 K; in that initial work the excitonic bistability was seen up to 120 K. At that temperature kT is more than twice the 4.2-meV binding energy, so it did not seem surprising when the bistability based upon the exciton resonance disappeared. On the other hand, Coulomb correlation certainly persists to room temperature, since it is responsible for the observed step-function band edge rather than the parabolic edge predicted for no correlation. Band tailing also increases with temperature, increasing α_B for the same laser-exciton detuning. Background absorption is very detrimental to cavity finesse and to peak transmission, making bistability more difficult (Sections 2.5.3, 7.1, and Appendix D). It is likely that the GaAs bistability disappeared for both reasons, i.e., weaker exciton nonlinear refraction and stronger background absorption. However, after room-temperature operation was achieved in the superlattice sample, the dewar was removed and microscope objectives were used for better focusing. Room-temperature operation of the pure GaAs device was then seen (Section 3.6.2).

The mechanism for excitonic bistability is described as follows. Figure 3.6-6 shows the optical excitation of a valence electron to form an electron and a hole which are bound to each other to form a free exciton that is not bound to any site within the crystal. Nonlinear absorption measurements show that the exciton feature can be saturated without appreciably affecting the band-to-band absorption (see Figs. 3.6-7 and 8). Note that in Fig. 3.6-8, the ratio of the saturable to the background absorption coefficients is $\alpha_0/\alpha_B = 6.5$, which was the largest ratio observed. Recall from Eq. (2.5-46) that bistability requires

$$\frac{\alpha_0 L}{T + \alpha_B(\Delta)L} > 4[1 + (1 + \Delta^2)^{1/2}] , \tag{3.6-1}$$

which reduces to $\alpha_0/\alpha_B > 8$ for on-resonance ($\Delta = 0$) and high reflectivity ($T \ll \alpha_B L$) conditions. This explains the failure to see purely absorptive bistability in GaAs. Equation (3.6-1) assumes the transition is homogeneously broadened. Figure 3.6-8 shows that a homogeneous saturation curve is consistent with the data, but so is an inhomogeneous saturation curve of the form $(1 + I/I_S)^{-1/2}$ from Eq. (2.4-43). The width of the exciton absorption was typically 5 Å, but widths as narrow as 3.5 Å have been seen. In contrast, Ulbrich (1981) has recently observed linewidths as narrow as 1 Å and a ratio of peak exciton to band-to-band absorption of about 20 (compared with 2.5 in the bistability sample and ≤ 6.5 in all other samples studied by Gibbs et al.). His sample was about 0.5-μm thick and had no AlGaAs windows or substrate. And his saturation energy/area was only 1 fJ/μm², giving more hope for improved GaAs devices. Ulbrich's narrow widths imply that samples with wider absorption widths, such as the bistable sample, are inhomogeneously broadened.

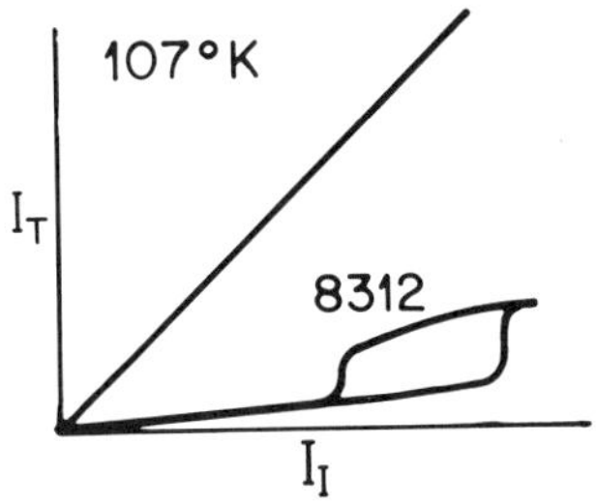

Fig. 3.6-5. Excitonic optical bistability in GaAs at 107 K. Gibbs, McCall, Passner, Gossard, Wiegmann, and Venkatesan (1979). © 1979 IEEE

If one goes far enough off resonance, i.e., $\Delta\omega = \omega_{EX} - \omega \gg \gamma_2^*$, then the inhomogeneities are no longer important and the resonance can be treated as a single homogeneous line again and Eq. (3.6-1) is applicable. Once the difficulty with α_B was understood, it was found by measurements such as Fig. 3.6-7 that α_B decreases with detuning below the exciton resonance. Equation (3.6-1) was then satisfied with the following approximate parameter values: $\Delta \simeq$ (15 Å detuning)/(4 Å exciton half width half maximum (HWHM) linewidth) = 3.75, $T + \alpha_B L \simeq 0.3$ from the measured finesse of 8, $\alpha_{TOTAL} L \simeq 10$ measured on the exciton peak of which $\alpha_{EX} L \simeq 7.5$ came from the exciton and $\alpha_B(0)L \simeq 2.5$ from the

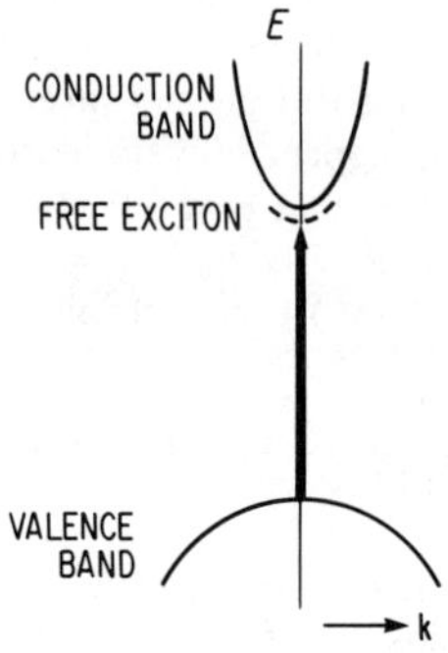

Fig. 3.6-6. Greatly simplified energy level diagram of GaAs showing the formation of a free exciton, i.e., of an electron-hole pair bound together (by 4.2 meV) but free to move through the crystal.

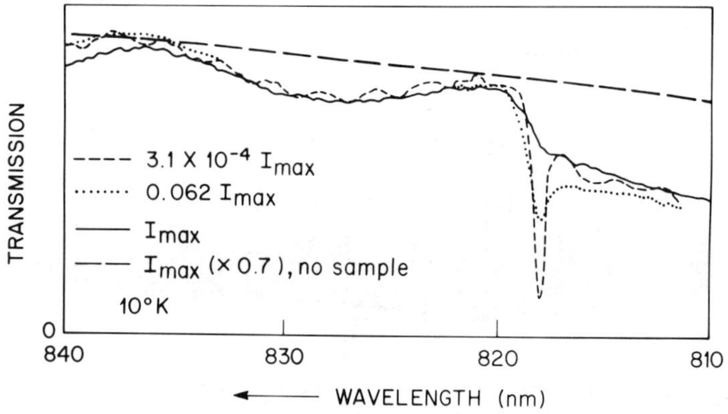

Fig. 3.6-7. Nonlinear transmission of 500-ns pulses through an antireflection-coated $Al_{0.24}Ga_{0.76}As$-GaAs-$Al_{0.24}Ga_{0.76}As$ heterostructure with layer thicknesses in micrometers of 2.38, 0.42, 3.33, respectively. $I_{max} \simeq 43$ kW/cm². Gibbs, Gossard, McCall, Passner, Wiegmann, and Venkatesan (1979).

unsaturable band tail. Equation (3.6-1) becomes 25 > 20, i.e., it is satisfied, and bistability was seen as described above. Of course, if the homogeneous linewidth is only 1 Å FWHM, $\Delta = 30$ but $\alpha_0 L$ must be increased to 60 and inequality Eq. (3.6-1) becomes 200 > 124. For large Δ and small T, Eq. (3.6-1) becomes $\alpha_{EX}\delta\lambda_{EX} > \alpha_B \Delta\lambda$, i.e., it is the area of the exciton absorption, not its height or

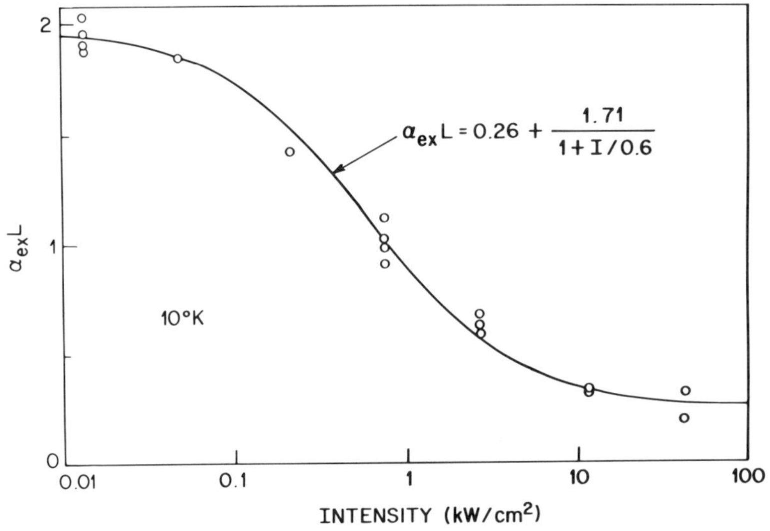

Fig. 3.6-8. Saturation of the free-exciton resonance in GaAs with 500-ns pulses of wavelength equal to the wavelength of peak exciton absorption. The solid curve is a least squares fit of $\alpha L = A[1 + B/(1 + I/I_S)]$, which yielded $A = 0.257 \pm 0.033$, $B = 6.65 \pm 0.99$, and $I_S = 0.605 \pm 0.064$ kW/cm². Gibbs, Gossard, McCall, Passner, Wiegmann, and Venkatesan (1979).

width alone, that is important. Likewise the maximum obtainable phase shift is proportional to $\alpha_{EX}\delta\lambda_{EX}/\Delta\lambda$.

The simple picture of excitonic bistability in GaAs can be viewed with the help of Fig. 3.6-9. At low input intensities there is a sharp free-exciton absorption peak and corresponding contribution to the refractive index or intracavity round-trip phase shift β (solid curves in Fig. 3.6-9a and b, respectively). The Fabry-Perot transmission profile is given in Fig. 3.6-9c showing the reduction of finesse at ν_{FP} due to absorption close to the exciton resonance at ν_{EX}. The laser frequency is tuned about one cavity width $\Delta\nu_{FP}$ away from ν_{FP} on the exciton side. At high enough intensities α and β decrease shifting ν_{FP} toward ν, until switch-on occurs in the usual fashion. Calculations of the expected phase shifts are in reasonable agreement with those observed and will be presented in Section 5.3. In fact, it will be found that if a density of $\simeq 10^{17}$ cm^{-3} excitations occurs whether by exciton, band-to-band, or below bandgap absorption, the exciton absorption disappears. The exciton is much like a hydrogen atom with a very large Bohr radius of 140 Å. With an excitation density of $\simeq 10^{17}$ cm^{-3} there is on the average one carrier per exciton volume to shield the Coulomb attractive force, destroying the exciton. Assuming an exciton lifetime of 10 ns and a 0.5-μm-thick GaAs sample, an intensity of 100 W/cm² is required to produce $\simeq 10^{17}$ cm^{-3} excitations. This is in good agreement with the value 150 W/cm² found by a numerical fit of the data shown in Fig. 3.6-8 but taking thick-sample and Gaussian-beam

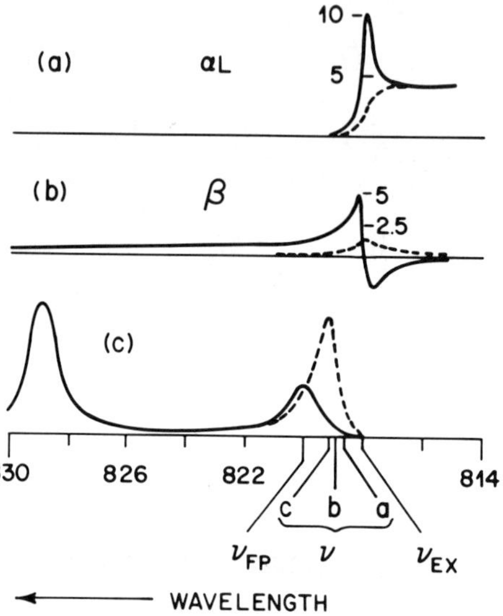

Fig. 3.6-9. Approximate (a) GaAs absorption αL, (b) round-trip phase shift β, and (c) Fabry-Perot transmission under conditions of optical bistability. The solid curves are for zero input intensity for which the exciton absorption is fully effective. The dashed curves are for intensities that are high enough to saturate the exciton feature, but low enough to leave the band-to-band contributions to α and β unaffected. Gibbs, McCall, Venkatesan, Gossard, Passner, and Wiegmann (1979).

effects into account. So again a nonlinear refraction mechanism is found that depends upon accumulated absorption over the last medium relaxation time, rather than upon the instantaneous value of the intensity as for a genuine n_2 process.

In the bistability and optical modulation (Section 5.3) experiments using the 4.2-μm-thick GaAs sample, index changes of $\delta n \simeq -10^{-2}$ required intensity changes of $\delta I \simeq 100$ kW/cm² = 1 mW/μm², yielding $\delta n/\delta I \simeq -10^{-4}$ cm²/kW = -10^{-2} μm²/mW. (Recent bistability with 50-μm-diameter spots to minimize carrier diffusion effects requires only $\delta I \simeq 3$ kW/cm², i.e., $\delta n/\delta I \simeq -0.003$ cm²/kW.) Of course, $\delta n/\delta I$ is not n_2. Those experiments were performed at intensities far above the $I_s = 0.6$ kW/cm² = 0.006 mW/μm² from Fig. 3.6-8. Presumably, that was necessary because the sample was 10 times thicker (requiring 10 times the absorption to maintain 10^{17} excitations/cm³) and because with $T = 0.1$ mirrors, the input required for switch-on may be 10 times the saturation intensity [see Eq. (2.4-21)]. One may represent the contribution of the exciton resonance to the refractive index as, from Eq. (2.1-30)

$$n_{EX}(\lambda) = \frac{\alpha_{EX}\lambda}{4\pi} \frac{\Delta}{1 + \Delta^2 + I/I_s} ; \qquad (3.6\text{-}2)$$

the wavelength λ exceeds the wavelength of peak exciton absorption by $\Delta\lambda$, and $\delta\lambda_{EX}$ is the width (HWHM) of the exciton resonance with peak absorption $\alpha_{EX}L$, and $\Delta \equiv \Delta\lambda/\delta\lambda_{EX}$. It is assumed that the exciton resonance is homogeneously broadened. Equation (3.6-2) predicts $n_{EX} \propto 1/\Delta$ for large Δ. The nonlinear index n_2 can be calculated as dn_{EX}/dI for $I = 0$:

$$n_2 = -\frac{\alpha_{EX}\lambda}{4\pi I_s} \frac{\Delta}{(I + \Delta^2)^2} , \qquad (3.6\text{-}3)$$

which falls off as $1/\Delta^3$ for large Δ. This has extreme values for $\Delta = 1/\sqrt{3}$: $|n_2|(\max) = 0.026\lambda\,\alpha_{EX}/I_s$; for the sample of Gibbs, Gossard, McCall, Passner, Wiegmann, and Venkatesan (1979) this gives a calculated value of n_2 of -0.57 cm^2/kW = -57 $\mu m^2/mW$ for GaAs with $\alpha_{EX} = 4/\mu m$ and I_s = 150 W/cm^2 = 1.5 $\mu W/\mu m^2$. Values of α_{EX} as high as 20/μm have been seen. An n_2 of -0.2 cm^2/kW has been measured using degenerate four-wave mixing by D. A. B. Miller, Chemla, Eilenberger, Smith, Gossard, and Wiegmann (1983). There is considerable hope that thinner GaAs etalons can be constructed with sharper exciton features and less background absorption permitting bistable operation much closer to the exciton resonance and with much lower input intensities.

At temperatures T above 100 K, thermal optical bistability (Section 3.5) was seen (Fig. 3.6-10) by detuning ν to the low-frequency side of ν_{FP} so that the positive dn/dT and dn/dI would restore coincidence between the two. Figure 3.6-11 shows that for excitonic bistability switch-on is easier as ν is reduced; therefore $\nu > \nu_{FP}$. Since excitonic bistability occurs, dn/dI must be negative, consistent with Fig. 3.6-9 (in fact a,b,c in Fig. 3.6-11 correspond roughly to a,b,c in Fig. 3.6-9c).

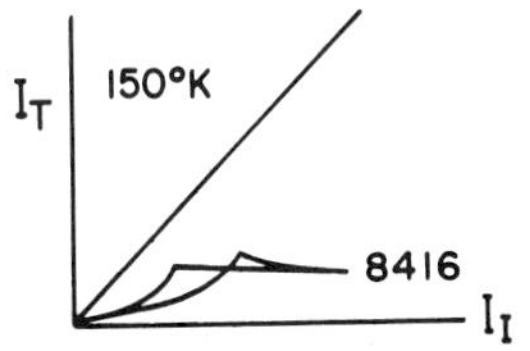

Fig. 3.6-10. Thermal optical bistability in the GaAs sample of Fig. 3.6-2 using 20-ms FWHM triangular pulses. Gibbs, McCall, Gossard, Passner, Wiegmann, and Venkatesan (1979).

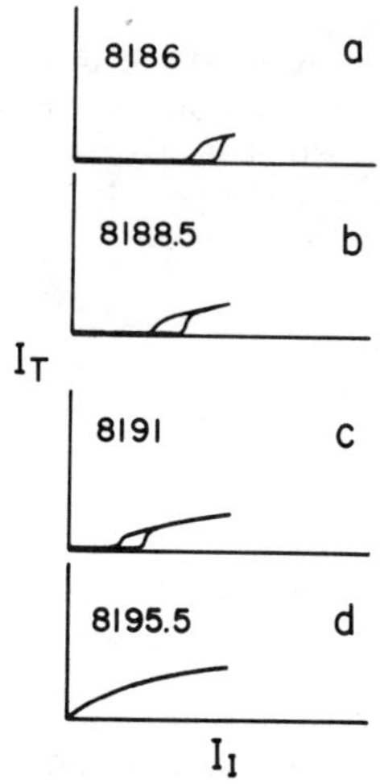

Fig. 3.6-11. Dependence of GaAs excitonic bistability on laser tuning for a fixed position on the sample at 10 K, using sample of Fig. 3.6-2. Gibbs, McCall, Gossard, Passner, Wiegmann, and Venkatesan (1979).

With 514.5-nm, 200-ps input pulses of ≃ 10 MW/cm², near-bandgap lasing of the GaAs bistability etalon occurred (Fig. 3.6-12) and was found to be tunable over tens of Ångstroms by tuning the etalon peak wavelength [Passner, Gibbs, Gossard, McCall, Venkatesan, Wiegmann (1980)]. Such above-bandgap pumping can be used to convert from a wavelength convenient for the pump laser to a longer tunable wavelength. Extensions of this lasing action have resulted in efficient room-temperature ribbon-whisker lasers [Stone, Burrus, and Campbell (1980)] and picosecond lasers [Damen, Duguay, Wiesenfeld, Stone, and Burrus (1980) and Duguay, Damen, Stone, Wiesenfeld, and Burrus (1980)].

In many ways the GaAs etalon is the most nearly practical optical bistable device so far. It is simple; it contains no wires (it is intrinsic). It is small, ≃ 5 μm³, with the potential of becoming very small (< 1 μm³). It is constructed of a convenient material already used for diode lasers and high-speed electronics. It has a fast turn-on of ≃ 1 ps. Its turn-off is a few nanoseconds or longer so far, much longer than desired, but impurities, radiation damage, surface recombination, or an electric field may improve that. The switching energy can be extrapolated to 1 fJ, i.e., only 4000 exciton-energy photons--about as low as statistics will allow for stable operation (Section 7.2). The short length means a free spectral range of about 100 Å, so a 1-Å laser linewidth and stability are sufficient. For example, bistability has been seen with multimode diode lasers (Section 3.6.2).

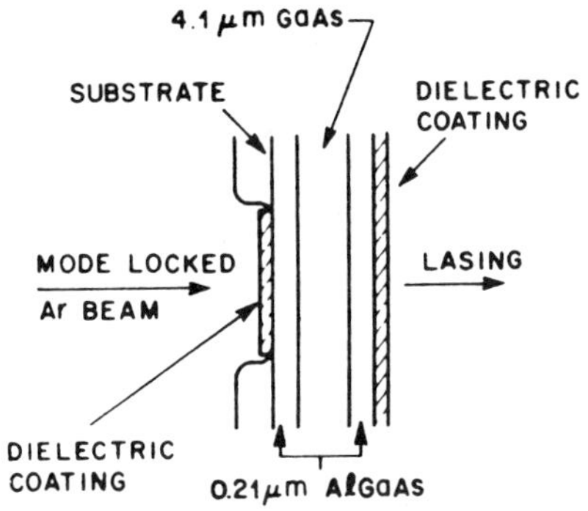

Fig. 3.6-12. Heterostructure sandwich grown by molecular beam epitaxy and used as a 5-μm longitudinally optically pumped laser. Passner, Gibbs, Gossard, McCall, Venkatesan, and Wiegmann (1980). © 1980 IEEE

3.6.2. GaAs-AlGaAs Multiple-Quantum-Well Device

For many potential applications of optical bistability, room-temperature operation is highly desirable if not essential. Jensen (1982a) achieved room-temperature operation in GaAs at 1.06 μm using 10-kW pulses and a valence-band distortion nonlinearity. Earlier it was suggested that excitonic bistability might be extended to room temperature using a multiple-quantum-well (MQW) crystal [Gibbs, McCall, Venkatesan, Gossard, Passner, and Wiegmann (1979)]. Multiple-quantum-well structures, consisting of thin (50 to 400 Å) alternating layers of GaAs and AlGaAs as in Fig. 3.6-13, can have exciton binding energies over twice that of pure GaAs [Dingle (1975)] suggesting a higher operating temperature. As the GaAs well thickness approaches the 280-Å diameter of free excitons in pure GaAs, the motion of the excitons and free carriers is essentially two-dimensional. The appropriate quantization within these multiple-quantum-well structures leads to an increased binding energy for the free excitons. From photoluminescence excitation spectra, R. C. Miller, Kleinman, Tsang, and Gossard (1981) have measured the difference in binding energies of excitons in the 1S and 2S states. From that difference they use the theory to calculate the binding energy B_{1S} of the ground-state exciton. The values of B_{1S} are plotted in Fig. 3.6-14 as a function of the GaAs well thickness L_W. For L_W = 100 Å, B_{1S} is almost 9 meV, compared with 4.2 meV for bulk GaAs. Greene and Bajaj (1983) calculate that the maximum binding energy depends upon the height of the barrier, i.e., upon the value of x in the $Al_xGa_{1-x}As$ barrier. For x = 0.30 they calculate that the heavy-hole binding energy peaks at 9.3 meV for a well thickness of 30 Å; see Fig. 3.6-14(b).

Figure 3.6-15 compares room-temperature exciton features in bulk GaAs with those in an ≃ 100-Å-well MQW sample. The exciton feature is clearly much more apparent in the MQW, but that may result from the larger binding energy and lower band-edge absorption rather than from a stronger free-exciton resonance.

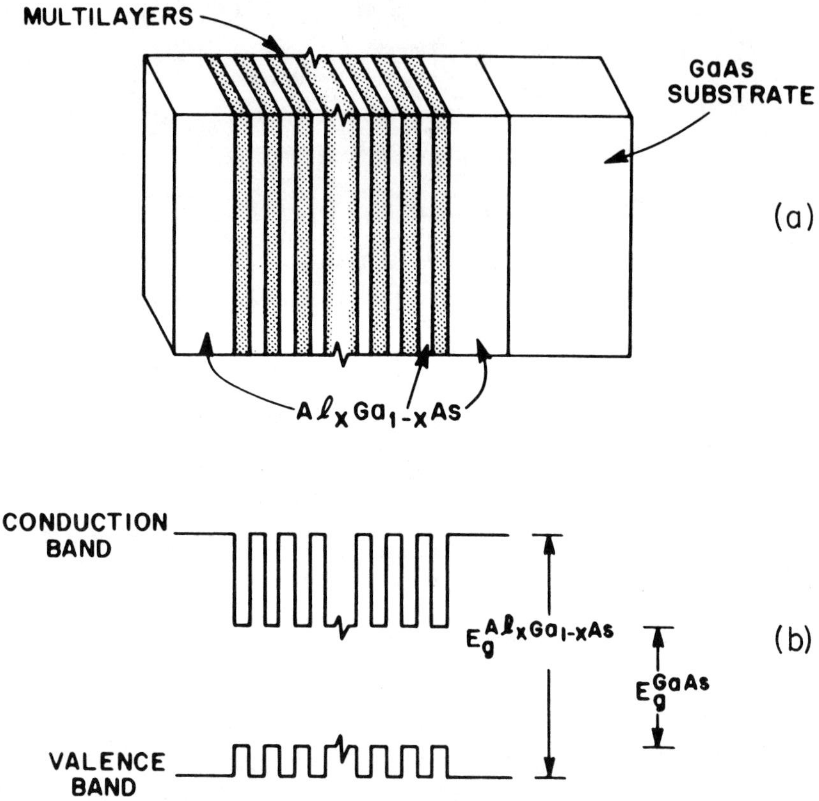

Fig. 3.6-13. Multiple quantum wells. (a) The sample consists of alternating layers of GaAs and $Al_xGa_{1-x}As$. (b) The energy gap of AlGaAs exceeds that of GaAs; free excitons and low-energy free carriers in the GaAs wells are unable to penetrate the AlGaAs barriers. Dingle (1975).

The saturation properties of the two samples are compared in Fig. 3.6-16: the initial saturation intensity of the MQW ($\simeq$ 500 W/cm²) is significantly smaller than that for bulk GaAs ($\sim$ 4.4 kW/cm²). From their Lorentzian analysis Chemla, Miller, Smith, Gossard, and Tsang (1982) estimated an $n_2 \geq 0.02$ cm²/kW near the excitonic resonance and later measured 0.2 cm²/kW by degenerate four-wave mixing. For further discussion of the MQW nonlinearity see Section 7.3 and Miller, Chemla, Eilenberger, Smith, Gossard, and Tsang (1982); D. A. B. Miller, Chemla, Smith, Gossard, and Tsang (1982a,b); D. A. B. Miller, Chemla, Eilenberger, Smith, Gossard, and Wiegmann (1983); Löwenau, Schmitt-Rink, and Haug (1982).

An artist's conception of the first MQW device which exhibited room-temperature optical bistability is presented in Fig. 3.6-17 [Gibbs, Tarng, Jewell, Tai, Weinberger, Gossard, McCall, Passner, Wiegmann (1982)]. The MQW etalon consisted of 61 periods, each containing a 336-Å GaAs layer and a 401-Å $Al_{0.27}Ga_{0.73}As$ layer. The total thickness of GaAs was only 2 μm, compared with 4.1 μm in the pure GaAs etalon of Section 3.6.1. The bistability is shown in Fig. 3.6-18; the wide loops imply light-induced cavity peak shifts of perhaps one quarter of an order. The intensities required are about 1 mW/μm², the switching times 20 to 40 ns, and the peak transmission > 10%. Overshoot (Section 3.3) is seen in the apertured transmission (Section 7.3) [Tarng, Tai, Moloney, Gibbs, Gossard, McCall, Passner, and Wiegmann (1982)].

From Fig. 3.6-14 one finds that the binding energy for 336-Å wells is only 6 meV, not very much larger than the 4.2-meV bulk value. This suggested searching for bistability in the pure GaAs sample at room temperature; Fig. 3.6-19 shows the resultant bistability. Room-temperature operation permitted the elimination of the dewar whose windows introduced losses and prevented good focusing. Without the dewar, high-quality microscope objectives yielded near-ideal focusing. Figure 3.6.-19 shows much better bistability in the MQW than the pure GaAs in spite of the fact that the latter was thicker. However, bistability in a 7-μm-thick bulk etalon is wide and comparable to Fig. 3.6-18a. Furthermore, two samples were grown without opening the molecular beam epitaxy (MBE) growth chamber: one was a bulk sample and one was a 300-Å-well MQW. The total GaAs thickness and the total sample thickness were made about the same in the two samples by adjusting the thickness of the AlGaAs windows. The bistability in the bulk sample was as good as in any MQW sample, i.e., wide loops were seen with < 100 mW (see Fig. 3.6-20) and the threshold was less than 10 mW. Apparently background absorption forces off-resonance operation, so that one cannot take advantage of the sharpness of the MQW exciton feature. Far off-resonance, the product of α_{EX} and $\delta\lambda_{EX}$, determines the maximum phase shift; that product is similar for the bulk and MQW. Elimination of background absorption might result in lower power bistability closer to a MQW free-exciton resonance. Further evidence that room-temperature GaAs bistability is excitonic in both bulk and MQW etalons is given in Ovadia (1984) and Ovadia, Gibbs, Jewell, Sarid, and Peyghambarian (1985): the free-exciton mechanism (Eq. (3.6-2) with the intensity dependence inserted through the measured $\alpha_{EX}(I)$ and $\Delta(I)$) yields a δn large enough for bistability; free-exciton computer simulations are consistent with the data; bistability is seen with less than 1 mW at ≃ 80K for which background absorption is much smaller.

The important point about room-temperature bistability in both bulk and MQW GaAs is that the binding energy is small (giving a large Bohr radius and low saturation intensity) but larger than the exciton linewidth. It is the weak optical phonon interaction of GaAs that makes free-exciton nonlinearities large at room temperature even though kT is 3 to 6 times the binding energy.

Room-temperature bistability in MQW devices has been seen with a sample consisting of 100 152-Å-thick GaAs wells sandwiched between 94%-reflection-coated microscope slides. Such sandwich structures eliminate the need to have each sample coated thus reducing the time between growth and observations.

Bistability has been seen with less than 10 mW [Jewell, Gibbs, Gossard, Passner, and Wiegmann (1983)], suggesting the use of a diode laser source. Since no single-mode diode laser was readily available with ≃ 870-nm wavelength needed for the 336-Å MQW, bistability was observed first with a multimode diode laser selected for good transverse-mode stability; see Gibbs, Jewell, Moloney, Tai, Tarng, Weinberger, Gossard, McCall, Passner, and Wiegmann (1983).

Coincidence was achieved at room temperature between a 53-Å MQW etalon and a Hitachi single-mode diode laser. Figure 3.6-21 shows the bistability that

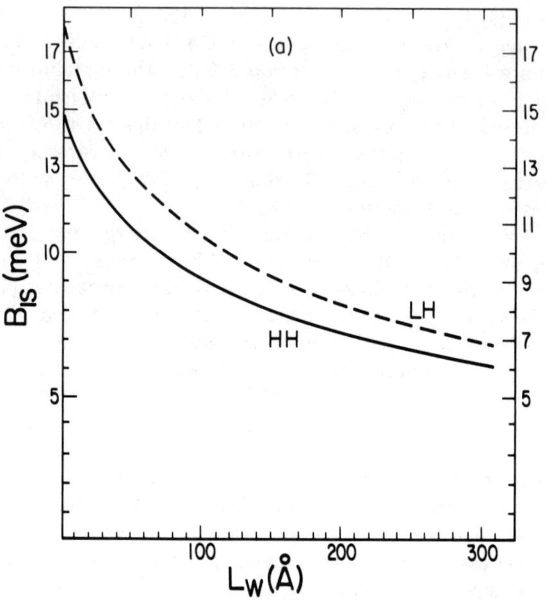
(a)
17
15
13
11
10
9
7
5
B_{1S} (meV)
LH
HH
100
200
300
L_W(Å)

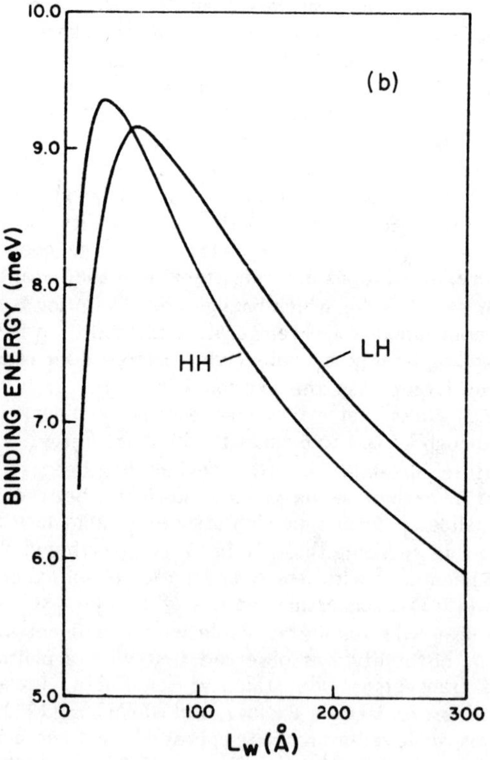
(b)
10.0
9.0
8.0
7.0
6.0
5.0
BINDING ENERGY (meV)
HH
LH
0
100
200
300
L_W (Å)

Fig. 3.6-14. (a) Exciton 1S-binding energy B_{1S} as a function of GaAs well thickness L_w for the light hole (LH) and heavy hole (HH) as calculated from data. Miller, Kleinman, Tsang, and Gossard (1981). (b) Variation of B_{1S} as a function of L_w for x = 0.30. Greene and Bajaj (1983).

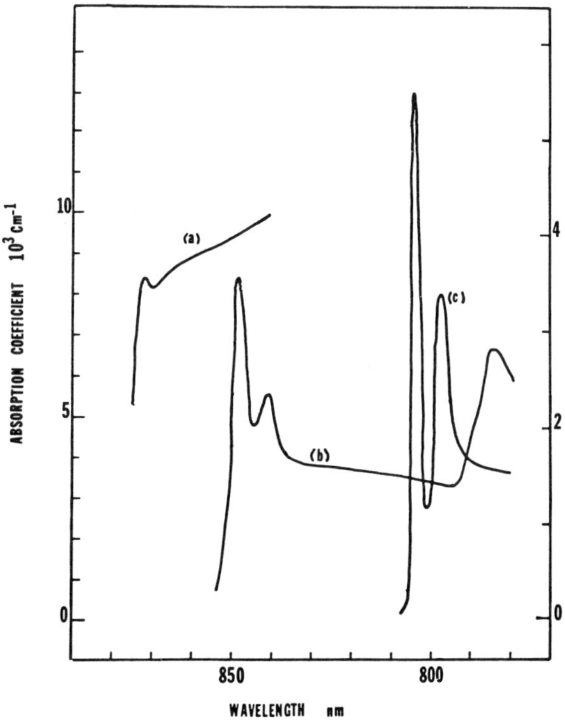

Fig. 3.6-15. Linear absorption spectra of (a) GaAs at 293 K, (b) MQW sample at 293 K, (c) MQW sample at 96 K. $L_w \simeq 100$ Å. Chemla, Miller, Smith, Gossard, and Tsang (1982).
© 1982 IEEE

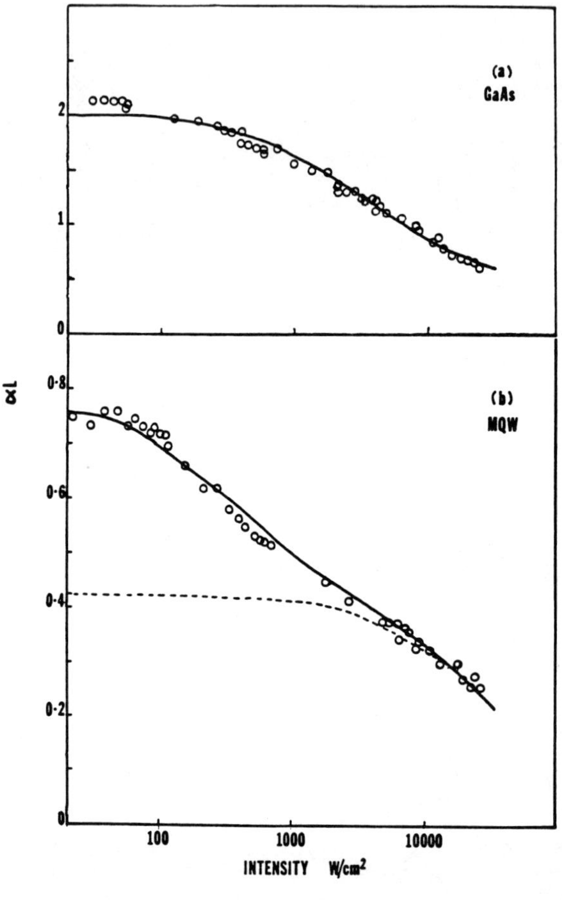

Fig. 3.6-16. Room-temperature absorption saturation in (a) the GaAs sample, (b) the MQW samples. Solid curves are (a) $\alpha L = 0.5 + 1.5/(1 + I/4400)$ and (b) $\alpha L = 0.35/(1 + I/500) + 0.43/(1 + I/44000)$ (αL is the optical thickness of the sample and I is the intensity in W/cm²). The broken curve in (b) is $\alpha L = 0.43/(1 + I/44000)$. Spot size ≃ 5 μm diameter. Chemla, Miller, Smith, Gossard, and Tsang (1982).

resulted, essentially indistinguishable from bistability using a dye laser and emphasizing the compatibility of the new devices with diode lasers. [For a cw degenerate four-wave mixing experiment on room-temperature GaAs-AlGaAs multiple-quantum-well material using a commercial semiconductor diode laser as the sole light source see Miller, Chemla, Smith, Gossard, and Wiegmann (1983).]

For pulses much longer than 1 μs, thermal effects turn off the GaAs devices; under certain conditions the device will then cool and turn back on, etc. These regenerative pulsations are discussed in Section 6.1. Thermal effects have prevented cw stable operation; presumably heat sinking will permit cw operation, but only many-millisecond operation has been reported so far.

Achievement of low-power room-temperature bistability not only increases the impetus toward all-optical systems but also simplifies greatly research in that direction; see Robinson (1984).

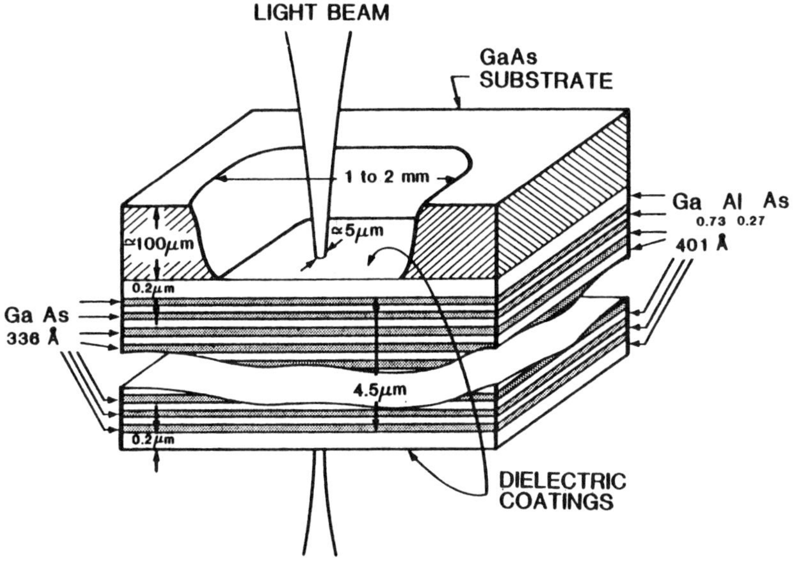

Fig. 3.6-17. GaAs-AlGaAs superlattice used for room-temperature optical bistability. Gibbs, Tarng, Jewell, Tai, Weinberger, Gossard, McCall, Passner, and Wiegmann (1982).

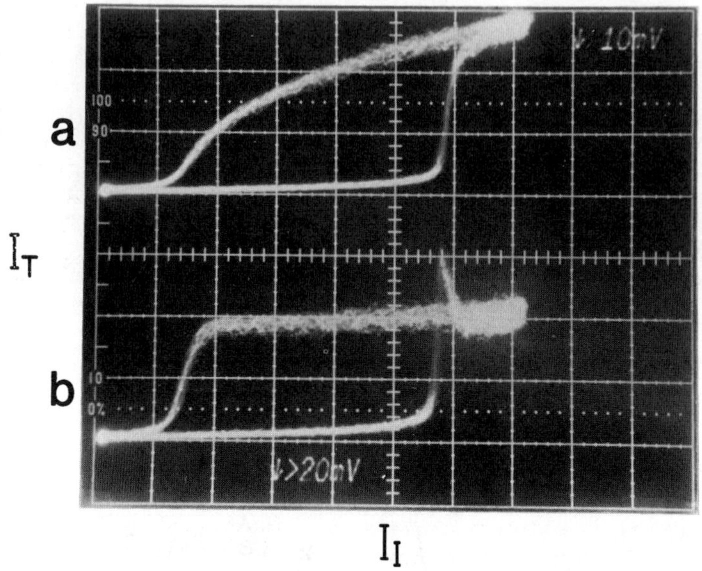

Fig. 3.6-18. Excitonic optical bistability in a GaAs-GaAlAs superlattice. (a) Entire beam. (b) Central part of output amplified 30X relative to (a). (The signal in (b) includes ≈ 1 μm diameter of the ≈ 5-μm diameter spot size at the sample.) Switch up (down) occurs later (earlier) for larger radii. 22°C, 881 nm. Gibbs, Tarng, Jewell, Weinberger, Tai, Gossard, McCall, Passner, and Wiegmann (1982b).

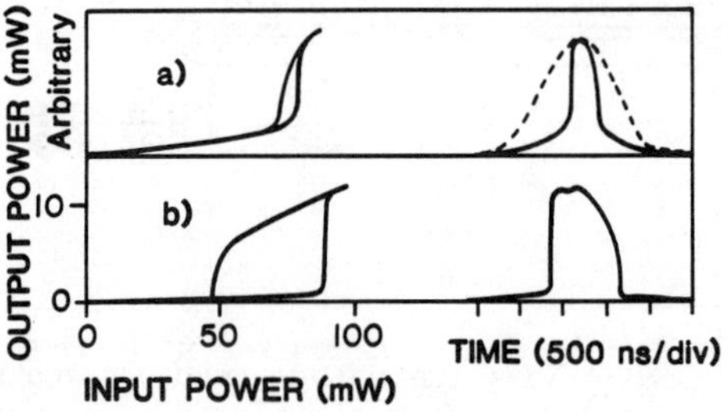

Fig. 3.6-19. Optical bistability in GaAs etalons at 22°C with a triangularly shaped input time dependence. (a) Bulk GaAs, 879.9 nm. (b) MQW, 880.7 nm. The dashed line is the input, normalized to the output. Gibbs, Tarng, Jewell, Weinberger, Tai, Gossard, McCall, Passner, and Wiegmann (1982a). © 1982 IEEE

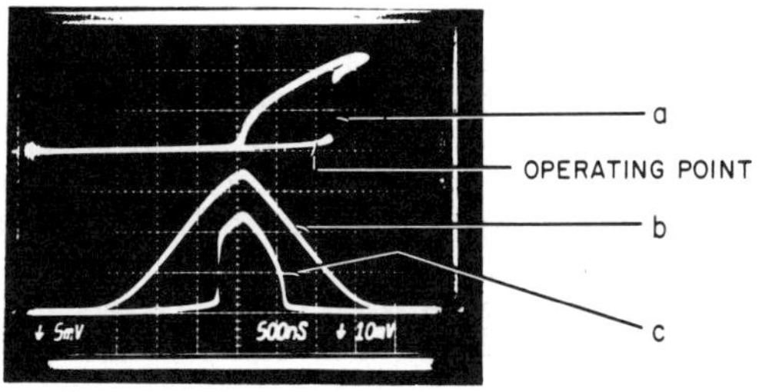

Fig. 3.6-20. Room-temperature optical bistability in a bulk GaAs etalon consisting of 1.5-μm thick GaAs with ≃ 0.2-μm AlGaAs stop layers sandwiched between R = 0.90 mirrors. (a) I_T vs I_I with maximum I_I about 56 mW. The "OPERATING POINT" was the holding value of I_I used in an experiment demonstrating switching through a 1-km fiber. (b) $I_I(t)$ and (c) $I_T(t)$; 500 ns/div. Venkatesan, Lemaire, Wilkens, Soto, Gossard, Wiegmann, Jewell, Gibbs, and Tarng (1984).

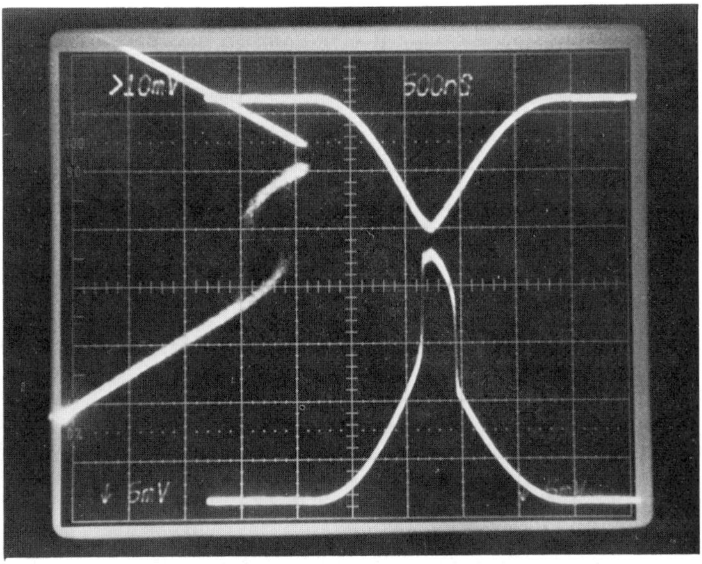

Fig. 3.6-21. Optical bistability in a GaAs MQW etalon using a Hitachi single-mode diode laser as the light source. Tarng, Gibbs, Jewell, Peyghambarian, Gossard, Venkatesan, and Wiegmann (1984).

3.7. InSb

Soon after optical bistability was reported in GaAs, it was reported in InSb [D. A. B. Miller, Smith, Johnston (1979)]. Several other references give further details: D. A. B. Miller and Smith (1979); D. A. B. Miller, Smith, and Wherrett (1980); Weaire, Wherrett, Miller, and Smith (1979); S. D. Smith and Miller (1981); S. D. Smith (1981a); D. A. B. Miller, Smith, and Seaton (1981a,b); D. A. B. Miller, Seaton, Prise, and Smith (1981). S. D. Smith is fond of pointing out the fact that the GaAs and InSb submissions to Applied Physics Letters arrived on the same day, June 15, 1979. But the GaAs work was received January 5, 1979 by the Conference on Laser Engineering and Applications [Gibbs, McCall, Passner, Gossard, Wiegmann, and Venkatesan (1979)], and the use of GaAs and InSb was suggested in Gibbs, McCall, and Venkatesan (1975). (S. D. Smith's rebuttal: "Our results were in fact reported at a conference in the U. K. in December 1978 and again in March 1979. Perhaps the key word is that the two approaches were quite independent.")

The principal InSb results are: a very low holding power of 8 mW; a very large nonlinear refraction of $n_2 \simeq 1$ cm²/kW (corresponding to an effective $\chi^{(3)} \simeq 1$ esu); and triode, transistor, or transphasor operation (see Section 5.4).

The first InSb bistable device consisted of a plane-parallel crystal etalon using only the natural reflectivity ($\simeq$ 36%) of its surfaces ($n_0 \simeq 4$). The sample was cooled to 5 K. A cw CO laser was operated at 1895 cm^{-1} photon energy (about 5-μm wavelength) just below the bandgap at 1899 cm^{-1}. About 500 mW were required to see bistability; see Fig. 3.7-1. It was seen in the fifth order; because of the low finesse ($\simeq$ 1) it could not be seen in lower orders. This can be understood from the graphical solution approach of Section 2.6. In Fig. 2.6-1, the slope of the transmission function must exceed the slope of the straight line for bistability, i.e., three intersections must occur. Figure 3.7-2 shows how bistability can occur in higher order when it cannot in lower order; the slope of the straight line decreases as it intersects the transmission curve in increasing orders. Of course, bistability will occur in higher order only if the intensity is sufficiently high and the nonlinear refraction does not saturate before the threshold order is reached. One can clearly see the abrupt increases in transmission in Fig. 3.7-1 corresponding to sweeping through each cavity peak; optical transistor action on the high-gain regions of Fig. 3.7-1 are discussed in Section 5.4. The intensity change required to sweep from one peak to the next increases, indicating some saturation of the nonlinear refraction.

In another sample of InSb, multiple bistable regions in successive orders were seen (Fig. 3.7-3)--the first multistability in an intrinsic bistable system. Also, bistability in reflection was seen (Fig. 3.7-3). Figure 3.7-4 shows bistability in a polished slice of inexpensive polycrystalline InSb of the kind normally used as a monochromator order-blocking filter. A simple two-layer Ge and ZnS coating yielded R = 0.70. The finesse measured with the Gaussian beam was $\simeq$ 3, allowing bistability to be seen in first order. The transverse profile of the transmitted beam changed discontinuously at the bistability transitions and, when observed in far field, exhibited a ring structure similar to that observed without Fabry-Perot action [D. A. B. Miller, Mozolowski, Miller, and Smith (1978); Weaire, Wherrett, Miller, and Smith (1979)]. Firth, Seaton, Wright, and Smith (1982) show spatial hysteresis far-field profiles. Accurate measurements of switching times have not been made because of slow detector and amplifier response times, but they were $\lesssim$ 500 ns for the device of Fig. 3.7-4.

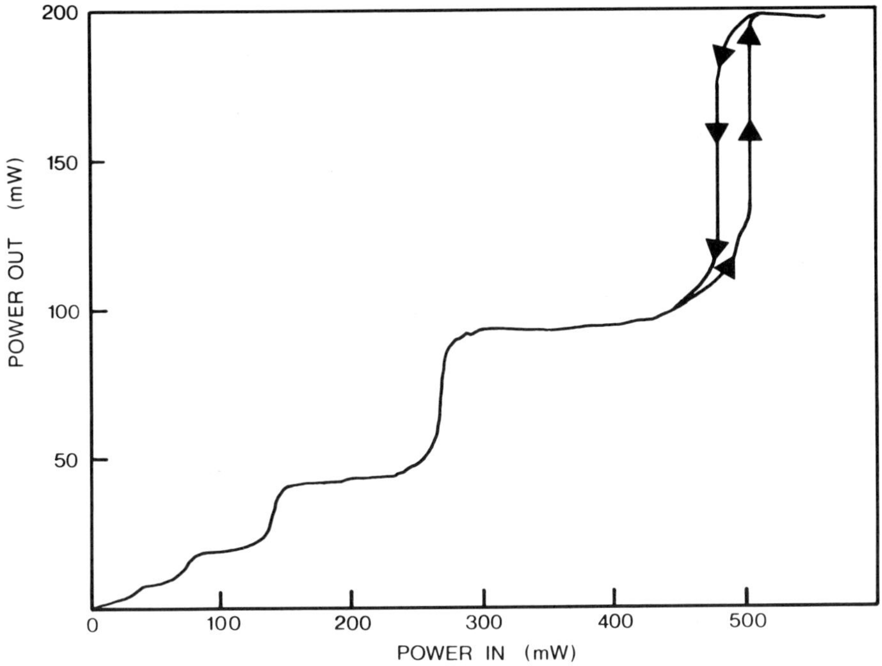

Fig. 3.7-1. Transmitted power plotted against incident power for a cw CO laser beam (wavenumber 1895 cm^{-1} and incident spot size 180 μm) passing through a plane-parallel InSb crystal (5 x 5 mm x 560 μm thick; difference between donor and acceptor densities ≃ 3 x 10^{14} cm^{-3} (n-type)) at ≃ 5 K. D. A. B. Miller, Smith, and Johnston (1979).

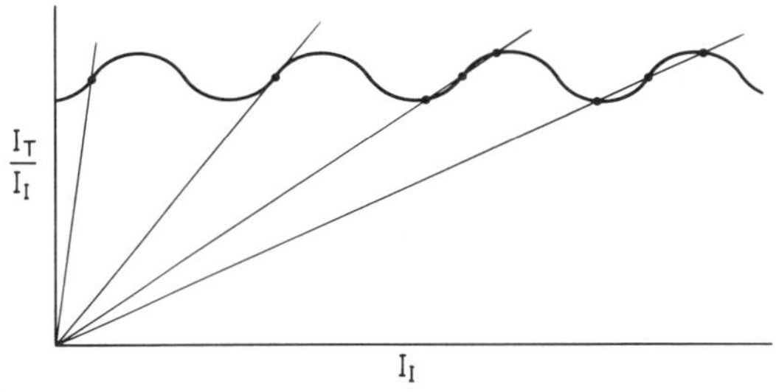

Fig. 3.7-2. Graphical construction for a low-finesse cavity, showing no bistability in first and second order and bistability for third and higher orders, completely neglecting any saturation effect.

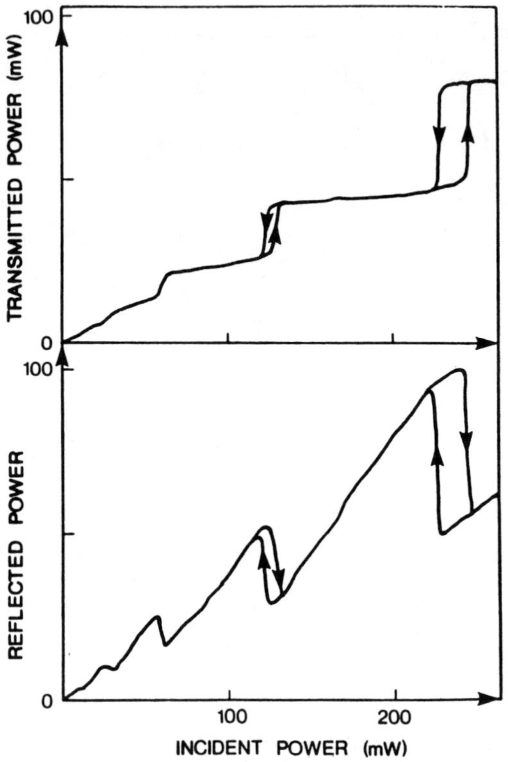

Fig. 3.7-3. Multistability in transmission and reflection. D. A. B. Miller, Smith, and Seaton (1981b). © 1981 IEEE

The nonlinear refraction has been measured through the self-induced distortion (self-defocusing) of an originally Gaussian beam passed through antireflection-coated InSb crystals [D. A. B. Miller, Seaton, Prise, and Smith (1981) and references therein]; see Section 7.3.2. It increases rapidly as the photon energy approaches the bandgap at both 5 and 77 K; see Fig. 7.3-4b. Miller, Smith, and Seaton (1981a) discuss various mechanisms for nonlinear refraction in semiconductors. The theory used in D. A. B. Miller, Seaton, Prise, and Smith (1981) is based on Kramers-Kronig relations which, although not generally valid for nonlinear optical processes, are applicable because the quantum-mechanical phases of the carriers have been randomized by thermalizing collisions. This band-filling mechanism can be understood as follows. A fraction of the optical absorption at the photon energy of interest (which may be below the bandgap) results in the creation of free electrons and holes in the semiconductor bands. The electrons and holes relax rapidly to thermal distributions, which then decay with a single interband recombination time. The thermal population of the band states partially blocks (i.e., saturates) the absorption which would otherwise exist in the spectral region above the bandgap energy; this alteration of the absorption spectrum in

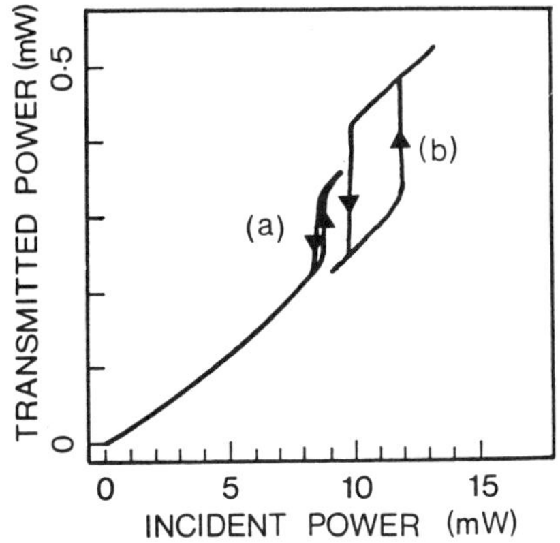

Fig. 3.7-4. Transmitted power plotted against incident power for cw CO laser beam (wavenumber 1827 cm^{-1}, spot size ≃ 150 μm) passing through a polished polycrystalline InSb slice (5 x 5 mm x 130 μm thick) coated to ≃ 70% reflectivity on both faces, held at ≃ 77 K. Onset of bistability is seen at ≃ 8 mW [trace (a)] with clear bistability at slightly higher powers with different cavity detuning [trace (b)]. D. A. B. Miller, Smith, and Seaton (1981b). © 1981 IEEE

turn affects the refractive index seen by the incident photons (which may be below the bandgap energy) through the Kramers-Kronig relations. Löwenau, Schmitt-Rink, and Haug (1982) have attempted a unified theory of the observed optical bistabilities in GaAs and InSb; see Section 7.3.4.

Room-temperature bistability in an uncoated 250-μm InSb etalon has been reported using two-photon nonlinearities from free-carrier generation and a CO_2 laser [Kar, Mathew, Smith, Davis, and Prettl (1983)]. The intensities were as low as 100 kW/cm^2, similar to those in the early GaAs work, and $\chi^{(3)} \simeq 10^{-4}$ esu or $n_2 \simeq 10^{-4}$ cm^2/kW. Of course, the 10-μm wavelength implies much larger interaction diameters and powers for two-photon InSb than excitonic GaAs.

Sarid, Jameson, and Hickernell (1984) have observed bistability on reflection from an InSb etalon (gold-coated on one face at 80 K with a 5.59-μm CO laser beam) and controlled the transmission by a beam coupled through the edge of the etalon.

3.8. OTHER SEMICONDUCTORS

Optical bistability in semiconductors has been discussed at some length already: excitonic bistability in GaAs (Section 3.6), band-filling bistability in InSb (Section 3.7), thermal bistability in ZnS, ZnSe, GaAs, Si, and glasses doped with semiconductors (Section 3.5), and increasing absorption bistability in several semiconductors (Section 2.10).

3.8.1. Te

Staupendahl and Schindler (1982) have tuned a 1.35-mm-thick uncoated ($R = 0.43$ at $\lambda = 10.6$ μm) plane-parallel slab of tellurium (Te) by refractive index changes caused by free carriers created by multiphoton absorption processes (Sections 5.2 and 5.3). They claim the first observation of optical bistability in an intrinsic semiconductor device at room temperature, but apparently Karpushko and Sinitsyn (1978) achieved that in ZnS. If the ZnS effect is thermal, then Staupendahl and Schindler may have achieved the first nonthermal semiconductor room-temperature bistability. They claim to see bistability on the basis of Fig. 3.8-1. To generate a sufficiently high density of electron-hole pairs, the sample was simultaneously irradiated by the 10.6-μm pulse shown in Fig. 3.8-1a (300-ns FWHM) and a slightly shorter 5.3-μm pulse (170-ns FWHM) at an angle of incidence large enough to prevent interference effects. They admit that "because of the actual temporal behavior of the 10.6-μm pulse and the 5.3-μm 'pump' the transmission behavior of the optical-optical modulator is slightly different from the ideal bistable case" (see Fig. 3.8-1c). We agree that the bistability is not very convincing; similar transient responses can occur without bistability if the relaxation time of the optically induced index change is comparable to the pulse lengths. Perhaps of more significance is the fact that etalon tunings approaching one order have been achieved, but the required intensities are several megawatts per square centimeter. Therefore, Staupendahl and Schindler concede, optical bistable devices with extremely low switching energies in the range of picojoules or femtojoules cannot be designed on the basis of a Te etalon. However, useful room-temperature applications include laser pulse shortening and fast switching in the infrared region.

3.8.2. CdS

Optical bistability has been seen in CdS using three different nonlinearities: a degenerate six-photon interaction, bound excitons, and renormalization of the bandgap by the formation of an electron-hole plasma.

Borsch, Brodin, Volkov, and Kukhtarev (1981) use a fifth-order nonlinearity to see a hysteresis in the reversed wave; 20 MW/cm^2 was required to see bistability in their 1-mm-thick sample. See also Borsch, Brodin, Volkov, Kukhtarev, and Starkov (1982) and Section 3.10.3.

Dagenais (1982, 1983) and Dagenais and Winful (1982, 1983a,b, 1984a,b) have used powers as low as 750 μW focused to 330 W/cm^2 to observe bistability in an uncoated ($n_0 \simeq 3$) plane-parallel platelet of 20-μm thickness. Weakly bound excitons in CdS have giant oscillator strengths many orders of magnitude larger than the oscillator strength per molecule of the free exciton. To achieve bistability the laser was tuned in the vicinity of the I_2 line of CdS at $\sim$ 487 nm, arising from an exciton bound to a neutral donor by 8.3 meV. The radiative lifetime is 500 ps, and a dephasing time T_2 of 40 ps is deduced from the 7.5-GHz absorption linewidth. Dagenais (1983) reports a saturation intensity of only 58 W/cm^2.

Far-field profile changes were observed similar to those accompanying self-focusing bistability (Section 3.9). At very low intensities, the far-field profile of the output beam is Gaussian. As the input intensity is increased, a circular fringe pattern develops due to the intensity-induced radial dependence of the index of refraction of the CdS platelet and interference in the far field. By further

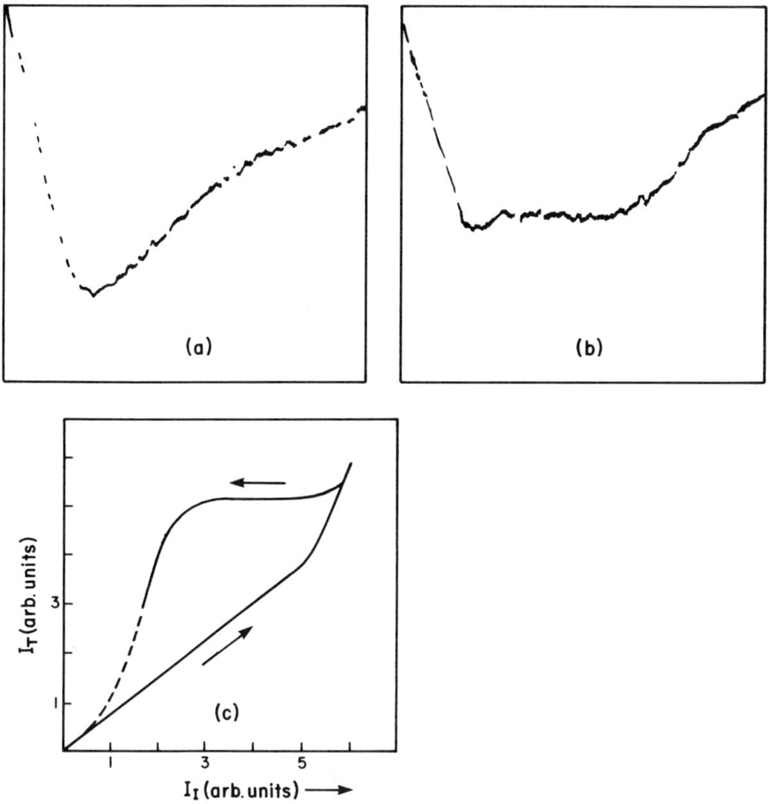

Fig. 3.8-1. Optical bistability in Te. (a) Incoming 10.6-μm pulse; (b) modulated pulse; (c) transmitted intensity I_T plotted against incident intensity I_I. 100 ns/division. Staupendahl and Schindler (1982).

increasing the input field, the number of rings increases as the fringe diameter decreases continuously. At some input intensity the ring pattern jumps discontinuously but with little change in total transmitted power. The bistability is observed using an aperture in the far field, so that only the intensity of the center fringe is monitored; see Fig. 3.8-2. The experiments have been performed at 2 K; thermal effects are seen unless the sample is immersed in liquid helium. Recent measurements on a 13-μm-thick CdS platelet by Sharfin and Dagenais (1984) and Dagenais and Sharfin (1985) show $\tau_\uparrow < 1$ ns, $\tau_\downarrow < 2$ ns, and < 8 pJ single-beam switching energy ($P_\uparrow \tau_\uparrow$ with $P_\uparrow = 8$ mW). Because bound excitons are unable to diffuse, it is hoped that switching powers and energies can be greatly reduced by tighter focusing.

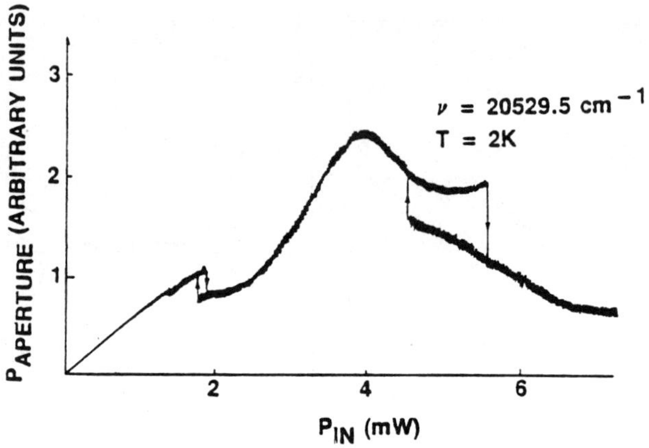

Fig. 3.8-2. Observation of transverse optical bistability in CdS using the optical nonlinearity in the vicinity of the bound-exciton transition. Dagenais and Winful (1984a).

For increasing absorption optical bistability in CdS see Section 2.10 and Klingshirn, Bohnert, and Kalt (1984).

3.8.3. SbSI

Intrinsic optical bistability has been observed in the semiconductor ferroelectric antimony sulfo-iodide (SbSI) using only 2 mW of power from a He-Ne laser [Uchinokura, Inushima, Matsura, and Okamoto (1981)]. SbSI is photoconductive, and photo-induced carriers cause a large refractive index change. The sample was an uncoated as-grown plate with a thickness of about 15 μm grown by a vapor transport method. Detailed time response studies have not been reported, but the observed switching times have been seconds.

3.8.4. CuCl

It has been suggested that the biexciton resonance in copper chloride (CuCl) close to 390 nm is a good candidate for optical bistability [Koch and Haug (1981); Hanamura (1981a,b); Haug, Koch, März and Schmitt-Rink (1981)]. A biexciton is a molecule consisting of two excitons; the biexciton in CuCl is bound by 30 meV. Biexciton bistability is a form of two-photon bistability; see Section 3.10. The biexciton resonance is a very efficient two-photon absorption process [Hanamura (1973)] in which excitonic molecules are created by a nearly resonant intermediate excitonic level. Hanamura (1981a,b) has pointed out that the third-order susceptibility can be divided into two parts, one due to real transitions and the other to virtual transitions. Of course, the decay time of a real excitation is just the lifetime of the excited state; the biexciton and exciton lifetimes are both about 300 ps. In contrast, the decay time of a virtual excitation is approximately the inverse of the frequency detuning, which can be less than a picosecond. Hanamura shows that operating conditions can be selected so that virtual contributions to the nonlinearity can be made to dominate. The hope of picosecond switching times has prompted several groups to look for bistability in CuCl in spite of the short wavelength and low temperatures required.

In the vicinity of the biexciton resonance, the complex dielectric function has been calculated to be given by [Koch and Haug (1981); Haug, März and Schmitt-Rink (1980); März, Schmitt-Rink, and Haug (1980); Haug, Koch, März and Schmitt-Rink (1981)]:

$$\varepsilon(k,E) = \varepsilon_\infty + \frac{(\varepsilon_0 - \varepsilon_\infty)E_x^2}{E_x^2 - E^2 - \Omega^2(k,E)} \tag{3.8-1}$$

where

$$\Omega^2(k,E) = \frac{4n_p|M|^2E_x(E_{xx} - E - i\Gamma_{xx})}{(E_{xx} - i\Gamma_{xx})(E_{xx} - 2E - i\Gamma_{xx})} . \tag{3.8-2}$$

The dielectric constants at zero and high frequency are $\varepsilon_0 = 5.6099$ and $\varepsilon_\infty = 5.59$. $E_x = 3.2022$ eV and $E_{xx} = 6.3722$ eV are the resonance energies at zero wavevector for the transverse exciton and the biexciton, respectively. Γ_{xx}, the full width at half maximum of the biexciton absorption curve, is 0.16 meV = $\Gamma_{xx,0}$ at low light intensities. The polariton density n_p is given by 1.794×10^{14} I/(MW-cm) where I is the intensity in megawatts per square centimeter. M is the matrix element for biexciton formation; $n_p|M|^2/I = 2.82 \times 10^{-8}$ (eV)^2 cm^2/MW. With $\varepsilon = |n|^2$ and $n = n' + in''$ one can calculate the nonlinear refractive index $n_2 \equiv \delta n'/I$ and the intensity-dependent absorption $\alpha = 4\pi n''/\lambda$. Or n can be used in the Fabry-Perot formula

$$\frac{I_T}{I_I} = \frac{(1 - R)^2}{(e^{\delta''} - Re^{-\delta''})^2 + 4R\sin^2\delta'} \tag{3.8-3}$$

where

$$R = \frac{(n' - 1)^2 + (n'')^2}{(n' + 1)^2 + (n'')^2} \tag{3.8-4}$$

and

$$\delta' + i\delta'' = \frac{2\pi Ln}{\lambda} \tag{3.8-5}$$

to search for conditions for optical bistability. Koch and Haug (1981) calculate bistability for L = 1 μm, no coatings, and I_I < 100 kW/cm^2. Henneberger and May (1982), using a theory found incorrect by Haug (1982), compute bistability for L = 10 μm, R = 0.9, and a holding intensity of "some MW/cm^2." Sarid, Peyghambarian, and Gibbs (1983a) conclude that Koch and Haug (1981) use some of the CdS parameters in their CuCl calculation leading to low values of intensities. Sarid et al. compute bistability for L = 10 μm, R = 0.9, and I_I = 10 MW/cm^2, including an intensity dependence in Γ_{xx} and background absorption from the tail of the exciton absorption.

Chase, Peyghambarian, Grynberg, and Mysyrowicz (1979) observed a broadening of the biexciton resonance with laser intensity which they attributed to collisions between excitonic particles. Further evidence was found by Masumoto and Shionoya (1981), Itoh and Katohno (1982), and Peyghambarian, Chase, and Mysyrowicz (1982). Schmitt-Rink and Haug (1981) offer a different explanation in terms of saturable absorption, but the broadening is observed even when the peak absorption is small. Peyghambarian (1982) and Peyghambarian, Sarid, Gibbs, Chase, and Mysyrowicz (1984) have developed a collision broadening model in which liquid-like collisions give rise to a scattering rate Γ_{XX} proportional to the ratio of the mean thermal velocity to the interparticle separation, yielding

$$\Gamma_{XX} = \Gamma_{XX,o} + \frac{\left[\left[\left[E - \frac{E_{XX}}{2}\right]^4 + 4(M_3\sqrt{I})^4\right]^{1/2} - \left[E - \frac{E_{XX}}{2}\right]^2\right]^{1/2}}{\sqrt{2}}, \tag{3.8-6}$$

where M_3 is a constant involving the two-photon absorption cross section and the mean thermal velocity. The value $M_3 = 0.2$ meV-cm/$\sqrt{1/MW}$ is obtained from the experimental result that the biexciton width is about 1 Å at ≃ 1 MW/cm².

Using Eq. (3.8-6) in Eq. (3.8-2), Sarid et al. (1983) calculate $\delta n'$ and find reasonable agreement with the measurements of Itoh and Katohno (1982). The inclusion of the correct CuCl parameters and collision broadening leads to the conclusion that much larger RLI_I products are needed to see bistability in CuCl than predicted by Koch and Haug (1981). Nonetheless, CuCl still has interesting possibilities: I_I = 10 MW/cm² and an area = $(\lambda/n_0)^2$ = 0.027 μm², imply a power of 2.7 mW. If the switching time is 1 ps, the power-speed product is 2.7 fJ or about 5000 photons, i.e., close to the statistical limit.

Haug, Koch, Neumann, and Schmidt (1982) have extended the biexciton bistability treatment to include extrinsic fluctuations by a classical noise source.

Evidence for biexciton optical bistability in CuCl has been presented by two different groups. Levy, Bigot, Hönerlage, Tomasini, and Grun (1983) use a 30-μm-thick CuCl platelet with R = 0.9 coatings and about 10 MW/cm² in 2.5-ns pulses with 0.5-ns detector resolution. Peyghambarian, Gibbs, Rushford, and Weinberger (1983) use a 10- to 12-μm-thick CuCl film grown on an R = 0.9 dielectric mirror with a second R = 0.9 mirror pressed against the film to form a Fabry-Perot etalon. They also use about 10 MW/cm² in 2-ns pulses with 0.6-ns detector resolution. The hysteresis curves of both groups seem to be detector limited, so switching that is fast compared with the input-pulse rise and fall times has not been seen. Limiting action has been seen convincingly (Fig. 3.8-3). The results are consistent with the calculations indicating the need for L ≃ 10 μm, R ≃ 0.9, and I_I ≃ 10 MW/cm² for CuCl biexcitonic bistability. It remains to be seen if the switching times are as short as 1 ps as predicted under these conditions. See also Peyghambarian, Gibbs, Rushford, Weinberger, and Sarid (1983) and Hönerlage, Bigot, and Levy (1984).

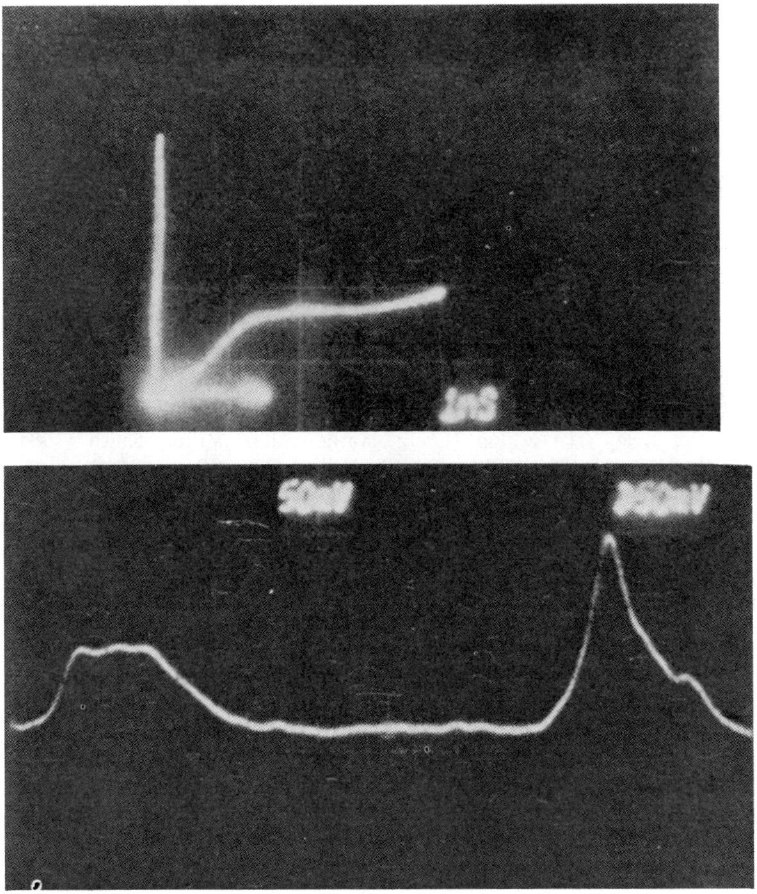

(a)

(b)

Fig. 3.8-3. Limiting action in a 10-μm-thick CuCl etalon. (a) Output vs input intensity. (b) Input pulse (right) and clipped output pulse (left) vs time. Time scale, 2 ns/division. Peyghambarian et al. (1983).

3.8.5. InAs

Poole and Garmire (1984b,c) have observed optical bistability in the reflected signal from a Fabry-Perot etalon consisting of polished n-type 170-μm-thick indium arsenide (InAs) with no coating on the front (R ≃ 31%) and a 800-Å silver film on the back (R ≃ 100%). The 3.096-μm line of an HF laser matches the bandgap at 77 K. Whole-beam bistable switching was seen with incident power levels as low as 3 mW corresponding to a peak intensity of 75 W/cm². The observed switch-on and switch-off times were limited by the 1-μs response time of the detector preamplifiers, but free-carrier lifetimes are typically 200 ns.

The nonlinear index is quite large, 0.03 cm^2/kW, and arises from band filling as in InSb [Poole and Garmire (1984a,d,e)]; see Sections 3.7 and 7.3.2. They extrapolate to a power of 10 nW by focusing to λ/n_0 on a 2-μm-thick etalon with front and back mirror reflectivities of 95% and 99%, respectively.

3.8.6. CdHgTe

The alloy semiconductor cadmium mercury telluride ($Cd_xHg_{1-x}Te$) has a bandgap that can be tuned from 0 to 1.5 eV by varying the composition. $Cd_{0.23}Hg_{0.77}Te$ has a bandgap close to twice the emission frequency of a CO_2 laser (10.6 μm). Nonlinear two-photon Fabry-Perot effects were observed in a 200-μm-thick uncoated sample using ≃ 100 kW/cm^2, 30-ns pulses [Mathew, Craig, and Miller (1984)]. Room-temperature two-photon optical bistability has been seen using 1.5-μs 500-kW/cm^2 pulses from a hybrid TEA CO_2 laser and an uncoated 400-μm-thick $Cd_{0.23}Hg_{0.77}Te$ etalon [Craig, Mathew, Kar, Miller, Smith (1984), Craig, Miller, Mathew, and Kar (1985), and A. Miller, Craig, Parry, Mathew, and Kar (1984)]. A 100-μm-diameter pinhole placed in contact with the exit face of the sample reduced the effect of etalon inhomogeneities.

At 77 K, the bandgap of $Cd_{0.23}Hg_{0.77}Te$ is within 3.5 kT of 10.6-μm radiation. A nonlinear response was seen with a 200-μm-thick etalon using 1 mW focused to 75-μm FWHM spot diameter [A. Miller and Parry (1984a); see also Craig, Mathew, and Miller (1984) and A. Miller, Parry, and Daley (1984)]. There are two principal difficulties in attaining nonthermal optical bistability with a cw laser. Auger recombination severely limits the size of the nonlinearity above about 10 W/cm^2, which causes higher order resonances to occur at relatively higher powers. The second difficulty is that the refractive-index temperature coefficient is negative, unlike that for most semiconductors. As a result, a cavity approaching resonance by a negative electronic nonlinearity will switch on thermally. When the electronic and thermal nonlinearities are of opposite sign, the thermal effect will tend to stabilize the operation on a steep nonlinear Fabry-Perot characteristic in the lower branch. An opposing thermal effect may cause switch-off in the upper state, however, resulting in regenerative pulsations (Section 6.1).

Khan, Kruse, Wood, and Park (1983) have also seen Fabry-Perot peak shifts of two or three orders in a 500-μm-thick uncoated etalon of $Cd_{0.21}Hg_{0.79}Te$ using a CO_2 laser.

3.8.7. GaSe

Bakiev, Dneprovskii, Kovaliuk, and Stadnik (1983a,b) report optical bistability in a 10-μm-thick uncoated etalon (R = 0.22) using ≃ 200 kW/cm^2 at 592.5 nm and observing 50-ns detector-limited switching times at 80 K. Using an uncoated 8-μm-thick etalon at room temperature, they observe switch-on at 620.7 nm but there is no hint of switch-off using 120-ns pulses. Gallium selenide (GaSe) is a layered semiconductor, so very flat etalons can be constructed by peeling apart the layers. They conclude that the nonlinear mechanism is probably not thermal but excitonic as in GaAs. The large free-exciton binding energy (20 meV) implies a small Bohr radius and large saturation intensity. One would also expect room-temperature operation, but the optical phonon interaction is so strong that the free-exciton resonance, though still quite apparent, is broadened at 300 K.

3.9. TRANSVERSE OPTICAL BISTABILITY

3.9.1. Self-Trapping, Self-Lensing, and Self-Bending in Extended Media

Bjorkholm, Smith, Tomlinson, Pearson, Maloney, and Kaplan (1981) have demonstrated a fundamentally new type of intrinsic dispersive optical bistability; see also Bjorkholm, Smith, Tomlinson, and Kaplan (1981). It uses self-trapping effects [See Bjorkholm and Ashkin (1974) and LeBerre, Ressayre, Tallet, and Mattar (1985)] and a single feedback reflector. The elimination of the optical cavity removes cavity build-up and decay time constants and relaxes the bandwidth requirements on the laser. These advantages are particularly pronounced in long cavities, which may be needed for instability studies requiring a medium response time shorter than the cavity round-trip time (Section 6.3).

The idea of transverse bistability based on the formation of a self-trapped filament is shown in Fig. 3.9-1. If the input Gaussian beam is focused to an appropriate diameter on the input face of an appropriate self-focusing medium, there is a critical value P_{cr} of the input power P_{in} for which the input laser beam passes through the medium with little change in diameter, i.e., it is self-trapped. For bistability, a lens L, aperture A, and mirror M are added such that there is good feedback for the self-trapped output but poor feedback for the low-intensity ($P_{in} << P_{cr}$) propagation. Hysteresis comes from the fact that once trapping is established, P_{in} can be reduced below P_{cr} and trapping will be maintained so long as $(1 + 2R_{eff})P_{in} \geq P_{cr}$ where R_{eff} is the effective feedback reflectivity into the self-trapped filament. [The factor of 2 arises because the counterpropagating field shifts the phase twice as much as the self-shift of the forward field (see Appendix D). Rapid diffusion can remove the grating between the forward and backward fields and reduce the factor of 2 to 1.] The essential phenomenon in self-trapping bistability is the movement of the apparent focus from the entrance to the exit of a long self-focusing medium by self-trapping.

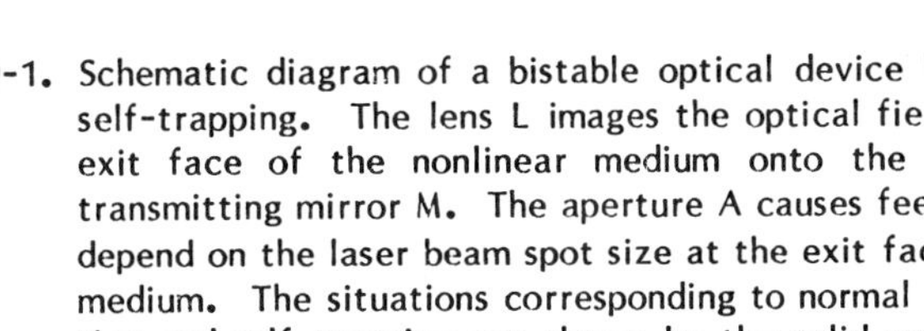

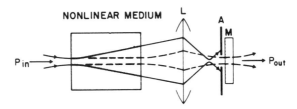

Fig. 3.9-1. Schematic diagram of a bistable optical device based on self-trapping. The lens L images the optical field on the exit face of the nonlinear medium onto the partially transmitting mirror M. The aperture A causes feedback to depend on the laser beam spot size at the exit face of the medium. The situations corresponding to normal propagation and self-trapping are shown by the solid and dashed lines, respectively. Bjorkholm, Smith, Tomlinson, and Kaplan (1981).

The Na vapor in Bjorkholm et al. (1981) was contained in a 20-cm-long heated cell constructed of Pyrex. The TEM_{00} mode from a cw ring dye laser was focused on the entrance window to an e^{-1} field radius of 80 μm corresponding to a confocal parameter of 6.8 cm. In the absence of self-focusing, the spot size on the exit face was 480 μm. A circular polarizer isolated the laser and cell. A 16-cm focal-length lens imaged the exit window with unity magnification onto a flat mirror (R = 0.94). An aperture of 100- to 200-μm diameter was placed several millimeters in front of the mirror.

Without feedback from the mirror, strong self-focusing and self-trapping were readily observed with approximately 150 mW of light tuned about 1.2 GHz above the resonance frequency of the $3S_{1/2}$, $F = 2 \rightarrow 3P_{1/2}$ Na transitions (589.6 nm). The Na density was about 2×10^{12} cm^{-3}. Bistability was observed when the mirror was aligned normal to the laser beam and the input light was amplitude modulated at ≃ 50 Hz with a spinning transmission grating composed of closely spaced fine wires; see Fig. 3.9-2. The switch-up and switch-down times were both about 20 μs. Probably the switch-up time is controlled by hyperfine optical pumping and the switch-down time by the time it takes unpumped atoms to enter and pumped atoms to leave the laser beam (see Section 3.2.4). The dashed lines in Fig. 3.9-2 show the calculated low- and high-transmission limits, based on the measurement of 45-mW power incident onto the aperture for an input power of 140 mW and no feedback. The high-transmission limit assumes 100% transmission by the aperture. At low intensities there is substantial absorption by the Na, but it is saturated for input powers exceeding 60 mW. Thus saturation of absorption is not the bistability mechanism. The switch-up occurred at a power level roughly equal to that needed for self-trapping without feedback. Visual observation of resonance fluorescence showed that when switching occurs, the laser beam abruptly changes from diverging propagation to what appears to be self-trapping.

A wide variety of behaviors was obtained depending upon the precise adjustments of the various parameters: laser frequency, Na density, and input and feedback optics. With higher Na density than that shown in Fig. 3.9-2 and a smaller input spot size, several switching levels and several hysteresis loops were obtained. Under certain conditions the output was noisy, appearing to switch back and forth between different transmission states (see Section 6.3).

Self-trapping bistability has been studied using liquid suspensions of dielectric particles as the nonlinear medium by P. W. Smith, Ashkin, Bjorkholm, and Eilenberger (1984). The combination of large nonlinear coefficient (calculable from first principles) and slow response time makes this medium attractive for cw and transient studies of complex nonlinear effects [P. W. Smith, Maloney, and Ashkin (1982)]. The nonlinear medium was contained in a cell with an antireflection-coated front face and with the interior surface of the exit window coated for 95% reflectivity. The cell length was 0.5 mm, and the 488-nm input beam was focused to a one-half-intensity diameter of ≃ 3.5 μm. Two liquid suspensions were used: an aqueous suspension of 80-nm quartz particles and a suspension of 70-nm quartz particles in o-dichlorobenzene. For the aqueous suspension, the refractive index of the particles (1.46) is higher than that of the liquid (1.33), and a Gaussian light beam induces gradient forces that pull particles into the light beam. By contrast, o-dichlorobenzene's refractive index (1.55) is higher than that of the particles, so the particles are pushed out of the laser beam. The data agree well with a simple theoretical model developed by Bjorkholm, Smith, and Tomlinson (1982) to provide a guide to the design of saturable-self-focusing devices. It predicts that the spot size of the self-focused beam will undergo oscillations as it propagates through

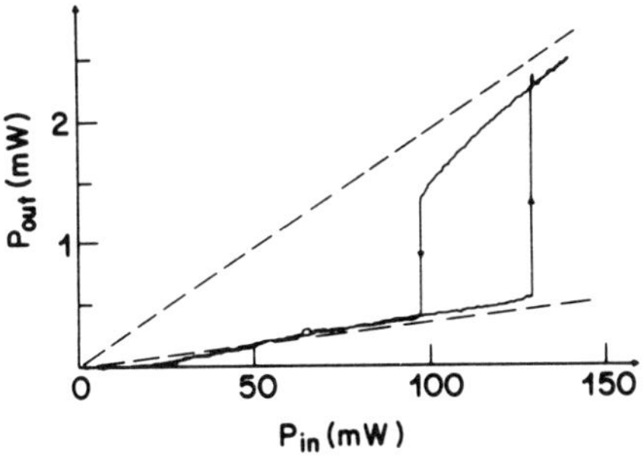

Fig. 3.9-2. Experimental curve of P_{out} versus P_{in} with parameters adjusted so that the device exhibits self-trapping bistable operation. The aperture diameter was 150 μm, and the arrows show the switching directions. Bjorkholm, Smith, Tomlinson, and Kaplan (1981).

the nonlinear medium and that, for a certain range of saturation parameters, this oscillation can result in multistable behavior. The observations are in good qualitative agreement.

Bjorkholm et al. (1981) extrapolate their self-trapping bistability to an input spot size of approximately one-wavelength diameter and estimate that a length $L \simeq 50\lambda$ would be adequate, yielding a 0.2-ps round-trip time ($n_0 = 1$). The index change $\delta n \simeq 0.05$ required for bistability would be hard to achieve. However, the potential for ultrafast switching [P. W. Smith and Tomlinson (1981a)] is clear. Note that the 50λ length needed is longer than the GaAs etalon device even when the physical length is multiplied by the finesse: $L_{eff} = (4\ \mu m)8 \simeq 40\lambda$. Granted the cavity may take many cavity response times to reach equilibrium, but the advantage, if any, of self-focusing, so far as the switching time is concerned, is slight. It will be exciting when bistable-etalon switch-off times are limited by cavity response times, for, by then, medium response times will have been reduced to a picosecond! The real challenge now is to identify very large nonlinearities that are very fast (Section 7.3); often "large" requires near-resonance, real transitions that are usually slow. Also note, in the comparison between self-trapping devices and etalons, that a lossless cavity has a peak transmission of 100%, whereas the self-trapping device's peak is the transmission of the feedback mirror, which is not likely to be high. The reflected beam might be used if it is not too badly distorted and can be separated. However, the self-trapping idea is clever and relatively new, so many more configurations and advantages will likely appear, perhaps in guided-wave designs (Section 3.10).

Gibbs, Derstine, Hopf, Kaplan, Rushford, Shoemaker, Weinberger, and Wing (1982) have studied self-trapping bistability in Na using an experimental arrangement similar to the original self-focusing bistability experiment except that no aperture was placed between the Na and the feedback mirror. A curved mirror, placed one radius of curvature away from the output end of the Na vapor region, makes self-trapping easier even though no aperture is used, emphasizing that the mechanism is a nonlinear wavefront effect. Use of an aperture outside the feedback mirror enhances the bistability signal because, with low absorption, the "on" and "off" stages are distinguished much more by changes in the diameter and divergence of the output beam than by changes in transmitted intensity.

Kaplan (1981b) has proposed the elimination of all feedback mirrors by using counterpropagating beams whose mutual influence by way of the nonlinear refractive index provides the feedback essential for bistability; see Fig. 3.9-3. His schemes can even work for defocusing, thus avoiding a catastrophic collapse of the beam in the case of an unsaturable nonlinear material. Defocusing schemes allow the use of semiconductors whose nonlinear refractive indices are large and negative just below the band edge where the absorption is sufficiently low. Kaplan also suggests bistability by using the self-bending of a light beam whose intensity profile possesses an asymmetrical shape; see Fig. 3.9-4.

Tai, Gibbs, Peyghambarian, and Mysyrowicz (1985) report cross-trapping optical bistability using the configuration of Fig. 3.9-3a but with no diaphragm. This is mirrorless bistability based on nonlinear index effects as opposed to the increasing absorption bistability of Section 2.10.

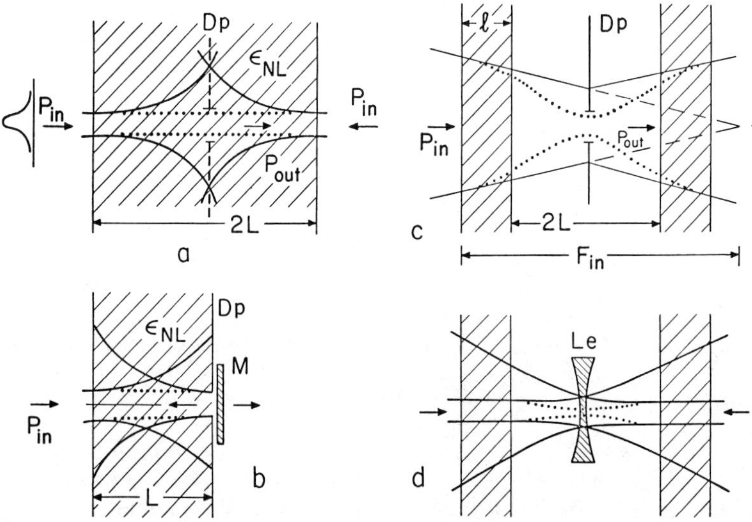

Fig. 3.9-3. Various configurations realizing optical bistability from mutual self-focusing (or self-defocusing) of counter-propagated beams. M, mirror; Dp, diaphragm; Le, lens. The solid lines show the spatial behavior of the low-power beams; the dotted lines correspond to the behavior of high-power beams after their switch into a high-transmittance state of the system. Kaplan (1981b).

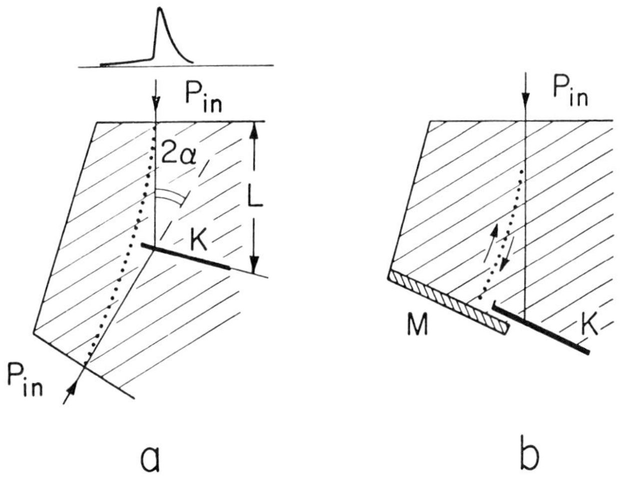

Fig. 3.9-4. Configurations realizing bistability from mutual self-bending of the light beams. M, mirror; K, knife edge. Kaplan (1981b).

3.9.2. Diffraction-Free Encoding in Short Media

Section 3.9.1 describes transverse optical bistability in a long medium in which self-lensing effects can change the beam diameter appreciably during propagation across the medium. Transverse bistability can also arise from thin-sample self-lensing for which the beam's diameter is altered very little during transit through the medium but is altered substantially by the time it is returned to the medium by a single feedback reflector. Kaplan (1981a,b) describes this situation by intensity-dependent lensing and emphasizes that it can work for self-defocusing as well as self-focusing.

This thin-sample-encoding (TSE: $|E|^2/E_S{}^2 \ll \alpha_0 L$ in Eq. (2.9-13)) bistability has been observed by Khoo and Shepard (1982), Khoo (1982), and Khoo, Yan, Liu, Shepard, and Hou (1984a,b) using self-focusing by nematic liquid crystals and by LeBerre, Ressayre, Tallet, Tai, Hopf, Gibbs, and Moloney (1984) using self-defocusing or self-focusing by Na vapor. Khoo (1982) and Khoo and Shepard (1982) focus an ≈ 0.5-W, 514.5-nm laser beam about 1.5 cm in front of a 60-μm homeotropically-aligned liquid crystal such that the optical field makes an angle of 70° with the director axis. A lens and a 90%-reflecting mirror in the near field provide the optical feedback. Bistability was observed in the near field through an aperture placed outside the bistable device. The usual counter-clockwise hysteresis loop was observed with the aperture on-axis, but reversed loops could be seen in off-axis positions. The feedback lens/mirror combination is positioned so that self-focusing by the crystal is required before the feedback becomes focused on the crystal; once that is established the input can be reduced resulting in a hysteresis. The switching times are 1 or 2 s. Khoo, Yan, Liu, Shepard, and Hou (1984a,b) analyze this and related experiments in greater detail and find good agreement with their analytic model.

The liquid-crystal TSE near-field bistability is accompanied by interference rings in the far field as a result of transverse phase modulation associated with the transverse intensity variation of the laser beam. TSE far-field bistability and rings have been studied both on the defocusing (with the input initially focused beyond the cell) and focusing sides of the Na D_2 resonance by LeBerre, Ressayre, Tallet, Tai, Hopf, Gibbs, and Moloney (1984). They also predict and see instabilities with a period of twice the round-trip time (Section 6.3).

In all of the transverse bistability cases described in Sections 3.9.1 and 3.9.2 so far, the mechanism leading to bistability has been a nonlinear refractive index effect. LeBerre, Ressayre, Tallet, Tai, Hopf, Gibbs, and Moloney (1984) predict bistability for pure amplitude encoding. If a cw on-resonance beam has the appropriate intensity and traverses an appropriate medium, the outer parts of the beam are stripped away and Fresnel fringes and on-axis enhancement occur in less than a Rayleigh length downstream [see LeBerre, Ressayre, Tallet, Tai, Gibbs, Rushford, and Peyghambarian (1984)]. If a plane feedback mirror is placed close to the cell, the on-axis enhancement can be used to yield bistability. Delayed feedback instabilities are predicted for this case, too.

3.10. OTHER OBSERVATIONS AND PROPOSALS

3.10.1 Two-Photon Optical Bistability

The Doppler-free two-photon dispersion near the Rb $5S_{1/2}$-$5D_{5/2}$ transition has been used to observe optical bistability and study transients (see Section 5.6) in beautiful experiments by Giacobino, Devaud, Biraben, and Grynberg (1980). Two-photon bistability was proposed in Arecchi and Politi (1978) and discussed theoretically in a number of other papers: Agrawal and Flytzanis (1980a,b; 1981b); Grynberg, Devaud, Flytzanis, and Cagnac (1980); Agarwal (1980); Hermann and Elgin (1981); Hermann and Thompson (1982); Hermann, Elgin, and Knight (1982); Walls, Kunasz, Drummond, and Zoller (1981); Tsukada and Nakayama (1982a); F. L. Li (1982, 1983a,b); Kobayashi, Kothari, and Uchiki (1984); Kothari and Kobayashi (1984). Koch and Haug (1981) and Hanamura (1981a,b) predicted two-photon bistability using the biexciton resonance in CuCl; this has been observed by Hönerlage, Bigot, and Levy (1983) and by Peyghambarian, Gibbs, Rushford, and Weinberger (1983); see Section 3.8.4. Room-temperature bistability in InSb using a CO_2 laser and two-photon nonlinearities has been reported by Kar, Mathew, Smith, Davis, and Prettl (1983); see Section 3.7.

3.10.2 Tristability, Polarization Effects, and Three-Level Systems

Optical tristability, i.e., three stable output states for the same input, can be achieved with a two-photon system as discussed by Hermann (1981); Walls, Kunasz, Drummond, and Zoller (1981); Hermann and Elgin (1981); Hermann and Thompson (1982); Hermann and Walls (1982); Tsukada and Nakayama (1982a); Walls (1982); F. L. Li (1983a,b); Parigger, Zoller, and Walls (1983); Agarwal (1982b); Kovarskii, Perel'man, and Averbukh (1982). Closely related are the discussions of three-level systems by Walls and Zoller (1980); Averbukh, Kovarskii, and Perel'man (1980b); Walls, Zoller, and Stëyn-Ross (1981); Kitano, Yabuzaki, and Ogawa (1981a,b); Savage, Carmichael, and Walls (1982); Adonts, Jotyan, and Kanetsyan (1982); Hassan, Tewari, and Abraham (1982). Lawandy and Rabinovich (1984) treat a three-level system interacting with two resonant fields, only one of which experiences cavity feedback.

Cecchi, Giusfredi, Petriella, and Salieri (1982) report optical tristability in Na vapor; for linearly polarized incident light, three stable states appear in the polarization of the transmitted light: σ_+ dominant, σ_- dominant, and linear polarization. Under the proper conditions, as the linearly polarized input intensity is increased to a threshold level, the output light switches from the linear to (in a random way) the σ_+ or the σ_- dominant state, keeping the acquired polarization when the incident intensity is further increased. This competition is similar to that between the counterpropagating beams in a ring dye laser close to threshold [Roy and Mandel (1980); Mandel, Roy, and Singh (1981); see Section 6.5]. Mlynek, Mitschke, Deserno, and Lange (1982, 1984) also report Λ-type three-level optical bistability in Na in a magnetic field.

Optical bistability without a cavity and without hysteresis of the light intensity has been seen in Na and analyzed by Yabuzaki, Okamoto, Kitano, and Ogawa (1984). The optical system is composed of a cell containing atoms with spin in the ground state, a $\lambda/8$ plate, and a mirror; the incident light is linearly polarized and nearly resonant with the atomic absorption line. The system has a positive feedback loop for the rotation of polarization, through competitive optical pumping by $\sigma_\pm$ circularly polarized components. When the input intensity exceeds a critical value, symmetry breaking occurs: the light polarization is rotated either clockwise or counterclockwise, and the atomic spin polarization is produced spontaneously parallel or anti-parallel to the incident beam propagation direction. By circularly polarizing the backward light, they see bistability of the transmitted intensity.

Polarization switching of a four-level $J = 1/2$ to $J = 1/2$ transition has been analyzed by Hamilton, Ballagh, and Sandle (1982). Areshev and Subashiev (1982) and Areshev, Murina, Rosanov, and Subashiev (1983) treat polarization and amplitude optical multistability in a ring cavity containing a nonlinear medium described by a susceptibility tensor of rank four $\chi^3_{ijkl}(\omega;\omega,\omega,-\omega)$, which has only two independent components (i.e., the medium is isotropic). Optical bistability and a reversed switching effect in a three-photon resonant medium are treated theoretically by Tewari and Hassan (1982).

3.10.3. Phase-Conjugation Optical Bistability

Optical bistability in degenerate four-wave mixing and phase conjugation has been proposed and discussed by Winful and Marburger (1980a); Flytzanis and Tang (1980); Agrawal and Flytzanis (1981a); Agrawal, Flytzanis, Frey, and Pradere (1981); Kukhtarev and Semenets (1981a,b); Kukhtarev and Starkov (1981); Flytzanis, Agrawal, and Tang (1981); and F. L. Li (1983b). Closely related is the work on counterpropagating beams and bidirectional ring cavities by Kaplan (1981a,b), Agrawal (1981a,b, 1982), Kaplan and Meystre (1981, 1982), and Silberberg and Bar Joseph (1982). The use of bistability to enhance the Sagnac effect has been proposed by Kaplan and Meystre (1981) and Kaplan (1982a, 1983), but a nonreciprocal optical nonlinearity leads to a false rotation signal if the counterpropagating beam intensities become unequal (they are assumed equal by Kaplan and Meystre). It remains to be seen if the proposed enhancement can actually increase the sensitivity of Sagnac gyroscopes.

One can make a distinction between using phase conjugation to monitor optical bistability and observing a bistability that does not exist without phase conjugation. Agrawal (1983a,b) calculates the case that phase conjugation is a nonperturbing monitor of a bistable Fabry-Perot cavity in which spatial hole burning effects of the pump-beam standing wave are treated. Mlynek, Köster, Kolbe, and Lange (1984) place Na vapor in a Fabry-Perot cavity and use a weak probe to generate a weak phase conjugation signal as a monitor of optical bistability (as a function of magnetic field). Tai, Gibbs, Peyghambarian, and Mysyrowicz (1985) use phase conjugation to monitor three types of transverse optical bistabilities. F. L. Li, Hermann, and Elgin (1982) calculate the phase conjugate signal used to monitor a Fabry-Perot cavity bistable for two-photon transitions. Agrawal (1981d) considers optical bistability in a Fabry-Perot cavity filled with a nonlinear medium and having one of its mirrors replaced by a phase-conjugate mirror. He finds that the bistability arises from saturated absorption alone: dispersive effects play no role even when the system is operated under off-resonance conditions. This is because of the phase-conjugate mirror: the dispersive phase shifts during a round trip cancel exactly.

Perhaps of greater interest is optical bistability without a cavity that results from phase conjugation, so that hysteresis is seen for the phase-conjugate signal versus the probe. Winful and Marburger (1980a) point out the similarity between the equations for four-wave-mixing phase conjugation and the equations for a Bragg reflector. In the latter (Section 3.10.5) there is a physical grating whose Bragg condition can be tuned by the intensity dependence of the refractive index. For four-wave-mixing, Winful and Marburger (1980a) show bistability neglecting pump depletion, but it occurs only for probes too large to be consistent with the no-pump-depletion assumption. A large phase-conjugate signal is included in some of the calculations by Winful and Marburger (1980a), Winful (1980), Kukhtarev and Semenets (1981b), Cronin-Golomb, White, Fischer, and Yariv (1982), and Borshch, Brodin, Volkov, Kukhtarev, and Starkov (1984). At low probe intensities the conjugate arises from diffraction of one pump off the grating formed by the probe and the other pump. At high intensities the phase-conjugate signal becomes so large that new holographic gratings are formed by its interference with the two-pump beams. This changes the phase of the pump waves and tilts the isophase surfaces of the grating and frustrates the Bragg diffraction condition for the formation of the conjugate. Cronin-Golomb et al. (1982) give an exact solution for the reflectivity of a four-wave-mixing phase-conjugate mirror with depleted pumps; Borshch et al. (1984) do the same for six-wave-mixing. There seems to have been no stability analysis of the proposed phase conjugation bistability systems. A small hysteresis loop of the phase-conjugate signal versus pump intensity ($\simeq 100\ \mathrm{MW/cm^2}$) for six-wave-mixing in CdS is attributed to phase conjugation bistability by Borshch, Brodin, Volkov, and Kukhtarev (1981, 1982) and Borshch, Brodin, Volkov, and Kukhtarev and Starkov (1984). Lytel (1984) predicts optical multistability in collinear degenerate four-wave mixing.

Feinberg (1982) has proposed a phase conjugation bistability in which light fed back to a fanning crystal discourages that fanning; Kwong, Cronin-Golomb, and Yariv (1984) report bistability and hysteresis in a photorefractive passive phase conjugate mirror (single domain crystal of barium titanate); they state the mechanism is quite different from Feinberg's but fanning and external mirrors are involved.

3.10.4. Nonlinear Interface Optical Bistability

Kaplan (1976, 1977) proposed bistability based on a plane-wave analysis of the intensity-dependent reflectivity of an interface between a linear and nonlinear medium. P. W. Smith, Hermann, Tomlinson, and Maloney (1979a,b) observed a hysteresis as a function of input intensity for the reflection coefficient of 1-ns,

7.5-GW/cm^2 ruby laser pulses at a glass-CS_2 boundary. It is now believed that, in steady-state, the reflection coefficient jumps abruptly but no hysteresis occurs. However, the lower branch of the reflection "hysteresis" curve is metastable, so it can persist for hundreds of relaxation times, thus explaining the glass-CS_2 hysteresis. This has been shown very convincingly using a cw argon-laser beam and a liquid suspension of dielectric spheres as an artificial Kerr medium [P. W. Smith, Maloney, and Ashkin (1982); P. W. Smith and Tomlinson (1984a,b)]. An incoming plane wave excites surface waves, so the plane-wave assumption breaks down [P. W. Smith, Tomlinson, Maloney, and Hermann (1981); Tomlinson, Gordon, Smith, and Kaplan (1982); see also Kaplan (1981d).] Boiko, Dzhilavdari, and Petrov (1975) also concluded that the plane-wave hysteresis is impossible because of the instability of a homogeneous plane wave. Of course, bistability is unnecessary for many signal processing applications; however the quasi-stability of the lower branch and the low "on"/"off" contrast may limit device applications. Regardless of the fate of the nonlinear interface with respect to practical devices, the tenacity of the researchers to reveal the beautiful physical effects contributing to the glass-CS_2 hysteresis is praiseworthy indeed.

Kaplan (1980; 1981c) suggested a hybrid version of the nonlinear interface device in which the reflected beam is detected and an electrical signal is fed back to an electro-optic element forming one of the interface media. P. W. Smith, Tomlinson, Maloney, and Kaplan (1982) have demonstrated bistability at an electro-optic interface consisting of a potassium dihydrogen phosphate (KD*P) crystal and an index-matching oil with a temperature-sensitive index of refraction.

Bosacchi and Narducci (1982, 1983) propose bistability for a frustrated-total-reflection optical cavity that can be thought of as a cross between a nonlinear interface and a Fabry-Perot etalon. Consider reflection (in this case the angle may be close to 45°) from a frustrated-total-reflection cavity consisting of a top spacer layer, a cavity layer, and an absorbing substrate. The cavity-layer refractive index is nonlinear and larger than that of the surrounding materials. The reflectivity as a function of the angle of incidence is characterized by a deep minimum at an angle determined by the thicknesses and optical constants of the structure. At resonance the intracavity intensity may be two or three orders of magnitude larger than the incident intensity, giving rise to a hysteresis by keeping the reflection frustrated with less input than required before the cavity build-up. Khoo and Hou (1982) have studied bistability characteristics of a planar nematic-liquid-crystal film sandwiched between two dielectrics under a total internal-reflection condition. Similar surface-plasmon and guided-wave structures are discussed in Section 3.10.5.

3.10.5. Guided-Wave Optical Bistability

Tremendous progress in electronics has been achieved by the development of planar integrated circuits with many electronic devices permanently and compactly constructed and interconnected on the same chip. Integrated optics is an attempt to achieve the optical analog. In certain applications, such as long-distance communication through fibers or single signal-processing devices, the advantage of guiding is often clear. However, there are tradeoffs in high-speed and parallel processing applications. Guided-wave devices permit the use of smaller nonlinearities because a high intensity can be maintained over a long distance; however, in some devices, a length of a millimeter or more implies a transit time of 10 ps or more, much too long for some applications. In fact, many guided-wave bistability papers make contradictory claims. They state that one advantage of guided waves is that they permit high intensities over distances much longer than the Rayleigh length possible for etalon devices. Then they state that another advantage is that by avoiding cavity build-up and decay times, guided-wave devices can switch faster.

Guided-wave interconnections are permanent and shock-resistant; however, they are the optical analog of electronic interconnections. They completely ignore the unique capabilities of light: a whole array can be imaged onto another array with only a lens in between; light beams can pass through each other with no interaction; even the direction of a beam can be changed. Lenses and beamsplitters can replace rats' nests of multiple guided-wave interconnections. No doubt waveguiding will play an important role in many optical control devices, but unguided devices that employ propagation of many beams perpendicular to the plane will likely dominate in parallel processing applications; see Section 7.4. Even so, it is very likely that unguided integrated devices, such as addressable holograms, will be used to manipulate parallel beams. For an overview of guided-wave approaches to optical bistability see Garmire (1981b), Sarid and Stegeman (1981), Sarid (1981d), and Stegeman (1982a).

Hybrid optical bistable devices employing an integrated electro-optic modulator or directional coupler are described in Chapter 4. Intrinsic bistable devices need optical feedback, which has been proposed to be provided by distributed Bragg reflectors [Okuda, Toyota, and Onaka (1976); Okuda and Onaka (1977a,b, 1978); P. W. Smith, Turner, and Maloney (1978); Okuda, Onaka, and Sakai (1978); Winful, Marburger, and Garmire (1979a,b); Winful and Cooperman (1982); Kowarschik and Zimmermann (1982)] or a ring-channel waveguide [Sarid (1981b)]. Okuda et al. propose structures with end-Bragg reflectors around a saturable absorber of either polyurethane film locally doped with cryptocyanine or rhodamine 6G or a GaAs-$Al_xGa_{1-x}As$ waveguide, with taper couplers or a Kerr medium of either CS_2 or a nematic liquid crystal. They study numerically the bistability properties as a function of the coupling coefficient of the corrugated waveguide, the background absorption, and the pulse width of the input light. Winful, Marburger, and Garmire (1979a,b) propose a nonlinear distributed feedback structure, and Winful and Cooperman (1982) study numerically its self-pulsing and chaos. Sarid (1981b,d) proposes two line channels coupled to a ring-channel waveguide having an optical-power-dependent refractive index; he calculates a switch-up power of 1.5 μW for a 1-cm-diameter InSb ring operating at 5 μm.

Distributed feedback in a nonlinear medium is shown in Fig. 3.10-1. In this case the distributed feedback can be considered as due to corrugations in the surface of a waveguide of nonlinear refractive index or as a plane sinusoidal refractive index modulation within a bulk nonlinear medium. In the distributed feedback medium the refractive index is assumed to be $n(z) = n_0 + n_1 \cos 2\beta_0 z + n_2 I$. This model does not include a sinusoidal perturbation on the nonlinear component of the refractive index. The transmission of a linear distributed feedback structure is shown in Fig. 3.10-2 and is seen to be very different from that for the nonlinear Fabry-Perot cavity. Its distinguishing feature is a single stop-band at the phase-matching condition. As a result of the different linear transmission function, the distributed feedback-bistable optical device transfer curves are very different from those for the nonlinear Fabry-Perot cavity. In principle it is possible to taper, chirp, or multiplex the distributed feedback grating to produce any desired transmission function. This additional degree of freedom makes it possible, in principle, to obtain any transfer function desired for optical signal processing.

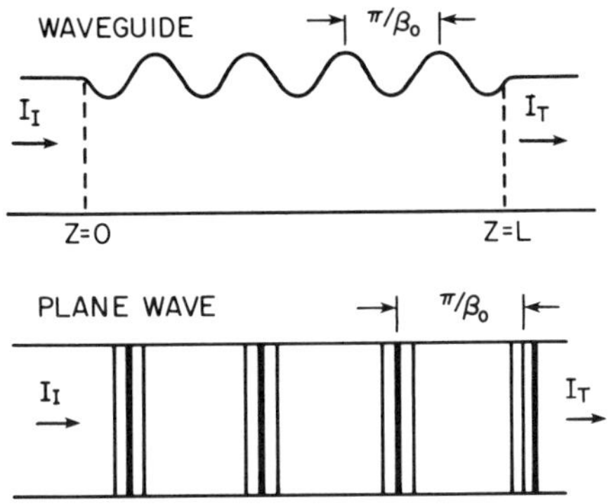

Fig. 3.10-1. Geometry for distributed feedback nonlinear bistable optical device: a) waveguide geometry, b) plane-wave geometry. Garmire (1981b).

Winful, Marburger, and Garmire (1979) have solved theoretically for the transfer function of the nonlinear distributed feedback structure with a uniform sinusoidal grating. The theory begins with Maxwell's equations for the forward and backward traveling waves. By making the plane-wave slowly varying envelope approximation, they obtain two coupled differential equations for the forward and backward waves. The solution to these equations are (analytic) Jacobian elliptic functions, which can be used to calculate numerically the transfer functions for the nonlinear distributed feedback bistable optical device. The results are shown in Figs. 3.10-3 and 3.10-4 for different values of the bias condition ($\Delta\beta L$) and the coupling constant of the grating (KL). By way of comparison, a nonlinear Fabry-Perot cavity with roughly the same reflectivity shows a very different transfer curve (Fig. 3.10-4).

Nonlinear properties of long-range surface plasmons have been studied for possible integrated optics applications [Wysin (1980); Wysin, Simon, Deck (1981); Sarid (1981a,c); Sarid, Deck, and Fasano (1982); Stegeman (1982a,b,c,d); Stegeman, Burke, and Hall (1982); Fasano, Deck, and Sarid (1982); Sarid, Deck, Craig, Hickernell, Jameson, and Fasano (1982); Craig, Olson, and Sarid (1983)]. A surface plasmon that is excited on the interface separating a metal and a nonlinear semiconductor has a propagation constant that depends on the power carried by the plasmon. Stegeman (1982a,b,c) compares the surface-plasmon approach with single and multilayer thin-film optical waveguides and finds the single-layer optical guide best. Sarid and Stegeman (1981) note that the nonlinearity may occur by means of the evanescent tail of the guided wave, so losses and propagation distances can be tailored. Furthermore guided wave devices circumvent transverse complications because the complete wavefront travels with a constant phase velocity, so distortions in both transverse dimensions can be eliminated by using channel waveguides.

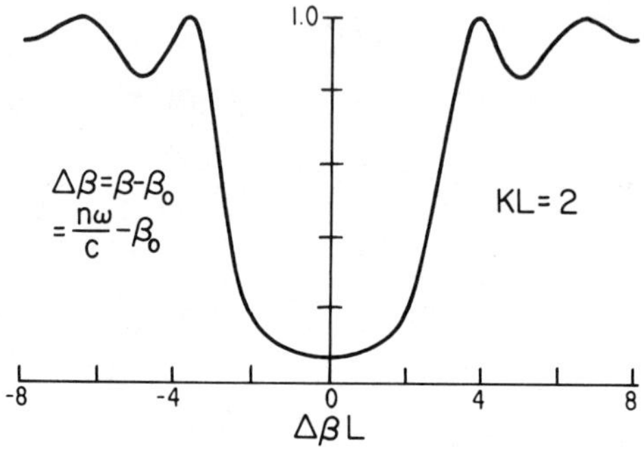

Fig. 3.10-2. Transmission of the linear distributed feedback structure as a function of wavevector mismatch times length. Garmire (1981b).

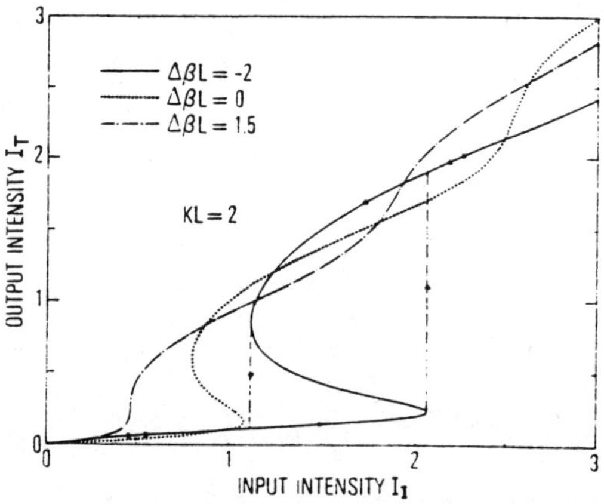

Fig. 3.10-3. Transfer curve for the nonlinear distributed feedback structure. Winful, Marburger, and Garmire (1979a).

Optical bistability from surface plasmon excitation has been reported by Martinot, Koster, Laval, and Carvalho (1982). See also Koster, Martinot, Pardo, Laval, Paraire, and Neviere (1984) and Martinot, Laval, and Koster (1984). When a surface plasmon wave is excited by attenuated total reflection in the Kretschmann configuration as shown in Fig. 3.10-5, the reflection coefficient exhibits a deep minimum for a given incidence angle corresponding to the resonance excitation of the plasmon wave. When the medium in which the excitation light propagates has a refractive index whose value is related to input intensity, bistability can be observed. Assume that the refractive index of the input medium is temperature dependent. If, for a low input light power, the incident angle is such that the

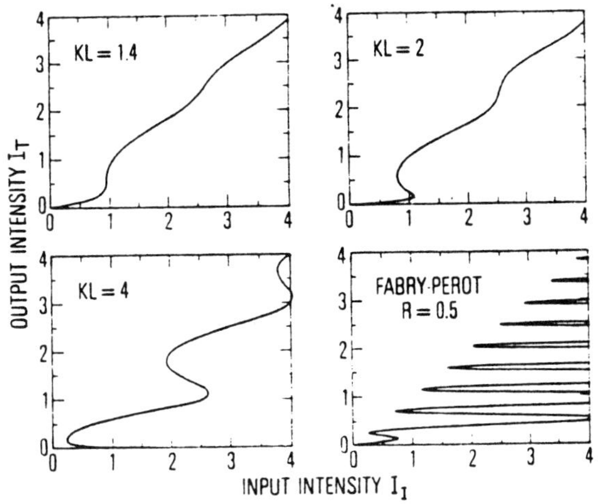

Fig. 3.10-4. Transfer curves for a nonlinear distributed feedback bistable optical device for different coupling efficiencies and for a nonlinear Fabry-Perot cavity. Winful, Marburger, and Garmire (1979a).

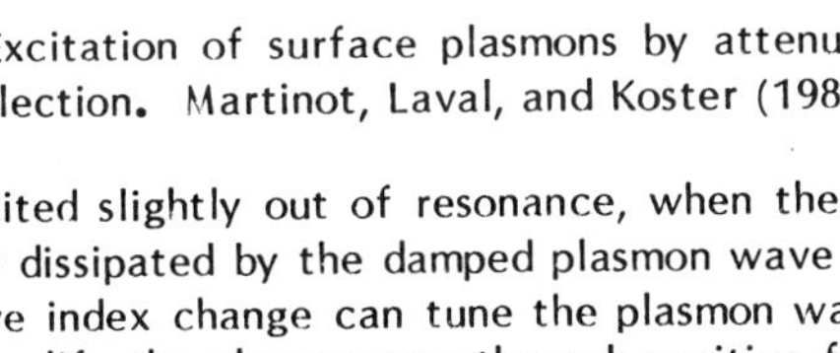

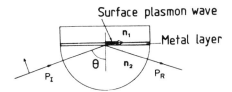

Fig. 3.10-5. Excitation of surface plasmons by attenuated total reflection. Martinot, Laval, and Koster (1984).

plasmon wave is excited slightly out of resonance, when the incident intensity is increased, the power dissipated by the damped plasmon wave heats the input medium. The refractive index change can tune the plasmon wave to resonance and this contributes to amplify the phenomenon through positive feedback, resulting in bistability. See Martinot, Koster, Laval, and Carvalho (1982) for a graphical construction and experimental demonstration of surface plamson bistability. They use 550 Å of silver as the metal layer and CS_2 as the temperature-dependent medium. The incident beam from a cw rhodamine 6G dye laser ($\lambda \simeq 600$ nm) is focused onto the silver layer at the center of a half-spheric cell that contains CS_2 and whose temperature is regulated at 15°C. When the incident power is varied, bistable characteristics are obtained; see Fig. 3.10-6. Bosacchi and Narducci (1982, 1983) propose bistability for a closely related case, the frustrated-total-reflection optical cavity. This is a resonant thin-film structure into which light is coupled by frustrated total reflection. See Section 3.10.4.

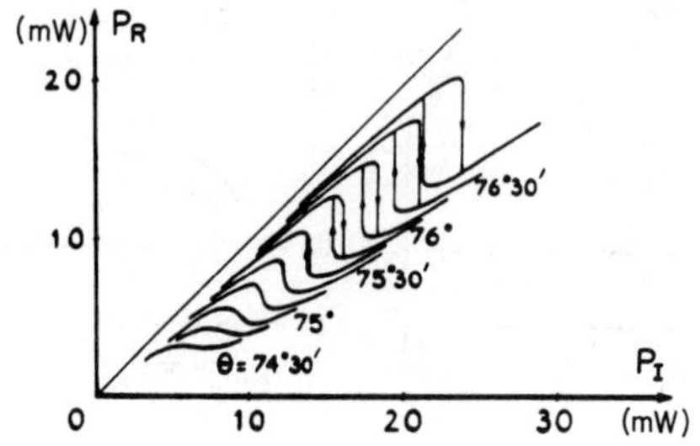

Fig. 3.10-6. Reflected vs incident power of a cw rhodamine 6G dye laser beam for different values of the incident angle θ. (The holder temperature is fixed at 15°C.) Koster, Martinot, Pardo, Laval, Paraire, and Neviere (1984).

An intensity-dependent hysteresis in the transmission of guided waves through a thin-film waveguide with a cladding having an intensity-dependent refractive index has been reported by Vach, Seaton, Stegeman, and Khoo (1984). A bead of liquid crystal MBBA was placed on the surface of a glass (n_f = 1.61, h = 1 μm)-on-glass (n_s = 1.52) thin-film waveguide. TE_0 and TE_1 modes were coupled in separate experiments into the waveguide with strontium titanate ($SrTiO_3$) prisms, guided through the region with the liquid crystal on top, and coupled out with a second $SrTiO_3$ prism. In both cases, the transmission through the system was typically 1% because of coupling inefficiencies, reflections at the two liquid-crystal boundaries, and scattering and absorption losses. For the TE_0 case, the transmitted power was linear in the incident power with some saturation effects evident at the highest power levels investigated (≃ 300 mW). In the TE_1 case, there was a pronounced saturation effect as well as hysteresis with ≃ 1-s time constant. The data are in good qualitative agreement with theory. For increasing input power in the TE_1 case the field is localized in the film giving higher transmission than when it is localized in the lossier liquid crystal. Liao and Stegeman (1984) examine just how efficient the coupling into the propagating modes of a power-dependent thin-film waveguide can be.

Bistability is not necessary, of course, for many logic operations. Jensen (1980b, 1982b) describes a nonlinear coherent coupler in which two waveguides are coupled by a nonlinear material. For pipelining, i.e., serial processing of two data streams, through an exclusive OR gate [Haus, Ippen, Lattes, and Leonberger (1982) and Lattes, Haus, Leonberger, and Ippen (1983)] or pulse compressor [Kitayama and Wang (1983)], the transit time is irrelevant; only the medium response time is significant. Maier (1982) proposes optical transistors and bistable devices using coupled waveguides by feeding back some of the light from one of the outputs into one of the inputs.

Normandin and Stegeman (1979, 1980) [also Stegeman and Normandin (1981)] demonstrate the convolution of picosecond optical signals by means of nonlinear mixing in a Ti in-diffused lithium niobate ($LiNbO_3$) waveguide. As further applications, they suggest signal convolution, time compression, inversion, and transient digitization on a picosecond time scale.

For optical logic and computation using guided-wave electro-optic devices see Taylor (1978). Chang and Tsai (1983) report an electro-optic analog-to-digital converter using a channel waveguide Fabry-Perot modulator array.

3.10.6. Mirrorless Optical Bistability

Bowden and Sung (1979, 1980, 1981) and Bowden (1981) propose mirrorless optical bistability for which the essential feedback is from the atomic pair correlation, i.e., from the inter-atomic correlation by means of the internal electromagnetic field (virtual photon exchange interaction). See also Drummond and Carmichael (1978); Narducci, Feng, Gilmore, and Agarwal (1978); Hassan, Bullough, Puri, and Lawande (1980); Puri, Lawande, and Hassan (1980). Abram and Maruani (1982) propose mirrorless semiconductor bistability due to the self-field changes in the dielectric function, but Sarid, Peyghambarian, and Gibbs (1983b) show that the branches of the dielectric function giving rise to the hysteresis are unphysical; see also Rozanov (1981a). Hopf, Bowden, and Louisell (1984) discuss the attainment of mirrorless bistability by way of local-field corrections; bistability is limited to cases in which the medium is so highly absorbing that it is opaque. This seems to be an example of the class of mirrorless bistability referred to as increasing absorption optical bistability (Section 2.10).

Winful (1982b) predicts mirrorless bistability for cholesteric liquid crystals arising from light-induced distortions of the cholesteric helix for waves in the Bragg regime. In distributed feedback bistability (Section 3.10.5), intense light changes the refractive index of a grating whose period remains fixed. For the cholesteric, the effect of the strong field is to alter its period.

Mirrorless optical bistability has been seen by cross trapping (Section 3.9.1) and phase conjugation (Section 3.10.3).

Emel'yanov and Zokhdi (1980) predict bistability and hysteresis when a laser causes changes in the static Lorentz field by redistributing the populations of the ground and excited states having different dipole moments.

3.10.7. Miscellaneous

Optical bistability has even been seen "without atoms"; that is, radiation pressure and gravity acting on a suspended lightweight mirror of a Fabry-Perot cavity can lead to dispersive optical bistability. See Dorsel, McCullen, Meystre, Vignes, and Walther (1983, 1984) and McCullen, Meystre, and Wright (1984).

Kaplan (1982b, 1984) has proposed bistable cyclotron resonance ($\lambda \simeq 1$ mm) based on a relativistic nonlinearity; the weak relativistic mass effect of an electron can result in bistability of the stationary orbit and kinetic energy of the electron. He suggests experiments using both free electrons and semiconductors (InSb in $\simeq 140$ kG using 240 W/cm^2 of 10.6-μm radiation). Cyclotron resonance bistability has recently been observed by Gabrielse, Dehmelt, and Kells (1985).

Bistability has also been studied with microwaves: observed in NH_3 by Gozzini, Maccarrone, and Longo (1982a,b), Arimondo, Gozzini, Lovitch, and Pistelli (1981), and Barbarino, Gozzini, Maccarrone, Longo, and Stampacchia (1982), and predicted in ferromagnets by Siva Kumari and Shrivastava (1982). Meier, Holzner, Derighetti, and Brun (1982) report bistability and chaos in nuclear magnetic resonance using ^{27}Al spins in ruby. The voltage across some suitably prepared Josephson junctions interacting with an external coherent radiation also shows hysteresis [Milani, Bonifacio, and Scully (1979); Shenoy and Agarwal (1980, 1981); Shenoy (1983)]. Wedding and Jäger have studied bistability of a radio-frequency electrical analog of a Fabry-Perot interferometer with a quadratic nonlinear medium.

Jain and Pratt (1976) propose an optical transistor based on the strong dependence of second harmonic generation on the phase-matching condition. A weak incoming signal can perturb the indices of refraction in a birefringent material so that the second harmonic power derived from a pump beam is significantly affected. They suggest Te as a suitable material, and Johnson and Pratt (1978) compute the termination of the phase-matching condition by free carriers generated by two-photon absorption to generate an ≃ 1-ps pulse at 5.3 μm with a peak intensity of 100 MW/cm². Quint, Johnson, Jain, and Pratt (1979) observe optical modulation of a 10.6-μm pump using a 5.3-μm signal; free carriers can be excited across the 0.33-eV bandgap of Te by absorption of two signal photons or of one signal photon and one pump photon. They observed a gain of 1.15 and conclude that an optical gate device employing a simple crystal of Te is not feasible.

Lugovoi (1979e) predicted bistability due to mutual mode quenching in optical second harmonic generation. Optical bistability, photon antibunching, spectra, and photon statistics for second and subharmonic generation in an intracavity crystal are treated theoretically by Drummond, McNeil, and Walls (1979, 1980).

Lugovoi (1978, 1979a,b,c,d, 1981, 1983) has discussed bistability, limiting, transistor action, and self-pulsing of a Fabry-Perot cavity containing quadratic nonlinear medium or a Cerenkov-type parametric oscillator. He obtained similar results much earlier for a resonator filled with a Raman medium (Lugovoi, 1969, 1977, 1983).

Armstrong (1979) and Baker and Armstrong (1981) extend the model of Bonifacio and Lugiato (1976) by adding a second light source able to ionize an atom in the upper level of the two-level atoms; since the ionization rate depends upon the excited state population, it can be used instead of fluorescence or transmission to monitor the bistability.

Litfin (1982) has operated both hybrid and intrinsic Fabry-Perot bistable devices based on orientational bleaching of F_A(II) color centers. "The ionic configuration of the center changes during the optical pumping cycle to form a saddle point configuration and back to its normal configuration." See Fig. 3.10-7. "However, there is a 50% probability that the halide ion separating the two wells will move into the original F_A center vacancy upon completion of the optical pumping cycle. If this happens the F_A center axis will have changed the orientation by 90°. Pumping the long wavelength absorption band with a pump polarisation along the z-direction, this center with its new orientation will no longer be able to absorb the pump light.... To demonstrate optical bistability the crystal was illuminated with two beams of the same wavelength (usually 632.8 nm) having different angles of incidence and perpendicular polarisation. The first beam that determines the initial distribution of the centers (distributing beam) is properly oriented to pump simultaneously the z and y directions but not the x direction. The second beam (probe beam) propagates in y direction and is polarised parallel to x, in order to excite only centers oriented in the x direction. If the crystal is exposed only to the distributing beam all centers will be oriented in the x direction. With increasing intensity of the probe beam the concentration of centers in the x direction is reduced and thus the extinction for the probe light is reduced resulting in the required nonlinear absorption."

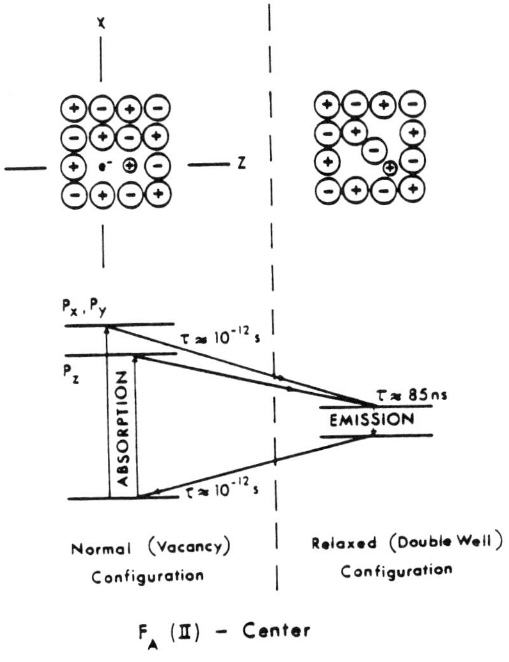

Fig. 3.10-7. Optical pumping cycle of F_A(II) centers. Litfin (1982).

CHAPTER 4

HYBRID OPTICAL BISTABILITY EXPERIMENTS

4.1. KASTAL'SKII'S PROPOSAL

Kastal'skii (1973) proposed an optoelectronic circuit consisting of a laser beam, Fabry-Perot cavity, detector, and electrical feedback to an intracavity electro-optic crystal; see Fig. 4.1-1. His analysis showed that this system could be used to generate a current instability or to stabilize the intensity of a laser beam. He pointed out that the frequency of peak transmission of a Fabry-Perot cavity can be shifted by changing the index of refraction by applying an electric field to various semiconductors and that high-speed modulation of a light beam by an electrical signal can be accomplished in this manner.

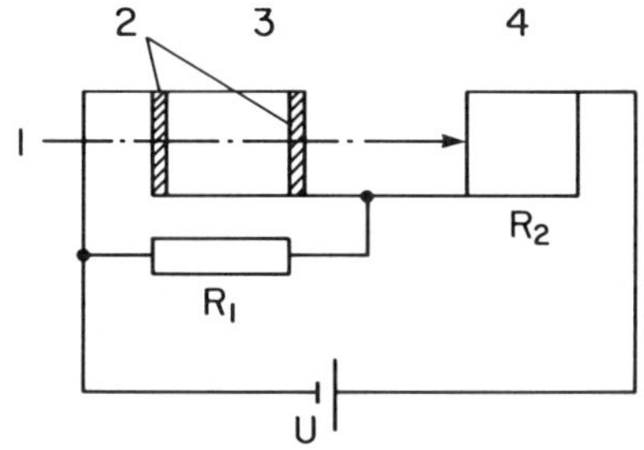

Fig. 4.1-1. Optoelectronic circuit. (1) Laser beam; (2) reflective coatings; (3) resonator with intracavity electro-optic crystal; (4) photoresistor. Kastal'skii (1973).

He assumed that the resistance of the photoresistor R_2 in Fig. 4.1-1 was determined completely by the transmitted light intensity. Starting out with the laser frequency coincident with a Fabry-Perot peak and applying a voltage, "the potential across R_1 alters the index of refraction n and thus, reduces the transmitted light intensity I_T. The photoresistor R_2 increases in value and takes up a large fraction of the applied voltage, so that the transition from $I_T = I_0$ to $I_T = I_{min}$ becomes extended in voltage. When the point $I_T = I_{min}$ is reached, the system becomes unstable and switching takes place in both current and voltage. The current-voltage characteristic $i(U_2)$ therefore has an S-type region of negative differential resistance.... This situation repeats periodically as the voltage is further increased." So Kastal'skii described purely dispersive hybrid optical bistability in which the hysteresis was mapped out by tuning the cavity rather than by modulating the intensity of the input light. He further predicted multistability. His multistable hysteresis curve is just a schematic plot of current versus voltage; he does not plot I_T versus I_I. He did analyze the optical limiter and gave parameters for which $\Delta I_T \approx \Delta I_I/20$. So Kastal'skii did not report any bistability experiments, but his proposed device was clearly an example of a purely dispersive hybrid bistable optical device. He did not propose intrinsic bistability, since the index changes were always by means of an electric field obtained from a light detector.

Kobayashi, Sueta, Cho, and Matsuo (1972) constructed a high-repetition-rate optical pulse generator that employs a Fabry-Perot electro-optic modulator as the output coupler of a laser/resonator. Bistability was neither predicted nor reported, but this hybrid cavity-dumper Fabry-Perot modulator would have functioned like the intrinsic saturable Fabry-Perot dumper of Bjorkholm (1967) and Szöke (1972) [see Section 1.4] if an output signal had been fed back to the dumper.

4.2. SMITH-TURNER HYBRID FABRY-PEROT BISTABLE DEVICE

Unaware of Kastal'skii's closely related theoretical work, P. W. Smith and Turner (1977a,b,c) and P. W. Smith (1978) proposed and demonstrated a hybrid bistable device consisting of a Fabry-Perot resonator with electrical feedback to an intracavity electro-optic crystal; see Fig. 4.2-1. The introduction of a hybrid bistable optical device was a very significant step because: it provided an ideal plane-wave purely dispersive and well-controlled device that could be used to compare theories and experiments (see, for example, Sections 5.6, 6.1, 6.3, 6.5); it made bistability accessible to almost any laboratory (see Section 4.3) since weak laser power is compensated by electrical power and any laser wavelength transmitted by the modulator can be used.

One can readily see that this device can be understood as an example of purely dispersive optical bistability (Section 2.3). Inside the Fabry-Perot cavity is an electro-optic crystal whose index depends linearly on the transmitted light intensity, i.e., $\beta = \beta_0 + \beta_2 I_T$ as in Section 2.3. The analysis is exactly that of Section 2.3 where it is assumed that the time retardation effects in the cavity can be neglected. In Section 2.3 all fields were treated as plane waves. In Fig. 4.2-1 the output detector integrates over the transverse spatial profile and feeds back a single voltage proportional to $\int I_T(x,y)dxdy$, which impresses the same phase shift on all parts of the electric field regardless of the amplitude of the field at that point. Therefore, the hybrid device behaves as an ideal plane-wave device regardless of the transverse spatial profile of the incident beam. This is a great simplification when careful comparisons between experiments and uniform plane-wave theories are desired. Of course, if the beam incident on a Fabry-Perot cavity is too small, the angular spread will cause an averaging of the transmission function over angle, which affects even a hybrid bistable device [Rozanov (1980a)].

Figures 4.2-2 and 4.2-3 show that the device does exhibit bistability, differential gain, and limiting as the laser-cavity frequency difference is changed while the input intensity is being modulated. Note that Fig. 4.2-3 shows the strong asymmetry with Fabry-Perot tuning characteristic of dispersive bistability (Section 3.2 and Figs. 3.2-4 and 5).

P. W. Smith and Turner suggested a device using a waveguide modulator that they predicted would lower the required voltage sufficiently to be supplied by a photovoltaic detector directly, permitting a self-contained device (no external wires or power sources); see Section 4.4.

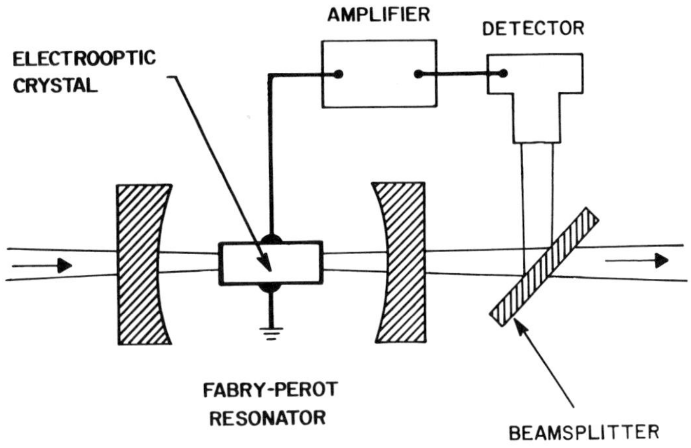

Fig. 4.2-1. Schematic of bistable Fabry-Perot device. The input was from a single-frequency He-Ne laser at 632.8 nm, the cavity was a 10-cm confocal resonator containing a KDP electro-optic modulator with half-wave voltage of 1200 V, the resonator mirrors had a transmissivity of 20%. The finesse of the entire system was ≃ 7. P. W. Smith and Turner (1977b).

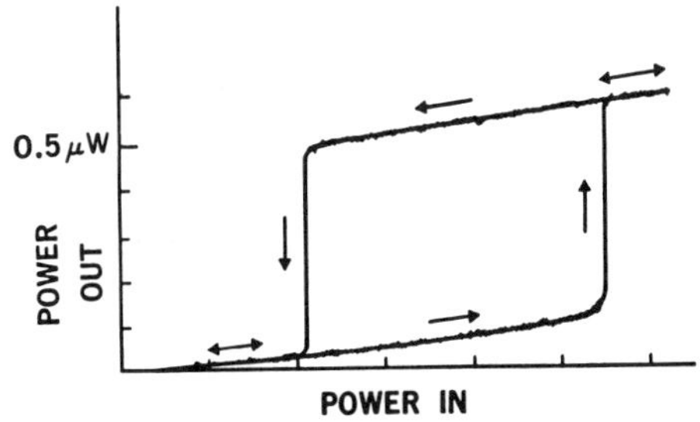

Fig. 4.2-2. Experimental data on optical power out vs optical power in for the device shown in Fig. 4.2-1. The arrows identify the portions of the plot followed for input power increasing and decreasing. The switching occurs with less than 1 μW incident on the detector. P. W. Smith and Turner (1977b).

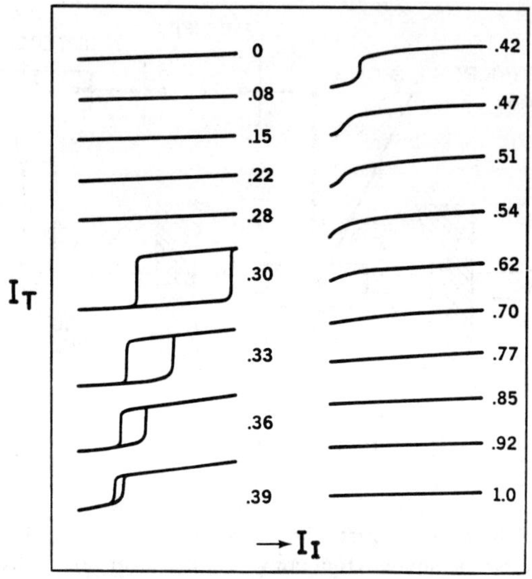

Fig. 4.2-3. Experimental data on output characteristic (power out vs power in) as a function of resonator tuning. The number indicates the tuning in units of a resonator period. P. W. Smith and Turner (1977b).

4.3. CAVITYLESS DEVICES; STUDENT EXPERIMENT

Marburger, Allen, Garmire, Levenson, and Winful (1978) and Garmire, Marburger, and Allen (1978) introduced bistability without mirrors. Feedback (intrinsic or extrinsic) is essential to bistability; in a hybrid device it is an external electrical signal. Any optical phenomenon that leads to an output signal that is a nonlinear function of a parameter that can be made proportional to the output will exhibit hybrid optical bistability. Elimination of the optical cavity removes the need for coherent light; any light source (multimode laser, LED, or incandescent bulb) with suitable stability characteristics can be used. This further simplification of bistable devices makes possible inexpensive simple systems for student use or for fundamental tests (Sections 5.6, 6.1, 6.3, 6.5) and may hasten the construction of practical waveguide devices.

Their incoherent mirrorless device is shown in Fig. 4.3-1; they observed the usual characteristic curves. Bistable operation and optical switch-on were dramatically evident. With a fixed laser intensity incident on the device between $I_\downarrow$ and $I_\uparrow$, switch-on was accomplished by shining a flashlight beam on the output detector and switch-off by interrupting the input.

To see that any device with a nonlinear transmission function $\mathscr{T}(V)$ can be used to construct a hybrid bistable device, refer to the graphical construction of Section 2.6. With feedback connected, $\mathscr{T}(V)$ must be satisfied, but so also must

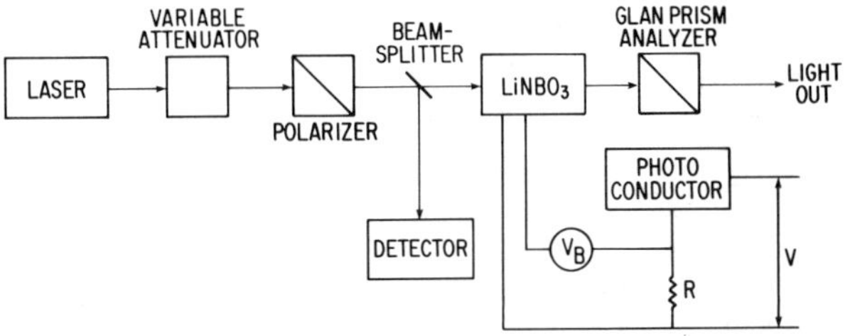

Fig. 4.3-1. Experimental apparatus for the incoherent mirrorless bistable optical device using an electro-optic polarization modulator. Garmire, Marburger, and Allen (1978).

$V = GI_T + V_B$, i.e., the voltage is related to the transmitted intensity by a gain G. So plot $\mathscr{T}(V)$ and the straight lines $I_T/I_I = (V - V_B)/GI_I$ versus V. Clearly, there exists, if $\mathscr{T}(V)$ is nonlinear, a bias voltage (fixing the intersection of the straight line with the V axis) and an input intensity or gain (determining the slope of the straight line) so that the straight line intersects $\mathscr{T}(V)$ at three places. That is the condition for bistability.

Garmire, Marburger, and Allen (1978) used a lithium niobate ($LiNbO_3$) c-cut crystal 1 x 1 mm^2 and 2-cm long with a half-wave voltage of 260 V. The transmission curve differed from the ideal form because of voltage-dependent nonuniformities of the local electric field. Commercial electro-optic modulators between crossed polarizers approach very closely the ideal sine-squared dependence.

In an experiment designed for use in a student laboratory, Greene, Gibbs, Passner, McCall, and Venkatesan (1980) use an electroded lead lanthanum zirconate titanate (PLZT) ceramic wafer bonded to glass-supported crossed linear polarizers (Motorola 9565). This wafer has the advantage of being commercially available and easy to use. Its disadvantages are its low 17% "on" transmission and its inherent hysteresis, which does not exhibit switching without feedback but does make highly quantitative comparisons with theories very difficult (Section 6.3). All of the components in the student apparatus of Fig. 4.3-2 are available commercially and inexpensively. Typical outputs are shown in Fig. 4.3-3. The flashlight beam switch-on and interrupt-input switch-off method is a fun way for the student to become convinced that two stable states exist for a constant input satisfying $I_\downarrow < I_I < I_\uparrow$. Feedback gain and bias controls allow one to completely simulate purely dispersive plane-wave intrinsic optical bistability. The low-power intrinsic bistability in a dye etalon or ZnS or ZnSe interference filter (Section 3.5) is an alternative student experiment.

4.4. DEVICES WITH WAVEGUIDE MODULATORS

P. W. Smith and Turner (1977b) proposed hybrid bistable devices employing waveguide modulators. Garmire, Allen, Marburger, and Verber (1978) and P. W. Smith, Kaminow, Maloney, and Stulz (1978, 1979) demonstrated such devices, and the latter group constructed a self-contained version free of electrical power. Cross, Schmidt, Thornton, and Smith (1978) constructed a hybrid bistable device

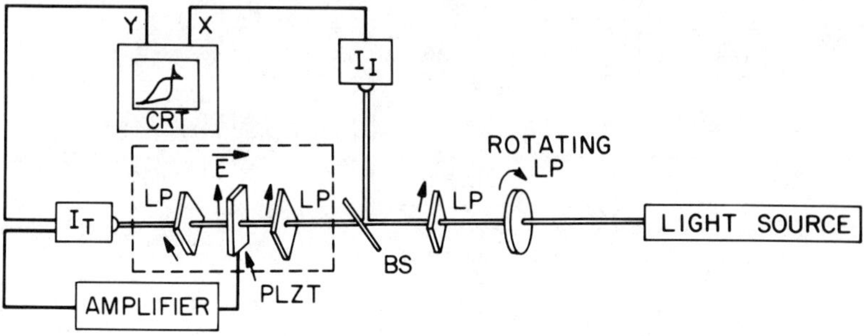

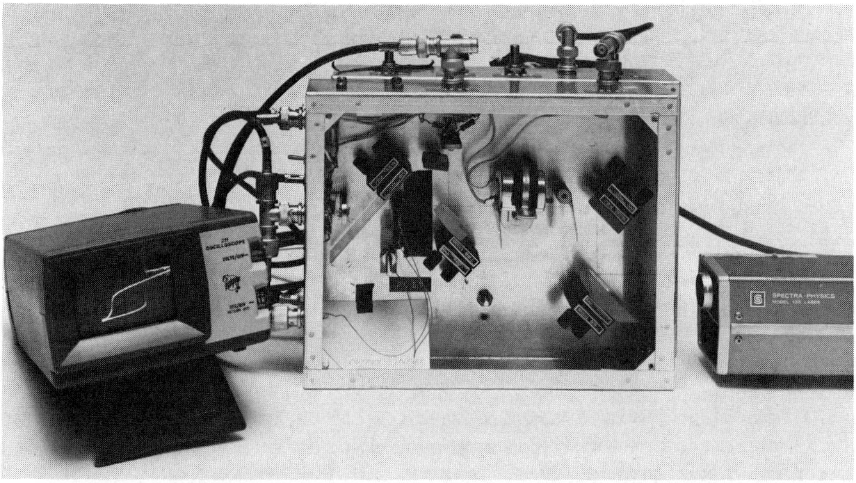

Fig. 4.3-2. Simple apparatus for observing optical bistability (top) and its schematic. The light source is a linearly polarized multimode He-Ne laser. LP is a linear polarizer, BS is a beamsplitter, and PLZT is a piezoelectric crystal in which an electrical feedback signal modulates the transmission through electric field-dependent index changes. The Motorola 9565 bonded wafer consists of the PLZT and LP's within the dashed box. I_I and I_T are the input- and transmitted-light detectors connected to the oscilloscope CRT. Greene, Gibbs, Passner, McCall, and Venkatesan (1980).

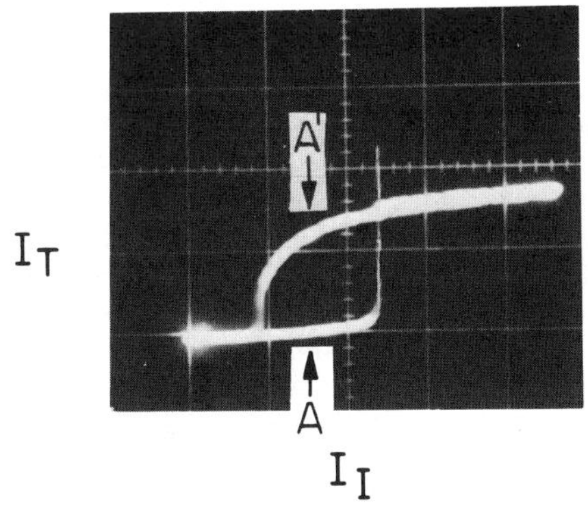

(a)

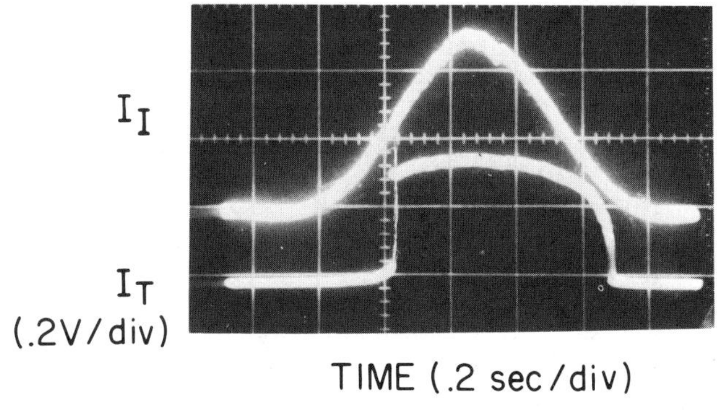

(b)

Fig. 4.3-3. Optical bistability observed with the apparatus of Fig. 4.3-2. Greene, Gibbs, Passner, McCall, and Venkatesan (1980).

that uses a waveguide directional coupler switch. These devices have been called integrated bistable optical devices, but they do not deserve that name. The modulator and coupler are waveguides, but nonplanar components (lens, detector, wires, etc.) are essential parts of the bistable device. All of these devices are huge in size (approximately centimeters) compared with the GaAs etalon (≃ 5 μm). The latter does not deserve to be called "integrated" either because the laser beam travels perpendicular to the plane rather than in the plane, but it shows that sufficient nonlinearity can occur in a very short length. Feedback and interconnections must occur in the plane before true integration can be claimed. These first two waveguide devices were significant in achieving low-voltage hybrid devices and in pointing toward truly integrated devices.

The devices of P. W. Smith, Kaminow, Maloney, and Stulz (1977, 1978, 1979) are shown in Figs. 4.4-1 and 2. In Fig. 4.4-1 the waveguide width was 4 μm, the length typically 1.5 cm, and the electrode separation 9 μm. The Ti in-diffused waveguide was oriented to be perpendicular (within 1/4°) to the cleavage plane of the x-cut $LiNbO_3$ crystal. The crystal ends were cleaved to produce parallel surfaces and coated to R = 0.5. They also report a hysteresis without electronic feedback, but without any hint of switching, arising from photorefraction in $LiNbO_3$ (relaxation time of seconds).

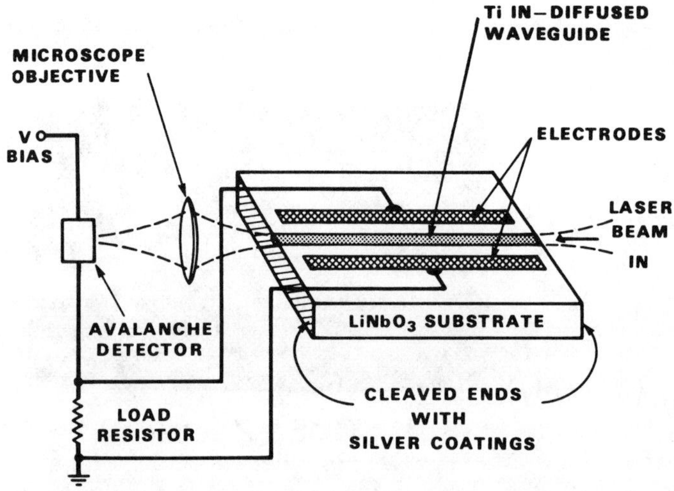

Fig. 4.4-1. Schematic view of device using waveguide modulator. P. W. Smith, Kaminow, Maloney, and Stulz (1978).

Figure 4.4-2 shows the self-contained device in which 96 miniature solar cells in series provide ≃ 40 V, enough to drive the 12-mm-long $LiNbO_3$ waveguide modulator directly. Typical operating conditions were 50 μW at 647.1 nm at the detector, R = 200 MΩ, C ≃ 50 pF. Thus RC = 10 ms is the switching time, and the 0.5 μJ is an indicator of the switching energy. By using an off-axis waveguide configuration to take advantage of the three-times-larger extraordinary-wave electro-optic coefficient, they achieved self-contained operation using a

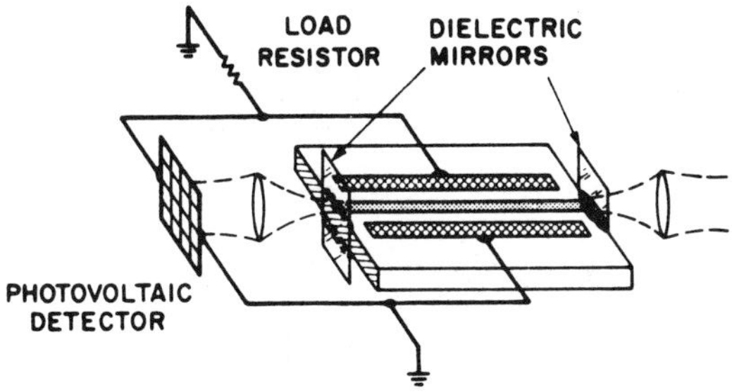

(a)

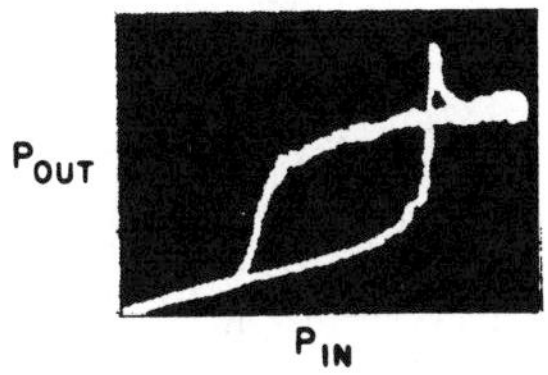

(b)

Fig. 4.4-2. (a) Schematic view of self-contained bistable device using a waveguide modulator. (b) Experimental optical hysteresis observed with the device shown in (a). P. W. Smith, Kaminow, Maloney, and Stulz (1979).

21-element photovoltaic detector for $\geq$ 2.5 μW at the detector and RC = 4 ms (RCP_T = 6 nJ). With an externally powered avalanche photodiode detector, RC = 50 ns and RCP_T = 0.5 pJ.

The directional coupler device is shown in Fig. 4.4-3a; the power incident on the two input ports exits from two distinct output ports, so that all-optical data processing or remote optical switching can be performed. Two single-mode strip waveguides forming a directional coupler are fabricated in $LiNbO_3$ by Ti in-diffusion. Multisection electrodes are then deposited on the waveguides and are connected so that the direction of the applied electric field reverses from section to section. The alternating electric field induces an alternating phase-mismatch between the waveguides that allows low-crosstalk switching to be achieved without stringent fabrication tolerances. The device used had an intersection length

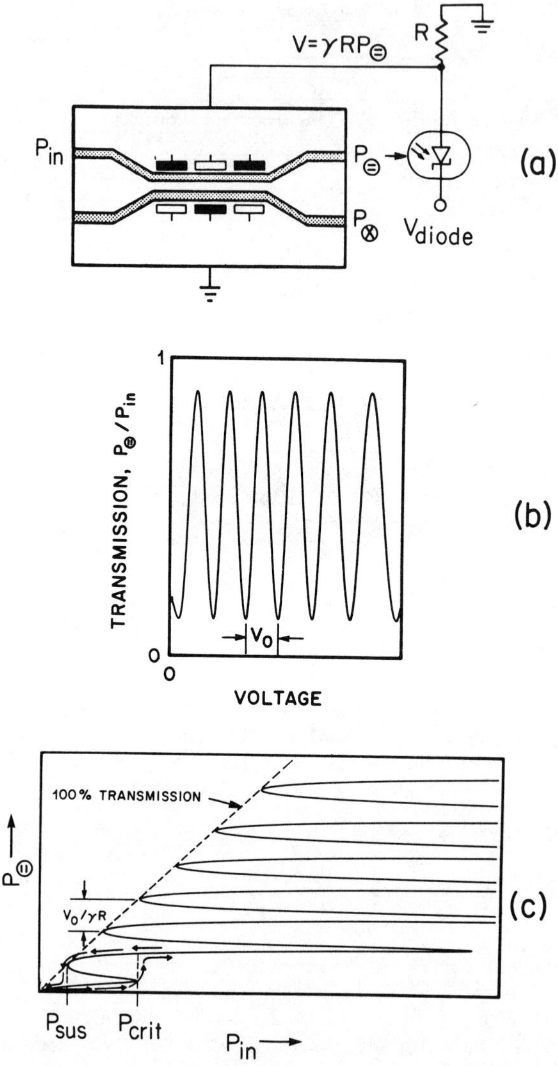

Fig. 4.4-3. (a) Waveguide directional coupler bistable device. (b) The calculated transmission as a function of drive voltage (no feedback) for a device with 12 sections of alternating $\Delta\beta$ and with intentionally poor zeros. (c) Transmission with feedback. Cross, Schmidt, Thornton, and Smith (1978). © 1978 IEEE

of ≈ 2 cm and 12 sections of alternating $\Delta\beta$. The voltage difference V_0 between adjacent transmission minima is less than 4 V; see Fig 4.4-3b. On/off ratios as high as 12 dB have been observed. A switching time of ≈ 300 μs and switching energy of ≈ 3 pJ were achieved for conditions yielding optical bistability. Figure 4.4-3c shows calculated hysteresis loops using Fig. 4.4-3b, and Fig. 4.4-4 shows experimental bistability and high gain. Schnapper, Papuchon, and Puech (1981) have analyzed and observed remote switching of an integrated directional coupler (Section 5.5). Tarucha, Minakata, and Noda (1981) have studied bistable state switching and optical triode operation using a directional coupler switch.

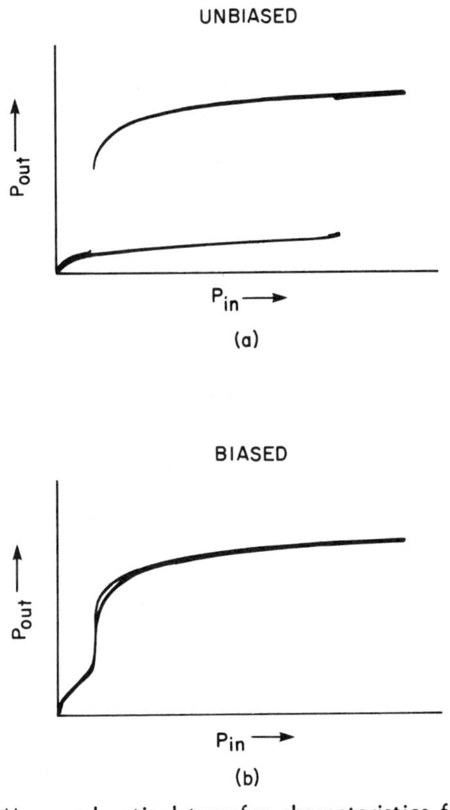

Fig. 4.4-4. Measured optical transfer characteristics for (a) a device with no electrical bias and (b) a device electrically biased past the first transmission minimum in Fig. 4.4-3b. Cross, Schmidt, Thornton, and Smith (1978). © 1978 IEEE

Hybrid bistability has been seen with an integrated optical modulator with an induced dielectric channel by Aksenov, Kukhtarev, Lipovskii, and Pavlenko (1983). The modulator operates by reflecting a light beam from a region of a planar optical waveguide in which the refractive index is reduced by an applied electric field. An acousto-optic waveguide modulator (sputtered tantalum pentoxide

(Ta_2O_5) on a glass substrate) has been used by Albert, Vincent, and Tremblay (1981) to observe hybrid bistability with less than 5 mW of acoustic power; also see Jerominek, Pomerleau, Tremblay, and Delisle (1984).

Carenco and Menigaux (1980) have observed bistability with the aid of an external amplifier using a directional coupler and a detector monolithically integrated in GaAs. Guglielmi and Carenco (1981), again with an external amplifier, see bistability with a monolithically integrated directional coupler and detector in $LiNbO_3$. One of the output channels of the directional coupler is made absorbing by using a metal cladding, thus forming a pyroelectric detector.

4.5. SURVEY OF OTHER HYBRID EXPERIMENTS

Hybrid bistable systems place few demands on the light source and optics; any sophistication is usually in the electronics or in the light-electronics interface device. Consequently, there have been many hybrid bistability experiments. They can be divided roughly into two types; those that study some bistability phenomenon such as regenerative pulsations or critical slowing down and those that extend the light-electronics interface systems. P. W. Smith (1980) has given a review of hybrid bistable optical devices and discussed their potential for optical signal processing.

As discussed before, a hybrid bistable system is an exact analog of a plane-wave purely dispersive intrinsic bistable system (neglecting intra-cavity time delay effects), so many of the hybrid experiments are very helpful in understanding the behavior of ideal intrinsic devices. P. W. Smith, Turner, and Maloney (1978) have shown pulse shaping (Section 5.2), differential gain (Section 3.2), optical triode (Sections 3.7 and 5.4), and analog-to-digital (multistable; see Fig 1.2-2) operations using the Smith-Turner apparatus of Fig. 4.2-1 or simple modifications of it. Multistability has been studied in mirrorless hybrids by Okada and Takizawa (1979a,b). P. W. Smith, Turner, and Mumford (1978) and P. W. Smith and Turner (1978) have used reflected light feedback (see Figs. 4.5-1 and 2). Feldman (1978a,b, 1979a,b) has used a Pockels cell mirrorless hybrid, sometimes in double-pass. With feedback from both the input and output laser beams, he has obtained an ultralinear modulator. Normally optimum differential gain occurs at a single value of the input intensity and is sensitive to changes in the dc components of the input. By using a linear combination of input and output intensities to determine the modulator voltage, the optimum operating point is achieved automatically, independent of the input power. Okada and Takizawa (1979c) have studied this scheme in further detail, both experimentally and theoretically. The intrinsic equivalent to obtaining part of the hybrid voltage from the input must be sending some of the input beam into the cavity from the side so that it affects the intra-cavity index constantly irrespective of what the cavity transmission is doing. McCall (1978) predicted regenerative pulsations and used a piezoelectric Fabry-Perot cavity with two feedback signals of opposite polarity and different time constants (see Section 6.1). Several other hybrid studies of regenerative pulsations have been made: Ito, Ogawa, and Inaba (1980, 1981); Okada (1980a,c); Okada and Takizawa (1980); Okada and Takizawa (1981a); Sohler (1980). Chrostowski and Delisle (1979a,b) also used a piezoelectric Fabry-Perot cavity and showed very asymmetrical nonlinear Fabry-Perot scans as in Figs. 3.2-5 and 3.2-15 and 3.5-2. Chrostowski, Delisle, Vallee, Carrier, and Boulay (1982) study multistability arising when the laser has several closely spaced modes. Garmire,

Marburger, Allen, and Winful (1979) have studied critical slowing down (Section 5.6); the response of the bistable system becomes sluggish in the vicinity of the phase-transition or switching points. Okada (1980b) has reduced the effect of critical slowing down by using two feedback voltages with different time constants to make the device appropriately unstable. Okada (1979) has studied the dynamical characteristics of the output of an electro-optic mirrorless hybrid device as a function of phase retardation (varied by applying a triangular voltage signal with an amplitude of one-wave voltage to the electro-optic crystal). This method of observing bistability is similar to scanning the Fabry-Perot cavity in the intrinsic case (Section 3.2). An electro-optic mirrorless hybrid with a delay added in the feedback has been used to observe Ikeda instabilities (Section 6.3) by Gibbs, Hopf, Kaplan, and Shoemaker (1981).

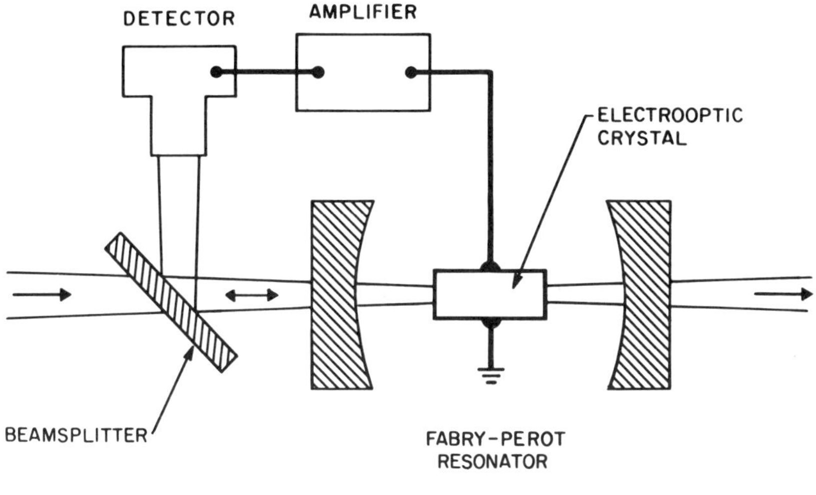

Fig. 4.5-1. Schematic diagram of hybrid bistable Fabry-Perot device using reflected light feedback. P. W. Smith, Turner, and Mumford (1978).

Several papers on hybrid optical bistability have extended the concept to new systems. Marburger, Allen, Garmire, Levenson, and Winful (1978) detect the second harmonic signal generated in $LiNbO_3$ with 1.5-μm laser light incident and feed it back electro-optically to control the phase mismatch. Mirrorless bistability was observed by Stark tuning NH_3 molecules by Pepper and Klein (1979) and Klein (1980). This work differs in two important respects from most hybrid experiments. Primarily, the light-electronics interface, the Stark cell, employs a resonance between the input light and a molecular transition; this requires a narrowband input and leads to molecular saturation effects at high intensities. Secondly, the mechanism is essentially absorptive; bistability can be seen even when the transmission is completely independent of the refractive index. This

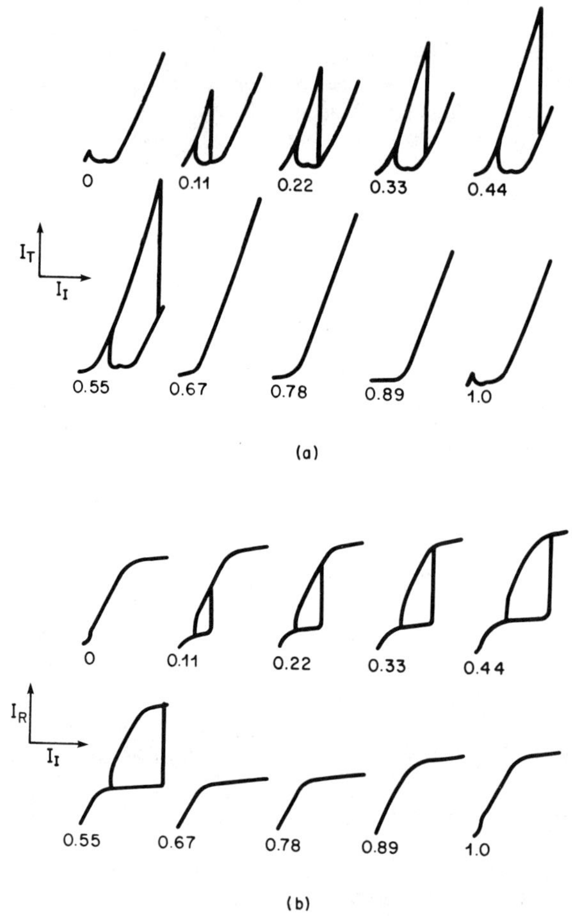

Fig. 4.5-2. Experimental plots of (a) transmitted and (b) reflected power versus incident power for the case in which the modulator is driven by the reflected power. The numbers refer to the resonator tuning in units of the resonator mode spacing. P. W. Smith, Turner, and Mumford (1978).

mirrorless bistability seems to be closely related to increasing absorption optical bistability even though their arrangement results in switching to lower absorption with increased input; see Section 2.10. Bourdet (1981) has performed similar experiments using two 200-MHz FWHM Stark components of monodeuterated ammonia (NH_2D). Papuchon, Schnapper, and Puech (1979) and Schnapper, Papuchon, and Puech (1979) have studied bistability in an integrated electro-optic two-arm interferometer. A bias voltage is placed across one arm and a voltage proportional to the output across the other. Ito, Ogawa, and Inaba (1979a,b, 1980, 1981) have studied bistability using similar integrated Mach-Zehnder interferometric switches. Sohler (1980) used an integrated optical cutoff modulator in $LiNbO_3$ that has a nonperiodic transmission characteristic as a function of applied voltage. Consequently bistability is seen, but no multistability. Hong-Jun, Jian-Hua, Jun-Hui, and Cun-Xiu (1981) use a liquid crystal (10-μm-thick phenylcyclohexane) as the electro-optic modulator for hybrid bistability; only a few volts were needed across the crystal. A magneto-optic bistable hybrid that employs a Faraday rotator has a high rotating power because of multiple reflections in a planar glass plate [Umegaki, Inoue, and Yoshino (1981)]; using the limiting action of bistability, they stabilize the output intensity of a He-Ne laser to 3.7×10^{-3} [Yoshino, Umegaki, Inoue, and Kurosawa (1982)]. Most of the hybrids have been operated in the visible, but Vincent and Otis (1981) use 10.6 μm from a cw CO_2 laser and a cadmium telluride (CdTe) electro-optic modulator. A novelty of their experiment is the way they obtain the essential nonlinear transmission $\mathcal{T}(V)$. The usual dc method would require several kilovolts across the CdTe modulator. Instead they connect a resonant circuit to the CdTe and use its tuning curve to produce the nonlinearity. The transmitted intensity is amplitude modulated; its envelope exhibits the bistability hysteresis when the output signal is demodulated and fed back to a voltage-controlled oscillator that drives the CdTe. Chrostowski and Delisle (1982) use an acousto-optic modulator and study the system's instabilities (Section 6.3). The use of electrorestrictive transducers for Fabry-Perot mirror control in a hybrid device has been demonstrated by Gomi, Uchino, Abe, and Nomura (1981), Nomura and Uchino (1982), and Gomi, Miyazawa, Uchino, Abe, and Nomura (1982). They found the operation of the electrorestrictive resonator to be more stable and reproducible than a piezoelectric device.

C. F. Li (1981) has used a silicon-controlled rectifier circuit as a nonlinear feedback amplifier in the crossed-polarizer device of Fig. 4.3-1 resulting in wider bistability loops. C. F. Li and Ji (1981) have demonstrated bistability using an ordinary Michelson interferometer with an electro-optic crystal in one arm. C. F. Li (1982) used the same principle in a waveguide device in which the output TE and TM modes interfered. Ogawa, Ito, and Inaba (1981, 1982) use the threshold characteristic of a diode laser and the saturation behavior of a photodetector to obtain bistable or differential gain depending upon the detector gain. They point out that their device is amenable to monolithic integration and could be as short as a few hundred micrometers compared with centimeter lengths of most hybrid devices. Monolithic integration of a bistable device consisting of a phototransistor and a light emitting diode or diode laser has been achieved by Grothe and Proebster (1983) and by Sasaki, Taneya, Yano, and Fujita (1984). Yajima, Sudo, Yumoto, and Kashiwa (1984) integrate an amorphous silicon photovoltaic detector and a $LiNbO_3$ Mach-Zehnder guided-wave modulator and modulate optically an external He-Ne laser with 5-μs response time. In related work, a train of 19-ps FWHM pulses has been produced at a 20-GHz repetition rate from the cw output of a He-Ne laser using a microwave-modulated optical Mach-Zehnder electro-optic waveguide interferometer fabricated in $LiNbO_3$ [Molter-Orr, Haus, and Leonberger (1983)].

Ryvkin (1981) proposes the use of the Franz-Keldysh effect to increase the absorption within an optical resonator consisting of a waveguide photojunction with reflecting ends, i.e., an ordinary heterolaser with a Fabry-Perot resonator. Light passing through the photojunction generates a photocurrent that changes the voltage across the photojunction. Under certain conditions an increase in the electric field and the corresponding increase in the absorption coefficient through the Franz-Keldysh effect can cause a decrease in the power absorbed and a corresponding decrease in the photocurrent proportional to this power. Bistability [Ryvkin and Stepanova (1982)] and oscillations [Ryvkin (1979)] can result.

A new type of optoelectronic device, a self-electro-optic-effect device (SEED) shown in Fig. 4.5-3, which uses the same GaAs/AlGaAs multiple-quantum-well material simultaneously as an optical detector and modulator, has been developed by D. A. B. Miller, Chemla, Damen, Gossard, Wiegmann, Wood, and Burrus

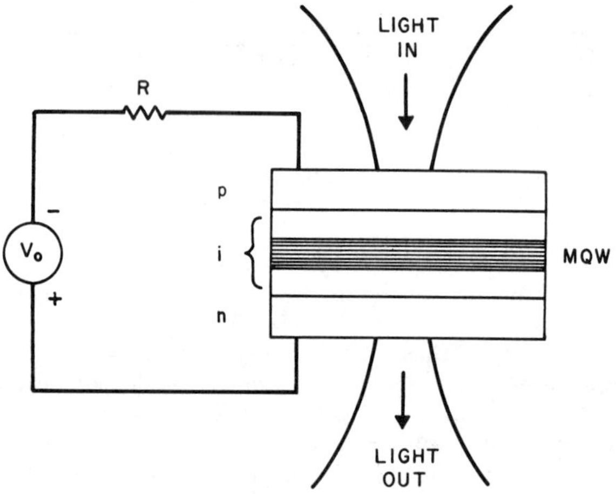

Fig. 4.5-3. Schematic of the quantum-well SEED. D. A. B. Miller, Chemla, Damen, Gossard, Wiegmann, Wood, and Burrus (1984).

(1984a,b); also see Robinson (1984) and D. A. B. Miller and Wood (1984). With a series resistor and constant voltage bias supply the SEED shows increasing absorption optical bistability (Section 2.10) that requires no mirrors. Bistability is seen at room temperature from ≃ 850 to 860 nm, at powers as low as 670 nW (1.5-ms switching time) or switching times as low as 400 ns (3.7-mW switching power) with ≃ 1-nJ optical switching energy in a 600-μm-diameter device. Total energies per unit area at ≃ 18 fJ/μm² were lower than for any previously reported bistable devices. The quantum-confined Stark effect permits the exciton resonances to remain resolved for shifts much larger than the zero-field binding energy and fields more than 50 times the classical ionization field.

P. W. Smith (1980) reviews hybrid bistable optical devices. Collins (1980b) edits a volume of Optical Engineering on Feedback in Optics, which contains several articles on hybrid bistability, especially on optical image processing. This chapter has dealt only with single discrete bistable devices; the great potential for parallel processing with two-dimensional arrays of bistable elements will be deferred to Section 7.4.

CHAPTER 5

OPTICAL SWITCHING: CONTROLLING LIGHT WITH LIGHT

One of the primary motivations for applied research in optical bistability is its potential for all-optical signal processing. Some of the functions already demonstrated experimentally were enumerated in Section 1.2 and Figs. 1.2-1 through 1.2-11. Clearly light beams cannot control one another directly. Some nonlinear optical material, i.e., a material whose transmission properties depend upon the light intensity of a control beam, is needed to transmit the information from a control beam to the main output beam. This chapter will describe various schemes for reshaping pulses and controlling optical beams, mostly by modifying the intracavity refractive index and scanning the transmission peaks of a Fabry-Perot interferometer.

The treatment of this book to this point has been largely steady-state. Only the steady-state solutions of the models in Chapter 2 were described. Some experimental switching times were reported in Chapters 3 and 4, and a comparison between an experiment and a theory of a transient absorptive Fabry-Perot cavity was given in Section 3.2.5b. Overshoot was discussed in Section 3.3. Instabilities, i.e., transient phenomena occurring with a constant input, will be discussed in Chapter 6. In this chapter the emphasis is upon the transmission changes resulting from a change in input intensity.

Figure 5.1-1 illustrates the basic concept of this chapter, namely the control of a light's beam intensity by scanning a Fabry-Perot interferometer. Scanning can be accomplished by a variety of external means, such as piezoelectric length control or refractive index control (by evacuating the space between the plates and leaking in a gas). Here the emphasis is upon scanning of the Fabry-Perot cavity by light. In a hybrid device this is done easily by applying the output voltage of a control-beam detector to an intracavity modulator or piezoelectric plate spacer. In intrinsic devices the tuning is by means of optical pathlength changes caused by the interaction of the signal beam itself or a second control beam with the intracavity nonlinear medium. This may be an instantaneous n_2 index effect or an accumulative index effect as in thermal or excitonic scanning. Or it might be a length change, for example by thermal expansion. Note that the control beam does not need to "see" a high-finesse cavity. Its wavelength can be selected so that the mirrors have no reflectivity for it. Or it can enter the Fabry-Perot resonator from the side and avoid the mirror altogether.

5.1. TRANSIENT NONLINEAR FABRY-PEROT INTERFEROMETER

When the Fabry-Perot cavity is scanned in a time shorter than or comparable to the cavity storage time, the steady-state treatment of Section 2.3 is no longer adequate. Here the treatment of Bischofberger and Shen (1978a,b) is adopted because of their very careful comparison and impressive agreement between

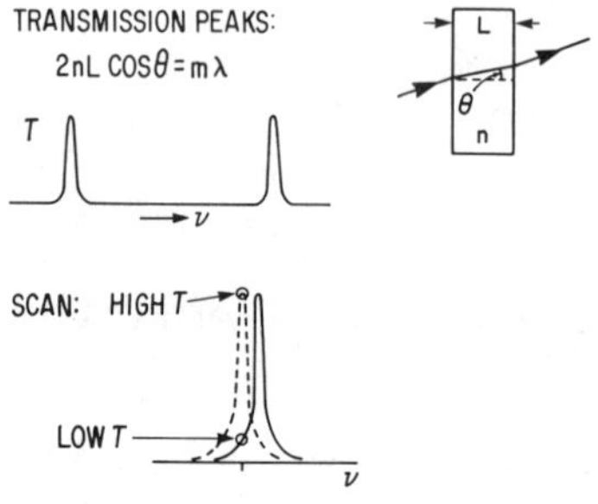

Fig. 5.1-1. Light transmission control by scanning a Fabry-Perot interferometer.

theory (described in this section) and experiment (described in Section 3.4). They use a purely dispersive model for their Kerr nonlinearities.

One of the ways to obtain the Airy function for the steady-state transmission of a Fabry-Perot cavity is to keep track of all the multiple reflections within the cavity and sum the infinite series for the transmitted electric field. In the transient case, the input light intensity and/or the intracavity optical pathlength may change before the infinite series can be approximated well by the closed-form sum. However, the same approach can be used numerically [Bischofberger and Shen (1979b)]:

$$E_T(t) = T e^{-i\omega t} e^{-\alpha_B L/2} \sum_{m=0}^{\infty} \xi_I\left[t - \left[m + \frac{1}{2}\right] t_R\right]$$

$$\times R_\alpha{}^m \exp\left[i\Phi\left[t, t - \left[m + \frac{1}{2}\right] t_R\right]\right], \qquad (5.1\text{-}1)$$

where α_B is the background (unsaturable) electric-field absorption coefficient and $R_\alpha = R \exp(-\alpha_B L)$; T and R are the intensity transmission and reflectivity coefficients; $\Phi[t, t - (m + 1/2)t_R]$ is the phase increment of the field entering the cavity at $t - (m + 1/2)t_R$ and leaving the cavity at t; and t_R is the cavity round-trip time. One can write

$$\xi_I\left[t - \left[m + \frac{1}{2}\right] t_R\right] = \sum_{\ell=m}^{\infty} \Delta \xi_{I\ell} \qquad (5.1\text{-}2)$$

$$\Delta \xi_{I\ell} = \xi_I\left[t - \left[\ell + \frac{1}{2}\right] t_R\right] - \xi_I\left[t - \left[\ell + \frac{3}{2}\right] t_R\right] \qquad (5.1\text{-}3)$$

where ℓ and m are positive integers. Then

$$\mathcal{E}_T(t) = T e^{-i\omega t} e^{-\alpha_B L/2} \sum_{\ell=0}^{\infty} \sum_{m=0}^{\ell} \Delta \xi_{I\ell} R_\alpha^{m} \exp\left[i\Phi\left[t, t - (m + \frac{1}{2})\, t_R\right]\right]. \tag{5.1-4}$$

If the variation of the field is small in t_R, then

$$\mathcal{E}_T(t) \simeq T e^{-i\omega t} e^{-\alpha_B L/2} \int_{-\infty}^{t} \left[\frac{d}{dt'} \xi_I(t')\right] \sum_{m=0}^{m_{max}} R_\alpha^{m}$$

$$\times \exp\left[i\Phi\left[t', t' - \left[m + \frac{1}{2}\right] t_R\right]\right] dt', \tag{5.1-5}$$

with m_{max} being the integer closest to but smaller than $(t - t')/t_R$

$$\Phi\left[t, t - \left[m + \frac{1}{2}\right] t_R\right]$$

$$= \sum_{\ell=0}^{m} \left[\frac{\omega}{c}\right] \int_0^L \left[n_0 + \delta n\left[t - \left[\ell + \frac{1}{2}\right] t_R + \frac{n_0}{c}\, z, z\right]\right] dz$$

$$+ \sum_{\ell=1}^{m} \left[\frac{\omega}{c}\right] \int_L^0 \left[n_0 + \delta n\left[t - \ell t_R + \frac{n_0}{c}\,(L - z), z\right]\right] dz\,. \tag{5.1-6}$$

If the variation of δn in a round-trip time t_R is negligible,

$$\Phi\left[t, t - \left[m + \frac{1}{2}\right] t_R\right] \simeq \left[m + \frac{1}{2}\right] \phi_0 + \frac{1}{2}\,\Delta\phi(t) + \sum_{\ell=1}^{m} \Delta\phi(t - \ell t_R)\,, \tag{5.1-7}$$

where $\phi_0 = \omega 2 n_0 L/c$ and $\Delta\phi(t) = \omega \oint \delta n(t,z)\,dz/c$ where $\oint$ denotes a round-trip integral. The spatial variation of δn arises from the interference of the forward and backward propagating waves in the cavity. Because it is sinusoidal it averages out in the round-trip integral of $\Delta\phi$. For Kerr liquids, δn and hence $\Delta\phi$ obey the Debye relaxation equation:

$$\tau_M \frac{\partial \Delta\phi}{\partial t} + \Delta\phi = \frac{\omega L}{c}\,(\delta n_F + \delta n_B), \tag{5.1-8}$$

where τ_M is the Debye relaxation time and δn_F and δn_B are the quasi-steady-state field-induced refractive indices seen by the forward and backward propagating waves E_F and E_B inside the cavity, respectively. The backward intensity has twice the effect on δn_F as does the forward intensity itself (see Appendix D):

$$\delta n_{F,B} = n_2|E_{F,B}|^2 + 2n_2|E_{B,F}|^2 .$$

Equation (5.1-8) then becomes

$$\tau_M \frac{\partial \Delta\phi}{\partial t} + \Delta\phi = \frac{3\omega L n_2}{c}(|E_F|^2 + |E_B|^2) . \qquad (5.1\text{-}9)$$

Also,

$$|E_T|^2 = n_0 T|E_F|^2 = \frac{n_0 T|E_B|^2}{R} . \qquad (5.1\text{-}10)$$

$$\Delta\phi(t) = \frac{3\omega L n_2(1 + R)}{n_0 c T \tau_M} \int_{-\infty}^{t} |E_T(t')|^2 \exp[-(t - t')/\tau_M]\, dt' \qquad (5.1\text{-}11)$$

is the integral form of Eq. (5.1-9).

The coupled set of Eqs. (5.1-4,7,11) describes fully the dynamical properties of the nonlinear Fabry-Perot interferometer. Bischofberger and Shen (1979b) have solved them numerically on a computer using iterative self-consistent loops.

When the input pulse duration τ_p is much longer than τ_M and t_R, the operation should be quasi-steady-state. Equation (5.1-1) reduces to the well-known result

$$E_T = \frac{T E_I \exp[i(\phi_0 + \Delta\phi)/2]}{1 - R_\alpha \exp[i(\phi_0 + \Delta\phi)]} , \qquad (5.1\text{-}12)$$

where from Eq. (5.1-11)

$$\Delta\phi \simeq \frac{3\omega L n_2(1 + R)|E_T|^2}{n_0 c T} . \qquad (5.1\text{-}13)$$

The experimental arrangement was summarized in Section 3.4, and bistability hysteresis curves were presented for nitrobenzene (Fig. 3.4-1) and CS_2 (Fig. 3.4-2). Bischofberger and Shen have made quantitative comparisons for various relative values of the input intensity pulse duration τ_p, medium response time τ_M, and the cavity round-trip time t_R, using $R_\alpha \simeq 0.78$ from the measured finesse of 13. Three different initial cavity-laser detunings, $\beta_0 = \Delta\phi_0$, have been selected corresponding to the steady-state limiting ($\Delta\phi_0 = 0$), ac gain ($\Delta\phi_0 = -0.1\pi$), and bistability ($\Delta\phi_0 = -0.2\pi$) modes of operation. Figure 5.1-2 shows transient hysteresis loops for the case $t_R \ll \tau_p \ll \tau_M$. The medium is affected quickly by the increase in field, but the medium relaxation is too slow to follow the decrease in input intensity. This is a cumulative effect where the phase change $\Delta\phi$ is proportional to the integral of the intracavity intensity. Note that the "bistability" loop shows no switching. The situation is not much improved in Fig. 5.1-3 where $\tau_p \approx \tau_M \gg t_R$; $\Delta\phi$ is beginning to follow the input somewhat better. However, even with $\tau_p \approx 4\tau_M$ and $\tau_M \gg t_R$ in Fig. 5.1-4, the response is far from the steady-state case. Reducing the medium response time to much less than τ_p helps considerably (Fig. 5.1-5); further reductions in τ_M change the response very little (Fig. 5.1-6). Almost no transient behavior remains when τ_p is lengthened to 62 ns, Fig. 5.1-7, so that $\tau_p/t_R \approx 600$. The transient response still evident in Figs. 5.1-5,6 is then primarily due to the finite buildup time of the cavity field. The cavity response time, $2n_0L/cT$, is only about 0.5 ns for an effective $T \approx 0.22$ corresponding to the measured finesse of 13. In Fig. 5.1-6, an input pulse length 30 times the cavity response time was insufficient for steady-state behavior. The experimental variations of τ_M using MBBA were accomplished by temperature control. In the simulations the values 2.2×10^{-11} esu and 1.1×10^{-11} esu were used for the nonlinear refractive indices n_2 of CS_2 and nitrobenzene, respectively.

The simulations have been carried further than the data because of experimental restraints. Figs. 5.1-8,9 show conditions for obtaining overshoot and damped oscillations; Venkatesan (1977) also found overshoot in his theoretical analysis. See Sections 3.2, 3.3, 3.5, and 3.6 for observations of overshoot.

Figure 5.1-10 shows that when $\tau_p \gg t_R \gg \tau_M$, the reflectivity of the Fabry-Perot mirrors sets the switching speed and affects overshoot and ringing. Lower mirror reflectivity slows down the switching speed and damps out the overshoot and ringing. The slower switching is attributed to the smoother and broader transmission curve, and the stronger damping to the lower finesse. They conclude that to have sharp optical switching, a cavity with a high finesse is needed, but then overshoot and ringing after switching will be more obvious.

A comparison is made in Fig. 5.1-11 between the steady-state theory and the quasi-steady-state theory and data. Note that the experimental intensities are considerably higher than the theoretical ones. This may be due, at least in part, to the fact that the theory assumes plane-wave inputs and the experiment uses Gaussian inputs. The agreement between theory and experiment in Figs 5.1-2 to 7 is remarkable considering that transverse effects were ignored. Also some of the data were taken under the conditions for Ikeda instabilities (Section 6.3), namely $\tau_M \ll t_R \ll \tau_p$. It is true that Bischofberger and Shen's detector was too slow to follow such instabilities, but Carmichael, Snapp, and Schieve (1982) point out that with those parameters, the ratio of minimum power to see Ikeda's instability to minimum power for bistability is 49. Since they barely had enough power for bistability, the Ikeda instability could not occur.

Figure 5.1-11c is quite surprising at first in that it shows experimental points on the negative slope portion of the S-shaped curve. Bischofberger and Shen (1978a) describe how this can be done using transient data. If the cavity response time is much shorter than the laser pulse, the Fabry-Perot transmission is still given by the steady-state formula Eq. (5.1-12) in which $\Delta\phi$ and $|\mathcal{E}_I|$ change slowly compared with the cavity. Also Eq. (5.1-8) leads to Eq. (5.1-13) for the maximum value of $\Delta\phi$ which occurs at time $t = t_{max}$ such that $\partial(\Delta\phi)/\partial t = 0$. Equation (5.1-13) is also the steady-state formula, but then $|\mathcal{E}_T|^2$ is time independent. From the analysis of transients such as Fig. 5.1-4c, t_{max} is deduced and the corresponding values of I_I and I_T are used to determine a point on Fig. 5.1-11c.

In the context of optical bistability, several other authors have considered the transient behavior of a nonlinear Fabry-Perot cavity. Venkatesan (1977) treats stability, overshoot, and phase evolution. Bonifacio and Lugiato (1976, 1978a)

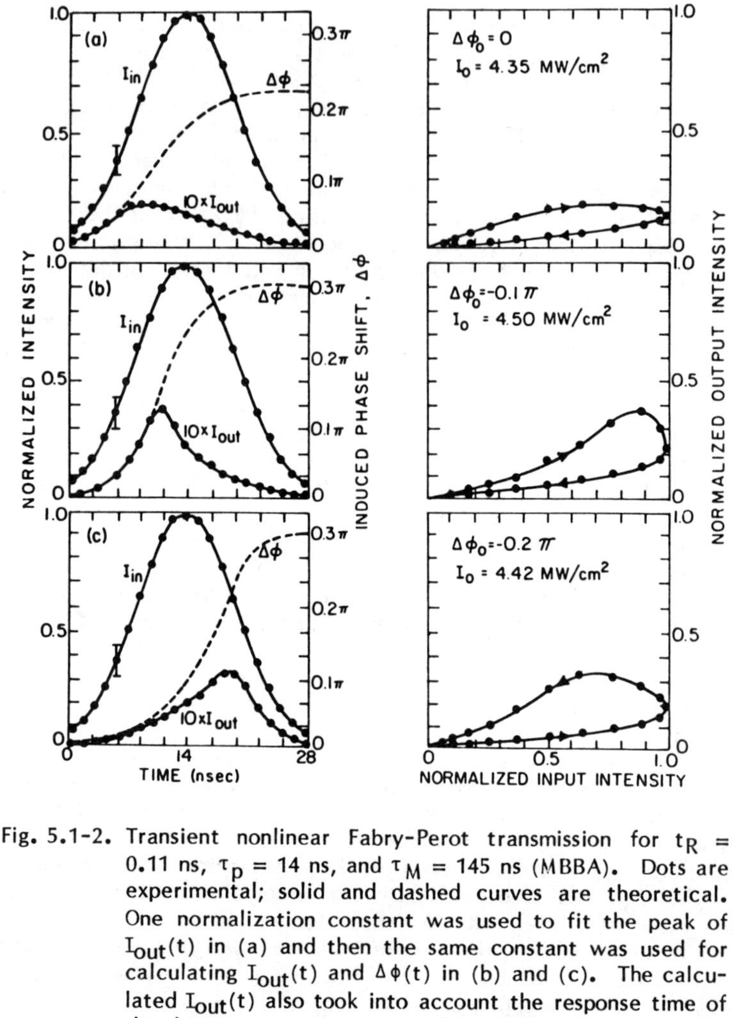

Fig. 5.1-2. Transient nonlinear Fabry-Perot transmission for t_R = 0.11 ns, τ_p = 14 ns, and τ_M = 145 ns (MBBA). Dots are experimental; solid and dashed curves are theoretical. One normalization constant was used to fit the peak of $I_{out}(t)$ in (a) and then the same constant was used for calculating $I_{out}(t)$ and $\Delta\phi(t)$ in (b) and (c). The calculated $I_{out}(t)$ also took into account the response time of the detector system ($\simeq$ 1 ns) through convolution with the detector response function. Bischofberger and Shen (1979b).

linearize around the equilibrium points (see Section 2.8) to discuss spectra and to note the line narrowing associated with critical slowing down (Section 5.6). Allen, Garmire, Marburger, and Winful (1978) and Garmire, Marburger, Allen, and Winful (1979) report studies of critical slowing down and other transient phenomena in hybrid bistable devices. Goldstone and Garmire (1981a) make numerical studies of dynamic response, particularly of switching phenomena (Section 5.5). Bonifacio and Meystre (1978) have calculated ringing and Fabry-Perot cavity round-trip effects when the input intensity is suddenly changed to a new value. Abraham, Bullough, and Hassan (1979) study the dynamics of absorptive bistability in a Fabry-Perot cavity using the Fleck-Meystre truncation discussed in

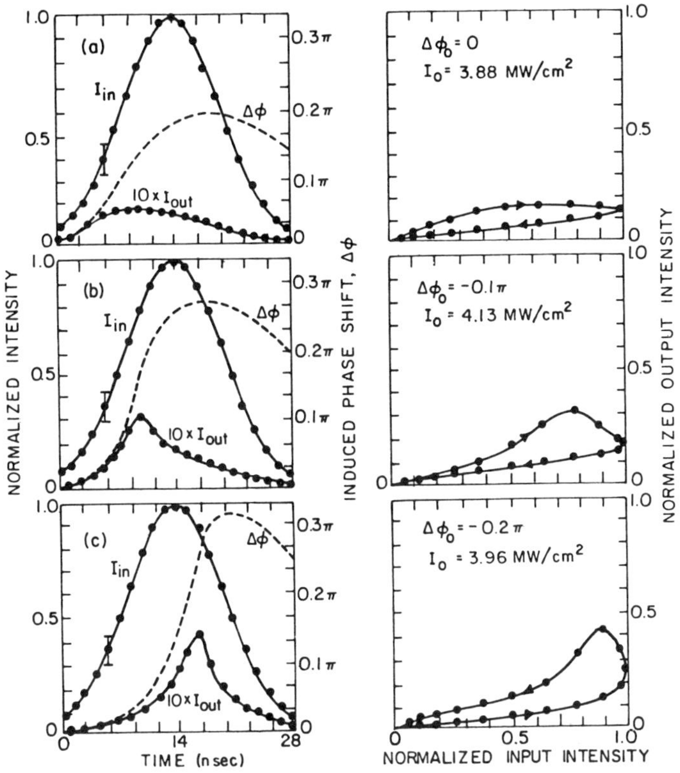

Fig. 5.1-3. Transient nonlinear Fabry-Perot transmission for t_R = 0.11 ns, τ_p = 14 ns, and $\tau_M \simeq$ 15 ns (MBBA). Bischofberger and Shen (1979b).

Section 2.5.2. They see transient effects as in Section 3.2 and amplification of the oscillatory part of the input signal envelope. Nishiyama (1980a) uses the same truncation and computes ringing and critical slowing down. Abraham and Hassan (1980) have included inhomogeneous broadening and dispersive effects and calculated Rabi ringing. They also allow for the inhomogeneous dephasing time T_2^* to change sinusoidally with time and find an oscillating output that they call thermal pulsing. Note that these pulsations differ from regenerative pulsations of Section 6.1 that arise from instabilities in the system itself, not from external pulsations. Hopf and Meystre (1979) study switch-up by an external pulse or by external noise (Sections 5.5 and 6.5). Drummond (1982) treats the limit that γ_L is larger than both γ_T and the cavity decay rate $cT/2n_0L$, which is the case of short semiconductor etalons. He finds switch-on transients with initial rates of order γ_L followed by a very rapid change in output intensity as the intracavity field builds up.

Noteworthy are the analytic transient studies of absorptive bistability in the large-C mean-field limit by Erneux and Mandel (1983) and Mandel and Erneux (1982a,b). They use a multiple-time-scale perturbation analysis to describe the

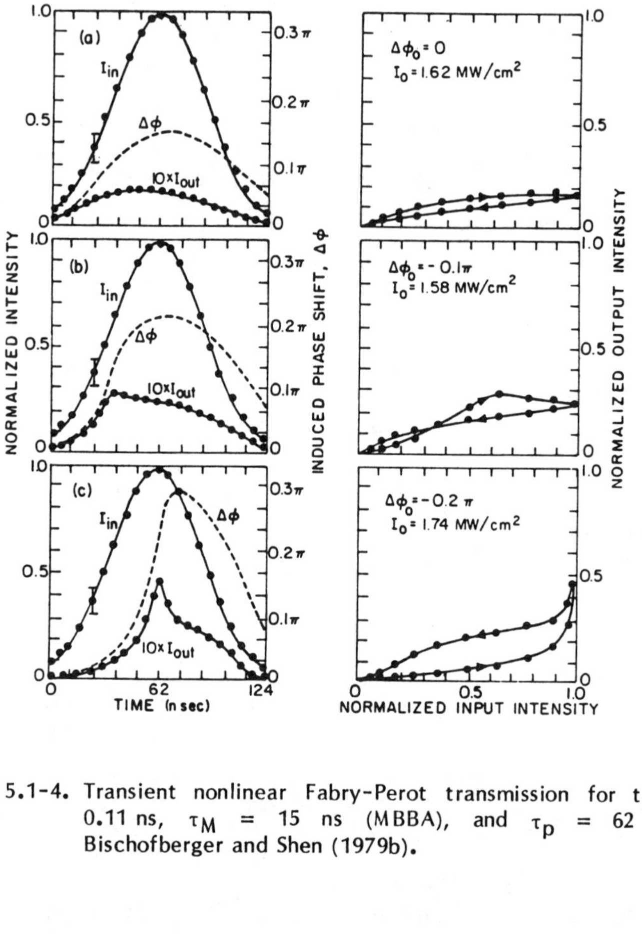

Fig. 5.1-4. Transient nonlinear Fabry-Perot transmission for t_R = 0.11 ns, τ_M = 15 ns (MBBA), and τ_p = 62 ns. Bischofberger and Shen (1979b).

transition between the two stable branches by simple equations. Near the low transmission branch the behavior of the system is governed by an equation for the atomic population, whereas near the high-transmission branch it is an equation for the field amplitude which determines the long-time evolution of the system.

Most of the remainder of Chapter 5 is concerned with transients in Fabry-Perot cavities, especially Section 5.6 on critical slowing down and Section 5.7 on phase switching.

5.2. PULSE SELF-RESHAPING AND POWER LIMITING

Pulse self-reshaping occurs whenever an input pulse of sufficient intensity is transmitted through a nonlinear Fabry-Perot resonator. Whenever optical

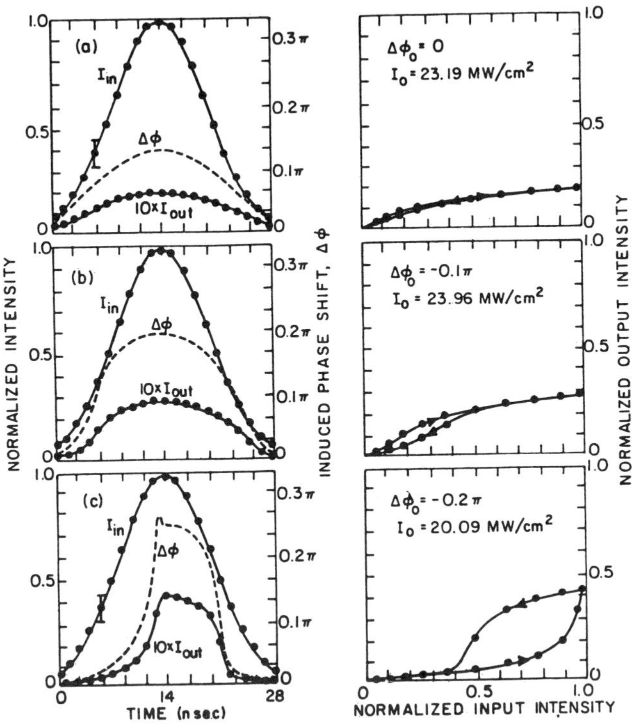

Fig. 5.1-5. Transient nonlinear Fabry-Perot transmission for τ_M = 0.045 ns (nitrobenzene), t_R = 0.1 ns, τ_p = 14 ns. Bischofberger and Shen (1979b).

bistability is seen by modulating the input intensity, the output is quite different from the input, i.e. self-reshaping has occurred. Of course, reshaping can occur without bistability. For example, if the laser and the cavity frequencies are coincident at low intensities, self-termination of a pulse can occur as the cavity-laser detuning increases with higher input. This case is shown in Fig. 1.2-7 in which a long tail is largely removed by self-detuning of the cavity. Optical discriminating is shown in Fig. 1.2-6 in which only the parts of the input exceeding a threshold are transmitted. Optical limiting action allows almost no change in transmitted intensity for large changes in input (Fig. 1.2-5). Section 5.1 contains numerous examples of pulse compression and squaring with good agreement between theory and data.

The simplicity of reshaping by self-sweeping of a Fabry-Perot cavity is illustrated by Figs. 5.2-1 to 3. In Fig. 5.2-1 a long square-shaped pulse is incident on an etalon initially detuned from the incident laser frequency. Absorption of part of the intracavity energy heats the etalon and sweeps its peak wavelength through the laser wavelength; a short pulse is generated. The pedestal could be

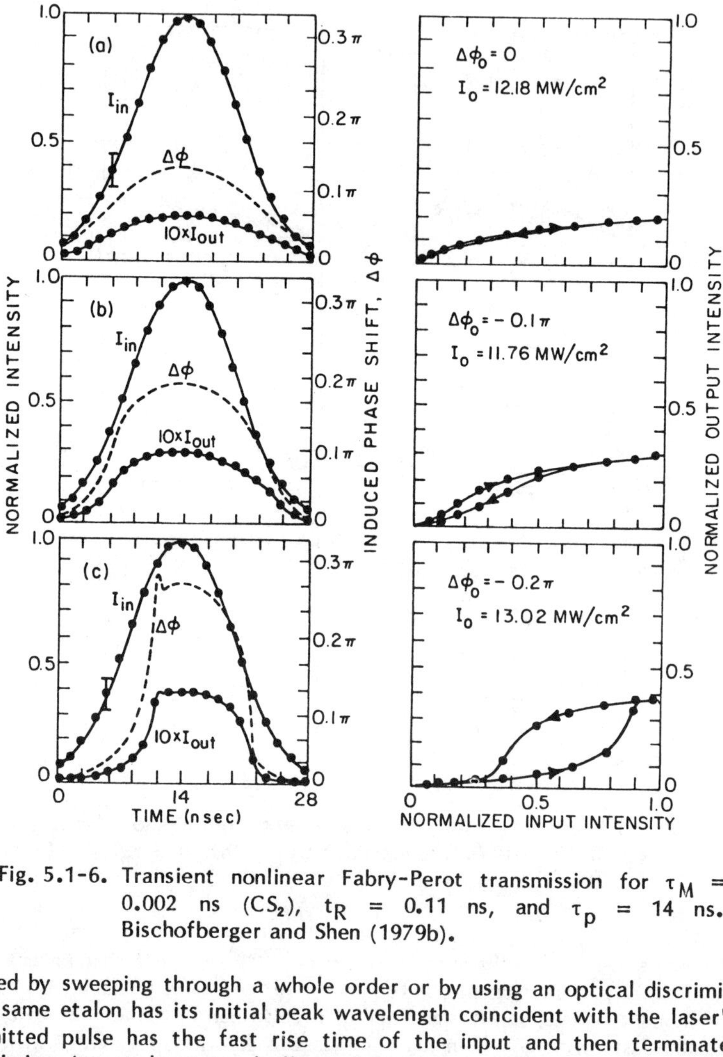

Fig. 5.1-6. Transient nonlinear Fabry-Perot transmission for τ_M = 0.002 ns (CS_2), t_R = 0.11 ns, and τ_p = 14 ns. Bischofberger and Shen (1979b).

removed by sweeping through a whole order or by using an optical discriminator. If the same etalon has its initial peak wavelength coincident with the laser's, the transmitted pulse has the fast rise time of the input and then terminates the transmission (or at least gradually reduces it) as self-sweeping destroys the etalon-laser coincidence (Fig. 5.2-2). Figure 5.2-3 shows self-sweeping through two etalon peaks using a thin GaAs etalon.

Kazberuk, Karpushko, and Sinitsyn (1978) proposed self-intensity modulation by a thin-film semiconductor etalon and showed some slight experimental pulse reshaping. They also show theoretical curves of pulse reshaping and the optically induced change in optical pathlength. Pulse self-reshaping has also been studied in Te [Staupendahl and Schindler (1982)] and Si [Eichler, Heritage, and Beisser (1981); Eichler, Massmann, and Zaki (1982); and Eichler, Massman, Zaki, and Heritage (1982)] etalon devices by refractive index changes via free-carrier production. In Si the cavity has been tuned by five orders with 300 mJ/cm^2.

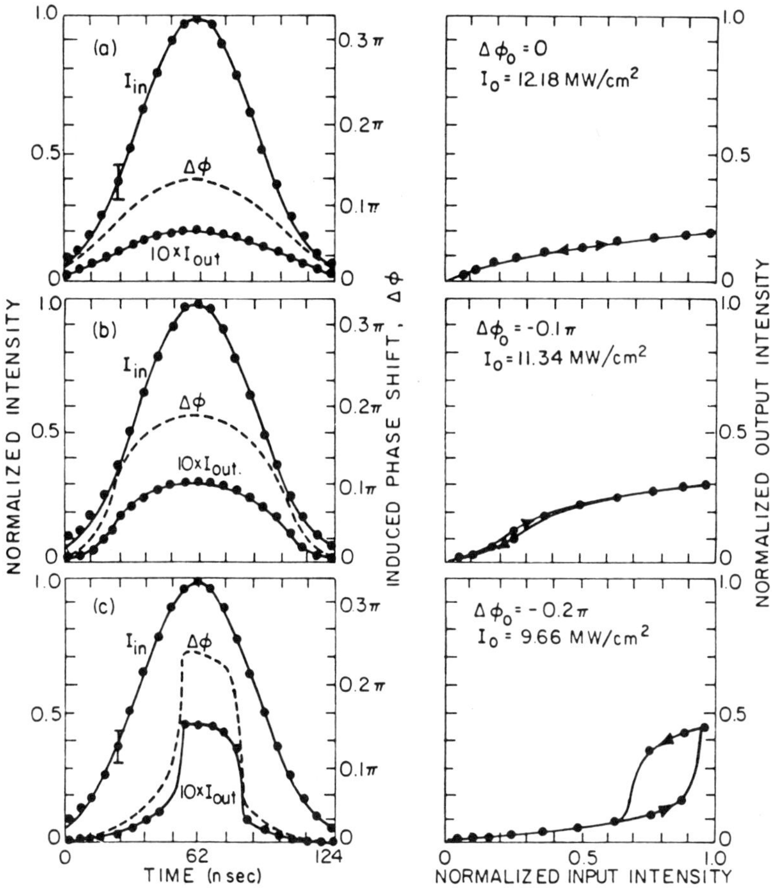

Fig. 5.1-7. Transient nonlinear Fabry-Perot transmission for τ_M = 0.002 ns (CS_2), t_R = 0.11 ns, and τ_p = 62 ns. Bischofberger and Shen (1979b).

Johnson and Pratt (1978) propose generating picosecond second-harmonic pulses in Te by generating free carriers by two-photon absorption, thereby destroying phase matching. Kitayama and Wang (1983) propose pulse compression using the power discriminating behavior of two waveguides coupled nonlinearly (Section 3.10.5).

There are also examples of pulse self-reshaping in the early attempts to see bistability (Section 1.4). Pulse shortening and tail reduction were seen by Szöke, Daneu, Goldhar, and Kurnit (1969). Spiller (1971, 1972) saw pulse shortening by more than a factor or two. He also saw optical discriminator action in the form of very nonlinear transmission. Input pulses differing by a factor of two differed

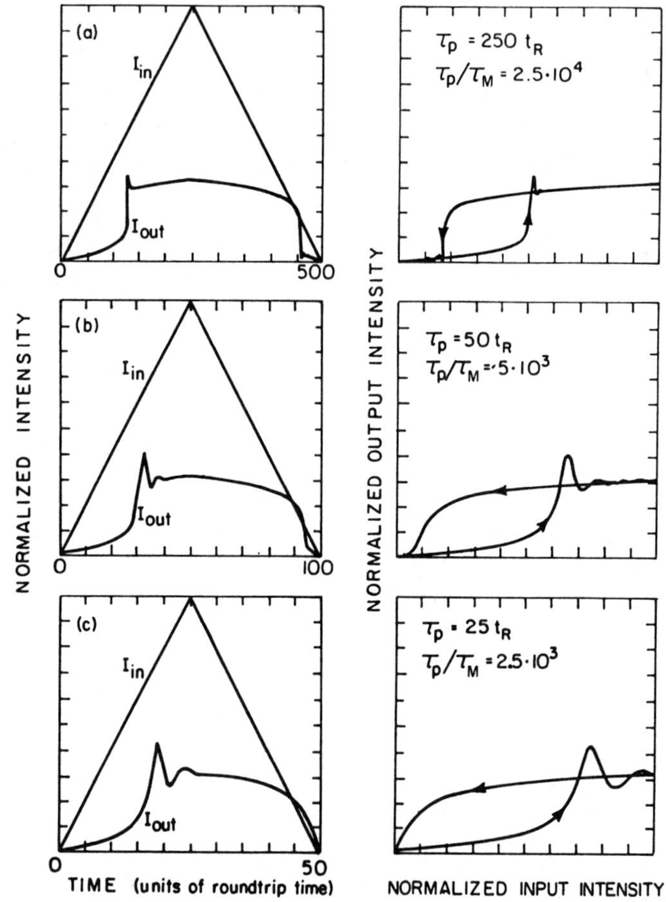

Fig. 5.1-8. Transient nonlinear Fabry-Perot transmission simulations for τ_M = 0.01 t_R, R_α = 0.7, and $\Delta\phi_0$ = -0.4π. Bischofberger and Shen (1979b).

by at least a factor of twenty upon transmission through Spiller's saturable absorber Fabry-Perot cavity; see Fig. 5.2-4. Spiller (1972) and Austin (1972) numerically studied pulse self-reshaping, noting that the reshaping is symmetric in normal operation and asymmetric under bistability conditions. Austin (1972) also saw a factor of two reduction in pulse duration.

Clearly, self-reshaping of optical pulses by sweeping a nonlinear Fabry-Perot cavity can be used to tailor light pulses. This should become increasingly useful as faster and stronger nonlinearities are discovered and utilized. Note that this reshaping technique is a direct tailoring of the input pulse by the cavity. The peak output power never exceeds the input peak power as it can in pulse compression techniques that redistribute the pulse energy in time, such as self-induced

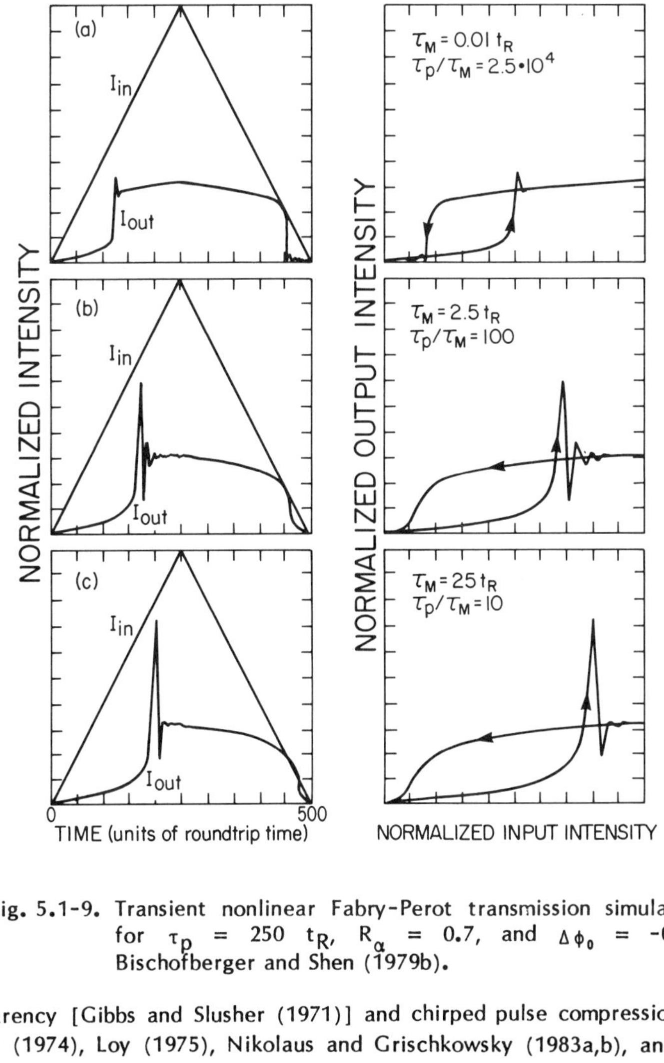

Fig. 5.1-9. Transient nonlinear Fabry-Perot transmission simulations for $\tau_p = 250\ t_R$, $R_\alpha = 0.7$, and $\Delta\phi_0 = -0.4\pi$. Bischofberger and Shen (1979b).

transparency [Gibbs and Slusher (1971)] and chirped pulse compression [Grischkowsky (1974), Loy (1975), Nikolaus and Grischkowsky (1983a,b), and Fujimoto, Weiner, and Ippen (1984)].

A special case of pulse self-reshaping is power limiting, which can be accomplished by a bistable device as shown in Figs. 1.2-5, 3.5-4, 3.6-10, 3.8-3, 4.2-3, and 4.4-3. Limiting action is achieved by initially tuning the peak of the nonlinear Fabry-Perot etalon to the laser frequency; as the input intensity rises the etalon detunes, keeping the transmitted power approximately constant. For example, Park (1984) has studied power limiting at 10.6 μm using a $Hg_{1-x}Cd_xTe$ nonlinear etalon. Many bistable devices are multistable, i.e., they will limit the power only over a limited range and then they switch up again, as in Fig. 1.2-2. A saturable nonlinearity, adjusted so that the next order cannot be reached, can be used to avoid multistability. Many power limiter proposals do not even use

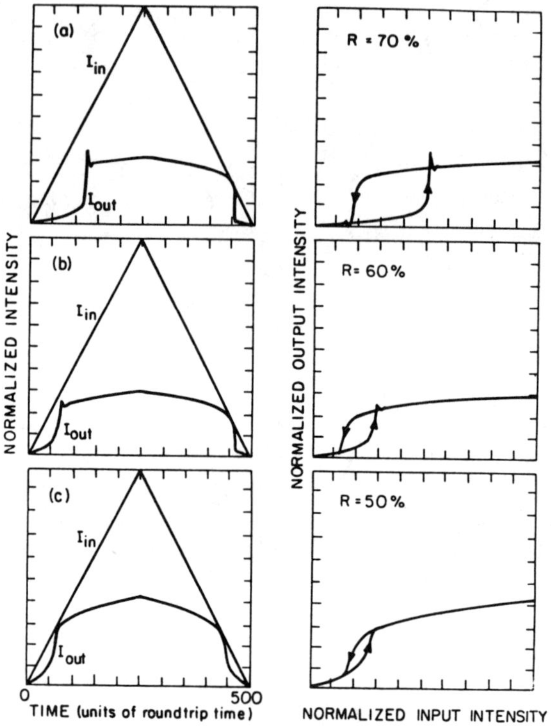

Fig. 5.1-10. Effect of cavity finesse in switching speeds and overshoot. The input peak intensity is kept constant, $\tau_M = 0.01\ t_R$, $\tau_p = 250\ t_R$, $\Delta\phi_0 = -0.4\pi$. Bischofberger and Shen (1979b).

bistability, but simply focus the input through a nonlinear medium and refocus the transmitted beam onto an aperture. The transmission of the system is very high at low input intensities, but at high intensities nonlinear self-lensing (whether focusing or defocusing) spreads out the beam by the time it reaches the aperture plane. Such power limiters generally have no multistable states, but interference rings and effects of multiple foci in self-focusing can lead to oscillations in the far-field on-axis intensity. Numerical calculations of on-axis oscillations are given in Hermann and Grigg (1984) for a Kerr medium and in LeBerre, Ressayre, and Tallet (1982) and LeBerre, Ressayre, Tallet, Tai, Gibbs, Rushford, and Peyghambarian (1984) for off-resonance two-level (saturable) atoms. Soileau, Williams, and Van Stryland (1983) and Williams, Soileau, and Van Stryland (1984) limit 1.06-μm and 0.53-μm pulses using CS_2. Boggess, Moss, Boyd, and Smirl (1984) limit 48-ps, 1.06-μm, $\leq$ 700-μJ pulses by a 1-mm-thick, antireflection-coated wafer of high-purity single-crystal Si.

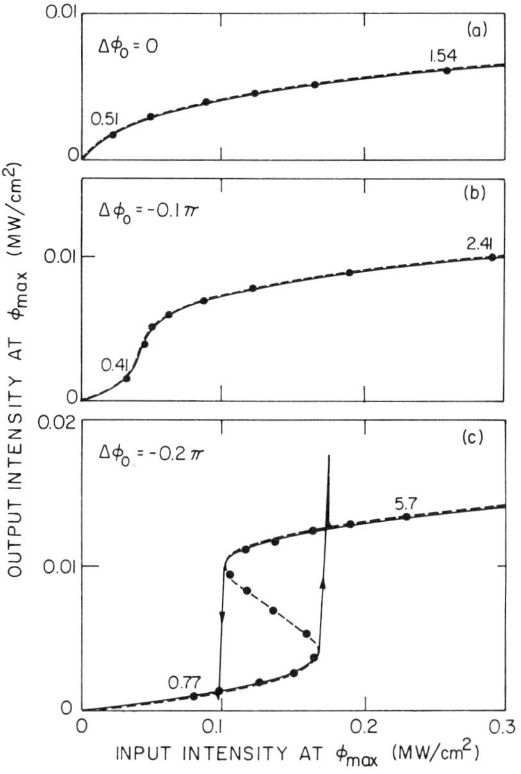

Fig. 5.1-11. Input versus output intensities at the maximum field-induced phase shift $\Delta\phi_{max}$ for three values of $\Delta\phi_0$. Each point corresponds to an input pulse with a certain peak intensity as indicated on a few points in megawatts per square centimeter (MBBA). Dashed lines are the steady-state characteristic curves of a nonlinear Fabry-Perot cavity. Solid curves are computer simulations of the characteristic curves for quasi-steady-state operations with τ_M = 2 ps assuming an input pulse with a peak intensity of 0.3 MW/cm². The bistable mode of operation in (c) shows overshoot immediately after switching. Bischofberger and Shen (1978a).

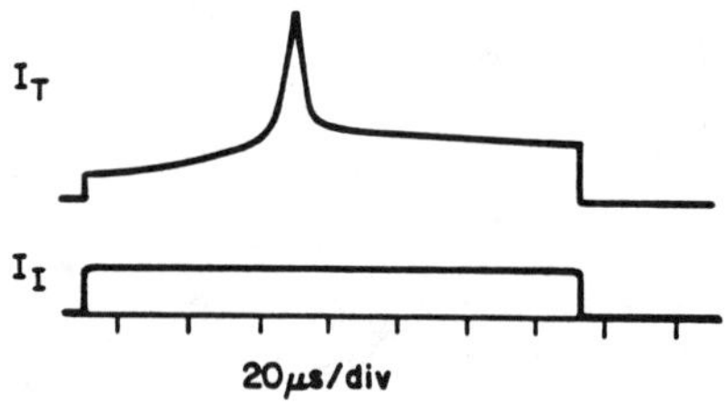

Fig. 5.2-1. Pulse production by thermal self-sweeping of a color filter etalon. At the beginning of the square input pulse, the etalon peak frequency and laser frequency are detuned by about one etalon instrument width, resulting in low transmission. Optical pathlength changes induced by the 11%/pass intracavity absorption sweep the etalon peak frequency through the laser frequency and beyond. Other parameters: Corning 4-74 color filter ground and polished into a 57-μm etalon and coated with R = 0.8; ≃ 600-mW, 647.1-nm incident beam focused to about 50 μm on the etalon. Venkatesan, Gibbs, McCall, Passner, Gossard, and Wiegmann (1979).

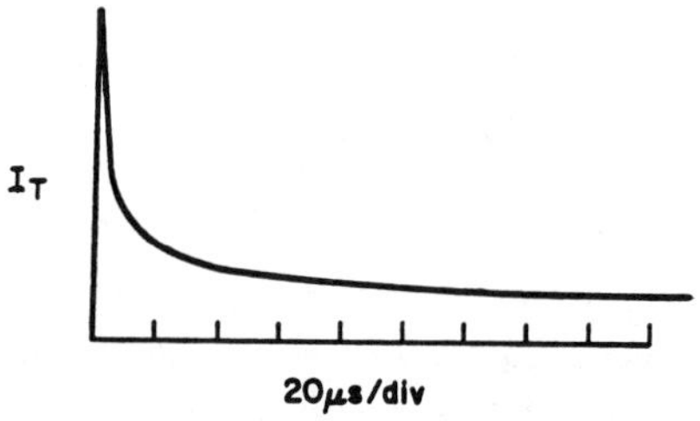

Fig. 5.2-2. Termination of a step function input pulse by thermal self-sweeping of a color filter etalon. Same conditions as Fig. 5.2-1 except etalon and laser frequencies are initially coincident, resulting in maximum initial transmission and maximum initial absorption.

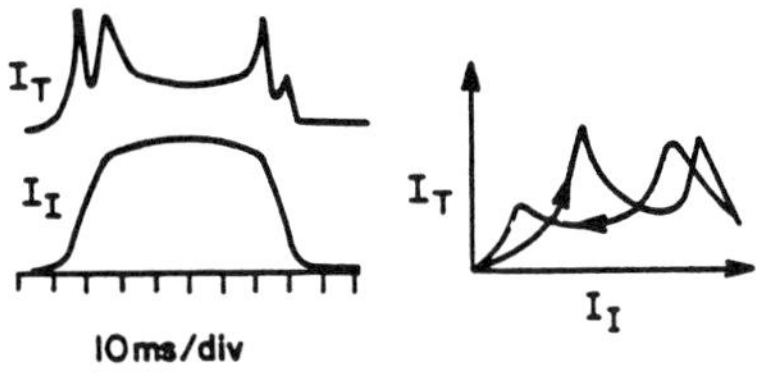

Fig. 5.2-3. Fabry-Perot sweeping by thermal effects in GaAs at room temperature. 0.5-μm GaAs between 2-μm GaAlAs windows with R = 0.9 and λ = 824.6 nm. The peak transmission was about 70%. The maximum input power was a few hundred milliwatts gated from a cw dye laser and focused to about 10-μm diameter. The heating shifts the band edge toward longer wavelengths, increasing the absorption at the operating wavelength and lowering the peak transmission. This is why the second peak is lower than the first even though the input is higher. Venkatesan, Gibbs, McCall, Passner, Gossard, and Wiegmann (1979).

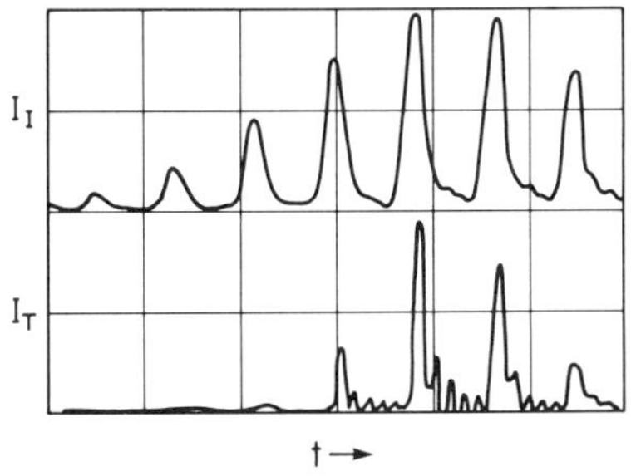

Fig. 5.2-4. Input pulse (upper trace) and output pulse (lower trace) for a saturable resonator filled with cryptocyanine in ethyl alcohol of unsaturated optical density $\alpha_B L = 0.75$. Mirror reflectivity R = 0.98, L = 125 μm. Time scale: 10 ns/div; vertical scale about 90 MW/div for the upper trace and about 70 kW/div for the lower trace; rise time of detector and oscillograph 0.5 ns. The ringing is due to the detection system. Spiller (1972).

5.3. CONTROL OF ONE BEAM BY ANOTHER

The wavelengths of peak transmission can be changed by a second beam as well as by the input beam itself as in Section 5.2. The optical pathlength change required to tune the Fabry-Perot cavity by half of an instrument width is

$$\delta(nL) = \frac{\lambda}{4\mathscr{F}}, \qquad (5.3\text{-}1)$$

where $\mathscr{F}$ is the cavity finesse. For a thermal device the required temperature change is

$$\delta T = \frac{\lambda}{4\mathscr{F}} \frac{1}{\partial(nL)/\partial T}. \qquad (5.3\text{-}2)$$

For the 4-74 Corning filter $n\partial L/\partial T \simeq 8.7 \times 10^{-6}$ L/K and $\partial n/\partial T = 3 \times 10^{-6}$/K, so that $L^{-1}\partial(nL)/\partial T \simeq 11.7 \times 10^{-6}$/K. For a length of 57 μm, a finesse of 8, and a wavelength of 632.8 nm, $\delta T \simeq 30$ K. To obtain this temperature change the absorbed energy required per unit area is

$$c_V \rho L \Delta T, \qquad (5.3\text{-}3)$$

which is 4.5 nJ/μm² for $c_V = 1.05$ J/Kg and $\rho = 2.5$ g/cm³. Figure 5.3-1 shows the control of a He-Ne beam by a 1-W Kr ion laser beam. A time of about 10^{-4} s was needed to scan the color filter etalon by half an instrument width with the control beam focused to 15 μm. This corresponds to 444 nJ/μm². However, the absorption coefficient at the control wavelength was only 0.12. Furthermore, the control wavelength was probably not at a Fabry-Perot peak. These two factors could easily reduce the 444 nJ/μm² incident to the 4.5 nJ/μm² calculated absorbed energy needed.

The control beam works more efficiently if the reflectivity is low and the absorption is high at its wavelength. Such is the case for a GaAs etalon controlled by 514.5-nm mode-locked Ar laser pulses: in Fig. 5.3-2, $R \simeq 0.9$ at $\lambda_s = 820$ nm and $R \simeq 0.25$ at $\lambda_c = 514.5$ nm. The 514.5-nm pulse was completely absorbed in the 4-μm-thick GaAs etalon, creating free carriers which screen out the free-exciton resonance, eliminating its contribution to the refractive index. For a finesse of 16, a $\delta(nL)$ of $\lambda/32$ is needed for the full-instrument-width shift. Elimination of the contribution to the phase from an exciton line with peak α_0 and HWHM $\delta\lambda$ results in a single-pass optical pathlength change of [Eq. (3.6-2)]

$$\delta(nL) = \left[\frac{\Delta}{1+\Delta^2}\right] \frac{\alpha_0 L}{2} \left[\frac{\lambda}{2\pi}\right], \qquad (5.3\text{-}4)$$

where the detuning factor Δ is the ratio of the laser-exciton wavelength difference to $\delta\lambda$. In Fig. 5.3-2a, $\alpha_0 L \simeq 7.5$, $\delta\lambda \simeq 4$ Å, and $\Delta \simeq 8$; then Eq. (5.3-4) yields $\delta(nL) \simeq 0.07\lambda$, about twice the $\lambda/32$ needed. In Section 3.6 it was noted that a density of $\simeq 10^{17}$ cm⁻³ free carriers is sufficient to remove the exciton's contribution to the index. Figure 5.3-2 was taken with a 2.5-W, 0.2-ns, 50-μm-diameter control pulse which has almost enough energy to provide 10^{17} cm⁻³ carriers throughout the 4.1-μm GaAs etalon.

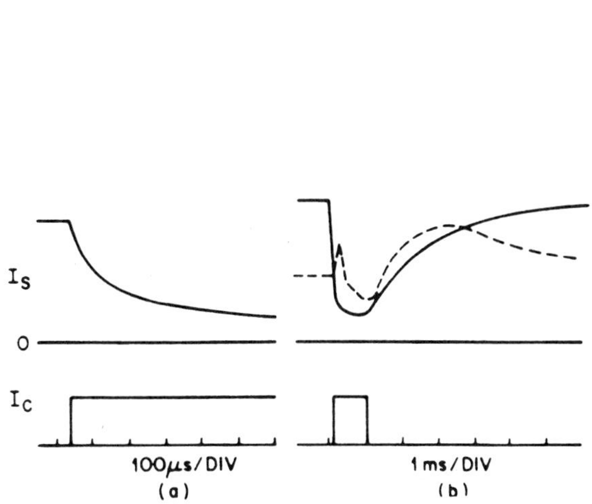

Fig. 5.3-1. Control of approximately a 1-mW He-Ne 632.8-nm beam I_S by approximately a 1-W Kr 647.1-nm beam I_C. A 57-μm 4-74 Corning filter with R = 0.8 is used. (a) Partial termination of the signal beam. (b) Solid curve: same as (a) with slower time scale. Dashed curve: control beam changed to ≃ 2-W Kr 647.1 and 676.4 nm and detuned so that the Fabry-Perot peak is swept through the laser frequency. Gibbs, Venkatesan, McCall, Passner, Gossard, and Wiegmann (1979).

These illustrations of optical modulation, using the color filter and GaAs etalons discussed in Sections 3.5 and 3.6 from the point of view of bistability, show that the calculated refractive index changes are consistent with the changes needed to explain the observed modulation. These examples illustrate the principles, although the control beams used were much more intense than the signal beams being controlled. The signal beams were deliberately made weak to avoid index changes by them. The optical transistor mode of operation allows control of a strong beam by a weaker one (Section 5.4). Clearly such techniques could be employed to gate an entire optical image with sub-picosecond resolution; recall the 1-ps GaAs NOR gate operating at 82 MHz (Section 1.2). If the sweep exceeds the instrument width, the gate can be open for a time much less than the duration of the control pulse.

There are many previous studies of controlling one light beam with another using similar ideas. Duguay and Hansen (1969) used ≃ 100 MW/cm^2 control beam to cause a Kerr liquid to become birefringent and change the polarization of the probe beam (see Section 5.8). Glass and Negran (1974) constructed a hybrid optical gate requiring no electrical power; a $LiTaO_3$:Cr detector absorbing 560-nm, 150-ns gating pulses was directly coupled to a lithium tantalum oxide ($LiTaO_3$) electro-optic modulator controlling a 632.8-nm beam. The transmission of a 10.6-μm beam along a GaAs waveguide can be altered by a 0.69- or 1.06-μm control beam, allowing a study of free carriers, etc.: Botineau, Gires, and Vanneste (1974); Azema, Botineau, Gires, Saissy, and Vanneste (1975, 1976a,b); Azema, Botineau, Gires, and Saissy (1978). Liao and Bjorklund (1976, 1977) used resonance enhancement by a two-photon resonance in atomic Na vapor to obtain large rotation angles with control intensities of 1 MW/cm^2. Grischkowsky (1977) has observed gating using the adiabatic following nonlinear index with the driving

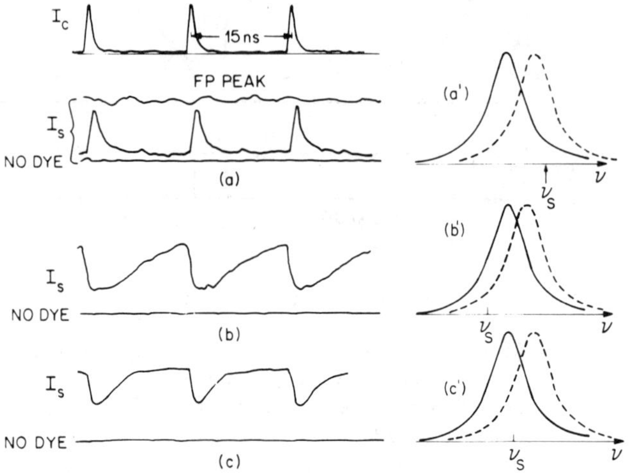

Fig. 5.3-2. Control of a 10-mW cw dye laser beam I_s by 2.5-W, 514.5-nm, 200-ps mode-locked Ar pulses I_c using a GaAs etalon. In all cases, there are about 15 ns between pulses. (a) Optical gate. Turn-on of the signal beam follows the integrated control pulse which sweeps the Fabry-Perot (FP) peak onto the signal wavelength; recovery occurs in 2 ns. "FP PEAK" is the transmission with Ar beam blocked and the Fabry-Perot cavity tuned to a peak by translating the sample. "NO DYE" is the detector output with dye-laser beam blocked. λ_s = 821.3 nm, 10 K ($\lambda_{EX} \simeq$ 818 nm). (b) Optical sawtooth. Turn-off follows the integrated control pulse which sweeps the Fabry-Perot peak off the signal wavelength. Detuning is such that recovery is not even complete by the next control pulse 15 ns later. λ_s = 821.3 nm, 10 K. (c) Optical modulation with faster recovery: λ_s = 821.3 nm, 50 K (λ_{EX} = 820 nm). (a')-(c') are pictorial representations of the Fabry-Perot instrument profile before (solid) and just after (dashed) the control pulse. The signal frequency ν_s is marked at a value which might generate the corresponding output (a)-(c). Gibbs, Venkatesan, McCall, Passner, Gossard, and Weigmann (1979).

and probing pulses in near resonance with the same single-photon transition. Gray and Casperson (1978) propose opto-optic modulation based on an orientational cross saturation effect in the amplifying medium of a laser oscillator or amplifier. Jamison and Nurmikko (1978) have switched 10-μm radiation on a picosecond time scale turning the pulse on by reflection and turning it off by free-carrier absorption (Section 5.8). Alcock and Corkum (1979) gated 10.6-μm radiation incident at Brewster's angle onto an n-type polycrystalline Ge sample using ≈ 15 MW/cm² at 0.7 μm to create a highly reflecting electron-hole plasma. The reflected pulse duration was 4 ns, but the transmission recovery took 50 ns by surface recombination.

An optically controllable millimeter-wave phase shifter has been constructed by DeFonzo, Lee, and Mak (1979) using 0.53-μm radiation to produce an electron-hole plasma in a silicon waveguide transmitting 94-GHz millimeter waves [also see Lee, Mak and DeFonzo (1980)]. Johnson and Auston (1975) demonstrated an opto-electronic picosecond switch for microwaves. Transmission of the switch is turned on by a surface layer of photoconductivity produced by a 0.53-μm pulse and is turned off by a volume photoconductivity produced by a 1.06-μm pulse, which shorts the device totally reflecting the incident microwave signal. Unlike conventional semiconductor junction devices, this switch does not depend on recombination or carrier sweepout to turn off. A 3-ps microwave pulse was produced by triggering a GaAs photoconductive switch driving a dipole antenna [Mourou, Stancampiano, Antonetti, and Orszag (1981)].

The frequency of a dye laser has been scanned by external optical control of an intracavity interference filter [Karpushko and Sinitsyn (1979)]; similar filters exhibited bistability (Section 3.5). Optical-optical modulation using Te etalons is reported by Staupendahl and Schindler (1982); see Section 3.5.

Tsukada and Nakayama (1982b) have given a detailed theoretical description of the modulation of the optical bistability of the main beam by a control beam. Both beams are assumed quasiresonant with a two-level atomic system, and the control is by means of changes in the dielectric susceptibility of the intracavity medium. Two-photon control mechanisms are discussed in Tsukada and Nakayama (1982a) and Hermann and Elgin (1981), the latter pointing out that the switching beam can undergo the reverse operation that it causes on the other beam. A recent proposal involving the simultaneous <u>coherent</u> propagation of two pulses in a three-level system might permit rapid switching [Mattar and Eberly (1979)].

5.4. OPTICAL TRANSISTOR OR TRANSPHASOR

One mode of operation of a nonlinear Fabry-Perot cavity is called differential or AC gain, i.e. $dI_T/dI_I > 1$ and the output is a single-valued function of the input. This analog operation is for all-optical inputs what transistor operation is for electrical inputs; hence the name <u>optical transistor</u>. Because its operation depends upon an intracavity phase shift in the dispersive devices, it is sometimes called a <u>transphasor</u>. In an age of digital electronics, the bistability mode has received more attention and is easier to observe than AC gain, which requires high stability to keep the laser and etalon on the high-gain mid-point [see Feldman (1978b) for a stabilization scheme]. Optical transistor characteristics can be seen by modulating the input beam or by using a second probe beam.

Optical transistor action was seen by modulating the input beam in the first optical bistability experiment; see Fig. 1.2-3 and Section 3.2 [McCall, Gibbs,

Churchill, and Venkatesan (1975) and Gibbs, McCall, and Venkatesan (1976)]. Further AC gain characteristics in Na can be seen in Figs. 3.2-10,11. The frequency range over which the gain was high was very narrow, making it difficult to obtain high linear gain. No special effort was made to stabilize on the high-gain point in order to study the gain.

A hybrid bistable device has been operated in the transistor mode as shown in Fig. 5.4-1. Differential gain using the arrangement of Fig. 4.2-1 is shown in Fig. 4.2-3 and Fig. 5.4-2. Characteristic curves with a small amount of AC gain were seen in the GaAs etalon of Fig. 3.6-2 as shown in Fig. 5.4-3.

D. A. B. Miller and Smith (1979) performed the first two-beam amplification experiment in an intrinsic device using InSb; see Fig. 5.4-4. The simplest view of the operation of their device is that the weak beam induces a small change in the phase thickness of the cavity which is transferred into the main beam. Hence, by analogy with the electrical transistor, they termed the device a transphasor. They observed gain not only with the two beams derived from one laser but also by using two different CO lasers. Tooley, Smith, and Seaton (1983) digitally amplify a 3-μW signal by a power gain of up to 10^4 with an InSb etalon operating at 77 K and 1819 cm^{-1}. D. A. B. Miller, Smith, and Seaton (1981a) emphasize that the distinction between AC gain of a signal impressed on the input and true two-beam optical transistor action can have important implications in a Fabry-Perot cavity. The two counterpropagating beams in a Fabry-Perot cavity and another input beam satisfy the conditions for degenerate four-wave mixing. Power may be transferred between beams and new beams may be generated.

Jain and Pratt (1976) proposed an optical transistor based on the strong dependence of second-harmonic generation on the phase-matching conditions, suggesting Te as a candidate material. Experimental results in Te by Quint, Johnson, Jain, and Pratt (1979) exhibited small gain, but they concluded that an optical gain device employing a simple crystal of Te is not feasible.

5.5. EXTERNAL OFF AND ON SWITCHING OF A BISTABLE OPTICAL DEVICE

A particular case of controlling one light beam with another (Section 5.3) is the use of one control beam to switch a bistable device "on" and another control beam to switch it "off." Clearly this can be accomplished easily in hybrid systems, and hybrids have been used to make interesting transient studies (see Sections 5.6 and 6.5). Switching of an intrinsic bistable device, the GaAs etalon of Section 3.6, has also been accomplished. A flat-top input pulse is used with an intensity above the switch-down but below the switch-up intensities of the bistable hysteresis loop (see Figs. 3.6-4,5). In the first GaAs external-pulse switch-on experiment [Gibbs, McCall, Passner, Gossard, Weigmann, and Venkatesan (1979)], a 200-ps, 590-nm, 0.6-nJ pulse switched on the GaAs etalon held at about 10 K; see Fig. 5.5-1. The observed time was just over 1 ns, limited by the resolution of the detection system. The 590-nm pulse illuminated a circle about 50 μm in diameter, so the switching energy per unit area was about 0.24 pJ/μm^2. This corresponds to a carrier density of about $1.6 \times 10^{17}/cm^3$, enough to saturate the free-exciton resonance as discussed in Section 3.6. Later switch-on was observed in a detector-limited time of 200 ps using a 10-ps, 600-nm, 1-nJ pulse focused to about 10 μm on the GaAs etalon held at about 80 K [Tarng, Tai, Jewell, Gibbs, Gossard, and Wiegmann (1981); Tarng, Tai, Jewell, Gibbs, Gossard, McCall, Passner, Venkatesan, and Wiegmann (1982); Gibbs, Jewell, Moloney, Tarng, Tai,

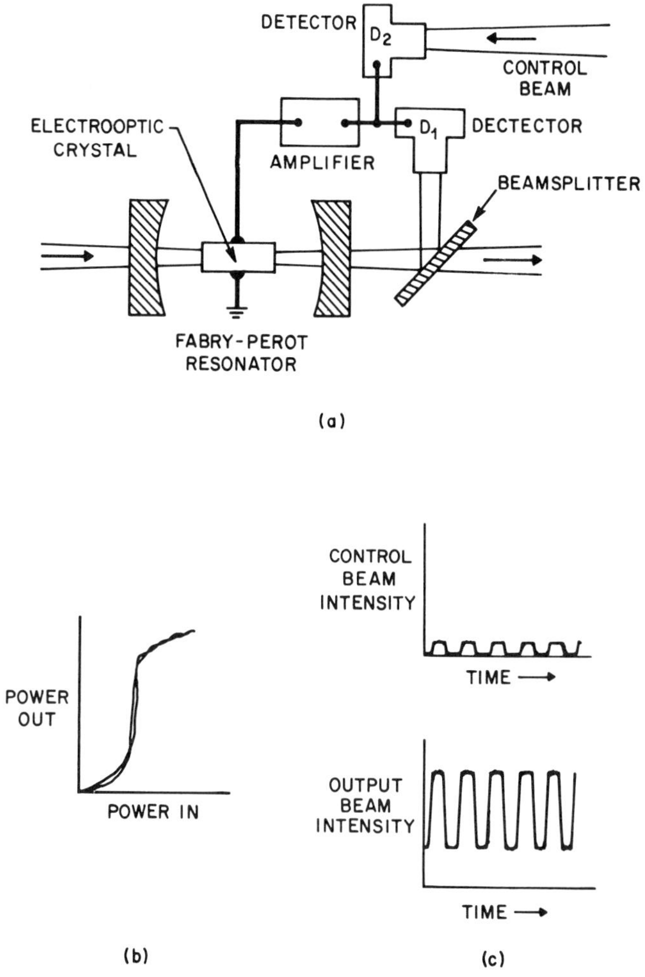

Fig. 5.4-1. (a) Setup for "optical triode" operation. (b) Experimental characteristic, (c) Input, and output for gain of 7X. P. W. Smith, Turner, and Maloney (1978).

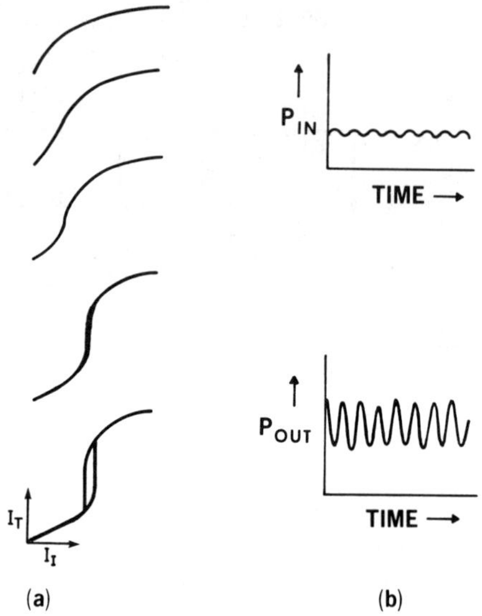

Fig. 5.4-2. (a) Experimental output characteristics as a function of resonator tuning. (b) Experimental input and output plots showing differential gain of 7X. P. W. Smith, Turner, and Maloney (1978). © 1978 IEEE

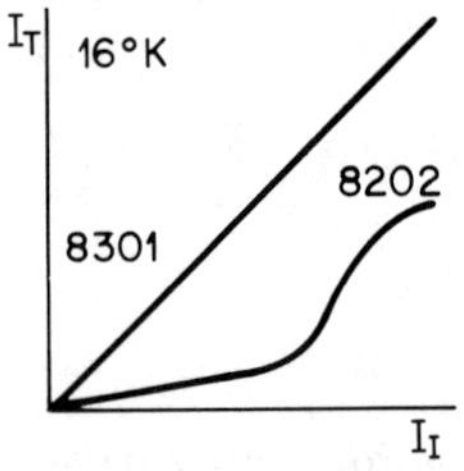

Fig. 5.4-3. AC gain characteristic in 4.1-µm-thick GaAs etalon with R = 0.9 coatings. Gibbs, McCall, Venkatesan, Gossard, Passner, and Wiegmann (1979).

Watson, Gossard, McCall, Passner, Venkatesan, and Wiegmann (1982)]. It is believed that 1-nJ switching energy was required even though the spot diameter was five times smaller because of rapid carrier diffusion. Shift of the Fabry-Perot peak in ≃ 1 ps for NOR-gate operation was described in Section 1.2. Other experiments suggesting 1-ps switch-on and saturation intensities of 1 fJ/μm^2 were presented in Section 3.6 [Ulbrich (1981)].

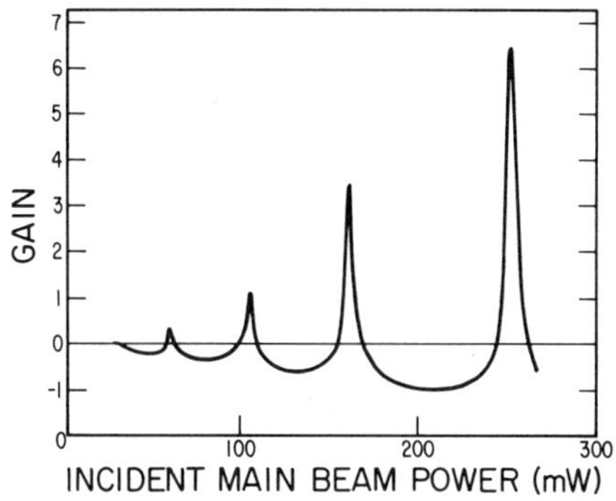

Fig. 5.4-4. Transphasor gain in InSb under conditions of Fig. 3.7-1. D. A. B. Miller and Smith (1979).

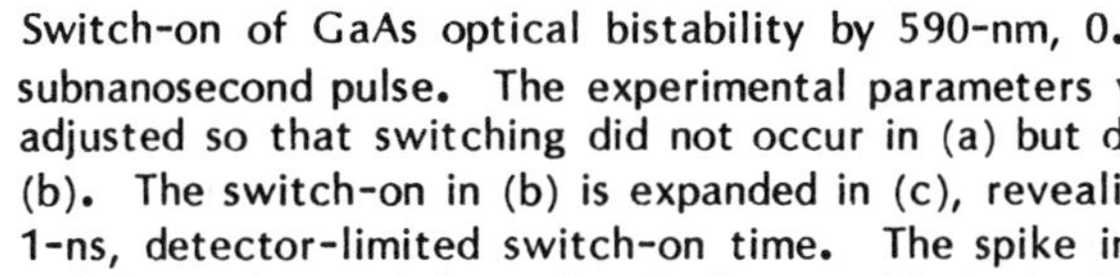

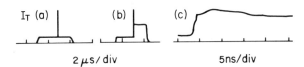

Fig. 5.5-1. Switch-on of GaAs optical bistability by 590-nm, 0.6-nJ subnanosecond pulse. The experimental parameters were adjusted so that switching did not occur in (a) but did in (b). The switch-on in (b) is expanded in (c), revealing a 1-ns, detector-limited switch-on time. The spike in (a) and (b) is absent if the red probe beam is blocked. Gibbs, McCall, Gossard, Passner, Wiegmann, and Venkatesan (1979).

S. D. Smith, Seaton, and Prise (1982) switch on a 200-μm-thick InSb bistable etalon with a 30-ps, 1.06-μm, 5-nJ pulse; they infer that the switch on should occur in the resonator build-up time of ≃ 8 ps, but they do not show it directly. Seaton, Smith, Tooley, Prise, and Taghizadeh (1983) use two 30-ps switch pulses, which together can switch on the InSb device but singly cannot, to demonstrate an AND gate or coincidence circuit. By delaying one pulse they measure 90 ns for the photogenerated carrier lifetime.

Switch-off was also reported in the fast switch-on GaAs experiment. Usually intrinsic bistable devices are switched off by reducing the input intensity below the switch-down value. For optical logic it would be preferable to accomplish switch-down with an external pulse. A 7-ns, 600-nm, 300-nJ pulse was used as the external switch-off pulse as shown in Fig. 5.5-2. The method is brute force in that the energy absorbed heats the etalon moving the hysteresis loop to higher input intensities. Switch-off occurs when the switch-down intensity exceeds the applied input intensity. Figure 5.5-3 shows external switch-on followed by external switch-off. Rapid switching back and forth would require rapid cooling between switch-off and switch-on pulses. Heat sinking through the etalon's thin (≃ 5-μm) dimension would accelerate cooling.

Hopefully, this first external switch-off of an intrinsic bistable device is just a beginning. Purely electronic mechanisms would be preferable. In a three-level system with one upper state, suppose the holding light is resonant with one transition and produces the nonlinearity by optical pumping of atoms to the other lower state. Then a pulse resonant with the other transition might be able to "unpump" the atoms causing switch-off to occur. Pumping with one polarization and switching off with another might also work. Rapid switching of the phase of the field incident on a dispersive device in which the medium response time is much shorter than the cavity response time τ_c has been shown numerically to switch the device off (Section 5.7). Since Ikeda instabilities (Section 6.3) sometimes occur when τ is shorter than the cavity round-trip t_R, the practicality or range of operation of phase switching needs to be determined. Switch-on and switch-off between stable states has always been found possible (Section 5.7) by intracavity phase shifting in computer simulations that do allow for Ikeda instabilities.

If fast external switch-off cannot be accomplished in a single device, it certainly can be accomplished with two. Jewell, Rushford, and Gibbs (1984) have proposed the use of a control etalon to switch-off a bistable etalon. The incident light passes through the control etalon on the way to the bistable etalon. Normally the control etalon is tuned to transmit the incident beam. The bistable etalon can be switched on by a picosecond pulse in the usual way. Switch-off is accomplished by sending a light pulse to the control etalon, detuning its peak from the laser frequency, and allowing the bistable etalon to switch off in the dark. If the recovery time of the control etalon is slightly longer than the switch-off time of the bistable etalon, the bistable etalon will stay in the "off" state when the incident light is restored. External switch-off using a control etalon has been demonstrated using dye etalons.

Schnapper, Papuchon, and Puech (1981) switch the output light from one channel to the other of a hybrid bistable directional coupler using switching information encoded in the incident light itself.

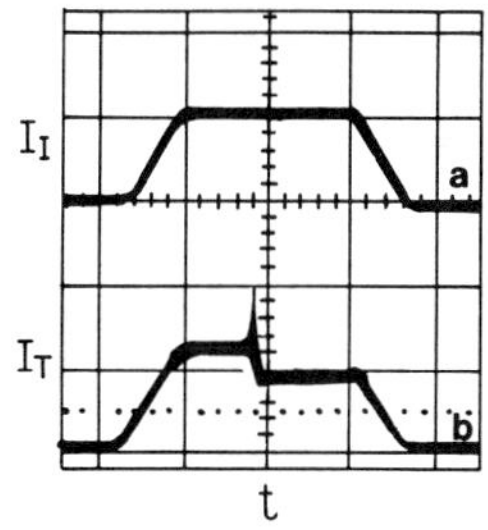

Fig. 5.5-2. Switch-off of GaAs optical bistability by a 7-ns, 600-nm, 300-nJ pulse. The etalon has high transmission for the duration of the pulse if the 300-nJ pulse is blocked (a), but it switches off if it is unblocked (b). 500 ns per major division. Tarng, Tai, Jewell, Gibbs, Gossard, McCall, Passner, Venkatesan, and Wiegmann (1982).

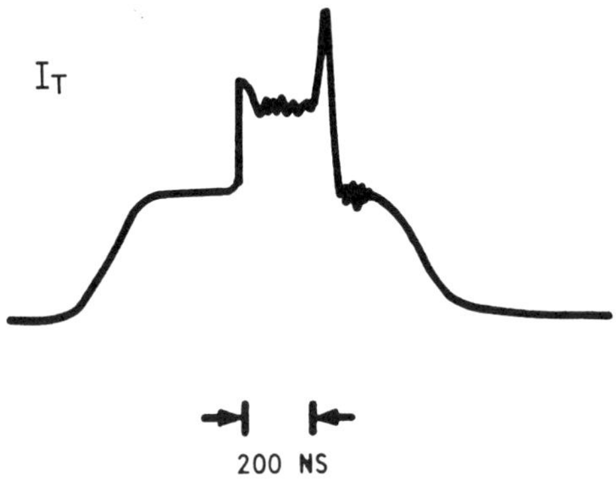

Fig. 5.5-3. Switch-on by a 10-ps, 1-nJ, 600-nm pulse followed by switch-off by a 7-ns, 300-nJ, 600-nm pulse. Tarng et al. (1982).

5.6. CRITICAL SLOWING DOWN

In Section 2.8 it was found that one of the eigenvalues λ_- of the matrix describing the evolution of an absorptive bistable system goes to zero near the switch-up point. Since the part of the evolution governed by that eigenvalue proceeds as $e^{+\lambda_- t}$, as $\lambda_- \to 0$ the response is very slow. This critical slowing down phenomenon is well known in phase transitions; it was first noted in optical bistability by Bonifacio and Lugiato (1976, 1978a) in an analysis of two-level atoms within a Fabry-Perot cavity; see also Narducci, Gilmore, Feng, and Agarwal

(1978), Agarwal, Narducci, Gilmore, and Feng (1978), and Bonifacio and Meystre (1978, 1979). Since the critical slowing down is the temporal counterpart of spectral narrowing (Section 2.8), it may be the best way to study the latter. Studying the time response is probably far easier than detecting a narrowing centered at the same frequency as the input beam.

Critical slowing down was first observed in bistability in a mirrorless hybrid device: Marburger, Allen, Garmire, Levenson, and Winful (1978); Allen, Garmire, Marburger,and Winful (1978); Garmire, Marburger, Allen, and Winful (1979). Grynberg, Biraben, and Giacobino (1981) have made an extensive study of transients in two-photon bistability in Rb and find excellent agreement with their theoretical analysis, which includes an analytic solution of the equations of motion for $\tau_M \gg \tau_c$; see also Giacobino, Cribier, Grynberg, and Biraben (1982); Cribier, Giacobino, and Grynberg (1983a,b). See Appendix H for a brief account of their work on critical exponents in dispersive optical bistability transients. Kimble and Grant (1982) report critical slowing down in intrinsic absorptive bistability in Na vapor. Mitschke, Deserno, Mlynek, and Lange (1983) see it in dispersive bistability using optical pumping of Na Zeeman levels. Barbarino, Gozzini, Maccarrone, Longo, and Stampacchia (1982) have seen critical slowing down in absorptive microwave bistability in NH_3. Harder, Lau, and Yariv (1982b) and Lau, Harder, and Yariv (1982a) report it in a laser diode containing a saturable absorber.

The experimental configuration of Garmire, Marburger, Allen, and Winful (1979), shown in Fig. 5.6-1a, consists of a Pockels cell modulator M driven by the output of a photoconductor PC. The operating characteristics without ($\mathscr{T}(V)$ curve) and with (V_{out} vs V_{in}) feedback are shown in Fig. 5.6-1b. They study the response when light from a 0.5-mW cw He-Ne laser is suddenly turned on to a value V_{in} in voltage units. A switching increment is defined as $\nu = (V_{in} - V_{\uparrow})/V_{\uparrow}$ where $V_{\uparrow}$ is the minimum input intensity (in voltage units) for switch-up in steady-state. Transient responses are shown in Fig. 5.6-2a, clearly demonstrating the critical slowing down phenomenon. The response time τ_R/τ_M, in units of the feedback modulator response time τ_M, is shown in Fig. 5.6-3 as a function of ν; τ_R is defined as the time required to reach 1/e of the steady-state value. In an intrinsic system τ_M corresponds to the response time of the nonlinearity. The response time τ_c of the output signal to the input signal is the response time of the detector current in their hybrid and corresponds to the cavity build-up time in a Fabry-Perot cavity. In the hybrid experiment $\tau_c \simeq 0.01$ s and $\tau_M \simeq 0.2$ s. The solid curve in Fig. 5.6-3 is obtained from the dynamical equation

$$\tau_c \tau_M \frac{\partial^2 V}{\partial t^2} + (\tau_c + \tau_M) \frac{\partial V}{\partial t} = V_{in} \mathscr{T}(V + V_B) - V \qquad (5.6\text{-}1)$$

by ignoring the first term. Equation (5.6-1) describes exactly the hybrid device of Fig. 5.6-1a, models approximately all types of nonresonant bistable devices, and gives the qualitative features observed and simulated by Bischofberger and Shen (Section 5.1). Time dynamics calculated using Eq. (5.6-1) are shown in Fig. 5.6-2b. Garmire et al. find that the switching energy tends to infinity as $\nu \to 0$ and conclude that to minimize the switching energy it is necessary to use devices with high extinction ratios which in turn require high switching intensities.

Equations (2.4-4a and 5b,c) are the mean-field coupled Maxwell-Bloch equations for purely absorptive optical bistability. In the so-called "good cavity" limit, $\gamma_c \ll \gamma_T, \gamma_L$, the cavity response is much slower than the medium response. Consequently, the medium polarization adiabatically follows the field, and v can be

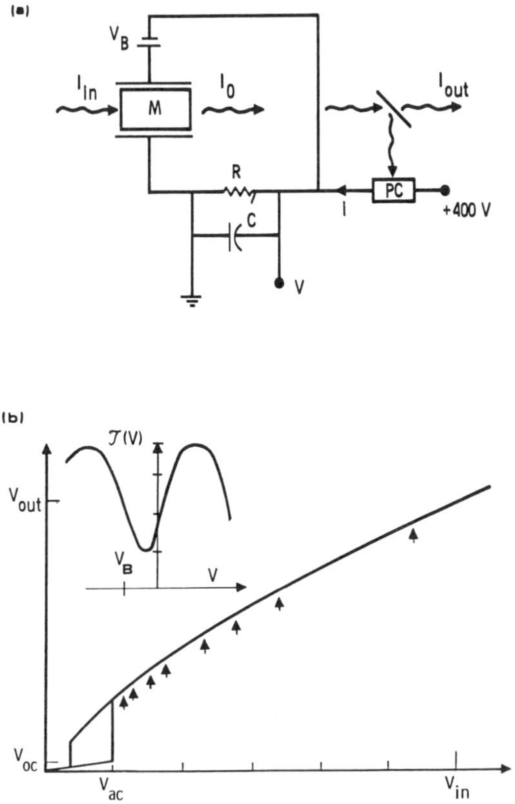

Fig. 5.6-1. (a) Simplified schematic of the hybrid bistable optical device. M = modulator, PC = photoconductor, and V_B = bias voltage. (b) Measured steady-state transfer curve using modulator whose transmission as a function of voltage is shown in the inset. Garmire, Marburger, Allen, and Winful (1979).

approximated by Eq. (2.4-14). Substitution of v into Eq. (2.4-4a), and using Eqs. (2.4-11, 12, 17) with $n_0 = 1$, one finds replacing $\partial\xi/\partial z$ by $\gamma_c(\xi - \xi_I/\sqrt{T})/c$ where $\gamma_c = cT/2Ln_0$ is the cavity decay rate,

$$\frac{\partial x}{\partial \gamma_c t} = -\left[x - y + \frac{2Cx}{1 + x^2}\right]. \qquad (5.6\text{-}2)$$

This equation has been solved numerically by Bonifacio and Meystre (1979) and analytically by Benza and Lugiato (1979b). This equation also describes the motion of a marble in the potential given by Eq. (2.7-1). For y above $y_\downarrow$ but less

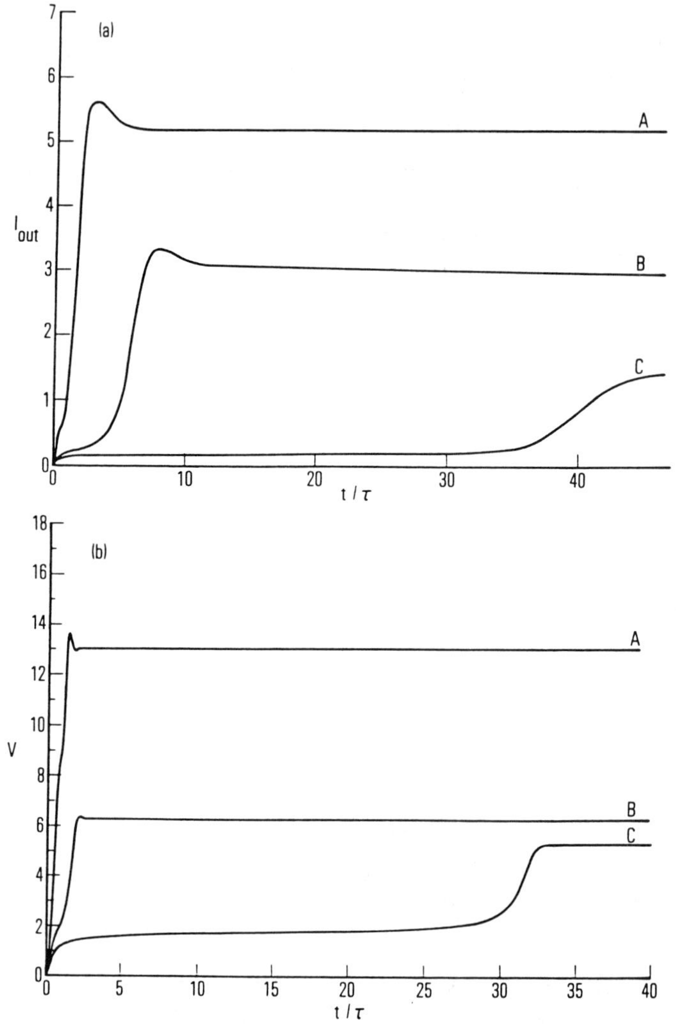

Fig. 5.6-2. (a) Experimental curves showing output as a function of time for three different step inputs. The values of the incremental switching intensity are (A) $\nu = 4.6$, (B) $\nu = 1.4$, and (C) $\nu = 0.03$. (b) Theoretical curves of output signal versus time, generated by solving Eq. (5.6-1) with incremental switching intensities of (A)$\nu = 4.6$, (B) $\nu = 1.0$, and (C) $\nu = 0.01$. Garmire, Marburger, Allen, and Winful (1979).

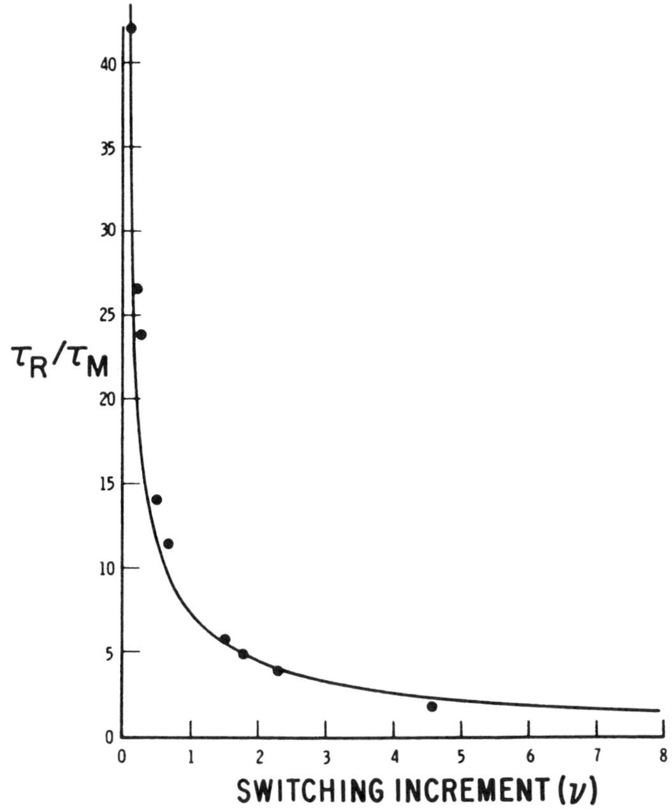

Fig. 5.6-3. Characteristic switching time versus increment of input switching signal beyond critical value. The theoretical plot was obtained from Eq. (5.6-1). Garmire, Marburger, Allen, and Winful (1979).

than the switch-up value $y_\uparrow$, the potential has two wells corresponding to two stable states of the bistable system. As $y \to y_\uparrow$, one of the wells becomes shallower and shallower, becoming a plateau for $y = y_\uparrow$. For $y > y_\uparrow$, the only minimum left corresponds to the upper state. For $y \simeq y_\uparrow$, the system, moving along a potential which is essentially flat, will take an extremely long time to reach its steady state. As y exceeds $y_\uparrow$ more and more, the potential steepens, and the delay decreases. Figure 5.6-4a shows the transient response given by Eq. (5.6-2) when y is suddenly changed from 0 to an operating value y_{op}. In general, $x(\gamma_c t)$ exhibits a plateau with a very slow time evolution, followed by a steep rise to the stationary value. As y_{op} approaches $y_\uparrow$ from above, the duration of the plateau increases very rapidly, whereas the slope of the steep part remains practically unchanged. Bonifacio and Meystre (1979) propose that this sensitivity to input field be used to design an electric-field-to-time converter. The approximate switching time is shown in Fig. 5.6-4b. Note that in this good cavity limit if y_{op} exceeds $y_\uparrow$ by only 10% (i.e., $I_I = 1.21\ I_\uparrow$), $\tau_R \simeq \gamma_c^{-1}$ so the switching time is approximately one cavity response time. In the bad-cavity limit of Fig. 5.6-3, $y_{op}^2 \simeq 2y_{\uparrow}^2$ resulted in $\tau_R \simeq 5/\gamma_2$.

Grant and Kimble (1983) have used their Na atomic-beam absorptive bistability apparatus (Section 3.2.5b) to study critical slowing down in an intrinsic system; see also Kimble and Grant (1982) and Rosenberger, Orozco, and Kimble (1984b). The input power P_I is turned on suddenly from 0 to some operating value P_{op} and the transmission $\mathscr{T}(t)$ is observed as a function of time. If $P_I >> P_{\uparrow}$, the time to reach within 90% of the steady-state transmission is close to the measured empty-cavity filling time $\tau_c = 633 \pm 100$ ns; see Fig. 5.6-5a. With $P_I = 1.04\, P_{\uparrow}$, Fig. 5.6-5b, the response is essentially undetectable for 3 μs followed by switch-on, resulting in $\tau_R \simeq 4$ μs. The results of many similar pulsed responses constitute Fig. 5.6-5c, clearly quite similar to Fig. 5.6-4b. In particular, their Na measurements confirm the predictions of Bonifacio and Lugiato (1978a) that the switch-up time can approach the empty-cavity response time if the input power is switched to a value exceeding $P_{\uparrow}$ by only 10 to 20% (in the good cavity limit, i.e., $\tau_M << \tau_c$).

Karpushko and Sinitsyn (1978) also saw critical slowing down in an intrinsic system, namely thin-film interference filters in which they observed thermal bistability (Section 3.5).

Hopf and Meystre (1979) have treated numerically the application of a switch pulse y_s of the form

$$y_s = A \exp[-(t - t_0)^2/\hat{t}^2] \qquad (5.6\text{-}3)$$

to a bistable system with steady input y_0 and switch-up value $y_{\uparrow}$. They assume purely absorptive bistability in the mean-field, good-cavity limits. Figure 5.6-6 shows the minimum pulse width $\hat{t}_m$ required for each value of A. For $A >> (y_{\uparrow} - y_0)$, they find that $A\hat{t}_m \simeq$ cst, indicative of a pulse area scaling law or energy per unit bandwidth scaling law. As $A \rightarrow (y_{\uparrow} - y_0)$, the scaling ceases to hold, and it takes extremely long pulses to activate the bistable memory (critical slowing down).

Critical slowing down can exacerbate the difficulty of finding sufficiently fast nonlinearities; the device may respond 5 to 10 times more slowly than the medium response time. The impact of critical slowing down upon semiconductor etalons has not been evaluated. Sections 1.2 and 5.5 clearly show very rapid switch-on in GaAs; in many cases the nonlinearity responds essentially instantaneously (<1 ps in GaAs) to the light. It is the response upon removal of the light which is slow; hot carriers must cool and recombine. So critical slowing down in switch-down is likely to be much more important for device considerations. The discussion here has been primarily for switch-on, but the situation is similar for switch-off [Grynberg, Biraben, and Giacobino (1981)]. Logic operations that permit switch-off in the dark, such as the NOR gate of Section 1.2, should eliminate critical-slowing-down limitations on switch-off.

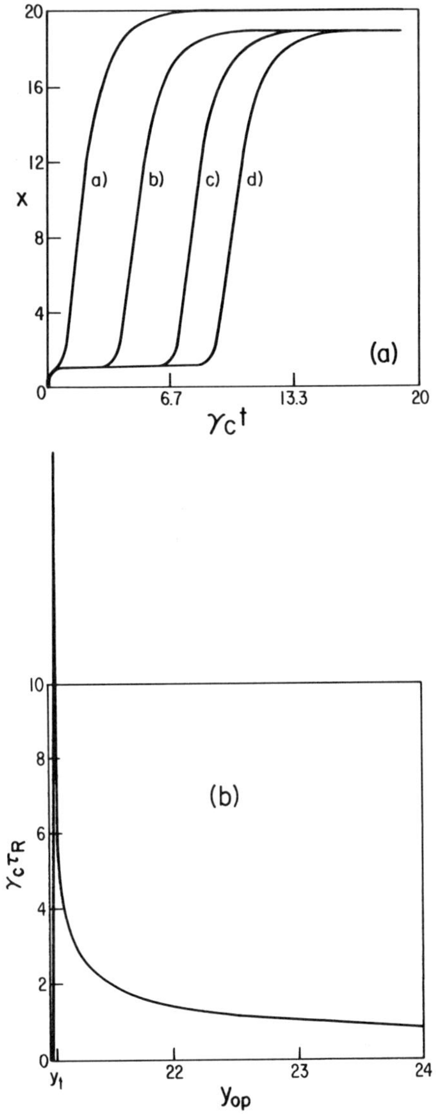

Fig. 5.6-4. (a) Time evolution of the transmitted field x in the mean-field approximation and good cavity limit for a) $y_{op} = 22$, b) $y_{op} = 21.1$, c) $y_{op} = 21.05$, d) $y_{op} = 21.04$. For $C = 20$ one has $y_{\uparrow} = 21.0264$. (b) Plot of the switching time vs the operating value y_{op} of the incident field for $C = 20$ for a good cavity and in the mean-field approximation. The switching time diverges for $y_{op} \to y_{\uparrow}$. Benza and Lugiato (1979b).

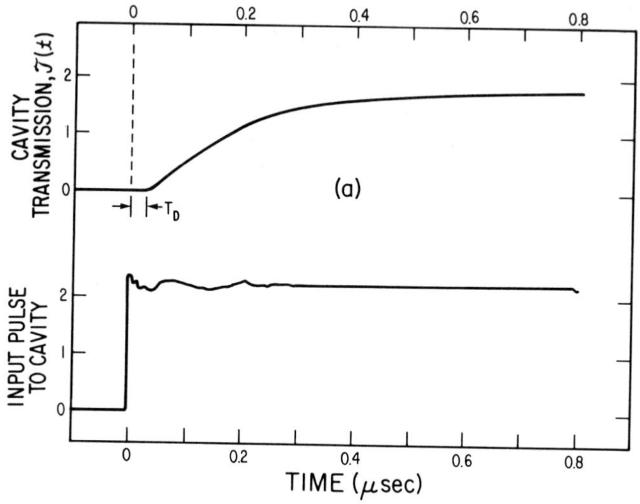
0
0.2
0.4
0.6
0.8
CAVITY TRANSMISSION, 𝒯(t)
2
1
0
(a)
T_D
INPUT PULSE TO CAVITY
2
1
0
0
0.2
0.4
0.6
0.8
TIME (μsec)

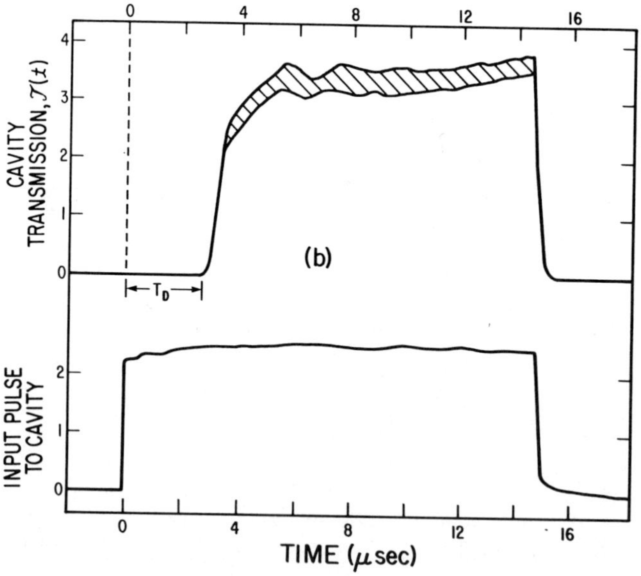
0
4
8
12
16
4
3
2
1
0
CAVITY TRANSMISSION, 𝒯(t)
(b)
T_D
INPUT PULSE TO CAVITY
2
1
0
0
4
8
12
16
TIME (μsec)

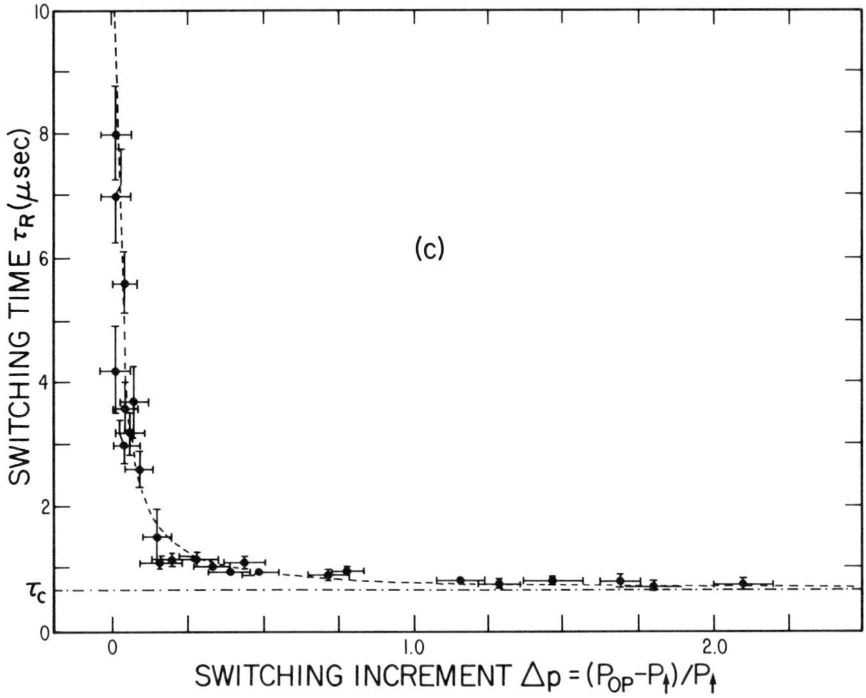

Fig. 5.6-5. (a) Cavity response $\mathscr{T}(t)$ to a near step-function change in incident power from zero to $P_{op} = 3P_\uparrow$ with $P_\uparrow$ as defined in Fig. 3.2-19. The small initial delay T_D is not present in the empty-cavity case. $\alpha_0 L = 0.94$. (b) As in (a), but with $P_{op} = 1.40\ P_\uparrow$. Note the change in time scale between (a) and (b). The shaded region denotes a noisy signal that is much the same for an empty cavity. $\alpha_0 L = 0.85$. (c) Time τ_R taken to reach 90% of steady-state output as a function of switching increment Δp above the switch-up power. Each point is obtained from a measurement such as in (a) or (b). The dashed line is the best fit of the function $\tau_c[1 + a/\Delta p]$, with $a = 0.5$ and is sketched only as a visual aid. The dot-dashed line is the empty-cavity response time τ_c. Grant and Kimble (1983).

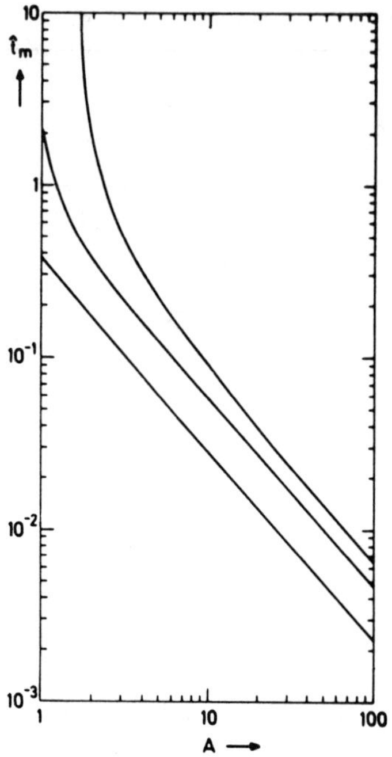

Fig. 5.6-6. Minimum pulse width $\hat{t}_m$ as a function of the pulse amplitude A for various values y_0 of the cw laser field. Upper curve: $y_0 = 20.8$, middle curve: $y_0 = 20.3$, lower curve: $y_0 = 19.6$. The other parameters are $C = 20$, $y_\uparrow = 21$, $\gamma_1^{-1} = 10$, $\gamma_2^{-1} = 20$, $\gamma_C^{-1} = 1$. Hopf and Meystre (1979).

5.7. PHASE-SHIFT SWITCHING

In absorptive optical bistability the electric field can be treated as a real variable; the approach to the stationary state after a perturbation is always monotonic. In dispersive optical bistability the electric field is necessarily a complex quantity which evolves in a two-dimensional phase space. Hopf, Meystre, Drummond, and Walls (1979) emphasize that intensity information is insufficient to predict switching behavior. Thresholds for switching between stable states do not occur at the same values as in the quasi-stable state; instead, anomalous thresholds are predicted.

5.7.1. Anomalous Switching and Input-Phase-Shift Switching

Hopf, Meystre, Drummond, and Walls (1979) consider a single mode interacting with a nonlinear medium described by $P = \chi^{(1)}E + \chi^{(3)}E^{3}$, which responds so rapidly that the medium can be adiabatically eliminated. They find that for the semiclassical equation of motion for the electric field, corresponding to Eq. (5.6-2),

$$\frac{\partial x}{\partial \gamma_C t} = -[1 + i(\omega_C - \omega)/\gamma_C]x - 2i\zeta|x|^2x/\gamma_C , \qquad (5.7\text{-}1)$$

where $\zeta \propto \chi^{(3)}$ and ω_C and ω are the cavity and laser angular frequencies respectively. They numerically solve Eq. (5.7-1) for the intracavity field when the input is switched from a steady-state y_0 to a new value y_{op}. They find an anomalous threshold such that, for $y_0 = 0$, switching sometimes occurs for $y_{op} < y_{\uparrow}$. Also, if the system initially is in the upper state, reduction of y to $y_{op} > y_{\downarrow}$ can result in switch-down depending upon y_0 and y_{op}. [This anomalous switching is called overshoot switching by Goldstone and Garmire (1981a); since the output often overshoots for normal switching, the name "anomalous switching" seems more definitive.] They calculate the trajectories of the field in the phase space Im(x) versus Re(x) and show that small changes in y_0 can change the final x values from the upper to the lower state or vice versa. They also find that when the system's trajectory goes close to one state but finally evolves to the other state, the switching can be very slow. This appears much like critical slowing down (Section 5.6), but in this case y need not be close to $y_{\downarrow}$ or $y_{\uparrow}$. Hopf and Meystre (1980) generalize the model to allow large mirror transmission and large phase shifts, appropriate for low-finesse multiple-order systems [uncoated GaAs (Section 3.6) and InSb (Section 3.7) etalons]. The anomalous threshold disappears under these conditions, but down-switching is performed more easily than in the $T \rightarrow 0$ case. Lugiato, Milani, and Meystre (1982) give analytical and graphical descriptions of anomalous switching.

Goldstone and Garmire (1981a) treat the dynamic response of a Fabry-Perot interferometer containing a nonlinear medium obeying a Debye relaxation equation

$$\tau_M \frac{\partial \Delta\phi}{\partial t} + \Delta\phi = \beta_0 + \beta_2|E_T|^2 \qquad (5.7\text{-}2)$$

similar to Eq. (5.1-8). They show that the field must satisfy

$$t_R \frac{\partial E_T(t)}{\partial t} + [1 - Re^{i\Delta\phi(t)}]\, E_T(t) = TE_I(t + t_R/2). \qquad (5.7\text{-}3)$$

Eqs. (5.7-2,3) give the steady-state curve in Fig. 5.7-1 and the transients in Fig. 5.7-2 in which anomalous switching and critical slowing down are apparent. The input is instantaneously switched from 0 to I_{op}. For step inputs $I_{op} < I_{\uparrow}$, if the overshoot value of $\Delta\phi$ exceeds $\Delta\phi_b$ (see Fig. 5.7-1), the cavity will anomalously switch to the high-transmission mode. This can occur at any input intensity as much as 30% below that required for steady-state switching; see Fig. 5.7-3. Goldstone and Garmire also find conditions under which the output alternates between high and low transmission states producing a subharmonic of the incident pulse train. This alternate switching has been observed using hybrid devices by Goldstone, Ho, and Garmire (1980) and by Martin-Pereda and Muriel (1982).

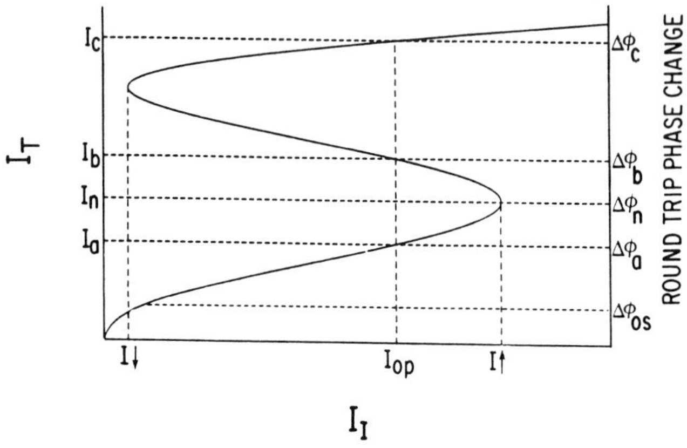

Fig. 5.7-1. A representative steady-state transfer curve, describing the dependence of both output intensity and round-trip phase change on the input intensity. Goldstone and Garmire (1981a). © 1981 IEEE

Goldstone and Garmire (1981a) also use their model to study switching by means of input phase shifts. In a multistable Fabry-Perot cavity they find level-by-level as well as two-level switches accomplished by anomalous switching by means of input phase shifts. When the Fabry-Perot cavity has only one effective time constant, rather than both t_R and τ_M, they find downward but not upward phase switching consistent with the work of Hopf and Meystre (1980).

Nishiyama (1980b) studies switching of a nonlinear absorptive Fabry-Perot cavity when the relative phase between two input beams incident from opposite sides is suddenly changed.

A rapid shift in the input phase may produce a sudden change in the interference between the cavity and input fields, which may enhance or reduce the cavity intensity until the cavity has time to adjust to the new input phase. The transient response of the nonlinear index may cause the cavity to switch to another steady-state output state for the same input intensity. Hopf, Meystre, Drummond, and Walls (1979) and Hopf and Meystre (1980) treat the response of a purely dispersive bistable optical device to an instantaneous change of phase of the driving field; instantaneous means fast compared to the cavity response time γ_c^{-1}. They show that, by repetitive calculations, one can divide the phase plane of E_i versus E_r into two regions. If the intracavity field is perturbed into one region it will evolve to the lower state; if perturbed into the other, it will evolve to the upper state. This approach is especially simple for intracavity phase shifts; see Figs. 5.7-4,5,6.

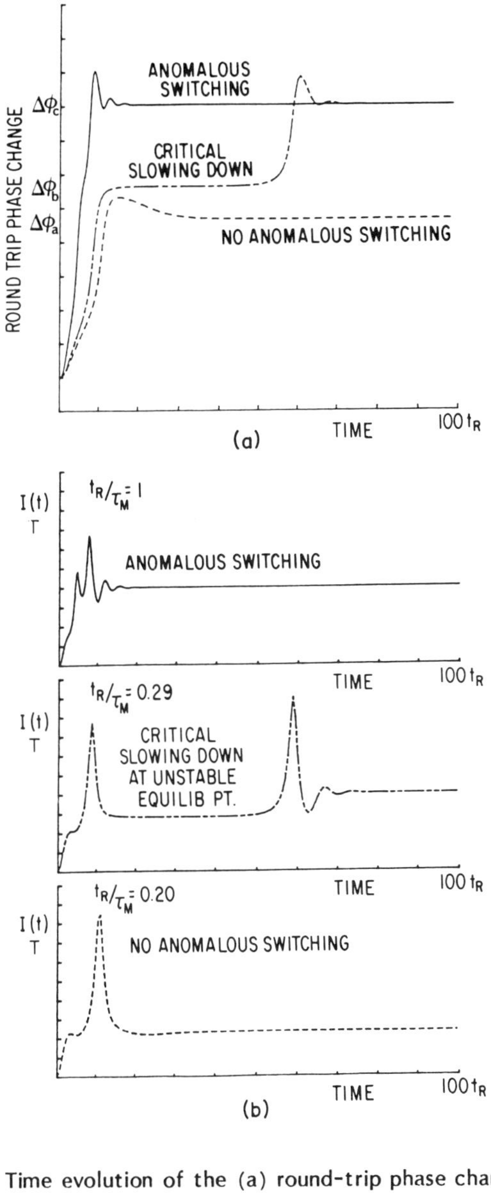

Fig. 5.7-2. Time evolution of the (a) round-trip phase change and (b) output intensity for a nonlinear Fabry-Perot cavity operating in the anomalous switching region for step inputs. Step input intensity is the same in all three cases. Goldstone and Garmire (1981a). © 1981 IEEE

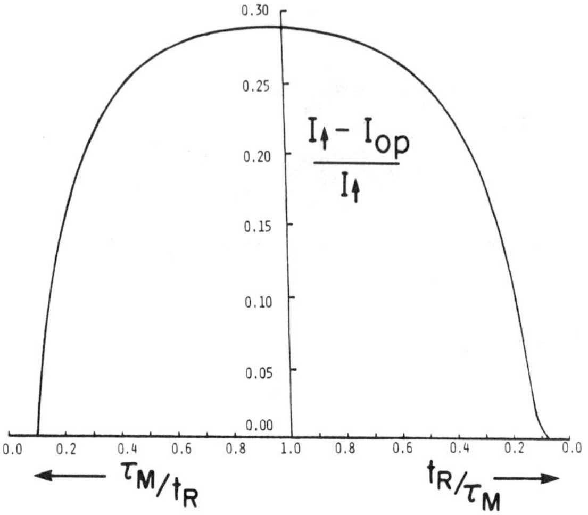

Fig. 5.7-3. Anomalous switching as a function of the ratio of system time constants. Goldstone and Garmire (1981a). © 1981 IEEE

5.7.2 Intracavity-Phase-Shift Switching

Section 5.7.1 deals with anomalous switching and with switching by a sudden change of the phase of the input field. In this section an external pulse shifts the laser-cavity detuning for a prescribed number of round-trip times by modifying the background refractive index of the medium [Tai, Gibbs and Moloney (1982) and Tai, Moloney, and Gibbs (1982b)]. This is called intracavity-phase-shift switching to distinguish it from the input-phase-shift switching of Hopf et al.

The bistable system is assumed to be initially in one of the two states with $I_\downarrow < I_I < I_\uparrow$. An external pulse is applied to change the cavity-laser detuning for a certain number of round trips and is then removed. One studies whether or not switching occurs as a function of the magnitude, sign, and duration of the change. The time for the system to return to equilibrium after the pulse is usually several times the cavity build-up time $\tau_C = \gamma_C^{-1} = L_T/cT$ where L_T is the total round-trip cavity length. Tai et al. allowed for Ikeda instabilities for calculations in the good-cavity limit ($\gamma_L^{-1}, \gamma_T^{-1} \ll t_R$). With operating parameters selected to avoid Ikeda instabilities, they found that intracavity phase switching can always cause switching. They included transverse effects in one dimension by numerically integrating Eq. (2.9-13) with the boundary condition

$$E(\tau,0) = \sqrt{T}\,E_I + R\exp[i(\beta_0 + \Delta\beta_0)]\,\eta E(\tau - t_R,0) \qquad (5.7\text{-}4)$$

where $\tau = t - z/c$ is the retarded time and

$$\eta = \exp\left[\int_0^{L_{NL}} \left[-\frac{\alpha_0}{2}\right] \frac{1 + i\Delta}{1 + \Delta^2 + |E(\tau - t_R, z)|^2/E_s^2}\, dz\right] \qquad (5.7\text{-}5)$$

and β_0 is the laser empty-cavity detuning, $\Delta\beta_0$ is the change in the intracavity phase shift induced by the external switching pulse, and L_{NL} is the length of the nonlinear medium.

Figure 5.7-4 compares the usual bistability loop (I_T versus I_I) with the phase-plane representation of the steady-state intracavity field (E_i vs E_r). Figure 5.7-5 shows the evolution from an initial upper state for 20 round trips under various detuning changes $\Delta\beta_0$. By properly choosing the magnitude and duration of the detuning change, the device can be set "on" (Fig. 5.7-6a) no matter which state it was in. Or with other length changes it can be switched to the other state (Fig. 5.7-6b) or switched off (Fig. 5.7-6c). These intracavity-phase-shift switching operations have been computed assuming a fast medium response, but the method works for a slow medium, too, if the sign, magnitude, and duration of $\Delta\beta_0$ can be selected at will. The switch-on and NOR-gate operations of GaAs etalons by switch pulses that see no cavity, but generate carriers that shift the cavity peak for beams being controlled, are basically intracavity-phase-shift switchings but with $\tau_M >> \tau_C$. Of course, switch-off and NOR-gate recovery are limited by τ_M; an intracavity phase shift can reduce the intracavity field for a few t_R, but if τ_M is much longer than t_R the medium's polarization will not change and the intracavity intensity will soon recover its initial high value.

5.8. PICOSECOND GATING

In Section 5.3 several techniques for controlling one light beam with another were presented that did not require that the nonlinear device be bistable. Those techniques capable of picosecond gating are of particular interest since one of the goals of all-optical devices is very fast (subpicosecond) switching [Smith and Tomlinson (1981a)].

Duguay and Hansen (1969) have developed an ultrafast shutter based on the optical (or AC) Kerr effect predicted by Buckingham (1956) and first observed by Mayer and Gires (1964). The ultrafast optical Kerr shutter uses optically induced birefringence to gate the light on and off on a picosecond time scale, much the way a conventional Kerr cell uses an electrically induced birefringence to gate light on and off on a nanosecond scale. A simple schematic of the arrangement is shown in Fig. 5.8-1. Ippen and Shank (1975, 1978) have operated a CS_2 Kerr shutter with subpicosecond pulses from a mode-locked cw dye laser allowing a very high repetition rate (≃ 100 MHz). Their gate resolution is limited by the 2-ps orientational response time of CS_2.

An ultrafast all-optical gate with 5-ps resolution has been demonstrated by Lattes, Haus, Leonberger and Ippen (1983) using a nonlinear $LiNbO_3$ Mach-Zehnder interferometer.

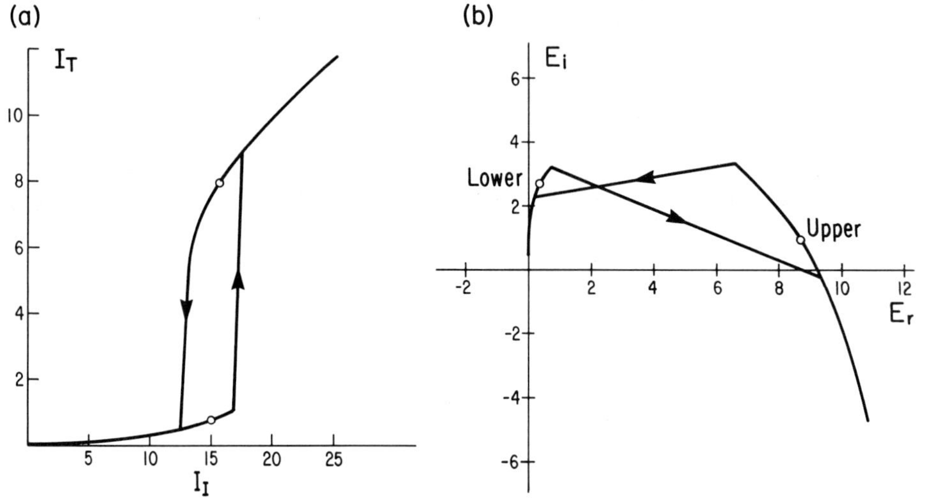

Fig. 5.7-4. The steady-state solution exhibited by the usual hysteresis loop. The parameters are as follows: R = mirror reflectivity = 0.9, $\mathcal{F}$ = cavity finesse = 30, instrument width = $2\pi/30$ radians, Δ = (laser-atom detuning)/(resonance half-width) = +5 (defocusing side), β_0 = laser empty-cavity detuning = +0.2 radians, β = low-intensity laser full-cavity detuning = -0.57 radians ($\simeq$ 2.7 instrument widths), $\alpha_0 L_{NL}/(1 - R)$ = bistability constant = 80, where L_{NL} = 0.3L is the length of the nonlinear medium. $I_{\uparrow}$ = switch-up intensity = 17, $I_{\downarrow}$ = switch-down intensity = 12.5. The same cavity parameters are used in Figs. 5.7-5,6. (b) The steady-state solution in the phase plane of the internal field at the output mirror. The arrows indicate the direction of switching as the input driving intensity goes above $I_{\uparrow}$ and below $I_{\downarrow}$. The two foci encircled in the phase plane are for constant input holding intensity of 15. The same holding intensity is used in Figs. 5.7-5,6. The upper branch focus has transmission = 52% and detuning = +0.007 radians ($\simeq$ 0.03 instrument widths). The lower-branch focus has a transmission equal to 5.6% and a detuning = -0.394 radians ($\simeq$ 1.89 instrument widths). Tai, Gibbs, and Moloney (1982).

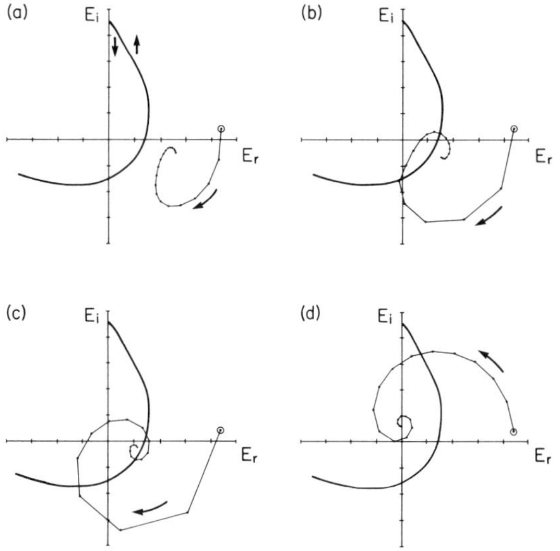

Fig. 5.7-5. Time evolution loci of the field in the phase plane driven by the external switching pulse for the first 20 round trips. The field starts at the upper-branch focus. The cavity parameters are the same as in Fig. 5.7-4. The dots correspond to the round trips. The critical boundary which separates the domains of stability for the two foci are superimposed in the figures. The detuning changes $\Delta\beta_0$ for (a), (b), (c), and (d) are -0.31, -0.62, -0.91, and 0.31 radians, respectively, i.e., about -1.5, -3, -4.5, and 1.5 times the instrument width. Tai, Gibbs, and Moloney (1982).

Low-power one-picosecond gating has been reported using a GaAs etalon by Migus, Antonetti, Hulin, Mysyrowicz, Gibbs, Peyghambarian, and Jewell (1985). As in the optical modulation in GaAs (Section 5.3), the coatings may be highly transparent to the control pulse and highly reflecting for the beam being gated. The semiconductor etalons can be scanned very rapidly (approximately picoseconds) through the laser frequency but recovery requires a few nanoseconds at present. See also Section 5.5; the picosecond semiconductor etalon lasers discussed in Section 3.6 are gates in a broad sense.

Subpicosecond gating has been achieved using optical phase conjugation by Bloom, Shank, Fork, and Teschke (1978). Gate times shorter than the molecular relaxation times are easily obtained. Optical phase conjugation [Fisher (1983)] is a form of degenerate four-wave mixing in which two pump (often counterpropagating) waves are mixed with an incident probe wave; see Fig. 5.8-2. A physical picture of the gating process can be described in the following manner.

In the Kerr medium the probe wave and each of the pump waves interact through the nonlinear refractive index to produce a phase grating, and the

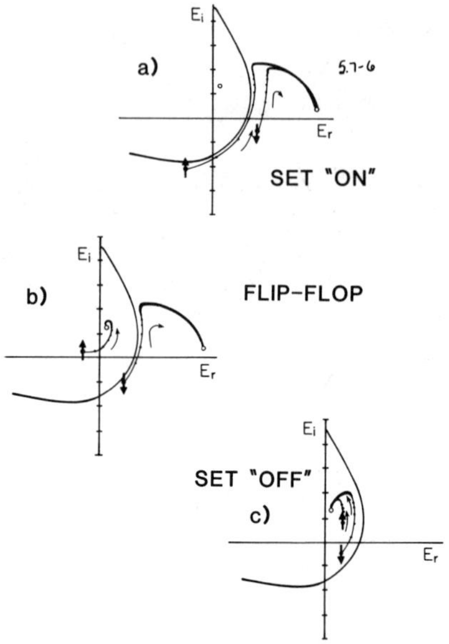

Fig. 5.7-6. The dynamical routes of recovery of the field after removal of a detuning change of $\Delta\beta_0$ = -0.93 radians impressed for (a) 3, (b) 5, and (c) 7 round trips. The ↓ (↑) arrows indicate that the device was initially in the lower (upper) state before the $\Delta\beta_0$ change. In (a), the final state is the "on" state regardless of the initial state. In (b), the final state is always the opposite of the initial state. In (c), the final state is the "off" state. Tai, Gibbs, and Moloney (1982).

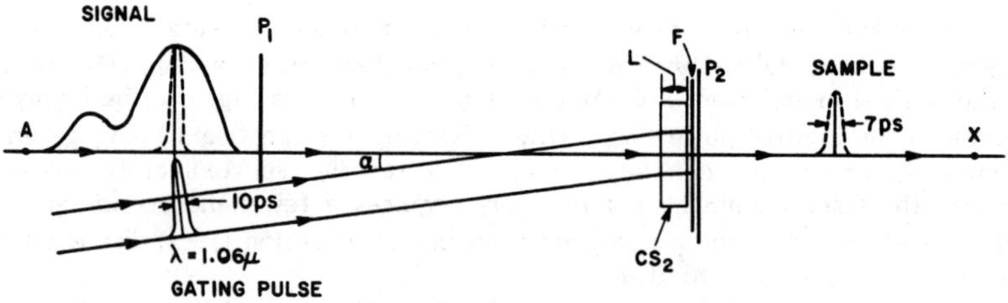

Fig. 5.8-1. The ultrafast optical Kerr shutter in its simplest form. Polarizers P_1 and P_2 are crossed and have their polarization axes at 45° to the plane of polarization of the gating laser pulse. Filter F greatly attenuates the gating pulse to prevent possible damage to P_2. In some cases F is unnecessary. Duguay (1976).

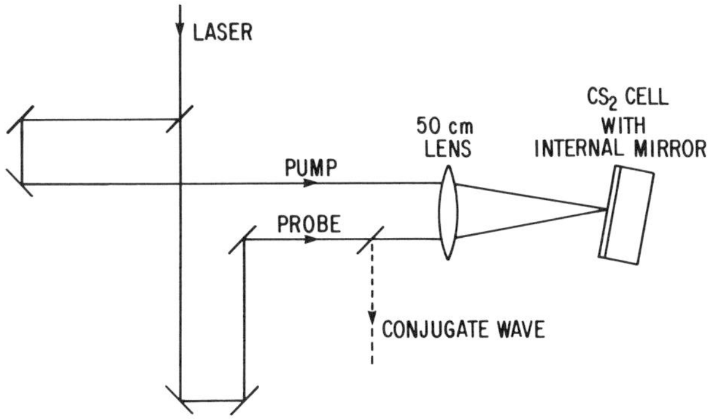

Fig. 5.8-2. Schematic of experimental arrangement for subpicosecond optical gating. Bloom, Shank, Fork, and Teschke (1978).

counterdirectional pump wave is diffracted by this grating to produce the backward-traveling conjugate wave. Furthermore, while the induced hologram will remain for the molecular relaxation time of the Kerr medium, the conjugate wave is generated only if both pump waves are present. Exact time coincidence between two pump pulses is not essential. It is possible to use a first pump pulse to "write" a hologram and a second delayed counterpropagating pulse to "readout" the hologram. However, the delay between the write and read pulses must be less than the molecular relaxation time. The noise background of this technique is limited only by scattering since the generated signal is produced in a direction far different from that of the incident beams. In addition, a conjugate wave will be generated only if the frequency of the object light is close to that of the pumps. Phase matching restricts the frequency bandwidth to $\Delta f = c/L$, where L is the interaction length. Since the interaction length is typically of order $c\tau_p$ (where τ_p is the pump pulse length), the acceptance bandwidth is approximately equal to the pump pulse bandwidth.

For other picosecond optoelectronic devices see Lee (1984).

CHAPTER 6

INSTABILITIES: TRANSIENT PHENOMENA WITH CONSTANT INPUT

Chapter 5 focused on transient phenomena initiated by a change in the input field amplitude and/or phase or by an external control pulse. This chapter examines transient phenomena that arise either from instabilities in the bistable systems themselves (Sections 6.1 to 6.4) or from fluctuations in the number of nonlinear atoms or in the number of intracavity photons (Section 6.5).

6.1. REGENERATIVE PULSATIONS BY COMPETING MECHANISMS

McCall (1978) proposed a relaxation oscillation type of instability for a bistable system in which the nonlinear refractive index has two contributions of opposite sign and with different time constants. For example, suppose the nonlinear optical susceptibility is $\sigma = \sigma_t - \sigma_e$, where σ_t is a thermal contribution and σ_e is an electronic or excitonic contribution. A regenerative pulsation mode can be heuristically understood (Fig. 6.1-1). Let σ_t be included in β, the detuning parameter, now time and intensity dependent. Imagine the thermal time constant to be relatively long, although in micrometer-size devices it could be quite short. Allow the input power to be turned on at $t = 0$, and suppose the device to be bistable and tuned so that the input power exceeds the $t = 0$ switch-up power. Then the device is in the highly transmitting mode, and a small amount of absorbed power begins to warm the device. Consequently, the detuning parameter β slowly changes, and the characteristic curve changes with the bistable region moving toward larger input powers. When the switch-down level crosses the input power level, the output switches to the lower level, and the device begins to cool. The bistable region then moves toward lower powers, until the switch-up level crosses the input power level; then the output switches to the upper level, and so forth. The output thus described will settle down to a periodic sequence of pulses.

McCall (1978) also demonstrated regenerative pulsations using a hybrid bistable device; see Fig. 6.1-2a. Light from a He-Ne laser was incident on a Fabry-Perot cavity after passing through an acousto-optic modulator. The light transmitted by the Fabry-Perot cavity was detected, and the resultant current after being amplified was applied to the piezoelectric crystals, moving one of the mirrors a distance proportional to the amplifier output. The amplifier output was related to the amplifier unit as $\tau_t\dot{\sigma}_t + \sigma_t = (a + 1)AP_T$ and $\tau_e\dot{\sigma}_e + \sigma_e = aAP_T$, with $a = 5$, $\tau_t = 1/3$ s, $\tau_e = 2/3$ s. Figure 6.1-2b shows the input and output of the Fabry-Perot cavity with an imprecisely determined value of detuning parameter β (however, the output frequency and modulation depth were not very sensitive to β). The fractional modulation exceeds 90%. The parameters a and τ_t/τ_e were adjusted to yield an insensitivity of regenerative pulsation frequency to input power. Such a system can be phase locked to a separate frequency standard.

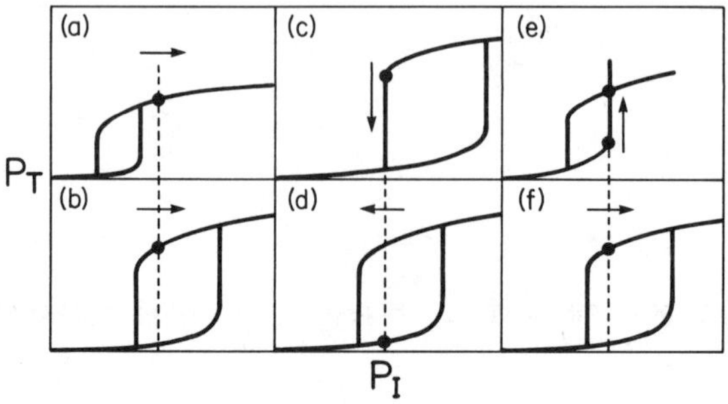

Fig. 6.1-1. Heuristic sequential display of regenerative pulsations. (a) Incident power turned on, the device is in a highly transmitting state, and a small amount of absorbed power causes the bistable region to move in the direction of the arrow. The heavy dots denote the operating output. The vertical dashed lines denote the input power. (b) The bistable region encloses P_I, but hysteresis dictates the indicated output, and the bistable region continues to move. (c) The bistable region just passed P_I, the device switched down, heating is greatly reduced, and motion to the right stops. (d) Cooling causes the bistable region to move left. (e) The switch-up level passes P_I, the device becomes highly transmitting and begins to warm. (f) The bistable region is moving to the right, as in (b), and the sequence (b) to (e) repeats. McCall (1978).

McCall analyzes the general theoretical conditions for obtaining regenerative pulsations.

Regenerative pulsations have been reported in several hybrid bistable systems: Allen, Garmire, Marburger, and Winful (1978); Sohler (1980); Okada and Takizawa (1980, 1981a); Okada (1980a); Ito, Ogawa, and Inaba (1981); Martin-Pereda and Muriel (1983c). These papers contain various configurations for obtaining competing signals with different time constants. Good agreement is found between the detailed circuit analysis and the observed characteristics. Note that the frequency of hybrid oscillators can be controlled electronically, simplifying the stabilization or locking of such devices. Such hybrids may perform important functions in the interface between electronics and all-optical devices.

Regenerative pulsations have also been observed in intrinsic bistable optical devices, first in a GaAs etalon [Gibbs, Jewell, Tarng, Gossard, and Wiegmann (1981); Jewell, Gibbs, Tarng, Gossard, and Wiegmann (1982)]. McCall (1978) outlined an intrinsic pulsation process in which switching is due to a fast electronic effect, but a slower thermal effect prevents either state from being stable. This case has been observed in a 4.2-μm-thick GaAs etalon at ≃ 80 K, with the free-exciton resonance providing the electronic nonlinearity. The hysteresis loop of Fig. 6.1-3a shows bistability in the GaAs device using the decrease in refractive index from saturation of the free-exciton absorption below resonance (Section 3.6). Optical bistability has also been seen in GaAs by another mechanism, namely the increase in optical path length with temperature (Section 3.6). Figure 6.1-3b illustrates the competition between exciton and thermal effects, with the result that the switch-down intensity is higher than the switch-up intensity for the particular input pulse used. In this case significant tuning of the Fabry-Perot transmission peak relative to the laser frequency occurs during the 40-μs "on" time. Figure 6.1-4a shows that during a long ($\geq$ 100 μs) flat-top input pulse, sufficient heating occurs to cause excitonic switch-down followed by cooling and excitonic switch-up, etc.

One would expect the oscillations to be perfectly periodic, but typically they are somewhat random in occurrence mainly due to noise from the input laser. Random oscillations can be obtained from a single nonlinear effect if the input intensity fluctuations exceed the difference between switch-up and switch-down intensities. This case is shown in Fig. 6.1-4b and the difference in appearance from Fig. 6.1-4a is readily apparent. The switching times in Fig. 6.1-4a are about 100 ns or less, consistent with excitonic switching.

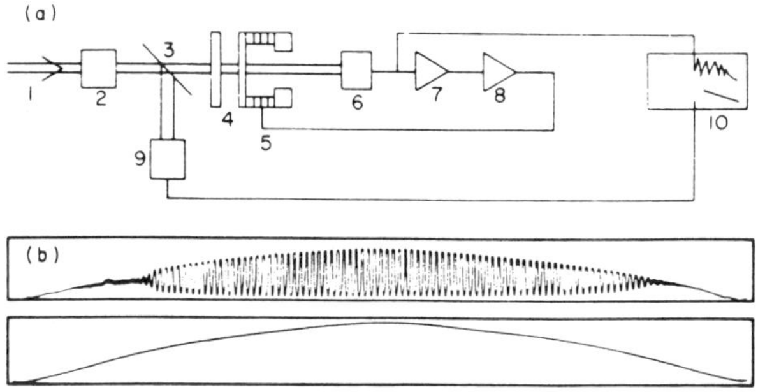

Fig. 6.1-2. (a) Experimental setup. (1) Incident He-Ne laser beam; (2) acousto-optic modulator provides triangular waves; (3) monitor beamsplitter; (4) Fabry-Perot plates; (5) piezoelectric drivers; (6) Fabry-Perot output detector; (7) amplifier with two adjustable time constants; (8) amplifier provides high voltage gain and bias (hysteresis effects in the piezoelectric crystals were too small to be consequential at a high bias voltage); (9) monitor; (10) chart recorder. (b) Upper recording: device output. Time increases to the right. The two maxima on each pulsation occur when the plate separation corresponds to maximum transmission. Lower recording: simultaneous device input. The total recording time is 490 s. The regenerative pulsation frequency averages 0.237 Hz. The oscillations cease at a lower input intensity than that at which they start because at values of the input intensity slightly less than the threshold for instability, the device is a high-Q narrowband amplifier of intensity modulation near 0.237 Hz. McCall (1978).

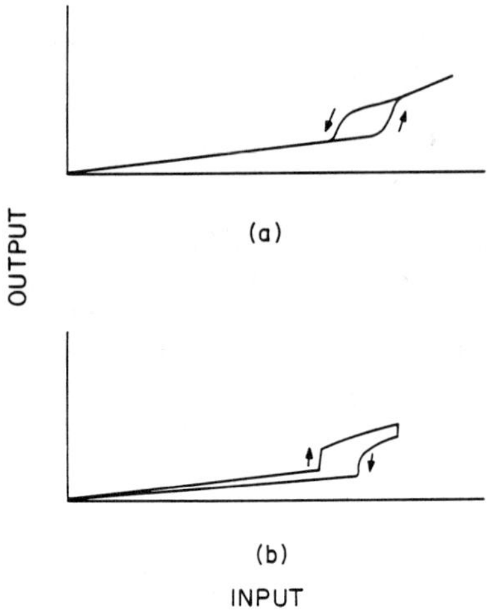

Fig. 6.1-3. Experimental GaAs hysteresis loops showing competition between exciton and thermal effects. (a) "Normal" bistability with 1-μs triangular input pulse, (b) "backward" hysteresis with 40-μs flat-topped input. Jewell, Gibbs, Tarng, Gossard, and Wiegmann (1982).

The GaAs regenerative pulsations are interpreted through the following model. Call the Fabry-Perot-cavity peak transmission frequency ν_c, the laser frequency ν, and the exciton frequency ν_{EX}. Initially $\nu_c < \nu < \nu_{EX}$. The input intensity shifts ν_c sufficiently for excitonic switch-on, yielding $\nu \lesssim \nu_c < \nu_{EX}$. Heating then decreases ν_{EX} toward ν resulting in increased intracavity absorption and thus lower etalon transmission and finesse. The heating also decreases ν_c toward ν until it is slightly less than ν. Rapid excitonic switch-down then occurs resulting in a jump of ν_c to a value less than its original value. In the "off" state, cooling brings ν_c back toward ν and decreases intracavity absorption until excitonic switch-up occurs again. The process repeats indefinitely, and "cw" pulsations have been observed at room temperature in a GaAs superlattice etalon (Fig. 6.1-5). This model is the basis for a computer simulation that assumes uniform plane waves and no defocusing effects [Jewell (1981)].

There is good agreement between the structure and the behavior of the simulated results computed from a noise-free input shown in Fig. 6.1-6 and the experimental data displayed in Fig. 6.1-4 and 5. With the sample in the "on" state, heating increases absorption in the sample, lowering the peak transmission, but at the same time ν_c moves closer to ν. Thus it is not qualitatively obvious whether throughput should increase or decrease with heating. Experiment shows a sharp decrease (Fig. 6.1-4). The simulation revealed this structure to be very sensitive to the slope of the absorption (assumed to be non-saturating at our intensities). Slopes less than about half the value used in Fig. 6.1-6 resulted in increased transmission with temperature. In both experiment and simulation, raising the intensity sufficiently results in the device remaining "on." Although the simulation employs a rather simplified model, its qualitative agreement with experiment suggests that it contains the essential theoretical aspects of the process. Quantitative agreement in areas such as power requirements, time scales, and throughput

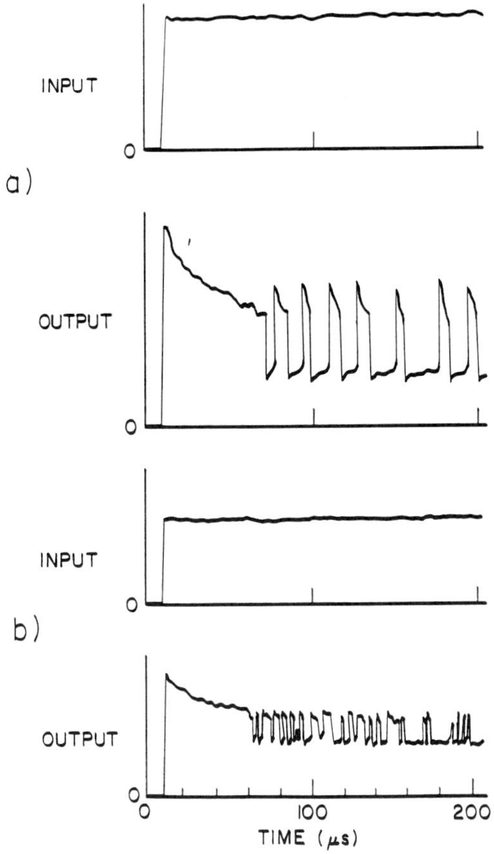

Fig. 6.1-4. Output vs time for a steady low-noise input to the GaAs etalon: (a) regenerative pulsations, (b) random noise switching. Peak transmission in the "on" state was ≃ 15 to 20%. Jewell, Gibbs, Tarng, Gossard, and Wiegmann (1982).

would require accurate detailed knowledge of the sample characteristics (e.g, band-edge structure).

Oscillations have been reported by Jain and Steel (1982) using HgCdTe in an external cavity coupled to a CO_2 laser cavity, but they cannot be associated with competition between thermal and electronic effects since they both act in the same sense [Miller, Parry, and Daley (1984)]. Their configuration is much like that of Spencer and Lamb (1972a,b) mentioned in Section 1.3, and it would be interesting to compare theory and experiment in that case. In contrast the GaAs etalon is decoupled from the laser by tilting; the short length of the etalon makes this a simple and effective way to achieve isolation.

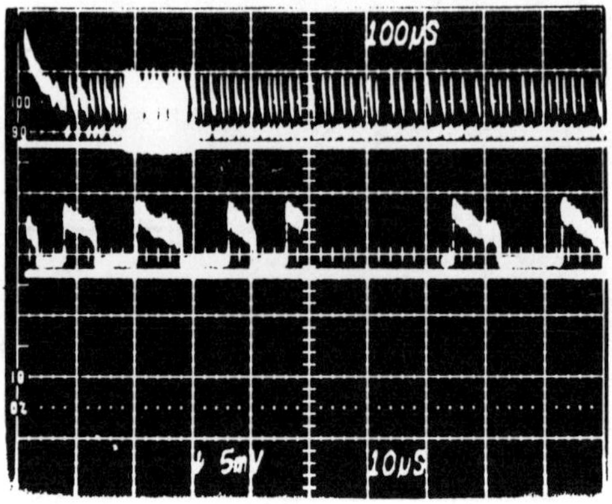

Fig. 6.1-5. Continuous-wave regenerative pulsations in a GaAs superlattice etalon at room temperature. Gibbs, Tarng, Jewell, Weinberger, Tai, Gossard, McCall, Passner, and Wiegmann (1982a). © 1982 IEEE

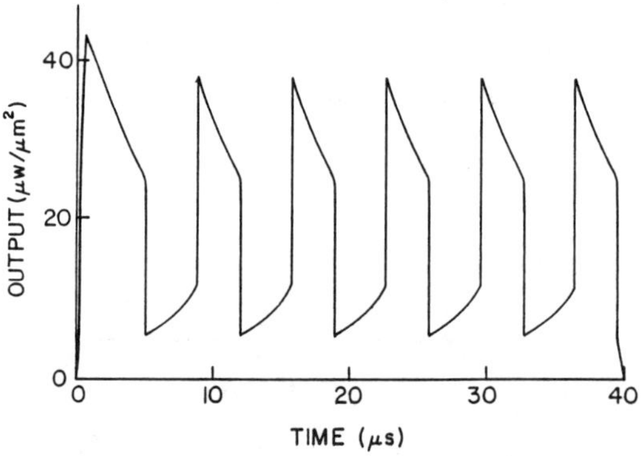

Fig. 6.1-6. Computer-simulated regenerative pulsations in GaAs. Jewell, Gibbs, Tarng, Gossard, and Wiegmann (1982). Jewell (1981).

Intrinsic regenerative pulsations have been seen in self-trapping optical bistability (Section 3.9.1) in Na [Gibbs, Derstine, Hopf, Kaplan, Rushford, Shoemaker, Weinberger, and Wing (1982)]. The experimental arrangement is similar to the original self-focusing bistability experiment except that no aperture was placed between the Na and the feedback mirror. Pulsations were clearly observed for two distinct detunings on the self-focusing (high-frequency) side of the $^2S_{1/2}(F=2) \rightarrow {}^2P_{1/2}$ sodium transitions at 589.6 nm; see Fig. 6.1-7. The pulsations depended critically upon optimum feedback, detuning from resonance, laser intensity, and Na density. No positive identification has been made of the competing mechanisms with different time constants, but the following scenario is a possibility. Assume that switch-on has just occurred by means of the nonlinear wavefront self-trapping mechanism. In the "on" self-trapped situation, the intensity is high inside the self-trapped filament, saturating the optical pumping, which is the origin of the wavefront nonlinearity. The saturated cylinder expands until there is no longer sufficient wavefront nonlinearity to keep the beam self-trapped, switch-off occurs, and the beam diverges from the input plane. The optically pumped atoms diffuse out of the region previously occupied by the filament while unpumped atoms diffuse in, and the wavefront nonlinearity gradually returns. Finally, switch-on occurs completing one cycle of the regenerative pulsation.

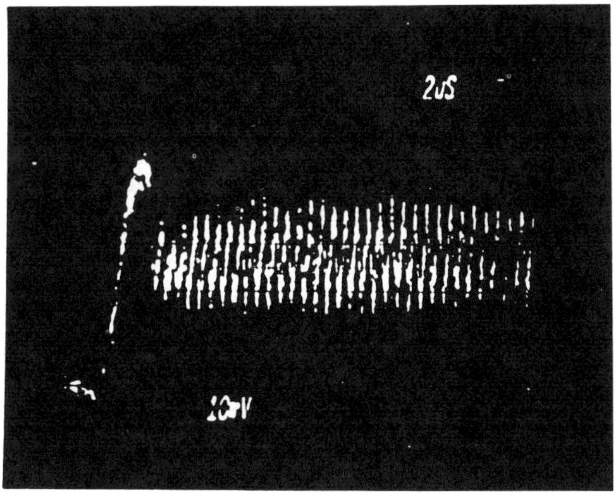

Fig. 6.1-7. Regenerative pulsations in Na vapor using self-trapping optical bistability. Derstine, Gibbs, Hopf, and Rushford (1982). Gibbs, Derstine, Hopf, Kaplan, et al. © 1982 IEEE

As the input intensity is increased, the regenerative pulsations, which begin as an almost sinusoidal oscillation, become more like a square wave. The spectrum evolves from a single spike to the multiple-peaked descending staircase of odd harmonics characteristic of a square wave. At still higher intensities the output becomes erratic and a broad background rises up in the spectrum; for waveforms and spectra see Hopf, Derstine, Gibbs, and Rushford (1984). This behavior is essentially identical to that observed in a hybrid bistable device with delayed feedback when noise terminates the period doubling sequence at period two (Section 6.3). Further study is needed to determine if this instability of the Na self-trapping bistability was the first observation of optical chaos in an intrinsic device; see Derstine, Gibbs, Hopf, and Rushford (1982). Instabilities with characteristic times much longer than a round-trip time were also seen in the original Na self-trapping experiment [Bjorkholm, Smith, Tomlinson, and Kaplan (1981)] and in the liquid-suspension-of-dielectric-particles self-trapping experiment [P. W. Smith, Ashkin, Bjorkholm, and Eilenberger (1984)].

Cheung, Durbin, and Shen (1983) have observed intrinsic regenerative pulsations in a Fabry-Perot interferometer filled with an 83-μm-thick nematic liquid crystal film (Section 3.4). The two counteracting mechanisms with different response times governing the nonlinear refractive index are optical-field-induced collective reorientation of molecules ($\tau_m \simeq 1$ s) and optical heating through residual absorption ($\tau_t \simeq 0.01$ s). Roughly 1-Hz oscillations were seen with an input intensity of 360 and 40 W/cm^2 in an external magnetic field of 0 and 1.7 kOe, respectively.

Intrinsic self-pulsations have also been reported in transverse optical bistability (Section 3.8.2) using an aperture in the far field in CdS with periods of 8 to 34 μs; the mechanism giving rise to them is unknown [Dagenais and Winful (1984a,b)]. Erratic pulsations at high intensity could be some form of optical chaos (Sections 6.3 and 6.4). On the other hand, if the whole transmitted beam is collected with a large-aperture lens, the pulsation tends to disappear.

Regenerative pulsations (at a few Hertz) of a different kind have been seen in germanium selenide ($GeSe_2$) evaporated films 6.4 μm thick using 1.4 to 2.7 kW/cm^2 at 632.8 nm [Hajtó, Zentai, and Somogyi (1977); Hajtó and Apai (1980); Fazekas (1981)] or 40 to 50 W/cm^2 using self-supported 6-μm films [Hajtó, Jánossy, and Forgács (1982); Hajtó and Jánossy (1983)]. Hajtó, Jánossy, and Firth (1983) propose a competition between a reversible photostructural change and a faster thermal heating which moves the absorption edge to shorter wavelengths for increased input intensity, cutting off the transmission; see Section 2.10.

Regenerative pulsations by competing mechanisms having unequal time constants have been studied mostly as an interesting instability of a bistable system. In principle, they could serve as all-optical oscillators, but they would probably need to be locked to a more stable frequency source. The demonstrated pulsations are relatively low frequency ($\lesssim$ 3 MHz), but much faster oscillations, perhaps approaching a gigahertz, could be obtained using the decrease in thermal response time with reduced size and/or temperature. Alternatively a device with two opposing non-thermal mechanisms might be extremely fast. There are other possible all-optical square-wave generators for non-gain media: ring-cavity self-pulsing (Section 6.2) resulting from mode competition in ring-cavity bistability, and the Ikeda instability (Section 6.3) where the medium response time is shorter than the cavity round-trip time. Another modulationless source of short optical pulses is the modelocked semiconductor laser that uses an external mirror and a saturable absorber near the output end of the laser [Ippen, Eilenberger, and Dixon (1980a,b); van der Ziel, Tsang, Logan, Mikulyak, and Augustyniak (1981); Harder, Smith, Lau, and Yariv (1983)]. Silberberg, Smith, Eilenberger, and Miller (1984) achieve 1.6 ps pulses using a GaAs/GaAlAs MQW structure as the saturable absorber.

Also Bjorkholm's (1967) "passive transmission switch" mentioned in Section 1.4 is a relaxation oscillator consisting of a laser with a saturable Fabry-Perot cavity as the output mirror [see also Basov, Morozov, Nikitin, and Semenov (1968); Szöke, Daneu, Goldhar, and Kurnit (1969); Harder, Lau, and Yariv (1981, 1982c); Satchell, Parigger, and Sandle (1983)].

Goldstone and Garmire (1981a) have studied numerically regenerative pulsations in the output of a three-mirror system consisting of two coupled Fabry-Perot cavities with a nonlinear refractive index in the second cavity [see also Spencer and Lamb (1972b), Goldstone (1984)]. The time constants of the second cavity must be less than the round-trip time of the first for oscillations to occur. Goldstone and Garmire (1981b, 1983a) have also computed regenerative pulsations in a lossless nonlinear Fabry-Perot cavity with a long medium response time compared with the cavity round-trip time, those two times competing in the McCall sense. See Goldstone and Garmire (1983b) for further analytical and numerical analysis of this kind of regenerative oscillation in a lossless nonlinear Fabry-Perot cavity.

6.2. STABILITY ANALYSIS; SELF-PULSING INVOLVING NONRESONANT MODES

In Chapter 2 the steady-state solutions of several models of absorptive and dispersive bistability were obtained, but the stability of those solutions was not tested. The perturbation expansion technique of stability analysis will be applied here to absorptive bistability following closely Gronchi, Benza, Lugiato, Meystre, and Sargent III (1981). Analytic expressions for the eigenvalues are found in the mean-field limit and numerically calculated figures summarize the exact eigenvalues. The negative-sloped regions of the x versus y "S" curve are found to be unstable as expected. Portions of the positive-sloped regions can also be unstable leading to oscillations now referred to as self-pulsing and first predicted by Bonifacio and Lugiato (1978f). They occur when the intracavity field is so high that the Rabi frequency exceeds the mode spacing and the neighboring modes experience gain [Gronchi et al. (1981) and McCall (1974)].

The stability analysis described below is a standard general approach and can reveal instabilities of various types. It is easier to discuss the instabilities in certain limits: regenerative pulsations (two opposing and unequal response times), self-pulsing ($T_2/(1 + X)^{1/2} < t_R$), Ikeda ($\tau_M < t_R$), etc.

From Eqs. (2.4-4a, 5, 13, 35), one finds

$$\frac{1}{c}\frac{\partial F}{\partial t} + \frac{\partial F}{\partial z} = -\frac{\alpha_0}{2}\left[\frac{\gamma_T}{\gamma_L}\right]^{1/2} v, \tag{6.2-1a}$$

$$\frac{\partial v}{\partial t} = -(\gamma_L\gamma_T)^{1/2}\,Fw - \gamma_T v, \tag{6.2-1b}$$

$$\frac{\partial w}{\partial t} = (\gamma_L\gamma_T)^{1/2}\,Fv - \gamma_L(w + 1), \tag{6.2-1c}$$

$$F(0,t) = Ty + RF(L,t - \Delta t), \tag{6.2-2}$$

where $\Delta t = (L_T - L)/c$ is the time taken by the light to propagate from mirror 2 to mirror 1 in Fig. 2.1-1. The stationary solutions of Eqs. (6.2-1) are given by Eqs. (2.4-14,15):

$$v_{st}(z) = \left[\frac{\gamma_L}{\gamma_T}\right]^{1/2} \frac{F_{st}(z)}{1 + F_{st}^2(z)}, \tag{6.2-3a}$$

$$w_{st}(z) = \frac{-1}{1 + F_{st}^2(z)}, \tag{6.2-3b}$$

and Eq. (6.2-1a) gives Eq. (2.4-33):

$$\frac{dF_{st}}{dz} = -\frac{\alpha_0}{2}\,\frac{F_{st}}{1 + F_{st}^2}. \tag{6.2-4}$$

Now introduce small deviations from the stationary values:

$$\delta F(z,t) = F(z,t) - F_{st}(z), \tag{6.2-5a}$$

$$\delta v(z,t) = v(z,t) - v_{st}(z), \tag{6.2-5b}$$

$$\delta w(z,t) = w(z,t) - w_{st}(z). \tag{6.2-5c}$$

Substituting Eqs. (6.2-5) into Eqs. (6.2-1) and keeping only the terms linear in the deviations, one obtains the following linearized equations by inspection:

$$\frac{1}{c}\frac{\partial \delta F}{\partial t} + \frac{\partial \delta F}{\partial z} = -\frac{\alpha_0}{2}\left[\frac{\gamma_T}{\gamma_L}\right]^{1/2} \delta v, \tag{6.2-6a}$$

$$\frac{\partial \delta v}{\partial t} = -(\gamma_L\gamma_T)^{1/2}\,[F_{st}(z)\,\delta w + w_{st}(z)\delta F] - \gamma_T\delta v, \tag{6.2-6b}$$

$$\frac{\partial \delta w}{\partial t} = (\gamma_L\gamma_T)^{1/2}\,[F_{st}(z)\delta v + v_{st}(z)\delta F] - \gamma_L\delta w. \tag{6.2-6c}$$

Using the boundary condition Eq. (6.2-2) and taking into account that $F_{st}(z)$ obeys Eq. (6.2-2), one obtains the following boundary condition for $\delta F(z,t)$:

$$\delta F(0,t) = R\delta F(L,t - \Delta t). \tag{6.2-7}$$

Now try solutions of the form

$$\delta F_\lambda(z,t) = \delta F_\lambda(z)\, e^{\lambda t} + \text{c.c.}, \tag{6.2-8a}$$

$$\delta v_\lambda(z,t) = \delta v_\lambda(z)\, e^{\lambda t} + \text{c.c.}, \tag{6.2-8b}$$

$$\delta w_\lambda(z,t) = \delta w_\lambda(z)\, e^{\lambda t} + \text{c.c.}, \tag{6.2-8c}$$

in Eqs. (6.2-6) and eliminate $\delta v_\lambda(z)$ and $\delta w_\lambda(z)$. Then

$$\frac{d\delta F_\lambda(z)}{dz} = -\left[\frac{\alpha_0}{2}\,\Gamma(\lambda,F_{st}(z)) + \frac{\lambda}{c}\right]\delta F_\lambda(z), \tag{6.2-9}$$

where

$$\Gamma(\lambda,F_{st}(z)) = \frac{\gamma_T}{1 + F_{st}^2(z)}\;\frac{\gamma_L(1 - F_{st}^2(z)) + \lambda}{(\lambda + \gamma_T)(\lambda + \gamma_L) + \gamma_T\gamma_L F_{st}^2}\,. \tag{6.2-10}$$

For $\lambda \to 0$, Eqs. (6.2-9,10) reduce to Eq. (A-3) evaluated for steady-state. Equations (6.2-8,9) give

$$\delta F_\lambda(z,t) \propto \exp\left[-\frac{\alpha_0}{2}\int_0^z dz'\,\Gamma(\lambda,F_{st}(z')) - \left[\frac{z}{c} - t\right]\lambda\right] + \text{c.c.} \tag{6.2-11}$$

From the boundary condition Eq. (6.2-7) one obtains an equation for λ:

$$1 = R \exp\left[-\frac{\alpha_0}{2} \int_0^L dz'\Gamma(\lambda, F_{st}(z')) - \frac{\lambda L_T}{c} \right], \qquad (6.2\text{-}12)$$

using $\Delta t = (L_T - L)/c$, where L is the absorber length and L_T is the total length of the ring cavity. Solving for λ

$$\lambda = \frac{c}{L_T} \ln R - \frac{\alpha_0 c}{2L_T} \int_0^L dz'\Gamma(\lambda, F_{st}(z')) - i\,\frac{2\pi nc}{L_T}, \qquad (6.2\text{-}13)$$

$$n = 0, \pm 1, 2, \ldots,$$

where the last term represents the cavity modes; note that adding multiples of $2\pi i$ to $\lambda L_T/c$ still satisfies Eq. (6.2-12). Define

$$\alpha_n = 2\pi nc/L_T, \qquad (6.2\text{-}14)$$

then for an empty cavity ($\alpha_0 = 0$)

$$\lambda = -i\alpha_n - \frac{c}{L_T} \ln \frac{1}{1 - T}. \qquad (6.2\text{-}15)$$

Note that $(c/L_T)\ \ln(1/1\text{-}T)$ is the cavity damping constant, which for $T \ll 1$ reduces to:

$$\gamma_c = \frac{cT}{L_T}. \qquad (6.2\text{-}16)$$

For $\alpha_0 = 0$ and $T \ll 1$, Eq. (6.2-11) becomes

$$\delta F_\lambda(z,t) = \text{const} \exp\left[-i\alpha_n\left[t - \frac{z}{c}\right]\right] \exp\left[-\gamma_c\left[t - \frac{z}{c}\right]\right] + \text{c.c.} \qquad (6.2\text{-}17)$$

Hence the deviation from the stationary state $\delta F(z,t)$ is a superposition of elementary solutions Eq. (6.2-17) composed of the cavity running modes $2\pi nc/L_T$ damped by mirror losses. The resonant mode corresponds to n = 0. The exact eigenvalue Eq. (6.2-13) can be solved numerically; a few figures will summarize some of those results. First, analytical solutions will be found and discussed in the mean-field limit.

In the mean-field limit, $F_{st}(z)$ becomes uniform in space and, since $F(L,t) \equiv x(t)$,

$$F_{st}(z) = x. \qquad (6.2\text{-}18)$$

Substituting Eq. (6.2-18) into Eq. (6.2-10) and putting Eqs. (6.2-10,16) into Eq. (6.2-13), one finds a cubic equation for λ. A stationary solution is stable if and only if $\text{Re}\,\lambda \leq 0$, so perturbations decay and the steady-state stationary values are regained. Label the eigenvalues as λ_{nj} where n refers to the nth running mode and j = 1,2,3 refers to the three solutions of the cubic equation. Consistent with the mean-field limit, i.e., $\alpha_0 L \ll 1$, $T \ll 1$, $\alpha_0 L/4T = C$ = constant, one can calculate λ_{nj} to first order in T (remember $\gamma_C \propto T$) following Benza and Lugiato (1979a). Since γ_C is proportional to T, for each n the three solutions are of the form

$$\lambda_{nj} = \lambda_{nj}^{(0)} + \gamma_C \lambda_{nj}^{(1)} + O(T^2) . \tag{6.2-19}$$

Now rewrite Eq. (6.2-13) as

$$\lambda = -i\alpha_n - \gamma_C\left[1 + \frac{2C\gamma_T}{1 + x^2} \frac{\lambda + \gamma_L(1 - x^2)}{(\gamma_T + \lambda)(\gamma_L + \lambda) + \gamma_T\gamma_L x^2}\right] . \tag{6.2-20}$$

Now try $\lambda_{n1} = -i\alpha_n + \gamma_C\lambda_{n1}^{(1)}$ as a good guess. Then to lowest order in T:

$$\lambda_{n1} = -i\alpha_n - \gamma_C\left[1 + \frac{2C\gamma_T}{1 + x^2} \frac{\gamma_L(1 - x^2) - i\alpha_n}{(\gamma_T - i\alpha_n)(\gamma_L - i\alpha_n) + \gamma_T\gamma_L x^2}\right] + O(T^2) . \tag{6.2-21}$$

Then from Eq. (6.2-20):

$$(\lambda + i\alpha_n + \gamma_C)\,[(\gamma_T + \lambda)(\gamma_L + \lambda) + \gamma_T\gamma_L x^2] + \frac{2C\gamma_T\gamma_C}{1 + x^2}\,[\lambda + \gamma_L(1 - x^2)] = 0 . \tag{6.2-22}$$

Since the cubic equation can be written as

$$(\lambda - \lambda_{n1})(\lambda - \lambda_{n2})(\lambda - \lambda_{n3}) = 0 ,$$

and anticipating that the leading terms in λ_{n2} and λ_{n3} will be zero order in T, approximate λ_{n1} by $-i\alpha_n$ and drop the γ_C terms in Eq. (6.2-22) to give:

$$(\lambda - \lambda_{n2})(\lambda - \lambda_{n3}) = \lambda^2 + (\gamma_T + \gamma_L)\lambda + \gamma_T\gamma_L(1 + x^2) , \tag{6.2-23}$$

therefore,

$$\lambda_{n2}, \lambda_{n3} = \frac{-1}{2}\left[\gamma_T + \gamma_L \pm [(\gamma_T - \gamma_L)^2 - 4\gamma_T\gamma_L x^2]^{1/2}\right] + O(T) . \tag{6.2-24}$$

For small T, $\mathrm{Re}\lambda_{n2}$, $\mathrm{Re}\lambda_{n3} \leq 0$ always, so the stability is determined by λ_{n1}

$$-\frac{\mathrm{Re}\lambda_{n1}}{\gamma_C} = 1 + \frac{2C\gamma_T}{1+x^2}\,\frac{\gamma_L(1-x^2)[\gamma_T\gamma_L(1+x^2)-\alpha_n^2]+\alpha_n^2(\gamma_T+\gamma_L)}{[\gamma_T\gamma_L(1+x^2)-\alpha_n^2]^2+\alpha_n^2(\gamma_T+\gamma_L)^2}\,. \tag{6.2-25}$$

For the resonant mode n = 0, the only one usually considered for bistability,

$$\frac{\lambda_{01}}{\gamma_C} = -\left[1 + 2C\frac{1-x^2}{(1+x^2)^2}\right] = -\frac{dy}{dx}\,, \tag{6.2-26}$$

where $y = x[1 + 2C/(1 + x^2)]$ is the stationary state Eq. (2.4-1). Then $\mathrm{Re}\lambda_{01}$ is positive if $dy/dx < 0$; therefore, all the stationary states lying on the part of the curve $x = x(y)$ with negative slope are unstable. Likewise the resonant mode is always stable where the slope is positive, but some off-resonance modes can become unstable.

Strictly speaking, Eq. (6.2-19) and the eigenvalues found therefrom are only correct for the mean-field single-mode limit. It does not give the correct first-order correction terms for λ_{n2} and λ_{n3} in the multimode case, although the more significant eigenvalues λ_{n1} are obtained correctly. The problem arises because the assumption of a uniform steady-state field [Eq. (6.2-18)] is only valid up to a correction of order (1-R). This correction is important when finding perturbative corrections to λ_{n2} and λ_{n3}; see Carmichael (1983a) and Lugiato (1983b).

Gronchi et al. (1981) have compared the mean-field and exact eigenvalues; see Figs. 6.2-1 to 3. They find that for C = 20, the results obtained in the mean-field limit are practically exact for T = 0.01. Figure 6.2-3 shows that the frequency of the unstable mode never differs from the frequency of the incident field by an amount larger than the intracavity Rabi frequency

$$\Omega_C = \frac{\kappa E_T}{T} = \kappa E\,.$$

In the mean-field limit they find an instability only when there is a hysteresis cycle [Bonifacio, Gronchi, and Lugiato (1981)],whereas in the case of dispersive bistability, Lugiato (1980b) finds an instability without bistability.

Figures 6.2-1 through 6.2-4 illustrate the origin of the multimode self-pulsing. Figure 6.2-1 shows that the resonant mode is stable for T = 0.1, $\gamma L/c = 0.2$, and x = 6 to 12 (of course, it is always stable). Figure 6.2-4 shows that for T = 0.1, $\gamma L/c = 0.2$, and x = 7, $\mathrm{Re}\lambda_{11}$ is positive, i.e., the gain exceeds the loss, so the adjacent mode is amplified and the single-mode operation is unstable. Part of the resonant mode's (mode 0) energy is transferred through the nonlinear response to the side mode, giving it gain. As in the usual laser problem, spontaneous emission provides the impetus for side-mode buildup [see Sargent, Zubairy, and DeMartini (1982)]. This gain without population inversion is discussed in Appendix A and was observed in the optical regime by Wu, Ezekial, Ducloy, and Mollow (1977). Hendow and Sargent III (1982a,b) give a population-pulsation derivation of laser and bistability instabilities.

The establishment of an instability does not reveal the time evolution. Gronchi et al. (1981) numerically integrate the exact coupled Maxwell-Bloch equations (6.2-1) and find that two different behaviors can evolve depending upon the precise values of the parameters [first found by Bonifacio, Gronchi, and Lugiato (1979b)]. In the first case the system precipitates to the (stable) stationary state on the low-transmission branch. In the other case the system evolves to a time-periodic state (limit cycle) in which the transmitted light consists of an undamped regular sequence of pulses. When the system is in the instability region, the dynamics involves a competition between the resonant mode and the unstable modes. When the unstable modes prevail, the system self-pulses; when the resonant mode dominates the dynamics, the system precipitates to a low-transmission branch. The bifurcation from a stable upper state to self-pulsing as the input is reduced is a first-order phase transition in that the self-pulsing appears abruptly. Analytical treatments of self-pulsing [Benza and Lugiato (1979a,b; 1981a,b; 1982); Benza, Lugiato, and Meystre (1980); Lugiato, Benza, Narducci, and Farina (1981, 1982, 1983)] find a first-order transition at the right boundary of the instability region and a second-order transition at the left boundary.

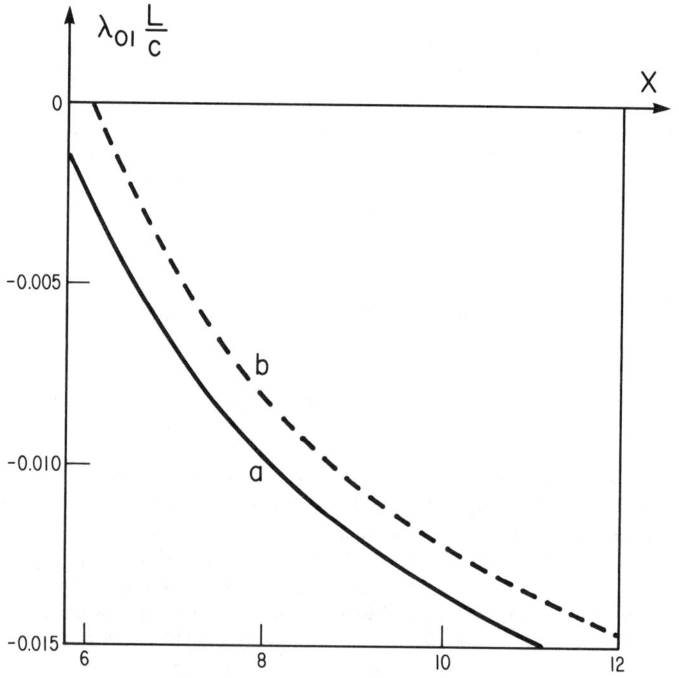

Fig. 6.2-1. Resonant mode eigenvalue λ_{01} as a function of x for C = 20, $L_T/L = 5$, and $\gamma = \gamma_L = \gamma_T$: (a) Exact as obtained numerically from Eq. (6.2-13) for T = 0.1, and $\gamma L/c = 0.2$. (b) Mean-field result for $T \ll 1$ (see Eq. (6.2-21)). Gronchi et al. (1981).

Inhomogeneous broadening reduces the side-mode gain [McCall (1974)]. Standing waves also reduce the effect [Casagrande, Lugiato, and Asquini (1980); Sargent (1980)]; Sargent concludes that an instability may exist when the medium is located at one of the ends of the cavity, but not if the medium fills the cavity. Asquini and Casagrande (1981) find a new instability originated by the counter-propagating field in a bidirectional ring cavity.

Risken and Nummedal (1968) predicted multimode operation in a homogeneously broadened unidirectional ring laser by integrating the coupled Maxwell-Bloch equations. This phenomenon is closely related to the present self-pulsing in optical bistability and corresponds to mode spacings that yield absorption, rather than gain, in the absorber formulae.

So far, our discussion of self-pulsing has been restricted to on-resonance purely absorptive optical bistability. Lugiato (1980b) has analyzed self-pulsing in dispersive bistability in a ring cavity for a homogeneously broadened atomic system in the limit $\alpha_0 L \ll 1$, $T \ll 1$, and $C = \alpha_0 L/4T$ fixed. He finds that the instability domain can be much larger than the bistability domain; an instability can occur even when the bistability is absent. This suggests that self-pulsing may be observed more easily in the dispersive case.

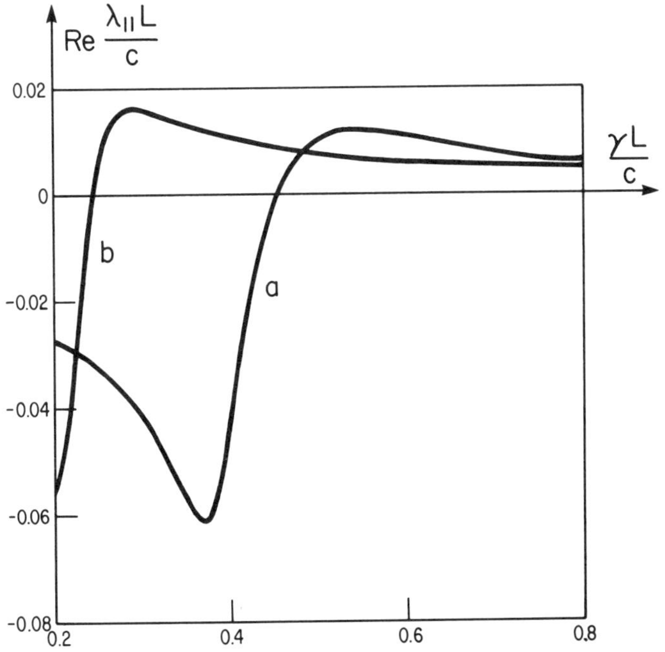

Fig. 6.2-2. Real part of the adjacent mode eigenvalue λ_{11} as a function of $\gamma L/c$ for $x = 6$ and $C = 20$, $L_T/L = 5$, and $\gamma = \gamma_L = \gamma_T$. (a) Mean-field result for $T \ll 1$ (see Eq. (6.2-25)). (b) Exact as obtained numerically from Eq. (6.2-13) for $T = 0.1$. Gronchi et al. (1981).

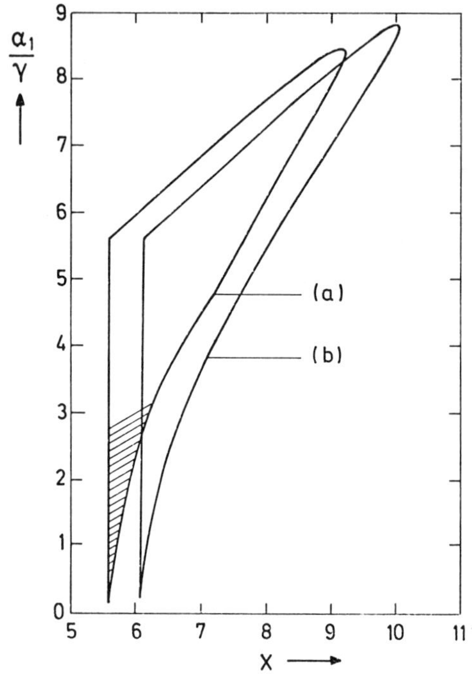

Fig. 6.2-3. Numerically determined instability region for T = 0.1, C = 20, $L_T/L = 5$, and $\gamma_T = \gamma_L = \gamma$. The shaded part indicates the region where both α_1 and α_2 (and sometimes other modes) are unstable. For comparison curve (b) shows the instability region in the mean-field limit. Gronchi et al. (1981).

At this writing, self-pulsing has not yet been observed. It should be characterized by a fundamental frequency of an inverse round-trip time, i.e. $\nu = 1/t_R$. If only adjacent side modes are unstable the output may be rather sinusoidal whereas it can be spiked if many modes are unstable. In contrast, regenerative pulsations (Section 6.1) are characterized by medium response times, and Ikeda periodic oscillations (Section 6.3) have frequencies of $1/2t_R$, $1/4t_R$, $1/8t_R$, ... The conditions for regenerative pulsations and Ikeda oscillations can be met easily by hybrid bistable systems, whereas it is not clear that there is a hybrid analog of Rabi flopping of two-level atoms resulting in side-mode gain. Transverse effects may also hamper or prevent the observation of self-pulsing since the radial dependence of the intensity would introduce a radial dependence to the Rabi frequency. Gaussian profiles did not prevent the observation of regenerative pulsations in GaAs; simulations (Section 6.3) predict that they substantially modify but do not eliminate Ikeda instabilities. Lugiato and Milani (1983, 1984) consider purely absorptive self-pulsing in a unidirectional ring cavity with spherical mirrors and an incident field matched to the TEM_{00} mode. In this one transverse-mode model they find that, when the radius of the cylindrical atomic sample is much larger

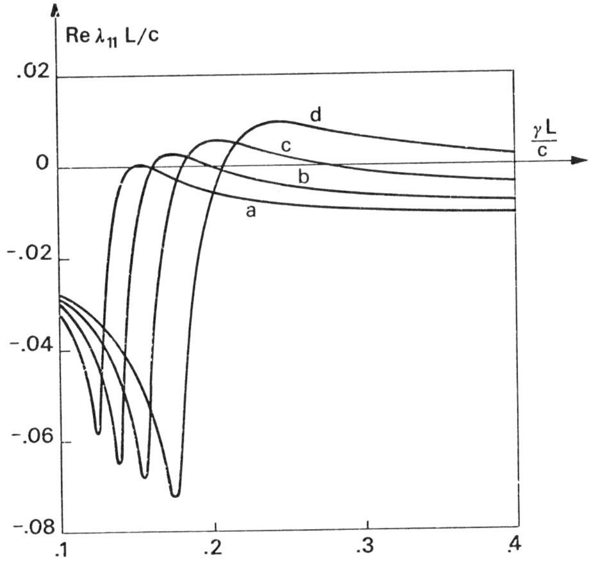

Fig. 6.2-4. Real part of the adjacent mode eigenvalue [exact as derived from Eq. (6.2-13)]. Re λ_{11} as a function of $\gamma L/c$ for T = 0.1 and C = 20, $L_T/L = 5$, and $\gamma = \gamma_L = \gamma_T$. (a) x = 6, (b) 7, (c) 8, and (d) 9. Gronchi et al. (1981).

than the beam waist, no self-pulsing occurs. Thus to observe self-pulsing in purely absorptive optical bistability, it is necessary to fulfill as closely as possible the plane-wave condition. For mixed absorptive and dispersive optical bistability, Carmichael, Asquini, and Lugiato (1984) find that there are conditions for which the instability domain is actually larger for a Gaussian than for a uniform-plane-wave input.

Self-pulsing involving nonresonant cavity modes also occurs in plane-wave analysis of a ring cavity containing two-photon homogeneously broadened absorbers, even when bistability is absent [Hermann (1982b)]. Ovadia and Sargent III (1983, 1984) find that side modes can build up for a centrally tuned strong mode for the two-photon laser but not for two-photon absorptive bistability, again assuming homogeneously broadened media. For strong-mode detuning, they find appreciable regions of side-mode gain for both problems, allowing multiwavelength instabilities to occur.

6.3. IKEDA INSTABILITIES: PERIODIC OSCILLATIONS, PERIOD DOUBLING, AND OPTICAL CHAOS

This chapter treats bistable systems that become unstable for constant input, i.e., the output intensity is time dependent even though the input is perfectly steady. Regenerative pulsations (Section 6.1) are relaxation oscillations in which two or more effects with different time constants compete, giving oscillation periods determined by the time constants. Self-pulsing (Section 6.2) occurs when two-level atoms are driven so hard that the Rabi frequency exceeds the mode spacing; gain at the next mode enables it to go unstable, giving an oscillation period equal to the cavity round-trip time t_R. Ikeda (1979) found another instability by showing that, in the limit that the medium response time $\tau_M = \gamma_T^{-1}$ is much shorter than t_R, a dispersive bistable ring cavity can exhibit oscillations with period $2t_R$. As the cavity is driven harder, the time required for the waveform to reproduce itself doubles, i.e. the period doubles with each bifurcation: $2t_R$, $4t_R$, $8t_R$, ... At high enough intensities the output becomes chaotic. Such behavior has been observed in various hydrodynamic experiments, and recently in optical bistable systems. From a device point of view, it is important to understand all instabilities of bistable systems so they can be avoided or utilized. The Ikeda instability is especially interesting because of the similarity between turbulence in these simple optical devices and in much more complicated situations such as hydrodynamics, population dynamics, biological feedback systems, etc.

Ikeda's essential step was to scrutinize time-delay effects in the dispersive bistability case. In Chapter 2, the time required to propagate the electric field from the output back to the input was properly ignored because steady-state solutions were being sought. The boundary conditions for a ring cavity (Fig. 2.4-1) are:

$$F(0,t) = \sqrt{T}\,F_I(t) + Re^{ikL_T}\,F\left[L, t - \frac{L_T - L}{c}\right] \tag{6.3-1a}$$

and

$$F_T(t) = \sqrt{T}\,F(L,t)\,e^{ikL} \tag{6.3-1b}$$

where the nonlinear medium fills the length L between the input and output mirrors, L_T is the total length of the ring cavity, and F is the complex slowly varying envelope of the intracavity electric field normalized as $F \equiv \kappa E/(\gamma_L\gamma_T)^{1/2}$. There are no medium contributions to $k \equiv \omega/c$.

The Maxwell-Bloch equations, for a two-level homogeneously broadened system are given by Eqs. (2.1-6,8); transforming to a retarded frame, z = z and $\tau = t - z/c$, one has

$$\frac{\partial F}{\partial z} = -\frac{\alpha_0}{2}\left[\frac{\gamma_T}{\gamma_L}\right]^{1/2} Q\,, \tag{6.3-2a}$$

$$\frac{\partial Q}{\partial \tau} = -(\gamma_T\gamma_L)^{1/2}\,Fw - (\gamma_T - i\Delta\omega)Q\,, \tag{6.3-2b}$$

$$\frac{\partial w}{\partial \tau} = (\gamma_L\gamma_T)^{1/2}\,\frac{FQ^* + F^*Q}{2} - \gamma_L(w+1)\,. \tag{6.3-2c}$$

(Ikeda's w is one-half the w in Eq. (6.3)). Equations (6.3-1,2) describe completely a system of two-level atoms in the semiclassical, slowly-varying-envelope, and uniform-plane-wave approximations. Clearly all of the instabilities of such a system are contained within these equations. The Bonifacio-Lugiato self-pulsing instability (Section 6.2) was first derived in the purely absorptive mean-field limit and later examined in the dispersive case, as well. The usual stability analysis (Section 6.2) applied to Eqs. (6.3-1,2) leads to the instability of Ikeda (1979) when the cavity round-trip time t_R is much longer than the medium transverse relaxation time $1/\gamma_T$, but it disappears in the mean-field limit [Lugiato, Asquini, and Narducci (1982)]. Direct numerical solution of these equations leads to periodic oscillations, period doubling, and chaos under appropriate conditions [Snapp, Carmichael, and Schieve (1981); Carmichael, Snapp, and Schieve (1982); Moloney and Hopf (1981); Moloney, Hopf, and Gibbs (1982 a,b)]. Here the derivation of Ikeda is followed in order to arrive at his difference equations, which facilitate the discussions of period doubling, the Feigenbaum (1980) universality relations, and the hybrid experiment that first demonstrated optical chaos.

Ikeda assumes rapid transverse relaxation, i.e., the polarization Q follows the field F adiabatically, and $\partial Q/\partial\tau \simeq 0$ in Eq. (6.3-2b). Therefore,

$$Q \simeq -\frac{(\gamma_T\gamma_L)^{1/2}Fw}{\gamma_T - i\Delta\omega} = -\left[\frac{\gamma_L}{\gamma_T}\right]^{1/2} Fw\left[\frac{1+i\Delta}{1+\Delta^2}\right], \tag{6.3-3}$$

where $\Delta \equiv \Delta\omega/\gamma_T$. Substituting Eq. (6.3-3) into Eq. (6.3-2a), one can write the electric field in integral form

$$F\left[z,\tau+\frac{z}{c}\right] = F(0,\tau)\exp\left[\frac{\alpha_0}{2}\left[\frac{1+i\Delta}{1+\Delta^2}\right]W(z,\tau)\right], \tag{6.3-4}$$

where

$$W(z,\tau) \equiv \int_0^z dz' w\left[z',\tau+\frac{z'}{c}\right]. \tag{6.3-5}$$

Substituting Eqs. (6.3-3) into Eq. (6.3-2c) and integrating over z, one finds

$$\frac{\partial W}{\partial\tau} = -\gamma_L(W+z) - \frac{\gamma_L}{1+\Delta^2}\int_0^z dz' w\left[z',\tau+\frac{z'}{c}\right]\left|F\left[z',\tau+\frac{z'}{c}\right]\right|^2; \tag{6.3-6}$$

now using Eq. (6.3-4)

$$\frac{\partial W}{\partial \tau} = -\gamma_L(W + z) - \frac{\gamma_L|F(0,\tau)|^2}{1 + \Delta^2} \int_0^z dz' w\left[z', \tau + \frac{z'}{c}\right] \times \exp\left[\frac{\alpha_0 W(z',\tau)}{1 + \Delta^2}\right]$$

$$= -\gamma_L(W + z) - \frac{\gamma_L|F(0,\tau)|^2}{1 + \Delta^2} \int_0^{W(z,\tau)} dW \times \exp\left[\frac{\alpha_0 W(z',\tau)}{1 + \Delta^2}\right]$$

$$= -\gamma_L(W + z) - \frac{\gamma_L|F(0,\tau)|^2}{\alpha_0} \left[\exp\left[\frac{\alpha_0 W(z,\tau)}{1 + \Delta^2}\right] - 1\right]. \tag{6.3-7}$$

Defining

$$\phi(t) \equiv \frac{W\left[L, t - \frac{L_T}{c}\right]}{L}, \tag{6.3-8}$$

$x \equiv t\gamma_L$, and $x_R = t_R\gamma_L$, Eq. (6.3-7) becomes, when evaluated at $z = L$ and $\tau = t - L_T/c$,

$$\frac{d\phi(x)}{dx} = -[\phi(x) + 1] - \frac{|F(0,x - x_R)|^2}{\alpha_0 L} \left[\exp\left[\frac{\alpha_0 L\phi(x)}{1 + \Delta^2}\right] - 1\right]. \tag{6.3-9}$$

From Eqs. (6.3-1a,4,8)

$$F(0,x) = \sqrt{T}\, F_I(x) + Re^{ikL_T} F(0,x - x_R) \exp\left[\frac{\alpha_0 L}{2} \left[\frac{1 + i\Delta}{1 + \Delta^2}\right] \phi(x)\right]. \tag{6.3-10}$$

For vanishingly small input the background refractive index n_0 is ($\phi(x) \to -1$):

$$n_0 = 1 - \frac{\alpha_0 \lambda}{4\pi} \frac{\Delta}{1 + \Delta^2}. \tag{6.3-11}$$

Therefore, Eq. (6.3-10) can be written as

$$F(0,x) = \sqrt{T}\, F_I(x) + RF(0,x - x_R) \exp\left[\frac{\alpha_0 L}{2} \left[\frac{\phi(x)}{1 + \Delta^2}\right]\right]$$

$$\times \exp\left[i\left[\frac{\alpha_0 L}{2} \left[\frac{\Delta}{1 + \Delta^2}\right] (\phi(x) + 1) - \beta_0\right]\right], \tag{6.3-12}$$

where

$$\beta_0 = -k(n_0L + L_T - L) + 2\pi M \qquad (6.3\text{-}13)$$

is the laser-cavity detuning for the linear absorber. Clearly the second term on the right-hand side of Eq. (6.3-12) is the intracavity field at the output mirror multiplied by R $\exp[ik(L_T - L)]$. The transmitted field at a time earlier by $(L_T - L)/c$ is then the product of $\sqrt{T}\exp[-ik(L_T - L)]/R$ and that term:

$$F_T\left[x - \gamma_L\left[\frac{L_T - L}{c}\right]\right] = \sqrt{T}\,F(0,x - x_R)\exp\left[\frac{\alpha_0 L}{2}\,\frac{\phi(x)}{1 + \Delta^2}\right]$$

$$\times \exp\left[i\left[\frac{\alpha_0 L}{2}\left[\frac{\Delta}{1 + \Delta^2}\right](\phi(x) + 1) - \beta_0 - k(L_T - L)\right]\right]. \qquad (6.3\text{-}14)$$

Equations (6.3-12,9,14) are Eqs. (8a), (8b), and (9) of Ikeda (1979) in our notation. He analyzed the stability of the multiple-valued stationary state in the limit of fast transverse relaxation, i.e., $x_R = t_R\gamma_L >> 1$, for which the above equations reduce to difference equations. He found that in the case of purely absorptive bistability the branch with positive differential gain is always stable. But for general detuning, instabilities occur on that branch. Ikeda plotted the sequence of complex electric fields and found that the points are attracted into a figure that appears to consist of an infinite set of one-dimensional curves. Almost identical figures were obtained when the initial field was changed over a wide range. To illustrate this situation we follow Ikeda, Daido and Akimoto (1980).

First assume that the absorption term in Eq. (6.3-12) is essentially independent of intensity; this requires $|F|^2 << \alpha_0 L$ [LeBerre, Ressayre, and Tallet (1982)]. Then Eq. (6.3-12) can be written as

$$E(t) = A + BE(t - t_R)\exp\left[i[\beta(t) - \beta_0]\right], \qquad (6.3\text{-}15)$$

where the intensity-dependent detuning is

$$\beta(t) \equiv \frac{\alpha_0 L}{2}\left[\frac{\Delta}{1 + \Delta^2}\right][\phi(t) + 1]. \qquad (6.3\text{-}16)$$

Similarly, Eq. (6.3-9) becomes

$$\gamma^{-1}\dot{\beta}(t) = -\beta(t) + |E(t - t_R)|^2, \qquad (6.3\text{-}17)$$

where $E(t - t_R)$ is proportional to $F(x - x_R)$. Equations (6.3-15,17) have been derived from the Maxwell-Bloch equations, but they apply to a Kerr medium with a response described by the Debye relaxation equation or to a hybrid device consisting of a Fabry-Perot cavity and intra-cavity phase shifter. The subscript on γ has been dropped to emphasize the generality.

Now consider the limit $\gamma t_R \to \infty$, so the medium responds to the electric field adiabatically:

$$\beta(t) \simeq |E(t - t_R)|^2 ; \qquad (6.3\text{-}18)$$

then E(t) satisfies a difference equation:

$$E(t) \simeq A + BE(t - t_R) \exp\left[i[|E(t - t_R)|^2 - \beta_0]\right] . \qquad (6.3\text{-}19)$$

Ikeda, Daido, and Akimoto (1980) have traced the time series $E_n \equiv E(nt_R)$ as A is increased and found that E_n undergoes successive bifurcations, forming a periodic set of 2^1 points, 2^2 points, 2^3 points, ..., successively, and finally wandering in an apparently erratic manner. The period doubling means that the first instability to appear as A is increased causes E_n to oscillate between two values, i.e., E_1, E_3, E_5, ... are all equal and E_2, E_4, E_6, ... are all equal but $E_1 \neq E_2$. When the next period doubling occurs, four t_R are required for the waveform to complete one period, and there are four different E values, etc. as illustrated graphically and experimentally for a hybrid crossed-polarizer device below. [Ruelle and Takens (1971) mention the period-doubling type bifurcation although their paper is concerned with the behavior on a torus resulting from successive Hopf bifurcations.] Figure 6.3-1 shows 5000 successive points of a computed series E_n in the chaotic regime. The series E_n itself is very sensitive to the choice of the initial point, but the "curves" on which E_n wanders, except for the first few points, are independent of it. In this sense these "curves" may be regarded as a strange attractor. Physically this erratic wandering of E_n may be attributed to large field-dependent phase shifts suffered by the electric field each time it travels around the cavity.

The difference equation approximation Eq. (6.3-19) is valid only over a limited initial range of time. This is because any fluctuation within t_R will develop into changes in E in shorter and shorter times until the adiabatic condition is violated. One must use the difference-differential Eqs. (6.3-15,17) to describe the long-term behavior.

In the limit $B \ll 1$ and $A^2B \simeq O(1)$, Eqs. (6.3-15,17) reduce approximately to the one-variable difference-differential equation for $\beta(t)$:

$$\gamma^{-1}\dot{\beta}(t) \simeq -\beta(t) + A^2\left[1 + 2B \cos[\beta(t - t_R) - \beta_0]\right] , \qquad (6.3\text{-}20)$$

where Eq. (6.3-19) has been approximated by:

$$E(t) \simeq A + B\, e\left[i[\beta(t) - \beta_0]\right] \left[A + BE(t - 2t_R) \exp\left[i[\beta(t - t_R) - \beta_0]\right]\right] + \ldots$$

$$|E(t)|^2 \simeq A^2 + 2A^2B \cos[\beta(t) - \beta_0] .$$

This limit corresponds to a Fabry-Perot cavity with very poor finesse since $B = R \exp[-\alpha_0 L/2(1 + \Delta^2)]$ must be small. Ikeda, Daido, and Akimoto (1980) solve Eq. (6.3-20) numerically; Fig. 6.3-2 shows the temporal behavior of the

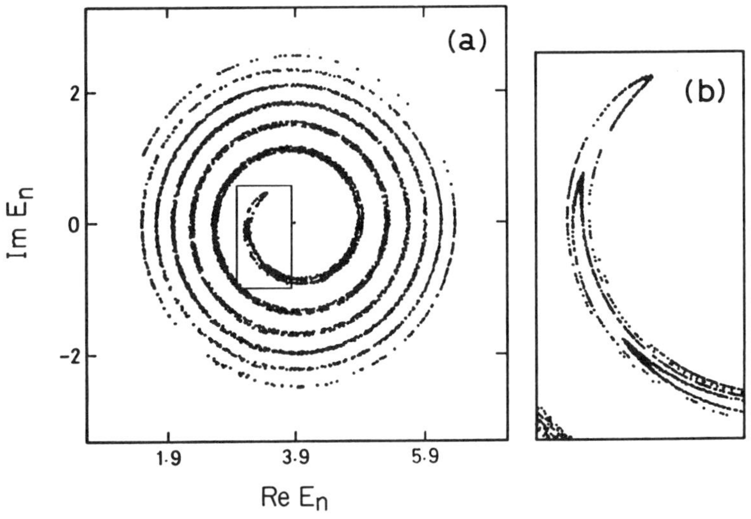

Fig. 6.3-1. (a) Plot of 5000 successive points of a series E_n on the complex E plane. The parameter values chosen are B = 0.4, A = 3.9, and $\beta_0 = 0$. (b) Enlargement of the rectangular region of (a). Ikeda, Daido, and Akimoto (1980).

output intensity and its power spectrum for two choices of the parameters. Period two is shown in Figs. 6.3-2 a,b and chaos in Figs. 6.3-2 c,d.

Ikeda, Daido, and Akimoto (1980) suggested that a hybrid bistable device (Section 4.3) be used to observe period doubling and chaos by inserting in the feedback a delay t_R much longer than the medium response time $\gamma^{-1} = \tau_M$. Rozanov (1982) points out that this was suggested earlier in Rozanov (1980b). Figure 6.3-2 shows period-two and chaotic outputs for that case; see also Murina and Rozanov (1981). Figure 6.3-3a shows an apparatus used by Gibbs, Hopf, Kaplan, and Shoemaker (1981) to make the first observations of period doubling and chaos in an optical bistable system. A TRS-80 digital computer was used to delay the transmitted voltage by t_R = 40 ms, much longer than both τ_M = 1 ms and the computer digitization step of 0.16 ms. [See also Bistrov, Gnedoy, Shipilov, and Shmanov (1982)]. Figure 6.3-3b shows ordinary bistability with t_R = 160 μs $\ll \tau_M$ = 1 ms; but for t_R = 40 ms the output has periodic and chaotic domains (Fig. 6.3-3c). The time dependences in those domains are shown in Fig. 6.3-4.

The transmission of crossed polarizers with an electro-optic crystal between (with electric field applied such that the slow and fast axes make angles of 45° with respect to the polarizer transmission axes) is given by

$$\frac{I_T(t)}{I_I} = \frac{1 - 2B \cos[\beta(t) - \beta_0]}{2}, \qquad (6.3\text{-}21)$$

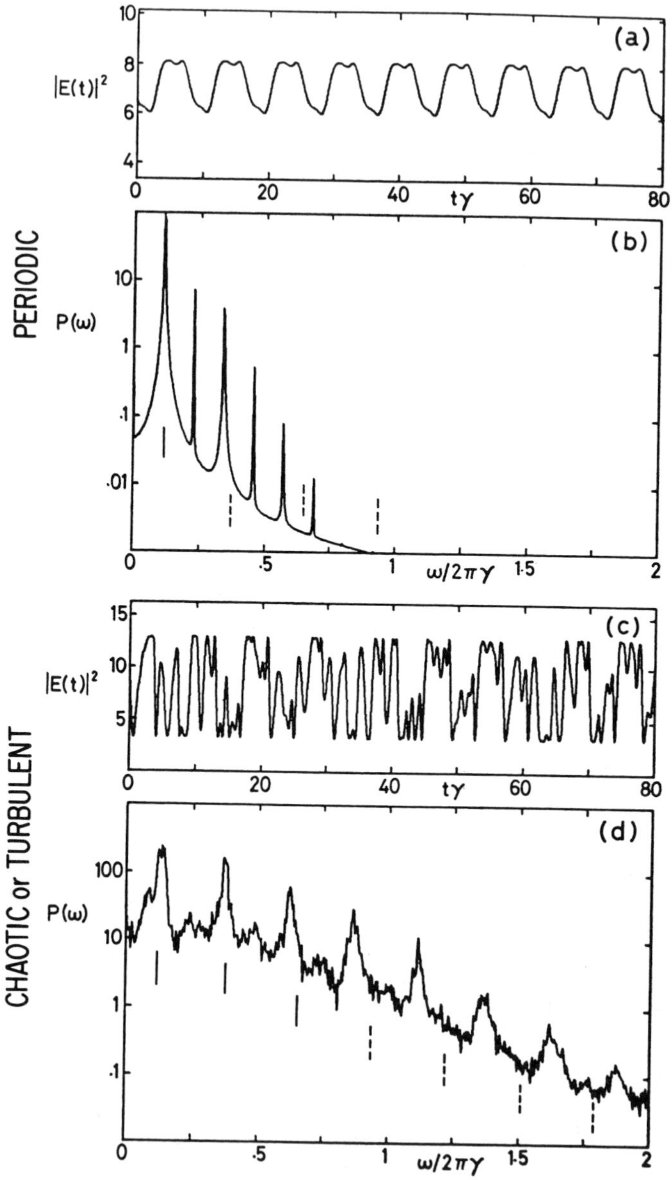

Fig. 6.3-2. $|E|^2$ vs time for B = 0.3, γt_R = 3.5, β_0 = 0, and (a) A = 2.17, and (c) A = 2.85; power spectrum of $|E|^2$ for B = 0.3, γt_R = 3.5, β_0 = 0, and (b) A = 2.17, and (d) A = 2.85. The solid and broken vertical lines in (b) and (d) indicate $|\mathrm{Im}(\lambda)|/2\pi\gamma$ of the unstable and stable modes, respectively. Ikeda, Daido, and Akimoto (1980).

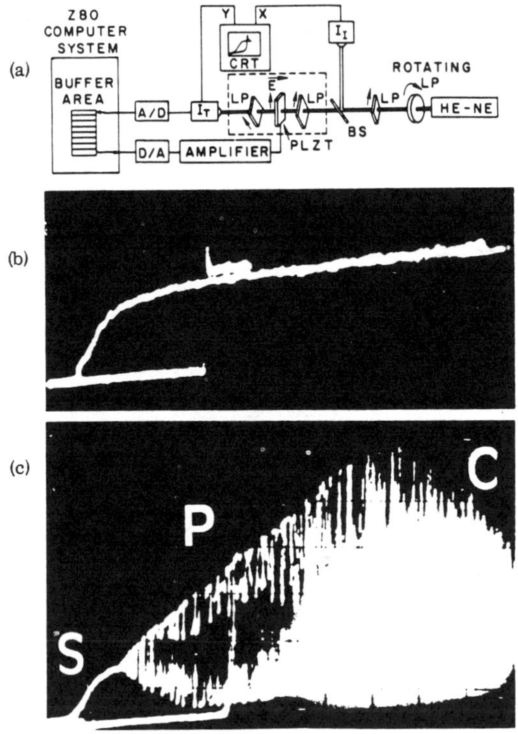

Fig. 6.3-3. (a) Block diagram of optical chaos apparatus. (b) Plot of output intensity (vertical axis) vs input intensity (horizontal axis) as the input intensity is cycled slowly from zero to some maximum and back over a time of 30 s. Here t_R = 160 μs << τ_M = 1 ms; β_0 = 0. (c) Same as (b) except t_R = 40 ms >> τ_M = 1 ms. The labels, S, P, and C indicate the stable, periodic, and chaotic domains of the upper branch. Gibbs, Hopf, Kaplan, and Shoemaker (1981).

where β is the phase shift introduced by the crystal and B = 0.5 for an ideal modulator. For an electro-optic device β is proportional to V_M, where V_M is the voltage across the modulator. The early hybrid employed a lead-based lanthanum-doped zirconium titanate (PLZT) modulator for which $\beta \propto V_M^2$; in that case the computer took the square root of the transmitted signal V_T as well as faithfully translated it in time by t_R. Therefore with the PLZT or later electro-optic modulators, one can write

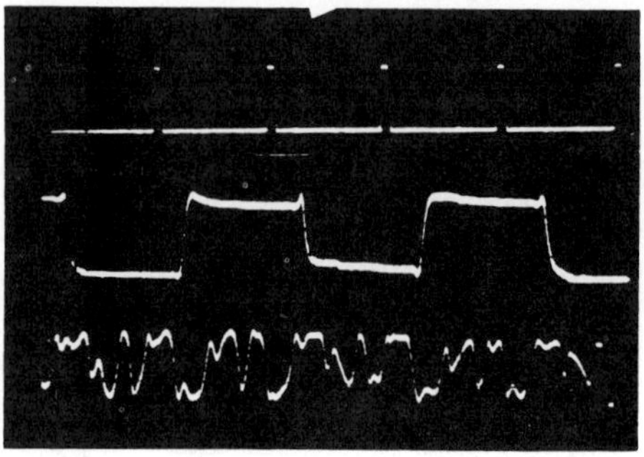

Fig. 6.3-4. Optical periodic oscillations and chaos. (a) Time calibration; one pulse every delay time $t_R = 40$ ms. (b) Intensity vs time in the period-two domain. (c) Intensity vs time in the chaotic domain. $\tau_M = 1$ ms; $\beta_0 = 0$. Gibbs, Hopf, Kaplan, and Shoemaker (1981).

$$\beta(t) = GV_T(t - t_R),$$

where G is a constant proportional to the feedback gain. If I_I is reduced to zero, $\beta(t)$ decays away in a time γ^{-1}, the detector-feedback-modulator composite relaxation time, i.e.

$$\gamma^{-1}\dot{\beta}(t) = -\beta(t) + GV_T(t - t_R)$$

$$= -\beta(t) + GI_IC \frac{\left[1 - 2B\cos[\beta(t - t_R) - \beta_0]\right]}{2}, \qquad (6.3\text{-}22)$$

where C converts I_T to V_T. Clearly Eq. (6.3-22) is of the form of Eq. (6.3-20). The use of parallel polarizers or increasing β_0 by π in Eq. (6.3-22) would make it identical to Eq. (6.3-20) used by Ikeda et al. (1980).

Hopf, Kaplan, Gibbs, and Shoemaker (1982) have studied the path to chaos using the apparatus of Fig. 6.3-3a but with a KDP modulator [see also Hopf, Kaplan, Shoemaker, and Gibbs (1981) and Hopf (1982)]. The phase was set at $\beta_0 = \pi/2$, so that for $I_I = 0$ the transmission was 50% and decreased as I_I increased. For all the waveforms studied $\beta < 3\pi/2$, so that the device was always in the lower branch. In other experiments a 1-km optical fiber was used for the delay, and the lower-branch path to chaos was studied with $\beta_0 = \pi$ and $B = 0.48\pm0.005$ as shown in Fig. 6.3-5. Period doubling through period eight was clearly seen. In the chaotic domain the reverse sequence was seen down to N_2 (noisy period two) and then higher harmonics appeared. For further discussion of frequency locking and "successive higher-harmonic bifurcations" see: Hopf et al. (1982); Ikeda, Kondo, and Akimoto (1982); Derstine, Gibbs, Hopf, and Kaplan (1982, 1983). Other interesting phenomena are predicted for different bias settings, for example, a "crisis," i.e., the destruction of a chaotic attractor by touching another

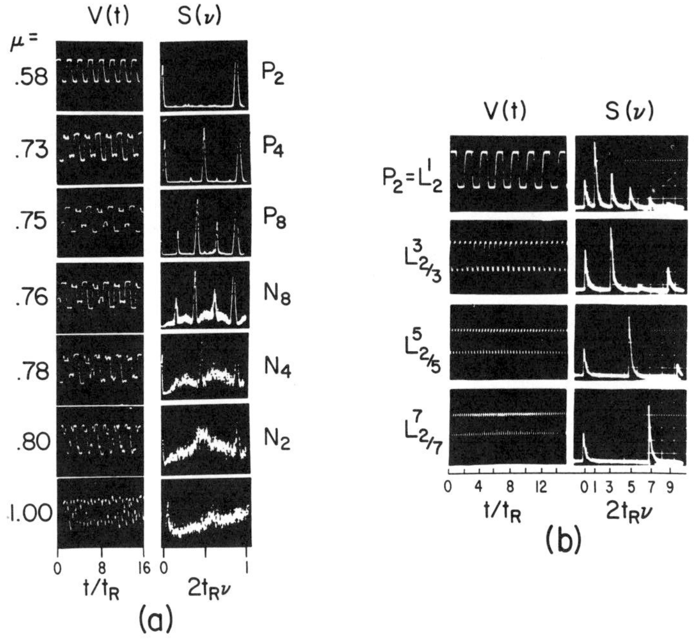

Fig. 6.3-5. Basic time-dependent outputs of the hybrid device with KDP modulator and optical fiber delay of 6 μs. $\gamma^{-1} = \tau_M$ ranged between 0.33 and 1 μs. (Left) P_n denotes an oscillation with period nt_R and N_n denotes period n chaos, for which there is no clear frequency component with $\nu = 1/(2nt_R)$. The $\mu = 1.00$ waveform is not identified; $\mu = GI_IC/2\pi$ in Eq. (6.3-22) can be altered via the gain or input intensity. (Right) Frequency locking waveforms in the chaotic regime between N_2 and N_1, showing successive higher harmonic bifurcations. $L^n{}_{2/n}$ denotes a periodic waveform where the subscript denotes the period $(2t_R/n)$ and the superscript denotes the n that defines the fundamental frequency: (a) $P_2 = L_2{}^1$, (b) $L^3{}_{2/3}$, (c) $L^5{}_{2/5}$, (d) $L^7{}_{2/7}$ [Hopf, Derstine, Gibbs, and Rushford (1984)].

manifold (in this case the unstable period one) [Mandel and Kapral (1983)], and explosions [Moloney, Hammel, and Jones (1984)].

The origin of periodic oscillations can be followed by a graphical solution [May (1976); Feigenbaum (1980)] in the limit $\gamma^{-1} \ll t_R$. Then with B = 0.5 and $\beta_0 = \pi/2$, Eq. (6.3-22) can be replaced by the difference equation

$$\beta_{(n+1)} = \pi\mu(1 - \sin\beta_n) , \qquad (6.3\text{-}23)$$

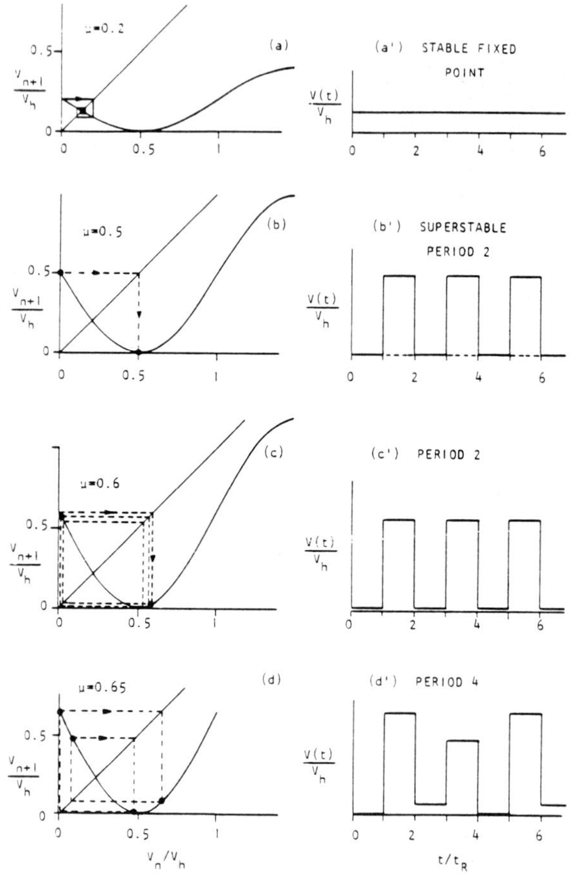

Fig. 6.3-6. Graphical origin of periodic oscillations. Left column [(a) to (d)]: graphical construction of V_{n+1} vs V_n. Right column [(a') to (d')]: V(t) plotted from left column after many cycles, i.e., under steady-state conditions. Gibbs, Hopf, Kaplan, Derstine, and Shoemaker (1981).

where the constant μ is proportional to the product of the feedback gain and input intensity. Figure 6.3-6 is a graphical plot of the delayed output voltage (proportional to β) for the case that μ is turned on abruptly at $t = 0$. The time evolution is found by applying in each t_R the voltage transmitted during the previous t_R. This is equivalent to translating from the nth point to the 45° line and then translating vertically to the (n + 1) point on the transmission curve. Superstable period two (Fig. 6.3-6b) is especially easy to follow: during the first t_R, no voltage has reached the modulator, but the transmission is 0.5; during the second t_R, a voltage of 0.5 on the modulator gives zero transmission; during the third t_R, zero volts are across the modulator again completing one period in two t_R's. Note the similarity between the period two and period four waveforms generated graphically from the difference equation (Fig. 6.3-6) and observed in the delayed-hybrid experiment (Fig. 6.3-5).

If one denotes the transmission function by f, then an intersection of f with the 45° line is a stable point if $|df/dV| < 1$, but is unstable otherwise. In Fig. 6.3-6a, $|df/dV| < 1$ at the intersection, so it is the stable operating point. In Fig. 6.3-6b, $|df/dV| > 1$ at the intersection, so it is unstable. If one goes through two iterations, i.e., $f^{(2)} \equiv f(f)$ or $V_{n+2} = f^{(2)}(V_n)$, two of the intersections have $|df^{(2)}/dV| < 1$; they are the two levels of the period two (Fig. 6.3-7a). Similarly the four levels of period four are found with $f^{(4)}$ (Fig. 6.3-7b), etc.

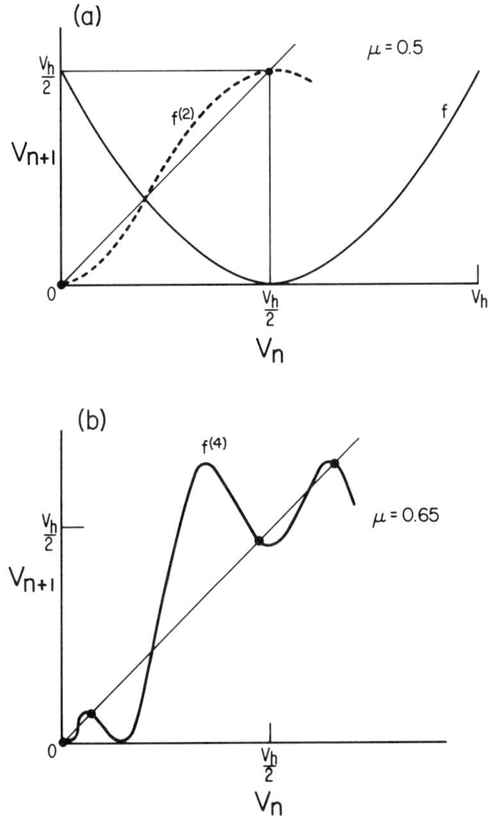

Fig. 6.3-7. Use of higher iterates to determine the steady-state levels. (a) The two intersections (•) of $f^{(2)}$ and the 45° line with $|df^{(2)}/dV| < 1$ define the two levels in period two of Fig. 6.3-6b. (b) The four intersections (•) of $f^{(4)}$ with the 45° line with $|df^{(4)}/dV| < 1$ define the four levels in period four of Fig. 6.3-6d.

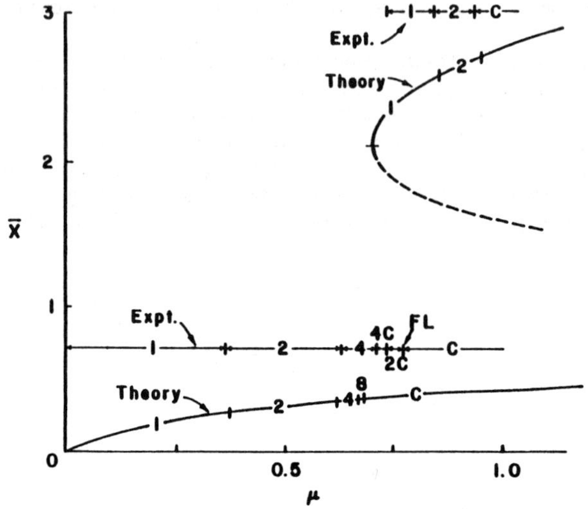

Fig. 6.3-8. The portion of the "S" curve of the bistable device over the domain $0 \leq \mu \leq 1$ of the experiment. $\overline{X}$ is the customary steady-state output (here mostly unstable) vs input laser intensity scaled as μ. The dashed portion is unconditionally unstable. The theoretical curve is labeled 1,2,4,8 to indicate period one (stable), period two, etc. Note that periods sixteen and higher on the lower branch and four and higher on the upper branch cover very small domains that are not shown. The label "C" indicates chaos. The experimental bifurcation structure is indicated along the horizontal lines (the experimental $\overline{X}$ is not shown). The domain of the frequency locked (FL) anomaly is too narrow to be shown ($\Delta\mu < 0.01$). Note that the uncertainty of locating the bifurcation points experimentally is quite large ($\simeq 5\%$) due to critical fluctuations. Hopf, Kaplan, Gibbs, and Shoemaker (1982).

Apart from possible important implications for practical bistable devices, the study of Ikeda instabilities is especially interesting because of the importance of turbulence in many diverse fields and the recent emergence of some universal properties of such phenomena [Lubkin (1981); Feigenbaum (1980); May (1976); Giglio (1982)]. The path to chaos observed in the computer-delay hybrid is compared with that expected in Fig. 6.3-8. Feigenbaum has predicted that the μ's in the period doubling sequence scale as

$$\delta \equiv \lim_{n\to\infty} \frac{\mu_n - \mu_{n-1}}{\mu_{n+1} - \mu_n} = 4.669 \,. \qquad (6.3\text{-}24)$$

Hence, the domains of stability vanish geometrically ($\mu_{n+1} - \mu_n \propto \delta^{-n}$), and the sequence of bifurcations stops at a finite value $\mu = \mu_c$, such that for $\mu > \mu_c$ the behavior is chaotic. The computer-delay-hybrid ratio of the period-two domain to period-four is 4.3 ± 0.3 in qualitative agreement with this remarkable prediction even though n is small.

Period eight was rarely seen using the computer-delay hybrid, but it is seen at will in the fiber-delay hybrid [Gibbs, Derstine, Hopf, Kaplan, Rushford, Shoemaker, Weinberger, and Wing (1982)]. The termination of the period-doubling sequence is now believed to be the effect of noise either from laser intensity fluctuations or photomultiplier shot noise [Derstine, Gibbs, Hopf, and Kaplan (1982); and Derstine, Hopf, Kaplan, and Gibbs (1982)]. This termination by noise was discovered by Crutchfield and Huberman (1980) in solving the problem of forced dissipative motion in an anharmonic potential with the aid of an analog computer and a white-noise generator. The effect of fluctuations in their problem was to produce a symmetric gap in the deterministic bifurcation sequence. The bifurcation gap has been observed in the fiber-delay hybrid. The doubling sequence terminated at period eight with lowest noise level attained, and at period four or period two with higher noise levels. However, the bifurcation gap was slightly asymmetric: a larger μ was required to terminate period-n chaos than period n. This difference may lie in the fact that the experimental noise was intensity dependent, whereas Crutchfield and Huberman used intensity-independent Gaussian noise. Zardecki (1982) has made further theoretical studies of the bifurcation gap for dispersive optical bistability. Chrostowski (1982) iterates a one-dimensional map for a hybrid acousto-optic bistable device and finds severe truncation of the bifurcation sequence with additive and multiplicative noise. Vallée, Delisle, and Chrostowski (1984) present beautiful figures of the effects of additive or of multiplicative noise upon the bifurcation diagrams calculated numerically and observed using an acousto-optic hybrid. See also Okada and Takizawa (1981b) and Gao, Narducci, Sadiky, Squicciarini, and Yuan (1984).

The analogy between the transition to chaos and critical point phase transitions is being pursued [Crutchfield, Nauenberg, and Rudnick (1981); Shraiman, Wayne, and Martin (1981)]. The Lyapunov exponent measures how fast two points in phase space separate in time. Once a period-doubling system becomes chaotic the Lyapunov exponent grows with a power-law dependence, just like an order parameter in a phase transition. How to distinguish noise from chaos in an actual experiment is nontrivial and a subject of high current interest. One definition of a chaotic system is that two points starting close together in phase space diverge exponentially in time, provided they start on a strange attractor [Ruelle and Takens (1971)]. This definition implies that chaotic systems should have continuous spectra, a property shared by nonchaotic systems such as incoherent light sources. Derstine, Gibbs, Hopf, and Sanders (1984) propose a new test for chaos, experimentally much simpler than the attractor test. They suggest that a chaotic system be treated like any other unstable system except that, since all trajectories are unstable, one must measure variances, i.e., error bars of trajectories, rather than the trajectories themselves. They repeatedly start their hybrid system in some prescribed state and examine its subsequent temporal evolution. In chaos, there is a finite delay between the initiation of the system at a known configuration and the time at which the system has developed such a large variance that the output can no longer be predicted deterministically. In contrast, uncertainties that are due to noise develop instantaneously in time. They find that the logarithm of the variances, $\ln[\delta X(t)]$, of various measurements $X(t)$ increases in time at a uniform rate independently of initial conditions and independent, to within experimental error, of the choice of X. The rate of increase of the variance is related to the metric entropy, which is a theoretical measure used to distinguish chaos from noise [Farmer (1982); Grassberger and Procaccia (1983)]. The metric entropy is positive and finite for chaotic systems and infinite for noisy systems. The linear rise in $\ln[\delta X(t)]$ implies that the slope of the rise (which is proportional to the metric entropy) is finite for all time and in particular for $t \to 0$, where the metric entropy should be measured. Noise develops temporally as $t^{1/2}$, so that the logarithm of the variance goes as $\ln(t)/2$. The derivative of the logarithm goes as $(2t)^{-1}$, and in the limit $t \to 0$ this quantity is infinite, giving infinite metric entropy. Therefore, the proposed test for chaos gives a finite metric entropy as expected for a chaotic system when the hybrid device is operated in the regime identified as chaotic from other indicators (period doubling to chaos, spectrum, etc.). In many cases, such as those noted in Section 6.1 or Gaussian-beam cavities, the bifurcation may occur from period two straight to chaos or along some other route. So a test for chaos independent of a detailed understanding of the bifurcation sequence is useful. The proposed test is roughly equivalent to the attractor test if the points all lie on the attractor; but the latter is unknown in the case studied.

Carmichael, Snapp, and Schieve (1982) and Snapp, Carmichael, and Schieve (1981) performed extensive plane-wave numerical studies of Ikeda ring-cavity instabilities and found them to be in excellent quantitative agreement with Feigenbaum's theory. Moloney (1984c) stresses that the nonlinear difference equation describing the plane-wave ring cavity is two-dimensional (or 1D complex) in contrast to Feigenbaum's one-dimensional logistic map. He has identified three distinct types of behavior: (i) a period-doubling cascade to chaos and return via the same route, (ii) a period-doubling cascade to chaos but return via a different route, and (iii) a completely new type of periodic cycle that appears abruptly via a tangent (saddle-node) bifurcation. The latter provides direct numerical evidence for the 2D behavior of the complex map through the identification of a stable six cycle and its associated subharmonic bifurcations. Numerically the 6-cycle is seen to undergo its own period-doubling cascade, giving support to the hypothesis that all instabilities in the plane-wave model of a bistable ring cavity are of the period-doubling type.

Stability analysis of a Fabry-Perot standing-wave cavity by Firth (1981) and Firth, Abraham, and Wright (1982) shows Ikeda instabilities much like those found by Ikeda for a one-way ring cavity. Instabilities of a Fabry-Perot cavity whose dispersive third-order nonlinearity has a finite response time τ_M have been studied by Abraham, Firth, and Carr (1982) and Abraham and Firth (1983). With $\tau_M/t_R = 5$ they still find an Ikeda period-two oscillation with a "lethargic" $2t_R$ period. Ikeda instability persists on high-order branches even for τ equal to many round-trip times t_R. Firth and Wright (1982b) consider a short Kerr medium located at the center of a plane Fabry-Perot cavity and study the effect of the nonreciprocal grating term [factor of 2 in last terms of Eqs. (D-8,9)]. They find that if the nonreciprocal terms are removed, t_R oscillations are forbidden to all orders. Their physical interpretation is to regard the cavity as containing two wave trains at any time, spaced $t_R/2$ apart, and interacting with each other (via $\chi^{(3)}$) twice per round trip. In steady state, these trains are identical in amplitude and phase. In t_R oscillation this symmetry is broken; the two trains have unequal amplitudes and impose unequal phase shifts on each other. Only the nonreciprocal interaction, i.e. the phase grating, forces such a splitting. This oscillation is thus the first dynamic instability attributable to nonlinear reciprocity. For a curved-mirror Fabry-Perot cavity pumped by a Gaussian beam matched to the fundamental mode, the profiles remain quasi-Gaussian but dilate and contract in the course of the t_R oscillation.

Carmichael, Snapp, and Schieve (1982) and Firth (1981) show that the internal intensity for Ikeda instability is minimized with a small-signal cavity detuning of π. This "off" condition for the cavity makes the two sideband frequency components of the $2t_R$ amplitude modulation (period two) resonant with the pair of cavity modes adjacent to the pump frequency (likewise for the odd-harmonics of the period-two square wave). The basic Ikeda instability can then be given an attractive interpretation as a resonant four-wave parametric instability: the cavity acts as a doubly-resonant four-wave optical parametric oscillator. See Abraham and Firth (1983), Firth (1983); Bar-Joseph and Silberberg (1983); Firth Wright, and Cummins (1984); Silberberg and Bar-Joseph (1984); Carmichael (1984). An attractive feature of the four-wave mixing interpretation is that it enables one to comprehend the behavior beyond the first bifurcation point [Silberberg and Bar-Joseph (1983,1984)]. At higher input powers both harmonics and subharmonics sometimes appear. Harmonics of an oscillating sideband E_1 are generated by the wave-mixing process $E_1E_1E_0^*$. This term describes a source for oscillations at $\omega_0 + 2(\omega_1 - \omega_0)$, and therefore harmonics are always present. On the other hand, subharmonics from period doubling appear only at certain threshold intensities. A field E_3 at $\omega_3 = (\omega_1 + \omega_0)/2$ is amplified through the process $E_0E_1E_3^*$; oscillations at ω_3 occur only if gain and resonance conditions are satisfied for E_3. Obviously, once E_3 is present, the beating period is doubled. This process, repeating itself, is the basic mechanism of successive period doubling in a nonlinear optical system.

Firth, Wright, and Cummins (1984) give a nice picture of period doubling due to "transphasing," i.e., four-wave-mixing causes the resonance frequencies of a cavity to tune with pump intensity. For an initial detuning of $\theta = -\pi$, the laser frequency lies halfway between two modes of the cavity; all of the odd harmonics in a Fourier decomposition of a period-two square wave are coincident with cavity modes.

> "Above the Ikeda threshold, the nonlinear resonator no longer has constant optical properties - the $2t_R$ oscillation forbids this. The standard cavity mode condition, that the field repeats itself (up to a factor) after a round trip, then loses its force, since the medium itself has changed after t_R. It follows that one can now only insist on repetition after two round trips. Then the effective cavity length is doubled, and so its free spectral range is halved. At the $2t_R$ threshold these cavity modes are coincident with the pump frequency and the $2t_R$ peaks, but increasing the input will cause the modes, by transphasing, to split off and move in frequency. We now have a rescaled version of the previous argument: when transphasing brings these "dressed modes" just halfway between the pump and Ikeda ($2t_R$) modes, we again have a double resonance and thus, if the gain is big enough, will get self-oscillation, this time at $4t_R$. This, in turn breaks the $2t_R$ symmetry, doubles the mode spectrum, and the "new" modes, through transphasing to double resonance, give $8t_R$ oscillation; and so on. The period-doubling cascade is thus a progressive spontaneous breaking of the time-translation symmetry of the system, leading in the end to no time symmetry at all, or chaos."

The computations and hybrid experiments described thus far in this section have been almost all plane-wave. Often the bifurcation structure is altered substantially by transverse effects that occur when the input beam has a Gaussian spatial profile. This is not surprising from the point of view that an infinite sum of plane-wave solutions is required to represent the Gaussian output for the case of no transverse coupling. One might even be surprised that periodic oscillations survive, and often the lower subharmonics with very narrow plane-wave domains do not. It is less surprising that chaos persists, since it is aided by additional degrees of freedom.

Studies of Ikeda instabilities with transverse effects included have been largely numerical [Moloney and Hopf (1981); Moloney, Hopf, and Gibbs (1982a,b); Firth, Abraham, and Wright (1982); Firth and Wright (1982b)]. Section 2.9 and Eq. (2.9-13) describe the inclusion of transverse effects in the wave equation by means of a complex electric field and the Laplacian diffraction term, in the limit of rapid transverse relaxation, so that the polarization adiabatically follows the field. Most of the simulations have been done with a one-transverse-dimension Gaussian input profile given by Eq. (2.9-14).

Plane-wave and Gaussian profile hysteresis loops for the on-axis intensity are compared in Fig. 6.3-9. The parameters were selected so that the plane-wave unstable domain consists solely of a period-two output in the lower branch (Fig. 6.3-9a). The Gaussian-beam hysteresis loop shown in Fig. 6.3-9b is quite different, possessing no steady-output upper state in the bistable domain. Some of the waveforms computed in the unstable domains A and B in Fig. 6.3-9b are shown in Fig. 6.3-10. Only one point is calculated per round-trip time; lines connect the points to help one visualize the sequence of points. The initial time transient is shown only in part (a) of each figure. Usually the first unstable output has period $2t_R$, just as in the plane-wave limit. For higher input intensities the sequence of outputs is usually very different from the period-doubling Feigenbaum sequence. See also Moloney (1982a) and Firth, Abraham, and Wright (1982). The quasiperiodic solutions are made up of two different Fourier components of incommensurate frequencies; see Figs. 6.3-10 [A (b) and B (c and e)]. Frequency-locked outputs are also seen; see Figs. 6.3-10 [A (c and d) and B (d)]. The sequence of bifurcations P2 $\rightarrow$ QP2 $\rightarrow$ L9 in region B is an almost exact duplicate of the bifurcation sequence in the hydrodynamic experiment of Gollub and Benson (1980). Perhaps the similarity is fortuitous, but it suggests that there may be universal features of the path to chaos in addition to those discovered by Feigenbaum for one-dimensional difference equations.

The range of parameters for which instabilities occur has not been well characterized for the transverse-Gaussian ring-cavity case. Note that a large detuning of 5.4 radians was used in Fig. 6.3-10. A detuning greater than π gives a negative slope to the transmission function as needed for periodic oscillation. A graphical example of this for a Fabry-Perot hybrid is shown in Fig. 6 of Gibbs, Hopf, Kaplan, Derstine, and Shoemaker (1981). In Fig. 2.9-1 a detuning of less than π gave no Ikeda instabilities, so instabilities can be avoided in designing devices.

Moloney (1984b) stresses the richness of the possible paths to chaos of a plane-mirror ring resonator with an incident cw laser beam having a Gaussian spatial profile in one transverse dimension. The nonlinear medium consists of saturable two-level atoms operated in the purely dispersive limit. He identifies transitions to turbulence induced solely by self-focusing or plane-wave conditions. Three typical transition sequences observed in fluid experiments occur: period

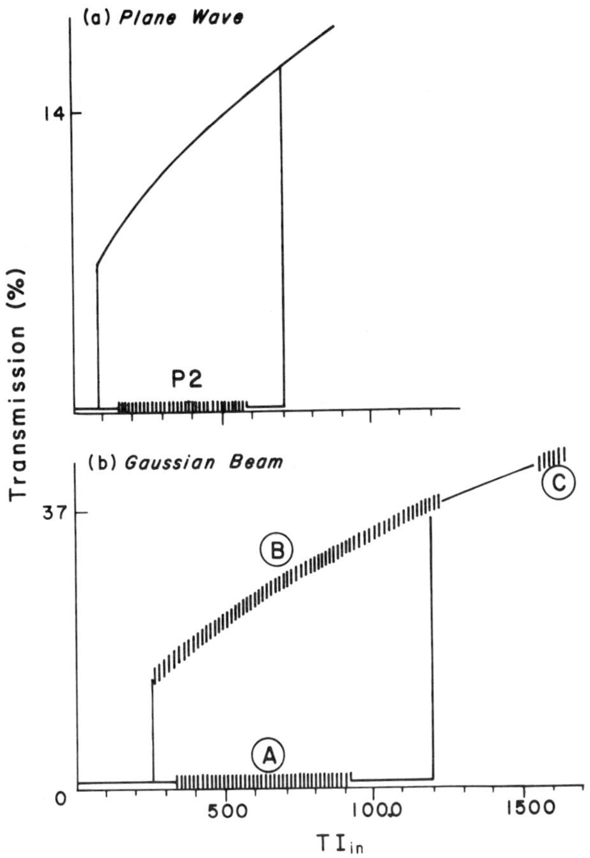

Fig. 6.3-9. Bistable loops associated with a detuned ring cavity. Parameters used in the calculation are detuning of β_0 = 0.4 radians between the laser and empty cavity, Δ = -20 (self-focusing), $\alpha_0 L_{NL}/(1 + \Delta^2) = 5$, $L_{NL} = 0.3L$, total length = 2L, T = 1 - R = 0.1, F = 0.54. The laser detuning from the peak of the full cavity at low light intensity is 5.4 radians. Input intensity I_{in} is in units of the saturation intensity I_s. (a) Plane wave (b) Gaussian beam. On-axis transmitted intensity is shown. Moloney, Hopf, and Gibbs (1982a).

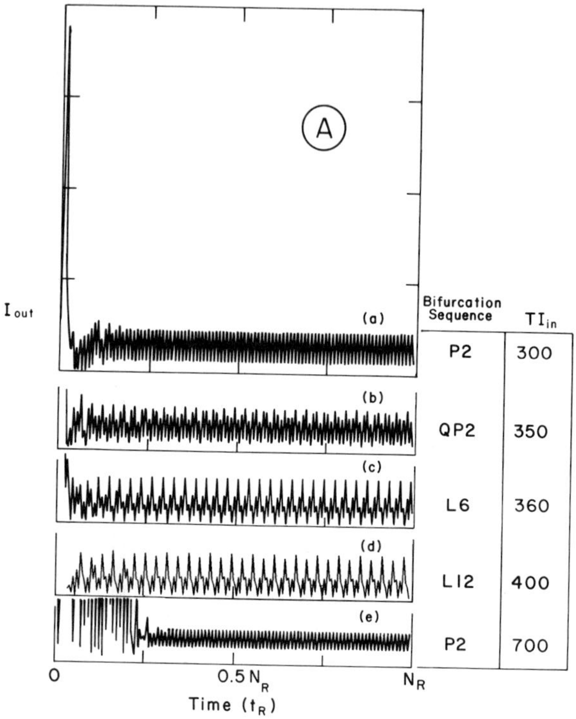
A
I_{out}
Bifurcation Sequence
TI_{in}
(a)
P2
300
(b)
QP2
350
(c)
L6
360
(d)
L12
400
(e)
P2
700
0
$0.5N_R$
N_R
Time (t_R)

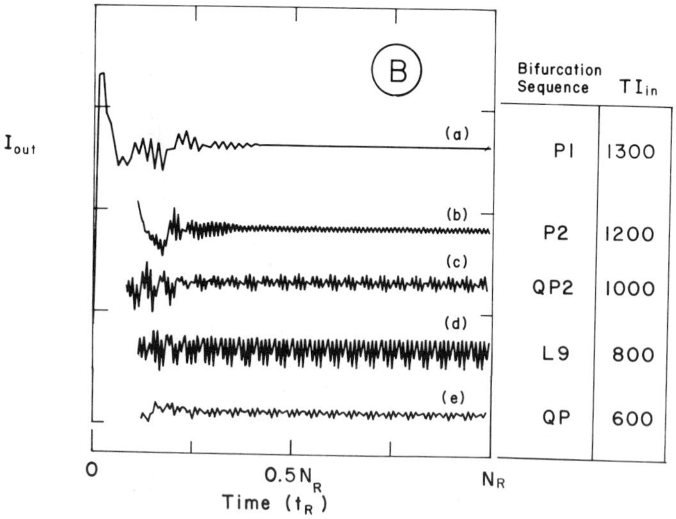
B
I_{out}
Bifurcation Sequence
TI_{in}
(a)
P1
1300
(b)
P2
1200
(c)
QP2
1000
(d)
L9
800
(e)
QP
600
0
$0.5N_R$
N_R
Time (t_R)

←

Fig. 6.3-10. Dynamic outputs representing a sampling of the unstable domains A and B of Fig. 6.3-9b. The time axis is in units of t_R, the cavity round-trip time. (A) Dynamic outputs for region A. Number of cavity round trips N_R = 200. (B) Dynamic outputs for region B. (a) N_R = 100, (b) - (d) N_R = 200, (e) N_R = 160. Moloney, Hopf, and Gibbs (1982a).

doubling to chaos; periodic-quasiperiodic breakdown of a two-torus to a chaotic attractor [Ruelle-Takens (1971) scenario]; and a periodic-intermittent chaos [Pomeau-Manville (1980) scenario)]. This richness might at first be viewed as an unwelcome escalation of complexity, but the similarities to cases where such complexities are unavoidable, such as fluids, are striking. It is suggestive that all-optical systems with one or both transverse degrees of freedom may help in the search for other universal features in turbulent systems. Optical systems afford the opportunity to control, to turn on and off, such complexities in experiments which can be performed very rapidly ($t_R \lesssim 1$ μs). Even though the outcome of such pursuits is unknown, such hopes motivate much of the research on systems that can exhibit all-optical chaos.

Periodic oscillations and erratic or chaotic outputs were described in Section 6.1 for self-trapping bistability in Na and etalon bistability in CdS. Those oscillations may be examples of intrinsic optical chaos, but the periodic oscillation frequencies are much less than $(2t_R)^{-1}$. The simple interpretation of the Ikeda delayed-feedback instability with characteristic periodic frequency $(2t_R)^{-1}$ makes that instability the preferred one to achieve experimentally. Three systems have exhibited it: modelocked pulses in an optical fiber interferometer, ring cavity containing ammonia, and a diffraction-free-encoding single-feedback-mirror Na system.

Bifurcation to chaos was observed using a single-mode optical fiber as the nonlinear medium ($n_2 \approx 10^{-13}$ esu) in a ring cavity by Nakatsuka, Asaka, Itoh, Ikeda, and Matsuoka (1983) [see also Nakatsuka, Asaka, Itoh, and Matsuoka (1983)]. The ring cavity consisted of an input and an output mirror having reflectivities of 0.60 and 0.20, respectively, in addition to two 10X microscope objectives and the 1.2-m optical fiber. The input pulses were the second harmonic of an actively modelocked and Q-switched YAG laser, yielding > 1 kW peak power and a 450-ns FWHM pulse train. The pulse width was 140 ps FWHM, the spectral width was 3 GHz (close to transform limited), and the separation between pulses was t_R = 7.6 ns. One transit time of the ring cavity was carefully adjusted to precisely equal the period of the modelocked pulses. This optical-fiber ring cavity is plane-wave-like as are hybrid devices, in this case because all parts of the guided wave are subjected to the same nonlinear phase shift. The round-trip losses give a value of 0.4 to 0.5 for the parameter B which characterizes the strength of the cavity feedback [Eq. (6.3-19)]. At peak powers of 50, 160, and 300 W the output pulse amplitudes were the same, alternated (period two), and jittered wildly (chaos) over some range, respectively. For the alternating pulses, A^2B, which characterizes the onset of chaos, was close to unity, and the nonlinear phase shift due to the propagation through the 120-cm optical fiber was calculated to be about π at the peak. Good agreement was found between numerical calculations of Eq. (6.3-19) and the output train of this clever all-optical bistable system.

The first observation of dispersive optical hysteresis in a ring cavity containing a molecular gas was made by Harrison, Firth, Emshary, and Al-Saidi (1984). Ammonia in a 1-m intracavity gas cell was used as the nonlinear medium; it was off-resonantly excited close to the aR(1,1) transition ($n_2 \simeq 2 \times 10^{-9}$ esu) using temporally smooth 100-ns pulses from a TEA CO_2 laser. Observations of $2t_R$ oscillation, with some indications of $4t_R$, were reported for this system by Harrison, Firth, Emshary, and Al-Saidi (1983). The time traces and frequency spectra show peaks with period of $2t_R \simeq 23.4$ ns and frequency $(2t_R)^{-1}$, respectively, and there is considerable similarity between the data and numerical simulations. Harrison, Firth, and Al-Saidi (1984a) used similar laser pulses but shortened the all-optical passive cavity. They used a Fabry-Perot resonator consisting of a single-surface Ge flat input coupler of reflectivity 36 to 85% and a single-surface Ge output coupler, of 2-m radius of curvature and reflectivity 76%. With 10 Torr of ammonia, a strong Ikeda oscillation (period $\simeq$ 13 ns, close to $2t_R \simeq 11.5$ ns) persistent throughout the pulse is evident in the neighborhood of minimum transmission. At lower pressures (4 to 8 Torr), where inhomogeneous broadening may be important, strong $4t_R$ oscillations are seen. At higher pressures (20 to 30 Torr) much more complex pulse shapes are seen. They report evidence for a $2t_R/3$ third harmonic oscillation which period doubles to $4t_R/3$. Aperiodic pulse shapes, characteristic of chaos, are also evident. "We suggest that the weight of evidence from the oscillatory pulse shapes supports the conclusion that we have driven an all-optical system through oscillation to chaos."

Optical bistability by means of diffraction-free-defocusing in a short cell and feedback by a single mirror was presented in Section 3.9.2. One-transverse-dimension numerical simulations show that this transverse bistability exhibits a rich bifurcation sequence [LeBerre, Ressayre, Tallet, Tai, Hopf, Gibbs, and Moloney (1984)]. An oscillation with a period of $2t_R$ has been seen in Na with $t_R \simeq 20$ ns and $\tau_M \simeq$ few nanoseconds using several Torr of Ar. This is apparently the first observation of a $2t_R$ instability in a passive absorber using a cw input. By varying the feedback distance, the frequency peak in the power spectrum shifted according to $(2t_R)^{-1}$. Under other conditions other peaks were seen in the power spectrum which were sensitive to alignment, laser detuning, and input power. Because most semiconductors have a self-defocusing nonlinearity below the band edge and must be physically thin for low absorption, this observation gives hope for studies of cw passive all-optical $2t_R$ instabilities using a semiconductor. A simple model, in which the backward field's waist is assumed to be the same as the input's at the cell, has been calculated analytically and evaluated numerically; it shows bistability, $2t_R$-oscillations, and more complicated outputs for more intense inputs.

Closely related to these all-optical intrinsic systems is the work on diode lasers with an external cavity [Otsuka and Iwamura (1983), Otsuka and Kawaguchi (1984a,b), and Kawaguchi and Otsuka (1984)].

A potential application of Ikeda periodic oscillations is to generate square-wave modulated optical beams [Gibbs, Hopf, Kaplan, Derstine, and Shoemaker (1981); Kaplan, Gibbs et al. (1982)]. The square-wave frequency is approximately $f_{SW} = (2t_R)^{-1} = c/(4Ln_0)$ for a Fabry-Perot cavity of length L and refractive index n_0. If the nonlinear medium in an intrinsic device occupies a very short length in the waist of the light beam in an otherwise empty cavity ($n_0 \simeq 1$), f_{SW} is 1, 10, 100 GHz for L = 75, 7.5, and 0.75 mm, respectively. Or for a semiconductor etalon with $n_0 = 3.75$, those frequencies are attained with L = 20, 2, and 0.2 mm, respectively. The difficulty, of course, is in finding highly nonlinear materials having a response time γ^{-1}, much shorter than the desired t_R. Such oscillators are likely to be faster and more stable than those based on regenerative pulsations or self-pulsing.

Even before the observation of periodic oscillations in a hybrid bistable device with delayed feedback, oscillations were observed in the intensity of lasers with delayed feedback. Gray and Casperson (1978) proposed square-wave modulation of a laser by delayed optical feedback that quenches the lasing by gain cross saturation; see Fig. 6.3-11. Damen and Duguay (1980) obtain 100-ps pulses using a feedback loop containing a semiconductor diode laser, a photodetector, an amplifier, and a time delay. Nakazawa, Tokuda, and Uchida (1981) use an optical-fiber electrical-amplifier feedback to a laser diode and observe sinusoidal and square-wave self-sustained oscillations. All of these devices can exhibit oscillations with periods of twice the round-trip time and to that extent they are related to the Ikeda oscillations. But these devices are essentially gain-control devices, whereas the oscillation labelled "Ikeda" here arises from an instability of a dispersive bistable system. Consequently, the nonlinear refractive index and changes in cavity-laser detuning are the fundamental parameters in the Ikeda case.

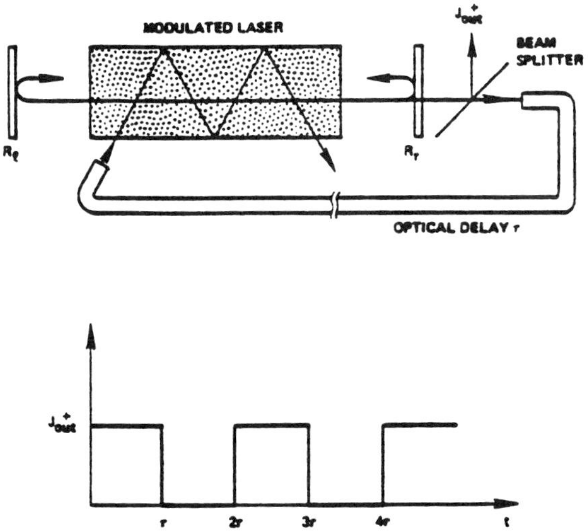

Fig. 6.3-11. Schematic drawing of a proposed delay-line oscillator using a modulated laser. The length and spacing of the output pulses are equal to the optical delay time τ. Gray and Casperson (1978). © 1978 IEEE

The field of instabilities in nonlinear optical systems is large and rapidly expanding. We have tried to summarize the work directly related to optical bistability, but a thorough coverage would be inappropriate in the present context.

Okada and Takizawa (1981b) and Petersen, Ravn, and Skettrup (1984) have used hybrid devices to study the case $\gamma^{-1} = t_R$, showing many waveforms. The corresponding theory has been treated by Narducci, Gao, and Yuan (1982), Gao Yuan, and Narducci (1983), Gao, Narducci Schulman, Squicciarini, and Yuan (1983), and Gao, Narducci, Sadiky, Squicciarini, and Yuan (1984). Chrostowski, Delisle, and Tremblay (1982) and Chrostowski, Vallee, and Delisle (1983) have studied Ikeda instabilities with an acousto-optic-modulator hybrid. Neyer and Voges (1982a,b) use a Mach-Zehnder interferometer on $LiNbO_3$ to study the dynamics of periodic and chaotic outputs. A twisted nematic liquid crystal modulator is used by Song, Lee, Shin, and Kwon (1983) with a computer delay. Murina and Rozanov (1981a,b) analyze the bifurcation sequence of electro-optic hybrids with delays.

Observations of special note in nonoptical systems include chaos, in a simple acoustic system [Kitano, Yabuzaki, and Ogawa (1983)], in a nuclear magnetic resonance (NMR) laser [Brun, Derighetti, Holzner, and Meier (1983) show clear phase-space portraits for P2, P4 plus erratic noise, P8, and chaos, and Meier, Holzner, Derighetti, and Brun (1982)], and in a Ge electron-hole plasma [Held, Jeffries, and Haller (1984)]. Observations of chaos in nonlinear electrical oscillators, the Belousov-Zhabotinskii chemical reaction, Rayleigh-Bénard convection, and Couette-Taylor flow are reviewed by Swinney (1983); see also Swinney and Gollub (1978,1981). A report on a meeting entitled "Testing Nonlinear Dynamics" summarizes the importance of deterministic nonlinear dynamics in many different fields [Abraham, Gollub, and Swinney (1984); see also Kadanoff (1983)]. For an introduction to laser instabilities and chaos see Abraham (1983), Gioggia and Abraham (1984), and references therein. For an extensive review of instabilities and chaos in a laser-driven nonlinear ring cavity see Englund, Snapp, and Schieve (1983). For other articles on optical chaos and laser instabilities see Mandel and Wolf (1984). Many fundamental papers on the mathematics of chaos are contained in Helleman (1980). Arecchi and Harrison (1985) is expected to contain many relevant review articles.

6.4. OTHER INSTABILITIES OF NONLINEAR CAVITIES

At the beginning of Section 6.3 it was noted that the coupled Maxwell-Bloch equations with boundary conditions, including time delays, describe completely the evolution of a two-level system in a semi-classical theory. The first instabilities of a bistable optical system to be predicted were regenerative pulsations (Section 6.1; to obtain two relaxation times a third level is necessary, or one medium time and the cavity time will suffice), self-pulsing (Section 6.2), and Ikeda periodic oscillations and chaos (Section 6.3). Subsequently several variations or limits have led to interesting predictions to challenge experimentalists.

Self-pulsing (Section 6.2) and the Ikeda instability (Section 6.3) involve more than one longitudinal mode. Under different conditions a single longitudinal mode can be unstable. Ikeda and Akimoto (1982a) have analyzed the case $t_R \ll \gamma^{-1} = \tau_M$, i.e., the opposite limit from Ikeda's first case. They assume that $\beta(t)$, β_0, and $1 - B$ are small quantities of the order of $t_R\gamma$ (mean-field assumptions), so Eqs. (6.3-15,17) are approximated by a set of ordinary differential equations:

$$E(t) \simeq E(t - t_R) + \frac{dE}{dt} t_R \qquad (6.4\text{-}1)$$

$$t_R \frac{dE}{dt} \simeq A - (1 - B)E + iBE(\phi - \phi_0) \quad (6.4\text{-}2)$$

$$\gamma^{-1}\dot{\phi} = -\phi + |E|^2 \quad (6.4\text{-}3)$$

or with $\tau \equiv t\gamma$, $\tau_R \equiv t_R\gamma$, $E \equiv \tau_R^{1/2}\bar{E}$, $\phi \equiv \tau_R\eta$, $A = \tau_R^{3/2}a$, $1 - B \equiv \tau_R b$, and $\phi_0 \equiv \tau_R\eta_0$

$$\dot{\bar{E}} = a - b\bar{E} + i(\eta - \eta_0)\bar{E} \quad (6.4\text{-}4)$$

$$\dot{\eta} = -\eta + |\bar{E}|^2 , \quad (6.4\text{-}5)$$

where $B \simeq 1$ was used in the last term of Eq. (6.4-4). Equation (6.4-4) has the same form as the optical Bloch equation, if $\mathrm{Re}\bar{E}$ and $\mathrm{Im}\bar{E}$ are regarded as two components of a Bloch vector. The total phase shift $\eta(\tau) - \eta_0$ plays the role of the Rabi nutation frequency and b that of damping. A characteristic of Eqs. (6.4-4,5) is that $\bar{E}$ exhibits a self-induced Rabi nutation, as can be seen by linearizing them:

$$\bar{E} \simeq \bar{E}_S + \delta\bar{E} , \quad (6.4\text{-}6)$$

$$\eta \simeq \eta_S + \delta\eta , \quad (6.4\text{-}7)$$

$$\delta\dot{\bar{E}} \simeq -b\delta\bar{E} + i\bar{E}_S\Delta n + i(\eta_S - \eta_0)\delta\bar{E} , \quad (6.4\text{-}8)$$

$$\delta\dot{\eta} \simeq -\delta\eta + \bar{E}_S\delta\bar{E}^* + \bar{E}_S^*\delta\bar{E} . \quad (6.4\text{-}9)$$

If $\eta_S >> 1$, $\delta\bar{E}$ oscillates rapidly with Rabi frequency η_S; $\delta\eta$ also oscillates with the same frequency, following $\delta\bar{E}$ adiabatically. They find that the lower branch of $|\bar{E}_S|^2$ vs a^2 is always stable, while the upper branch can be unstable. Period-doubling bifurcations to weak turbulence are seen (Fig. 6.4-1). Chaos occurs only for values of a for which the $|\bar{E}_S|^2$ vs a^2 relation is threefold. The chaos can be interpreted as a self-induced Rabi nutation disturbed by the unstable stationary state on the intermediate (negative-slope) branch. This new instability requires that the relaxation time of the medium be shorter than the lifetime of the cavity, i.e., $(1 - R + \alpha L/2) < t_R\gamma$ for a ring cavity (replace L by 2L and n_2 by $3n_2$ for a Fabry-Perot cavity). For $a >> 1$ and $\eta_0 >> 1$, the fundamental frequency is $a^2b^{-1/3}\ \gamma/2\pi$, which is much smaller than $1/2t_R$, the fundamental frequency in the delay-induced Ikeda instability.

Lugiato, Narducci, Bandy, and Pennise (1982a,b), Narducci, Bandy, Pennise, and Lugiato (1983), and Lugiato and Narducci (1984) have treated essentially the same single-mode instability as Ikeda and Akimoto, except that the latter's Kerr nonlinearity is replaced with the more general Bloch equations in the limit of very rapid transverse relaxation ($\gamma_T >>$ all other rates). For C = 70,000, Δ = 374, and θ = 340 [in Eq. (2.1-31) with $g(\Delta\omega) = \delta(\Delta\omega)$], they find in the upper branch a sequence of period-doubling bifurcations to chaos as the incident field y is decreased. They show attractive phase-space plots of the oscillations; see Fig. 6.4-2. The smearing out of these phase-space portraits by noise has been studied by Neumann, Koch, Schmidt, and Haug (1984) using Maxwell-Bloch equations with stochastic Gauss-Markoff noise sources (essentially Eqs. (6.5-18-20)).

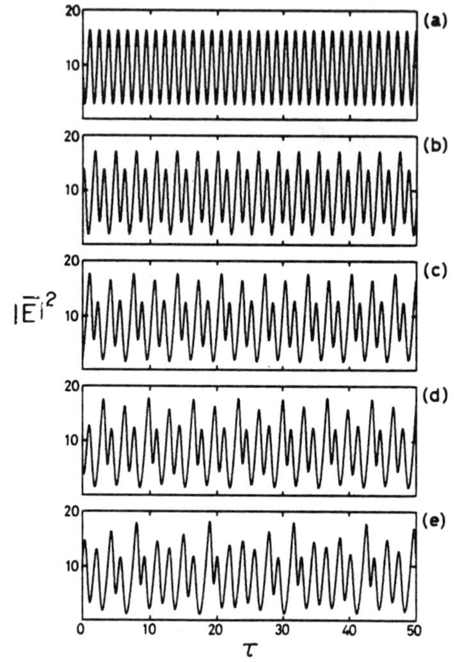

Fig. 6.4-1. Period-doubling bifurcations of Eqs. (6.4-4,5) for b = 0.3 and η_0 = 5: (a) period one (a = 5.4), (b) period two (a = 4.6), (c) period four (a = 4.0), (d) period eight (a = 3.84), and (e) chaotic (a = 3.5). The time parameter is $\tau = t\gamma = t/\tau_M$ with $\tau_M >> t_R$ assumed so the oscillation frequency is much lower than for self-pulsing or Ikeda period-two oscillations. Ikeda and Akimoto (1982a).

Lugiato, Horowicz, Strini, and Narducci (1984a,b) have investigated the effect of a Gaussian transverse intensity profile upon this single-wavelength instability. Unstable behavior is predicted to occur for experimentally accessible values of the bistability parameter C, and to be favored by the absence of bistability and by the selection of atomic (Δ) and cavity (θ) detunings having opposite signs. In fact, the instability domain is larger than that for the plane-wave case, quite unlike the disappearance of the multimode on-resonance self-pulsing for a Gaussian input.

Rosenberger, Orozco, and Kimble (1984a) and Orozco, Rosenberger and Kimble (1984) observe this on-resonance single-wavelength oscillation in the predicted region of parameter space: $60 \leq C \leq 175$, $\Delta = -0.75$ to -1.5, $\theta = +5$ to $+26$. They use a high-finesse confocal ring resonator, modified so that the light beam passes through the ten highly collimated beams of optically pumped Na atoms four times per round trip. The oscillations occur on the upper bistable branch and also when the detunings are too large for bistability. The oscillations were nearly sinusoidal, ranged in frequency from 20 to 60 MHz as the detunings were varied, exhibited a modulation depth as great as 7 to 1, and had a mean value above the steady-state value. Precipitation to the lower branch occurred occasionally.

Winful and Cooperman (1982) [see also Cooperman and Winful (1981)] show that the light transmitted by a nonlinear distributed feedback structure can be steady, periodic, or chaotic depending on the intensity of the input cw beam. Because the feedback mechanism is distributed continuously throughout the nonlinear medium, the system is accurately described by a set of partial differential equations instead of difference equations. The coupling between the input and the backward field occurs by means of the periodic perturbation of the linear refractive index of the form $n_0 + n_1 \cos 2k_0 z$ (where $n_1 < n_0$) and through the nonlinear phase term of the form $\beta_i \propto I_i + 2I_j$ (see Appendix D). They find that the instability begins with a period less than $2t_R$, and the period decreases continuously with input intensity, unlike the period doubling of an Ikeda instability. Apparently their instability is a relaxation oscillation or regenerative pulsation in which the grating stop band is alternately in and out of resonance with the laser wavelength. But when the input intensity is sufficiently large, the output becomes quite noisy.

Silberberg and Bar-Joseph (1982) show that self-oscillations and chaos can be obtained for an optical system without any external feedback. They consider two monochromatic waves interacting in a third-order dispersive nonlinear medium obeying a Debye relaxation equation [Eq. (6.3-17)]. The mutual influence of the two fields is only through the phase term, of the form $\beta_i \propto I_i + 2I_j$, which is identical for equal inputs. They find that the nonlinear phase shift required for an instability to occur is minimized for a transit time approximately equal to the medium response time τ; see Fig. 6.4-3. Their preliminary studies show that as the intensity is increased, bifurcations and period doubling are encountered just as in the Ikeda instability in a device having an external feedback. This instability's effect on the proposed enhancement [Kaplan and Meystre (1981)] of the sensitivity of Sagnac interferometers should be investigated.

Singh and Agarwal (1983) have analyzed the path to chaos for coherent two-photon transitions in a ring cavity and found Feigenbaum's scenario in spite of the very complicated two-dimensional map that characterizes this system. Period doubling and chaos have been found numerically in three- and four-level systems exhibiting optical tristability (Section 3.10.2) by Savage, Carmichael, and Walls (1982) and by Carmichael, Savage, and Walls (1983); Arecchi, Kurman, and Politi (1983) find periodic oscillations.

Mitschke, Mlynek, and Lange (1983) and Lange, Mitschke, Deserno, and Mlynek (1984) observe magnetically induced optical oscillations in a Fabry-Perot resonator containing Na vapor. This is the self-sustained spin precession proposed by Kitano, Yabuzaki, and Ogawa (1981a,b and 1984). When a magnetic field perpendicular to the light axis (z) exceeds a critical value, the magnetization can rotate sufficiently far to change the sign of its z component and thus to reverse the spin orientation. This causes the polarization state of the light field to become unstable, and the system jumps to the complementary state. The oscillation frequency is thus close to the Larmor frequency. The dynamical Stark effect can cause periodic oscillations, even in the single-mode case and in the mean-field approximation, for two-photon single or double-beam optical bistability or for optical double resonance [Averbukh, Kovarskii, and Perel'man (1982) and Perel'man, Kovarskii, and Averbukh (1981)].

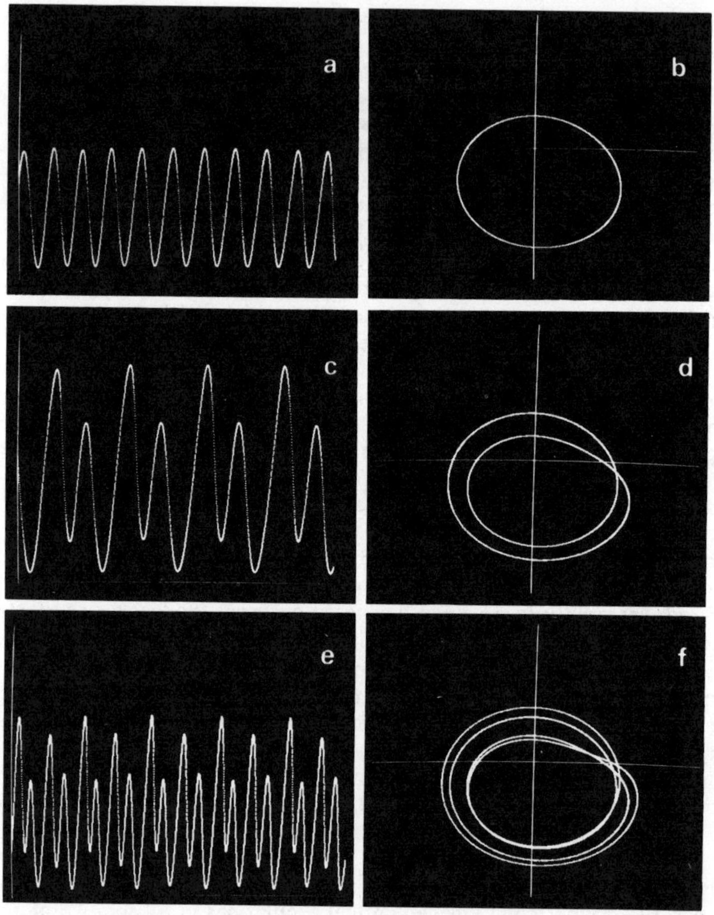

Fig. 6.4-2. (Left) Ring-cavity oscillations and (right) phase-space representations on the complex x plane. $\gamma_C/\gamma_L = 0.25$; (a,b) $y = 2000$, (c,d) $y = 1350$, and (e,f) $y = 1225$ with $y_{\uparrow} \simeq 1500$ and $y_{\downarrow} \simeq 300$. Lugiato, Narducci, Bandy, and Pennise (1982a).

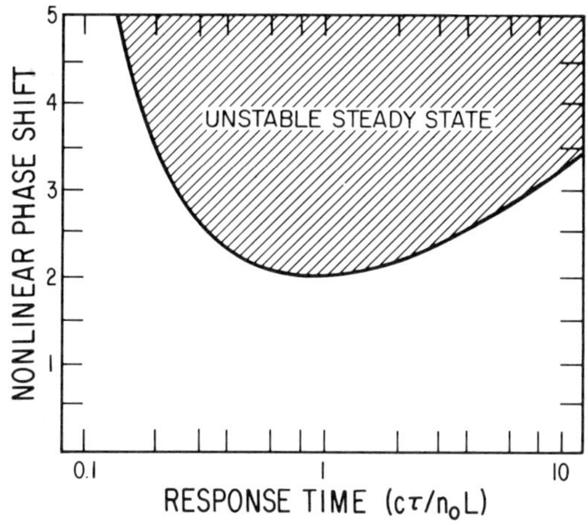

Fig. 6.4-3. Instability threshold intensity vs the ratio of the response time of the medium τ to the medium transit time n_0L/c. The stationary solution is unstable in the shaded region. Silberberg and Bar-Joseph (1982).

Sargent, Zubairy, and DeMartini (1982) present the first fully quantum mechanical theory of optical bistability instabilities that shows how the instabilities grow from spontaneous emission. Hillman, Boyd, and Stroud (1982) show that the self-consistent treatment of a probe beam propagating through a driven two-level medium is described in terms of bichromatic natural modes that are then the proper modes for analyzing the stability of homogeneously broadened ring lasers and bistable ring cavities. Maeda and Abraham (1982) and Abraham, Coleman, Maeda, and Wesson (1982) have studied self-pulsing and subharmonic bifurcations of a single-mode, Fabry-Perot, 3.51-µm helium-xenon (He-Xe) laser. Brunner and Paul (1982) numerically study instabilities of multimode gas lasers.

Arecchi, Meucci, Puccioni, and Tredicce (1982) report multiple subharmonic bifurcations, eventually leading to chaos, in a Q-switched CO_2 laser operating at the 10.6-µm P(20) line. A time-dependent perturbation is introduced into the cavity by an electro-optical modulator driven by a sinusoidal frequency, adding to the photon population and molecular population inversion the crucial third degree of freedom essential for chaos in a system described by differential equations. They give evidence of the coexistence of independent basins of attraction in the phase space (generalized multistability), which may lead in suitable situations to the appearance of low-frequency divergences in the power spectrum. Arecchi and Lisi (1982) had already demonstrated the generation of low-frequency noise by hopping between attractors (random intermittency) using a suitable electronic oscillator with a cubic nonlinearity and driven by an external modulation.

Heartbeat phenomena occur when a laser beam travels vertically upward through a shallow pool of absorbing liquid (Gouesbet, Rhazi, and Weill (1983)) or horizontally near and under its free surface (Anthore, Flament, Gouesbet, Rhazi, and Weill (1982)). The latter makes a transition to 2D optical turbulence.

The period-doubling picture of the path to chaos is that, when some bifurcation parameter μ is increased to a certain critical value, a periodic orbit in phase space is replaced by an orbit that has twice the period. This continues with ever-decreasing changes in μ required to reach the next doubling until at a critical value the period becomes infinite and the motion aperiodic. The system is quite sensitive to the initial conditions. Mathematically chaos occurs if two trajectories in phase space that are initially very close will separate at an exponential rate with time. This occurs in a purely deterministic way; classical noise-free equations describe the motion. Most of the simulations and experiments of Sections 6.3 and 6.4 have not been put to this test, but the obvious noisy appearance of the output and the growth of a broad background in its spectrum justify the use of "chaos" in describing the time dependence that supplants the periodic output. Apparently any system described by one or more difference equations or by three or more differential equations will exhibit chaotic behavior if driven appropriately. Obviously this includes most systems. From the point of view of designing practical devices, these various instabilities may restrict the range of operation; but every bistable system exhibiting an instability seems to have a comfortable range of "stable" bistable operation. Under other conditions periodic oscillations are stable and can be utilized.

6.5. FLUCTUATIONS AND NOISE

In Section 7.2 a gedanken experiment is described using a single atom in an optical cavity of length λ and cross-sectional area λ^2. Semiclassically, optical bistability is predicted, but quantum mechanically one expects spontaneous switching back and forth between the two states. In the potential-well description of Section 2.7 one can say that the potential is rapidly changing, so the marble rattles all around. Or the potential is more or less fixed but the marble has enough energy to easily surmount the barrier between the walls corresponding to the two bistable states. The probability distribution is not two well-separated peaks but two very wide overlapping peaks. Since bistability using a single atom is not ruled out by fundamental considerations, statistical limitations will likely determine the minimum number of atoms and photons per device in practical devices some day. Consequently, quantum statistical studies are of potential practical importance to bistability, and they are always of fundamental interest in systems simple enough to serve as a testing ground for theories and approximations.

The first discussion of the effects of statistical fluctuations upon optical bistability was apparently at the Fourth Rochester Conference on Coherence and Quantum Optics in 1977 [Bonifacio and Lugiato (1978b); Agarwal, Narducci, Feng, and Gilmore (1978)]. There the source of the fluctuations was spontaneous emission and the emphasis was upon determining the spectrum of the transmitted and fluorescence light (see Section 2.8) rather than upon switching between the two states. A study of the fluctuations requires a fully quantum mechanical analysis. However, as discussed in Section 2.8, the quantum regression theorem claims that the time dependence of the polarization fluctuations in the stationary state can be obtained from the dynamics of the polarization itself. Other early publications on fluctuations are: Willis (1977, 1978); Bonifacio and Lugiato (1978e); Narducci, Gilmore, Feng, and Agarwal (1978); Bonifacio, Gronchi and Lugiato (1978); Bulsara,

Schieve, and Gragg (1978); Schenzle and Brand (1978, 1979a,b); Chrostowski and Zardecki (1979); Willis and Day (1979); Agarwal, Narducci, Feng, and Gilmore (1980).

Willis emphasizes that the fluctuation behavior is strongly dependent upon the relative values of the polarization relaxation rates γ_L and γ_T, the cavity damping rate γ_c, and the collective field damping rate γ_{CF}. If $\gamma_{CF} \gg \gamma_L$, γ_T, and γ_c, the master equation is fundamentally non-Markovian; radiation and matter have to be treated on an equal footing, and the damping terms have only a weak effect on the dynamics. If $\gamma_c \gg \gamma_{CF} \gg \gamma_T, \gamma_L$, then the field follows the polarization adiabatically and the single-mode master equation of superfluorescence applies. If $\gamma_L, \gamma_T \gg \gamma_{CF}$, then the polarization adiabatically follows the field, and the laser master equation with an external field is appropriate. The usual approach is to obtain the Fokker-Planck equation from the appropriate master equation or Langevin equations with various assumptions about the noise sources. First we will describe an experiment that illustrates many of the concepts but utilizes controllable shot noise rather than spontaneous emission noise.

6.5.1. Shot-Noise Fluctuations in a Hybrid Experiment

Consider a crossed-polarizer hybrid device as shown in Fig. 6.5-1 and described in Section 4.3 [McCall, Gibbs, Kaplan, and Hopf (1981); McCall, Ovadia, Gibbs, Hopf, and Kaplan (1984)]. The transmission curve of the device is shown in Fig. 6.5-2:

$$\mathscr{T}(V_M) = \frac{I_T}{I_I} = [1 - D\cos(\pi V_M/230)]/2 , \qquad (6.5\text{-}1)$$

where $D = 0.965 \pm 0.002$ was measured. The voltage V_M across the modulator can also be written as

$$V_M = GV(230/\pi) , \qquad (6.5\text{-}2)$$

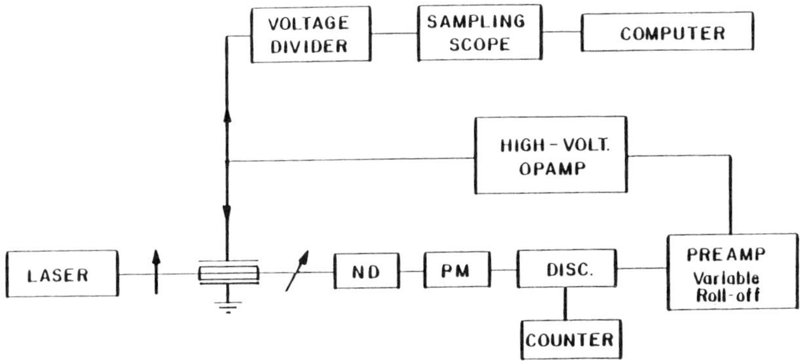

Fig. 6.5-1. The experimental apparatus for studying shot-noise fluctuations in a crossed-polarizer hybrid device. McCall, Gibbs, Hopf, Kaplan, and Ovadia (1982).

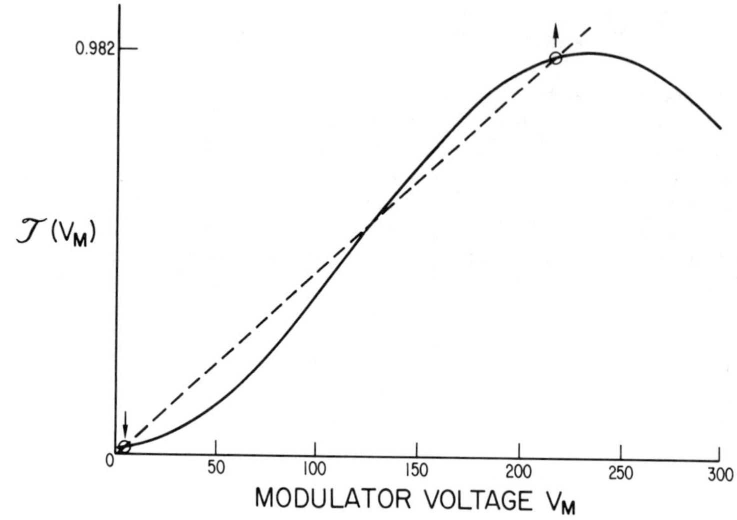

Fig. 6.5-2. Modulation transmission $\mathcal{T}(V_M)$ and feedback load line. McCall, Ovadia, Gibbs, Hopf, and Kaplan (1985).

where V is output voltage of the preamplifier and G is the gain of the high-voltage operational amplifier. V, of course, depends upon the maximum counting rate C_0 (for $V_M = 230$ volts), the modulator transmission $\mathcal{T}$ at voltage V_M, the standard number of electrons delivered by the discriminator per detected photon, and the gain and bandwidth (or decay time τ) of the preamplifier. Since V is proportional to $\mathcal{T}$, Eq. (6.5-2) defines a load curve of $\mathcal{T}(V_M)$ versus V_M which is a straight line of slope proportional to G^{-1}. According to the graphical approach to steady-state optical bistability (Section 2.6), G can be chosen to have a holding value G_H exceeding the switch-off value $G_\downarrow$ and less than the switch-on value $G_\uparrow$. If $C_0\tau$ is very much larger than 1, the device will remain indefinitely in the upper $\uparrow$ or lower $\downarrow$ state depending upon whether G was last above $G_\uparrow$ or below $G_\downarrow$.

Fluctuations are introduced by increasing the neutral density attenuation until $C_0\tau$ is of order unity or less; G must be increased correspondingly to maintain the same average operating point in the bistable loop. The stochastic arrival of photons and the statistical nature of their photoelectric detection will then cause V to fluctuate. Large-enough fluctuations of the slope of the straight line in Fig. 6.5-2 will cause switching between the previously stable bistable states. This is shown in Fig. 6.5-3 in which the device spontaneously switches back and forth between the "on" and "off" states. Note that the absolute magnitude of the noise depends upon the device's transmitted intensity, I_T, i.e., it depends upon the state of the device. This can also be seen by the widths of the peaks in Fig. 6.5-4, which is a histogram of the number of occurrences of a given voltage versus the voltage (labeled as a channel number). The "on" distribution peaked at channel 42 is clearly broader than the "off" distribution peaked at channel 8 (close to zero voltage at channel 6). To obtain the distribution in Fig. 6.5-4, the feedback gain G (or equivalently the input intensity) was adjusted so that the device spent roughly the same amount of time in each of the states. This operating point ($G \simeq 530$) is closer to the switch-down point ($G_{\downarrow} \simeq 400$) than the switch-up point ($G_{\uparrow} \simeq 1000$) as shown in Fig. 6.5-5; G values are given in arbitrary units. Experimentally, it was found that if the probabilities in the two states were made equal at one preamp bandwidth setting, they would become quite unequal if nothing but the bandwidth was changed. To study fluctuation switching away from the equal-probability point, it was necessary to externally force the device into the lower probability state and observe its return to the high-probability state; see Figs. 6.5-5,6.

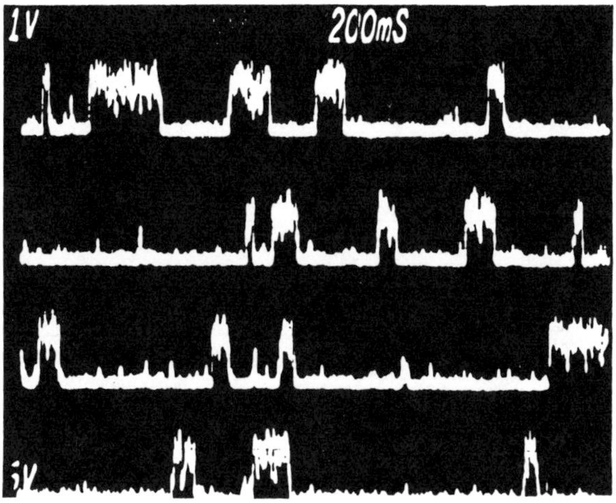

Fig. 6.5-3. Noise switching between the two bistable states. The absolute voltage fluctuations are larger in the upper state. McCall, Ovadia, Gibbs, Hopf, and Kaplan (1985).

An exact theory of the shot-noise experiment has not been solved, but good agreement with approximate models is found if one allows the counting rates to be adjusted [McCall, Ovadia, Gibbs, Hopf, and Kaplan (1984); Ovadia (1984)]. The high-voltage amplifier output in Fig. 6.5-1 can be written as

$$V(t) = \zeta \sum_i H(t - t_i) e^{-(t - t_i)/\tau} \tag{6.5-3}$$

where ζ is the voltage increment per detected photon, $H(t)$ is the Heaviside step function, the times t_i denote when photons are detected, and the decay time τ is set by the preamplifier roll-off frequency. The rate of detecting photons is $C_0\mathscr{T}(V)$. From Eq. (6.5-3), one has

$$\frac{dV}{dt} = -\frac{V}{\tau} + \zeta \sum_i \delta(t - t_i) , \tag{6.5-4}$$

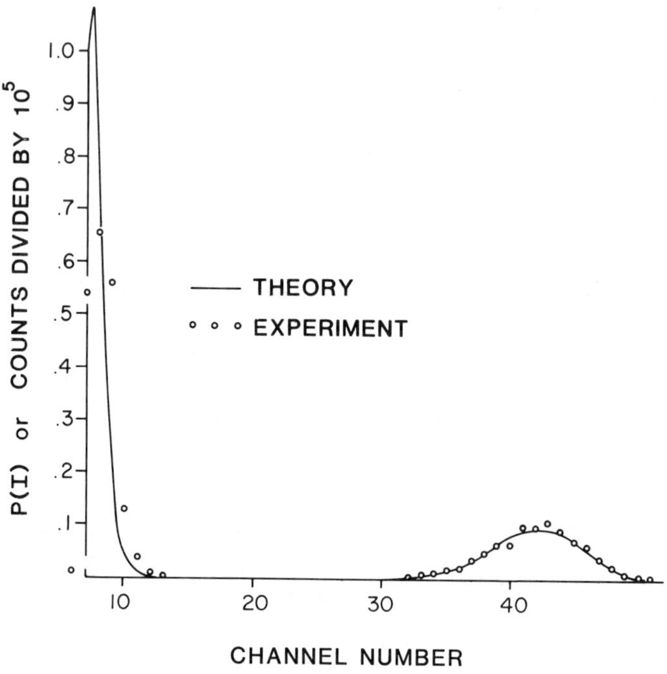

Fig. 6.5-4. A histogram of output voltage occurrences (open circles) for C_0 = 37 kHz and 100-Hz bandwidth. Zero voltage is at about channel 6, the lower state is about channel 8, the upper state about channel 42. The computer-simulated histogram from Eq. (6.5-7) is shown as the solid curve with G = 660 and A = 61.83 (effectively 62 atoms, slightly less than half being excited). McCall, Ovadia, Gibbs, Hopf, and Kaplan (1984).

a Langevin-type equation for the fluctuating voltage. Since the photons are detected at random times, the process is Markovian (no memory).

Let P(V,t)dV be the probability for V to be between V and V + dV at time t then

$$\frac{\partial P(V,t)}{\partial t} = \frac{1}{\tau}\frac{\partial}{\partial V}[VP(V,t)] + R(V-\zeta)P(V-\zeta,t) - R(V)P(V,t)\,, \quad (6.5\text{-}5)$$

where $R(V) = C_0\mathscr{T}(V) = A[1 - D\cos(GV)]/2\tau$ and $A \simeq \tau C_0$. Eq. (6.5-5) is the difference-differential equation which is considered exact; the first term describes the exponential decay process, while the last two terms take into account the detection of photons. Expanding the second term to second order in ζ, one finds

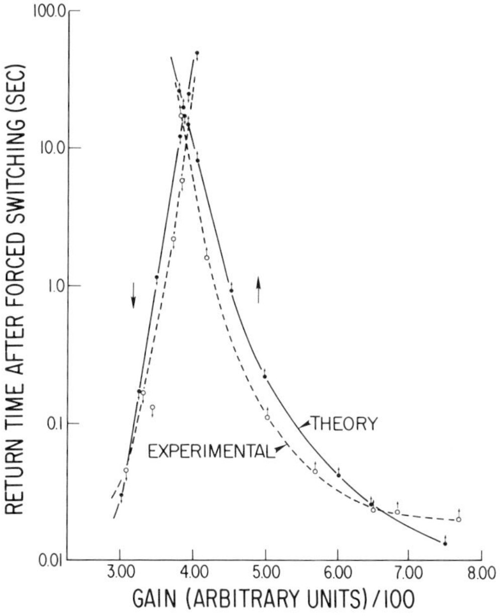

Fig. 6.5-5. Return time after forced switching to the lower state (↑ points) and to the upper state (↓ points) as a function of feedback gain setting (in arbitrary units). A = 61.83. McCall, Ovadia, Gibbs, Hopf, and Kaplan (1985).

$$\frac{\partial P(V,t)}{\partial t} = \frac{\partial}{\partial V}\left[\left[\frac{V}{\tau} - \zeta R(V)\right]P(V,t)\right] + \frac{\zeta^2}{2}\frac{\partial^2}{\partial V^2}[R(V)P(V,t)] , \qquad (6.5\text{-}6)$$

the corresponding Fokker-Planck-like equation. The steady-state solution to Eq. (6.5-5) is zero for $V < 0$, whereas the solution to Eq. (6.5-6) does not have this property. Equation (6.5-6) can be compared with a simple ladder model, somewhat similar to the birth-death macroscopic description of Gragg, Schieve, and Bulsara (1979), that fits the data well. Suppose the device can be modeled by a collection of two-level systems in which only a few, namely N, are excited, so that induced emission can be neglected. Let $1/\tau$ be the decay rate due to spontaneous emission, and let R_N be the rate for photon absorption and excited atom creation. Let P_N be the probability that the bistable device has exactly N excited systems. If N increases by K for each detected photon, the voltage from the preamplifier is $V_N = N\zeta/K$ giving GV_N across the modulator. Then P_N evolves according to the master equation

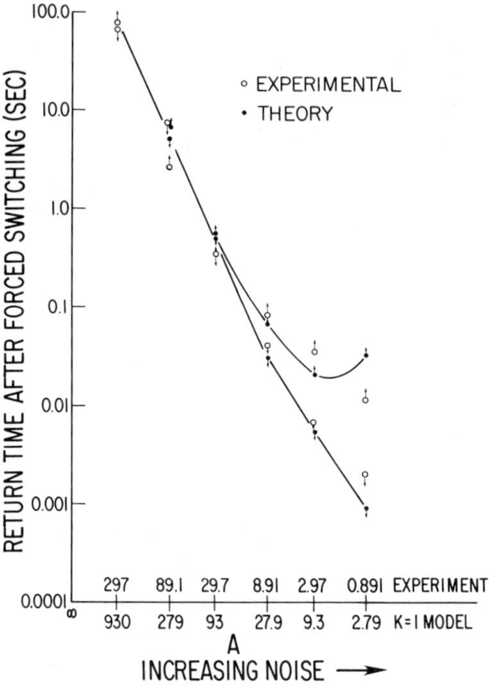

Fig. 6.5-6. Return time after forced switching to the lower and upper states as a function of the amplifier upper roll-off frequency. G = 396 (840) for return times $\tau_{\uparrow}(\tau_{\downarrow})$ after forced switching to the lower (upper) bistable state. A = 27.87(100 Hz/bw) where bw is the amplifier upper roll-off frequency in Hz, corresponding to systems with 279 atoms at 10 Hz to 3 atoms at 1 kHz. McCall, Ovadia, Gibbs, Hopf, and Kaplan (1984).

$$\frac{dP_N}{dt} = -\frac{NP_N}{\tau} + \left(\frac{N+1}{\tau}\right) P_{N+1} - R_N P_N + R_{N-K} P_{N-K}, \qquad (6.5\text{-}7)$$

corresponding to a simple ladder diagram. The first two terms on the right-hand side of Eq. (6.5-7) describe decay out of and into the Nth level, while the last two represent absorption out of and into the Nth level. This model discretizes both the excitation, as in the experiment, and the decay, contrary to the experiment. The unit of photon excitation is K, and the unit of phonon decay is 1, therefore, as $K \to \infty$, Eq. (6.5-7) approaches the exact model. Expanding Eq. (6.5-7) to second order in $K/N = \zeta/V$, one has

$$\frac{\partial P}{\partial t} = \frac{\partial}{\partial V}\left[\left[\frac{V}{\tau} - \zeta R\right]P\right] + \frac{\zeta^2}{2}\frac{\partial^2}{\partial V^2}\left[\left[R + \frac{V}{\zeta K\tau}\right]P\right] \tag{6.5-8}$$

which clearly reduces to Eq. (6.5-6) for $K \to \infty$.

To find the steady-state probability distribution for $K \to \infty$, set $\dot{P}_N = 0$; let $P_0 = 1$, $P_1 = R_0$, and apply Eq. (6.5-7) for $N \geq 2$ repeatedly:

$$P_N{}^{S.S.} = \frac{\tau^N}{N!}\prod_{i=0}^{N-1} R_i \Big/ \sum_{N=0}^{\infty}\frac{\tau^N}{N!}\prod_{i=0}^{N-1} R_i \,. \tag{6.5-9}$$

Equation (6.5-9) is plotted as the solid curve in Fig. 6.5-4. For $K > 1$, the steady-state solution is not simply expressed.

For finite K, the last term in Eq. (6.5-8) represents an increase in noise; the relaxation process introduces uncertainty. For infinite K the relaxation process is perfectly noise-free as in the experiment and Eq. (6.5-5). The ratio $V/\zeta K\tau : R$ is about $1/K$ where it counts, since the peaks of the bimodal distribution occur approximately where $V = \zeta R\tau$ according to the argument of the first derivative term.

The fitting of the models to the distribution of Fig. 6.5-4 was done by forcing A and G to be such that two experimental ratios were fit: the ratio of the peak value of the upper-state histogram to the minimum in between the peaks (1184 experimentally) and the ratio of the upper-state area to the lower-state area (0.4576 experimentally). The latter ratio is, of course, the ratio of the time spent in the upper state to the time in the lower. The histogram was fit for K = 1, 2, 3,...10. The values of A thus obtained fit well the relation $A = 31.54 + 30.28/K$. The value of 31.54 for $K = \infty$ matches poorly the experimental value of 80.1, which means there was 2.5 times as much noise in the system as they calculated. The variation of A with K is what one expects by requiring the coefficient of P in the second derivative term in Eq. (6.5-8) to remain constant. G, which determined the noise-free operating curve, hardly varied with K at all. The eigenvalues yielding $\tau_\uparrow$ and $\tau_\downarrow$ were model independent. Consequently the K = 1 case is taken for the eigenvalue discussion here.

For nonsteady-state, one assumes $P_N \simeq P_N{}^{S.S.} + P_N{}^{(\lambda)} e^{-\lambda t/\tau}$ in Eq. (6.5-7) which gives the following recurrence formula:

$$P_{N+1}{}^{(\lambda)} = \frac{1}{N+1}\left[\tau R_N P_N{}^{(\lambda)} - \tau R_{N-1} P_{N-1}{}^{(\lambda)} + (N - \lambda) P_N{}^{(\lambda)}\right] . \tag{6.5-10}$$

The eigenvalue λ is obtained from Eq. (6.5-10) by using the following property of P_N:

$$1 \equiv \sum_{N=0}^{\infty} P_N = \sum_{N=0}^{\infty} P_N{}^{S.S.} + e^{-\lambda t/\tau}\sum_{N=0}^{\infty} P_N{}^{(\lambda)}$$

where $\Sigma P_N(\lambda)$ is zero since

$$-\lambda \sum_{N=0}^{\infty} P_N(\lambda) = -\sum_{N=0}^{\infty} N P_N(\lambda) + \sum_{N=0}^{\infty} (N+1) P_{N+1}(\lambda)$$

$$-\sum_{N=0}^{\infty} R_N P_N(\lambda) + \sum_{N=0}^{\infty} R_{N-1} P_{N-1}(\lambda) \equiv 0$$

summing over Eq. (6.5-7). Thus

$$\sum_{N=0}^{\infty} P_N(\lambda) = \frac{0}{\lambda} = 0 \text{ if } \lambda \neq 0. \tag{6.5-11}$$

Therefore, if one assumes an initial value for λ and iterates Eq. (6.5-10) with the constraint on $P_N(\lambda)$ according to Eq. (6.5-11), the solution for λ converges smoothly depending upon the gain, decay time, and counting rate.

The tunneling times are obtained from the eigenvalue in the following way. Starting with the equations of motion for the probability that the device is in the lower or upper state one has,

$$\frac{dP_\uparrow}{dt} = \frac{-P_\uparrow}{\tau_\downarrow} + \frac{P_\downarrow}{\tau_\uparrow} \tag{6.5-12a}$$

$$\frac{dP_\downarrow}{dt} = \frac{P_\uparrow}{\tau_\downarrow} - \frac{P_\downarrow}{\tau_\uparrow} \tag{6.5-12b}$$

with $P_\uparrow(t) + P_\downarrow(t) = 1$, and $P_\uparrow$ and $P_\downarrow$ are the probability distributions in the "on" and "off" states, respectively. If one assumes $P_i = A_i + B_i e^{-\lambda t/\tau}$ where $i = \uparrow$ or $\downarrow$, then the eigenvalue must satisfy:

$$\begin{vmatrix} \frac{\lambda}{\tau} - \frac{1}{\tau_\downarrow} & \frac{1}{\tau_\uparrow} \\ \frac{1}{\tau_\downarrow} & \frac{\lambda}{\tau} - \frac{1}{\tau_\uparrow} \end{vmatrix} = 0$$

or

$$\lambda^2 - \lambda\left[\frac{1}{\tau_\uparrow} + \frac{1}{\tau_\downarrow}\right]\tau = 0, \tag{6.5-13a}$$

which has the trivial solution $\lambda = 0$ and

$$\lambda = \left[\frac{1}{\tau_{\uparrow}} + \frac{1}{\tau_{\downarrow}}\right]\tau. \tag{6.5-13b}$$

Thus, the eigenvalue is just the product of the spontaneous decay time and the sum of the tunneling rates. Using the fact that in steady state Eq. (6.5-12) implies $P_{\uparrow}/P_{\downarrow} = \tau_{\downarrow}/\tau_{\uparrow} \equiv \gamma$, one can solve for the tunneling times

$$\tau_{\downarrow} = (1 + \gamma)\,\frac{\tau}{\lambda} \tag{6.5-14a}$$

$$\tau_{\uparrow} = (1 + \gamma)\,\frac{\tau}{\lambda\gamma}\,. \tag{6.5-14b}$$

In the lower state $\gamma \equiv 0$ and in the upper state $\gamma \equiv \infty$. In Fig. 6.5-4, the solid curve is plotted for $\gamma \equiv 0.4575$, which means that the lower state is slightly more likely to be occupied than the upper state.

From Eqs. (6.5-14) the return times after forced switching versus the gain in arbitrary units, with C_0 = 37 kHz and 300-Hz bandwidth, are shown in Fig. 6.5-5. The simulated curve shows the same basic functional dependence as the experimental data within the experimental error of 10% to 20% in the counting rate, 25% in the bandwidth, and a factor of almost 2 in the tunneling times.

The calculated "averaged" escape time $(\tau_{\uparrow}\tau_{\downarrow})/(\tau_{\uparrow} + \tau_{\downarrow})$ is plotted vs the gain in arbitrary units from 10 Hz to 1 kHz amplifier bandwidth in Fig. 6.5-7 for C_0 = 5.6 kHz, i.e. a noisier system. From those curves one can determine two gain settings within the bistable loop, G = 840 and G = 396, such that $\tau_{\uparrow}$ and $\tau_{\downarrow}$ are equal for the 10-Hz and 30-Hz bandwidths. (In the same units the model's noise-free switch-up and switch-down points are 1058 and 380.) Using Eqs. (6.5-14), the return times after forced switching to upper and lower states are plotted versus amplifier upper roll-off in Fig. 6.5-6.

Again the simulated curve agrees as well as can be expected with the experimental data in Fig. 6.5-6. From Fig. 6.5-7 the apparent shift in the peak of the return time toward higher gain as the amplifier bandwidth is increased causes the splitting of the straight line into two branches in Fig. 6.5-6. In other words, at a given gain setting, changing the amplifier bandwidth causes the system to prefer one state to the other. In addition, according to the ladder model, the return time to the upper state will start to increase above some critical bandwidth. Since at a larger bandwidth the decay time τ is faster, it becomes less probable, for a given counting rate, to get more than one photon within τ. If achieving $V > V_{\uparrow}$ requires the detection of several photons within τ, switch up becomes less probable the smaller C_0 (provided C_0^{-1} is comparable to or longer than τ).

Although the ladder model agrees well with the data of the hybrid experiment, Gragg, Schieve, and Bulsara (1979) find that their similar birth-death model does not agree with the microscopic approach of Bonifacio, Gronchi, and Lugiato (1978). In particular, the birth-death approach gives a much larger width and a different value of the input field for the phase-transition region. The ladder model should be an accurate description of the increase in the modulator voltage because that always occurs in quantized units of the discriminator's standard output pulse. However, the decay of the modulator voltage is continuous, so the larger N the better the ladder model.

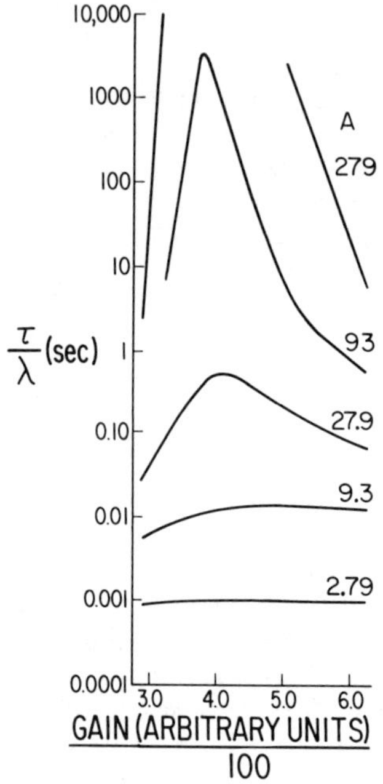

Fig. 6.5-7. The response of the "average" escape time after forced switching, as a function of the feedback gain setting (in arbitrary units), at different amplifier bandwidths. A = 27.87(100 Hz/bw). McCall, Ovadia, Gibbs, Hopf, and Kaplan (1985).

6.5.2 Theories of Optical Bistability Fluctuations

Beginning with a quantum statistical theory, Bonifacio, Gronchi, and Lugiato (1978) have derived a Fokker-Planck equation describing absorptive optical bistability in the good-cavity limit:

$$\frac{\partial \overline{P}(x,t)}{\partial(\gamma_c t)} = -\frac{\partial}{\partial x}[K(x)\overline{P}(x,t)] + \frac{1}{2}\frac{\partial^2}{\partial x^2}[D(x)\overline{P}(x,t)] , \tag{6.5-15}$$

where $\overline{P}(x,t)$ is the conditional probability distribution of finding x at time t given x_0 at time t_0. The drift and diffusion coefficients, K(x) and D(x), are given by

$$K(x) = y - x - \frac{2Cx}{1 + x^2} , \tag{6.5-16}$$

$$D(x) = \frac{C}{N_s}\left[\frac{x}{1 + x^2}\right]^2 . \tag{6.5-17}$$

The saturation photon number N_S is equal to $N_T \gamma_L/(8C\gamma_C)$ where N_T is the total number of two-level atoms. The Fokker-Planck equation (6.5-15) can be obtained from semiclassical coupled Maxwell-Bloch equations (2.1-6,8) by adding noise terms and making appropriate approximations. One has the following system of Langevin equations:

$$\dot{Q} = -(\gamma_T - i\Delta\omega)Q - (\gamma_T\gamma_L)^{1/2} Fw + F_Q(t), \qquad (6.5\text{-}18)$$

$$\dot{w} = -\gamma_L(w + 1) + (\gamma_L\gamma_T)^{1/2} \frac{FQ^* + F^*Q}{2} + F_w(t), \qquad (6.5\text{-}19)$$

$$\frac{\partial F}{\partial z} + \frac{1}{c}\frac{\partial F}{\partial t} = -\frac{\alpha_0}{2}\left[\frac{\gamma_T}{\gamma_L}\right]^{1/2} \int_{-\infty}^{+\infty} Q(z,t,\Delta\omega)g(\Delta\omega)d(\Delta\omega) + \frac{F_F(t)}{c}. \qquad (6.5\text{-}20)$$

Incoherent interactions such as spontaneous emission and collisions are included in γ_T and γ_L. It is assumed that the different dissipative processes are adequately represented by the white-noise driving terms F_Q, F_w, and F_F. These random forces are assumed to be mutually uncorrelated Gaussian random processes with zero mean and with the following autocorrelation functions:

$$\langle F_Q(t_1)F_Q(t_2)\rangle = D_Q\delta(t_1 - t_2), \qquad (6.5\text{-}21)$$

$$\langle F_w(t_1)F_w(t_2)\rangle = D_w\delta(t_1 - t_2), \qquad (6.5\text{-}22)$$

$$\langle F_F(t_1)F_F(t_2)\rangle = D_F\delta(t_1 - t_2), \qquad (6.5\text{-}23)$$

where the constants D_Q, D_w, and D_F characterize the strength of the fluctuations. The usual procedure is to make simplifying assumptions about the various time constants so that the Langevin equations reduce to a single equation for the field or the inversion. Then the corresponding Fokker-Planck equation is derived and solved.

As an illustrative example, consider again the good-cavity absorptive bistability. In the mean-field approximation, F and $c\partial F/\partial z$ in Eq. (6.5-20) can be replaced by x and $+\gamma_c(x-y)$, respectively. Furthermore if one assumes purely absorptive bistability ($\Delta\omega = 0$) in a homogeneously broadened medium ($g(\Delta\omega) = \delta(\Delta\omega)$) and $\gamma_c \ll \gamma_T, \gamma_L$, one can adiabatically eliminate the atomic variables to obtain the following field evolution equation:

$$\frac{dx}{d(\gamma_c t)} = y - x - \frac{2Cx}{1 + |x|^2}$$

$$+ \frac{2CxF_w}{\gamma_L(1 + |x|^2)} - \frac{2CF_Q}{(\gamma_L\gamma_T)^{1/2}} + \frac{C(xF_Q^* + x^*F_Q)}{(\gamma_T\gamma_L)^{1/2}(1 + |x|^2)} + \frac{F_F}{\gamma_c}. \qquad (6.5\text{-}24)$$

For $F_w = F_Q = F_F = 0$, Eq. (6.5-24) is just Eq. (5.6-2). In order to obtain the microscopic Fokker-Planck equation (6.5-15) of Bonifacio, Gronchi, and Lugiato, one can make use of the prescription [Schenzle and Brand (1979b)] that the

Fokker-Planck equation corresponding to the Langevin equation with multiplicative noise term $G(x)$

$$\dot{x} = K(x) + G(x)F_W \qquad (6.5\text{-}25)$$

is

$$\frac{\partial \overline{P}}{\partial t} = -\frac{\partial}{\partial x}\left[\left[K(x) + \frac{D_W}{2}\frac{\partial G^2}{\partial x}\right]\overline{P}\right] + \frac{D_W}{2}\frac{\partial^2}{\partial x^2}(G^2\overline{P}). \qquad (6.5\text{-}26)$$

Then with $D_Q = D_F = 0$ and small D_W, the drift term in D_W can be neglected and an equation of the form of Eq. (6.5-15) results [Zardecki (1980)].

The moments of the distribution $\overline{P}^{(st)}(x)$ of Eq. (6.5-15) are given by

$$\langle x^n \rangle = \int_0^\infty dx\, x^n\, \overline{P}^{st}(x). \qquad (6.5\text{-}27)$$

Figure 6.5-8 shows that $\sigma^2 = \langle x^2 \rangle - \langle x \rangle^2$ is largest for y just where $\langle x \rangle$ begins to increase on its way to the high-transmission state. Figure 6.5-9 shows the semi-classical S-shape curve, the Maxwell-rule transition (for which the two cross-hatch areas are equal), and the lower quantum mean value transition. The sharpness of the distribution $\overline{P}^{st}$ is determined by C/N_s; for $C/N_s < 0.2$ the distribution is peaked narrowly on the upper and/or lower state and transitions between them are very rare.

Willis and Day (1979), Graham and Schenzle (1981), Drummond and Walls (1980b, 1981), and Chrostowski, Zardecki, and Delisle (1981) have treated the quantum theory of dispersive bistability.

Lugiato, Farina, and Narducci (1980) find for a bistable absorber in the good-cavity limit that the time evolution of the system evolves in two stages. The first consists of a local relaxation around the minimum (or minima) of a suitably defined potential; see Fig. 6.5-10. The second state, which occurs over a much longer time scale, results from tunneling from one minimum to the other across the local maximum of the same potential. Such long-term decay of a metastable state or tunneling has been studied by a combination of analytical and numerical techniques by Bonifacio and Lugiato (1978a,e), Hopf and Meystre (1979), Schenzle and Brand (1979a), Farina, Narducci, Yuan, and Lugiato (1980), Drummond, McNeil, and Walls (1980a,b), Hanggi, Bulsara, and Janda (1980), Hanggi, Grabert, Talkner, and Thomas (1984), Talkner and Hanggi (1984), Hanggi, Marchesoni, and Grigolini (1984), Bonifacio, Lugiato, Farina and Narducci (1981) and Walls, Drummond, and McNeil (1981). See also Sections 5.1, 5.5, and 5.6. Brand and Schenzle (1978) treat a stochastic model for diffusion in a nonsymmetric bistable potential and solve exactly and analytically the underlying time-dependent Fokker-Planck equation: such a solution can test the limits of validity for various approximation schemes. Numerical calculations of time dependencies have been carried out in various limits: good-cavity [Zardecki (1980)] and bad-cavity [Zardecki (1981)] absorptive bistability; Chrostowski, Zardecki, and Delisle (1981): good-cavity

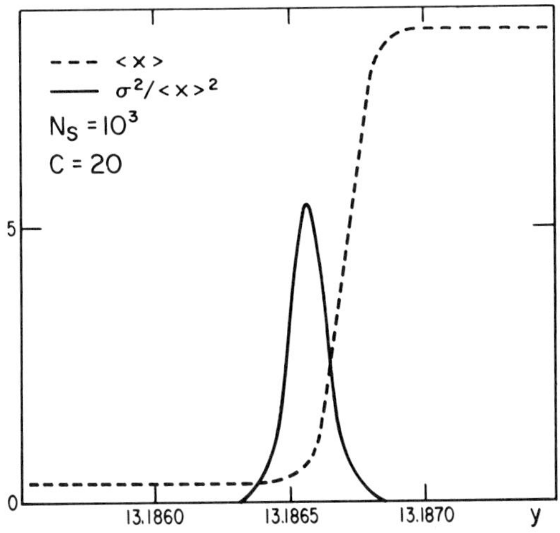

Fig. 6.5-8. Mean value and relative fluctuations of the transmitted field. Bonifacio, Gronchi, and Lugiato (1978).

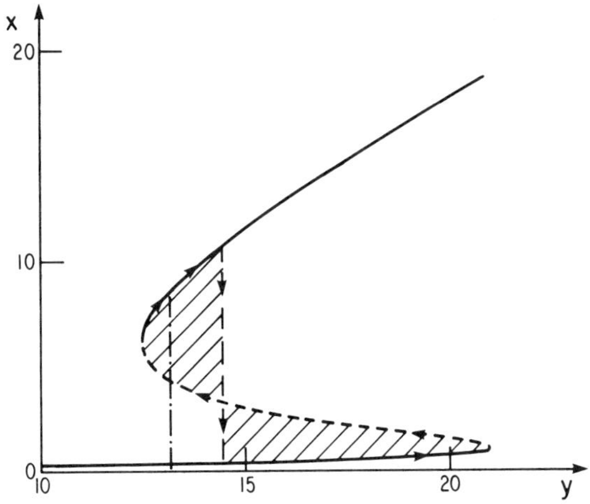

Fig. 6.5-9. Semiclassical stationary solutions, Maxwell rule, and mean value of the normalized field amplitude. $N_S = 10^3$, C = 20. (Solid line) Semiclassical (stable); (dashed) semiclassical (unstable); (dash-dot) $\langle x \rangle$. Bonifacio, Gronchi, and Lugiato (1978).

dispersive; Drummond (1982): bad-cavity arbitrary detuning and $\gamma_C, \gamma_T >> \gamma_L$, obtaining an exact steady-state solution valid on or off resonance. Chrostowski and Zardecki (1979) study numerically the transmission of noise by an absorptive bistable system and find that in quasi-steady state the transmitted light shows enhanced intensity fluctuations; see Fig. 6.5-11.

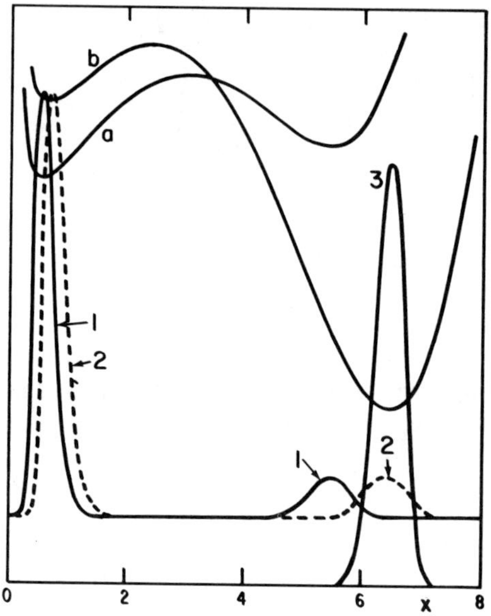

Fig. 6.5-10. Schematic representation of the local and global relaxation processes. Curves a and b represent the initial and final free energy functions $U(x,y_0)$, $U(x,y_{op})$, respectively. Curve 1 shows the initial P function corresponding to $U(x,y_0)$; curve 2 is a representation of the solution $P(x,y_{op},t)$ at the completion of the local relaxation (note that the peaks of the initial distribution have been displaced to coincide with the new minima of the $U(x,y_{op})$, while the respective areas of the two peaks have remained constant). Curve 3 shows the $P(x,y_{op},\infty)$ after completion of the tunneling process. Lugiato, Farina, and Narducci (1980).

Lugiato, Casagrande, and Pizzuto (1982), Drummond and Walls (1980, 1981), Casagrande and Lugiato (1980) and Lugiato and Strini (1982a,b,c) concentrate on nonclassical effects in optical bistability, namely antibunching and squeezing. Squeezing is defined as follows. Suppose a cavity mode is described by the annihilation (creation) operator $A(A^{\dagger})$ obeying the boson commutation relation $[A,A^{\dagger}] = 1$. Consider two components in quadrature:

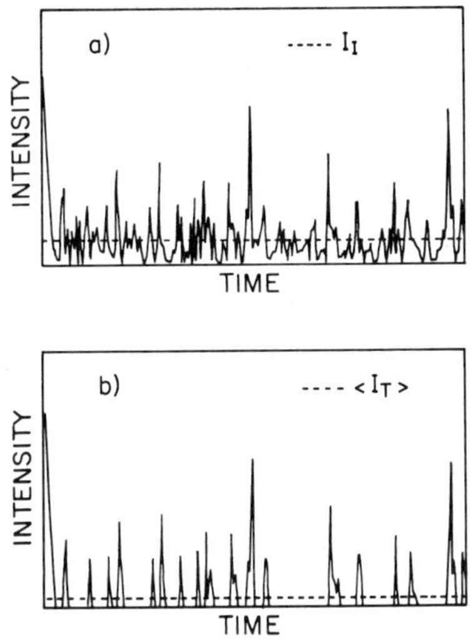

Fig. 6.5-11. Instantaneous intensity of the input (a) and output (b) of an absorptive bistable system. $C = 20$, $I_\downarrow = 155.9$, $I_\uparrow = 442.1$, $\langle I_I \rangle = 200$, $\langle I_T \rangle = 69.87$, $g^{(2)} = \langle I_T^2 \rangle / \langle I_T \rangle^2 = 8.14$. Chrostowski and Zardecki (1979).

$$x_1 \equiv \frac{A + A^\dagger}{2}, \qquad x_2 \equiv \frac{A - A^\dagger}{2i} . \qquad (6.5\text{-}28)$$

Squeezing [Yuen (1976)] is said to occur when the variance of either of the two components is smaller than in the usual coherent Glauber states, i.e.

$$\langle \delta x_1^2 \rangle < \frac{1}{4} \quad \text{or} \quad \langle \delta x_2^2 \rangle < \frac{1}{4} . \qquad (6.5\text{-}29)$$

Squeezing by a factor of about 2 is predicted for degenerate parametric amplifiers and two-photon optical bistability. Lugiato and Strini (1982a,b,c) find that increasing the order of an n-photon process only decreases $\langle \delta x_1^2 \rangle$ to 0.06 from the value $\simeq$ 0.1 for two-photon optical bistability. Walls and Milburn (1981) analyze the potential of a dispersive bistability system as a "squeezed-state" detector with reduced quantum fluctuations and find the magnitude of the squeezing to be too small to be of interest as a concept for improving the detection of gravitational radiation.

Lugiato, Casagrande, and Pizzuto (1982) show that when the fluctuations are normal, the second-order correlation function $g^{(2)}(0)$ is given by

$$g^{(2)}(0) = \langle I_T^2 \rangle / \langle I_T \rangle^2 = 1 + \frac{4}{N_s x_{st}^2} \left[\langle \delta x_1^2 \rangle - \frac{1}{4} \right] , \qquad (6.5\text{-}30)$$

where N_s is the saturation photon number ($\gamma_L N/8\gamma_c C$) and x_{st} is a solution of $y = x_{st}[1 + 2C(1 + x_{st}^2)^{-1}]$. Thus whenever squeezing occurs, there is also antibunching ($g^{(2)}(0) < 1$); antibunching occurs because an atom that has just spontaneously emitted must be re-excited before emitting again. For bistability, where N_s is usually very large, antibunching will be very difficult to see. Drummond and Walls (1980b, 1981) also discuss antibunching in the cases for which the nonlinearity arises from a cubic nonlinearity (Kerr medium) or Bloch equations (two-level atoms), respectively.

A detailed discussion of the ordering prescription for the operators in the characteristic function of the system atoms-plus-field seems out of place here. Perhaps it is sufficient to point to the detailed discussion by Lugiato, Casagrande, and Pizzuto (1982). It is precisely with respect to the nonclassical effects that the ordering matters. They point out that the two hierarchies of equations for the moments obtained from the two Fokker-Planck equations (Glauber-Sudarshan normal ordering and Wigner symmetrical ordering) are perfectly equivalent and correct. But if nonclassical effects are present, the Glauber-Sudarshan representation leads to a diffusion matrix that is not positive definite. Thus if one wants to obtain the quasiprobability distribution one must use the Wigner function formulation, because the Glauber-Sudarshan function does not exist. One can use the Glauber-Sudarshan function formulation by adopting the formalism of the complex P representation, devised by Drummond and Gardiner (1980). Their method has led to exact solutions to a nonlinear Fokker-Planck equation in which detailed balance holds. Using the complex P representation, the distribution itself is complex and one must calculate the moments at the cost of integrations along suitable contours in the complex plane. For a thorough discussion see Drummond, Gardiner, and Walls (1981). In other situations, the Wigner function formalism is more suitable because it gives directly the quasiprobability distribution.

In most quantum statistical calculations, the theory is eventually linearized. Carmichael, Walls, Drummond, and Hassan (1983) show that in this approximation the positive P-representation is justified; i.e., one can use standard results for moments and spectra, naively ignoring the fact that the diffusion matrix is not positive definite, and the results so obtained are correct.

Fluctuations, bistability, and antibunching in subharmonic and second harmonic generation are examined by Drummond, McNeil and Walls (1979, 1980), Brand, Graham, and Schenzle (1980), and Walls, Drummond, and McNeil (1981). Reid, McNeil, and Walls (1981) treat fluctuations in multiphoton lasers and mutiphoton bistability when thermal noise dominates the quantum noise.

Mandel, Roy, and Singh (1981) and Roy and Mandel (1980) have measured the probability distribution of photoelectron counting for each of the two counter-propagating traveling-wave modes of a ring dye laser. Each distribution is found to be double-peaked similar to Fig. 6.5-4. The system exhibits quasi-bistability, with spontaneous switching between states, and the relative intensity fluctuations of the slightly weaker mode is greater than unity. They find qualitatively good agreement with a Fokker-Planck theoretical description, but they believe the discrepancies are statistically significant and remain as a challenge to a more complete theory of the ring dye laser. Casagrande and Lugiato (1978) have studied a laser with saturable absorber for an arbitrary ratio between the saturation parameters of the passive and active atoms. They find three main effects of the saturation in the amplitude diffusion term of their renormalized Fokker-Planck equation: (i) an enhancement of the sharpness of the first-order-like phase transition, (ii) a large shift in the threshold of this bistable laser toward lower values of the pump of the amplifier, (iii) a notable narrowing of the peak of the intensity fluctuations as a function of the pump parameter.

Other papers relevant to or concerned with fluctuations and optical bistability can be found in Mandel and Wolf (1984) and Bowden, Gibbs, and McCall (1984). In particular, Schenzle and Thel (1984) treat the case of colored (finite bandwidth) noise.

CHAPTER 7

TOWARD PRACTICAL DEVICES

Most of the material in Chapters 1 through 6 is devoted to an analysis of the properties of bistable optical systems and to a description of experimentally successful bistable systems. Some of the all-optical logic operations possible with bistable devices are summarized in Section 1.2. In this chapter the emphasis is upon the characteristics of an ideal device (Section 7.1), fundamental limitations (Section 7.2), the prospect of improving present devices by identifying giant nonlinearities (Section 7.3), and the utilization of the full power of optics by parallel processing (Section 7.4).

7.1. DESIRABLE PROPERTIES; FIGURES OF MERIT

The characteristics of an ideal discrete bistable optical device are easily enumerated. The device should be small, i.e., of characteristic dimension of a micrometer, so that millions of them can be placed in a small volume. To avoid excessive heat problems and to minimize the operating power, the holding power should be less than 1 mW and the switching energy should be less than 1 fJ (1 fJ = 10^{-15} J). It should be fast ($\lesssim$ 1 ps) so that high-speed operations can be accomplished and the picosecond-pulse advantages of all-optical systems utilized. It should be operable at room temperature to eliminate cryogenics and to increase its range of applicability. Finally, it should be integratable to permit large numbers of interconnections insensitive to external perturbations.

Clearly, none of the demonstrated devices is a good approximation to the ideal device. The GaAs superlattice device (Section 3.6) is the best so far: $\lesssim$ 3-μm length and $\lesssim$ 10-μm diameter, < 10-mW holding power at 300K and <1 mW at 85K, 3-pJ switching energy, ≃ 1-ps external switch-up time, 40-ns triangular-input switch-down time (few ns with $\lesssim$ 10-ns input), and room-temperature operation. It is not integrated, but it is constructed of materials used for laser diodes and high-speed electronics. Measurements (see Section 3.6) of exciton saturation energies of only 1 fJ/μm^2 at ≃ 2 K and < 100-ps recovery time by Ulbrich (1981) suggest that 1-μm devices may someday be efficient and fast. The CS_2 (Section 3.4) recovery time is ≃ 1 ps, but the n_2 is so small that a long cavity is required resulting in switch-off times approaching 1 ns. Thus in all respects (except for true cw operation) the GaAs device is superior to all other intrinsic devices. Hybrid devices are relatively large and slow so far.

The use of bistable devices for two-dimensional image processing (Section 7.4) has different requirements. Large distortion-free devices are needed, but response times as slow as 1 ms are acceptable for some parallel-processing applications. Hybrid devices such as liquid-crystal light valves may be satisfactory for such applications if the cost can be made reasonable, but two-dimensional intrinsic devices may switch faster and be cheaper to construct. The best materials need to be identified or engineered and transverse coupling effects understood.

In evaluating discrete bistable optical devices it would be useful to have a meaningful figure of merit for comparison. For a plane wave incident on a plane Fabry-Perot cavity at normal incidence, constructive interference occurs for $2nL = m\lambda$. The index change needed for bistability is $\delta n \simeq \lambda/(2L\mathcal{F})$ where $\mathcal{F}$ is the finesse and a Δm of $1/\mathcal{F}$ corresponds to an initial detuning of one instrument width. If $n = n_0 + n_2 I$ then for $R \simeq 1$,

$$\frac{\delta n}{n_2} = I = \frac{\lambda}{n_2}\left[\frac{1}{2L\mathcal{F}}\right] \simeq \frac{\lambda}{n_2}\left[\frac{T}{2\pi L}\right]; \tag{7.1-1}$$

if the criterion for merit is minimum intensity then n_2/λ is a measure of the medium's merit [Gibbs, McCall, and Venkatesan (1980)]. However, in most devices to date a limit is placed upon L by unsaturable background absorption, so the inequality to be satisfied for bistability is Eq. (2.5-46):

$$\frac{\alpha_0 L}{T + \alpha_B L} > 4\,[1 + (1 + \Delta^2)^{1/2}]\,, \tag{7.1-2}$$

for a homogeneously broadened two-level transition. The $\alpha_0 L/T$ part of Eq. (7.1-2) corresponds to $n_2 L/T$ in Eq. (7.1-1). In a lossy etalon $\alpha_B L$ can dominate T, so $n_2/(\lambda\alpha_B)$ is a better measure of the material merit [D. A. B. Miller (1981a)]; this is derived in Appendix D; see Eq. (D-27).

The above measures of merit apply to plane-wave intensities. From a device viewpoint it is the power, not the intensity, that is of primary concern. Inputs with shorter wavelengths can be focused and guided in smaller cross sections, yielding higher intensities for the same input power. For a plane Fabry-Perot cavity, the effective Fresnel number F_{eff}

$$F_{eff} = \frac{n_0 a^2}{\lambda L \mathcal{F}} \tag{7.1-3}$$

needs to be of order unity to avoid strong diffraction losses; a is the HWHM beam radius at the etalon. For $(\alpha_B L + T) \ll 1$,

$$\mathcal{F} \simeq \frac{\pi}{T + \alpha_B L}\,. \tag{7.1-4}$$

The intensity is related to the power P by

$$I = \frac{P}{\pi a^2} \tag{7.1-5}$$

which by Eq. (7.1-1) becomes with $F_{eff} = 1$

$$P > \frac{\pi\lambda^2}{2n_2 n_0}\,. \qquad \text{Plane-parallel cavity} \tag{7.1-6}$$

This suggests n_2n_0/λ^2 as a figure of merit where the λ/n_0 factor reflects the focusing capability. This figure of merit is independent of α_B (of course, $\alpha_B L \ll 1$ was assumed) because the smaller α_B the less δn is needed but also the less tightly the beam can be focused. This simple analysis has ignored any frequency dependence of n_2 or α_B.

In a confocal Fabry-Perot cavity the curved mirrors compensate for diffraction, so $a \simeq \lambda/n_0$ is possible:

$$P > \frac{\lambda^3}{n_2 n_0^2}\left[\frac{\pi}{2L\mathscr{F}}\right], \tag{7.1-7}$$

or

Diffraction-compensated or guided-wave devices

$$P > \frac{\lambda^3}{n_2 n_0^2}\left[\frac{T + \alpha_B L}{2L}\right]. \tag{7.1-8}$$

The figure of merit $n_2n_0^2/\lambda^3$ emphasizes the advantage of short wavelengths from the guided-wave viewpoint. The need for a large n_2/α_B is, of course, still present; n_2/α_B must decrease more rapidly than λ^3 so that the guided-wave-bistability power will increase with decreased wavelength.

Equation (7.1-6) gives $P > 30$ mW for GaAs with $\lambda = 0.827$ µm, $\alpha_B = 0.05$ µm^{-1}, $|n_2| \geq 10^{-4}$ cm^2/kW, $n_0 = 3.5$, $L = 4.2$ µm, and $T = 0.1$. These values are those of the first bulk GaAs experiment (Section 3.6), so they are not the optimum values. Bistability has been seen with as little as 60 mW in that etalon. Equation (7.1-8) predicts 170 µW for guided-wave ($a \simeq \lambda/n_0$) operation with those same GaAs parameters. Bistability has been seen in multiple-quantum-well devices with less than 10 mW at room temperature and less than 1 mW at 80 K in spite of carrier diffusion over tens of micrometers. The measured value of n_2 is indeed very encouraging for microwatt-power, carrier-confined nanosecond devices or milliwatt-power subnanosecond devices. For InSb, Eq. (7.1-6) gives $P > 290$ µW with $\lambda = 5.47$ µm, $\alpha_B = 10$ cm^{-1}, $n_2 \simeq 0.4$ cm^2/kW, $n_0 = 4$, $L = 115$ µm, and $T = 0.3$. Bistability was seen with as little as 8 mW. Equation (7.1-8) predicts 0.46 µW for guided-wave operation. Of course, P in Eqs. (7.1-5 to 8) is intracavity power, not input power; one has, $P_T = TP$, and in the "on" state for a perfect dispersive device $P_I \simeq P_T$ so that $P_I \simeq TP$. However, in some devices $P_T \ll P_I$ in the "on" state. In any passive (absorbing) device, $P_T < P_I$.

Clearly, minimizing the power alone is inadequate. For many applications speed is important, so $\tau_\uparrow$ and $\tau_\downarrow$ should be minimized as well. Cavity build-up times and critical-slowing-down phenomena restrict the length to a few micrometers for $\simeq$ 1-ps response time. Even a cavityless one-feedback-pass device cannot be longer than about 50 µm for 1-ps round-trip time ($n_0 \simeq 3$).

The discussion in this section has assumed that $\delta n = n_2 I_{max}$, but in many cases δn depends upon the energy absorbed in the last relaxation time τ of the nonlinearity. For example, the δn in GaAs depends upon the density of carriers N_c that "saturates" the exciton transition by shielding the Coulomb attractive force between the electron and the hole. Or, in thermal bistability the temperature rise is proportional to the energy absorbed in the last thermal conduction time. In such cases

$$\dot{N}_C = -\frac{N_C}{\tau_M} + \frac{\alpha_B I}{\hbar\omega}, \qquad (7.1\text{-}9)$$

if absorption arises from an unsaturable α_B with $\alpha_B L << 1$. Then

$$N_C(t) = \frac{\alpha_B}{\hbar\omega}\int_{-\infty}^{t} \exp\left[-\frac{t-t'}{\tau_M}\right] I(t')dt'; \qquad (7.1\text{-}10)$$

if τ_M is much longer than the pulse width τ_p of $I(t)$, then

$$N_C(t) = \frac{\alpha_B}{\hbar\omega}\int_{-\infty}^{t} I(t')dt', \qquad (7.1\text{-}11)$$

assuming only nonresonant absorption. In the other limit $\tau_M << \tau_P$,

$$N_C(t) \simeq \frac{\alpha_B}{\hbar\omega}\tau_M I(t), \qquad (7.1\text{-}12)$$

so that $\delta n \sim N_C \sim I$, making the expansion $n = n_0 + n_2 I$ meaningful. Reducing τ_M increases the required intensity. With limited input intensity, a sample is likely to be labeled "good" if τ_M is long, so I_{max} can be lower. This tradeoff between power and speed is usually the case. Equations (7.1-11,12) also emphasize that absorption is essential when δn depends upon absorbed energy. If δn arises from an instantaneous n_2 effect, $\alpha_B L << 1$ is desirable; but if δn arises from absorbed energy, then $\alpha_B L \simeq 0.01$ to 0.1 should be better. A saturable absorption is preferable to a nonresonant one, since it produces carriers but permits high finesse in the "on" state.

7.2. FUNDAMENTAL LIMITATIONS

Several authors have addressed the question of what physical limits are placed on the possibility of attaining an ideal device: Keyes (1975, 1981), Fork (1982), McCall and Gibbs (1981, 1982b), P. W. Smith and Tomlinson (1981a), Kogelnik (1981), and P. W. Smith (1982, 1983, 1984b).

For bistability $2C = \alpha_0 L/T$ must exceed 8 by the semiclassical analysis of Chapter 2. Suppose a single atom is confined to an optical cavity of area λ^2 and length λ. Practically, choose $T = 0.05$, so that $\alpha_0 L \simeq 1$ should certainly yield bistability. A single atom whose absorption line is broadened only by lifetime effects has an absorption cross section σ_a of about λ^2. Then $\alpha_0 L \simeq \sigma_a NL \simeq \lambda^2\lambda^{-3}\lambda \simeq 1$, so bistability should be possible semiclassically. This single-atom bistability [McCall and Gibbs (1981)] will "glitch" too often, and the semiclassical approach is invalid; see Section 6.5. For a reasonable-width bistability loop to remain in the desired state, one needs about 1000 two-level atoms or the statistical equivalent. This amounts to 0.25 fJ for 1.5-eV photons. As mentioned above there is some indication that this can be approached in GaAs. Already GaAs devices (Section 5.5) have been switched at 10 K by 0.6 nJ spread over a 50-µm diameter, corresponding to 15 fJ if the diameter is reduced to $\lambda/n_0 = 1/4$ µm. Later measurements at 77 K required $\simeq$ 1 nJ for $\simeq$ 10 µm diameter. This may be because diffusion is rapid out of the smaller diameter; see Olsson et al. (1982). Room-temperature switching of a higher finesse superlattice device required less than 3 pJ for $\simeq$ 10µm diameter [Sections 1.2 and 3.6.] Clearly reducing the switching energy toward 1 fJ is very important.

The 1000-quanta switch-energy limit applies to semiconductor electronic and Josephson devices also, yielding limits of about 0.25 fJ and 10^{-3} fJ, respectively. The switch energy must be several factors of ten times the operating temperature, however. So room-temperature operation is possible in principle for semiconductor and optical bistable devices with energy gaps of order 1 eV but not for $\simeq$ 1-meV Josephson-junction devices.

Switching times are limited by cavity build-up time, medium response time, and cavity dynamical effects (Section 5.6). A cavity 1-µm long with T = 0.1 and $n_0 = 3$ has a build-up time τ_c of about 0.2 ps. Switching conditions can be selected to minimize critical slowing down. The medium response time is the present practical limitation. There is no such limit in principle since picosecond response times are common in semiconductors with relaxation enhanced by impurities, electric-field sweepout, radiation damage, surface recombination (very thin, 0.5-µm-thick samples), or loss of crystal structure (amorphous). Identifying fast optical nonlinearities that are still strong, i.e., require low intensities, is a major objective at present (Section 7.3).

P. W. Smith (1983, 1984b) has summarized fundamental limits in Fig. 7.2-1. Thermodynamic arguments conclude that a single yes-no switching operation must dissipate a minimum of about 1 kT of energy, where k is Boltzmann's constant and T is the absolute temperature. Quantum mechanical considerations lead to a minimum dissipation of h/τ_s of energy, where h is Planck's constant and τ_s is the switching time. The kT and h/τ_s limits are much lower than the statistical 1000-quanta limit discussed above. The region to the left of the "thermal transfer" line has been ruled out by Keyes (1975) for repetitive switching. A higher switching rate would result in an unacceptable temperature rise in the device. A device can be operated in this region provided that it is operated at less than the maximum repetition rate or that not all of the switching power is dissipated in the switching element. Keyes (1975) also shows that a device operating by absorbing light and saturating an optical transition needs to operate to the right of the "absorptive nonlinearity." He assumed a dipole moment ea_0 (product of electron charge and Bohr radius); in a multiple-quantum-well GaAs/AlGaAs structure it can be seven times larger, shifting Keyes' curve to the left. Use of a Fabry-Perot resonator can shift it farther to the left [Fork (1982)]. P. W. Smith (1982) calculates the "reactive nonlinearity" line for an optical Kerr effect medium along the lines of Section 7.1 using $n_2 = 6 \times 10^{-9}$ cm^2/kW for polydiacetylene PTS which has the largest known n_2 with $< 10^{-13}$ s response time. [See also Carter, Chen, and Tripathy (1983)]. In Fig. 7.2-2 is a similar chart with some additional points labeled by various materials. P. W. Smith (1983, 1984b) obtained those points by <u>extrapolating</u> reported experimental results to λ^2 cross-section devices with a length adjusted for minimum switching energy. He assumed a finesse of 30 for the Fabry-Perot resonator and considered the response time to be the switch-off time of the device (in some cases the reported switch-on times are much faster). Figure 7.2-2 shows that, with presently

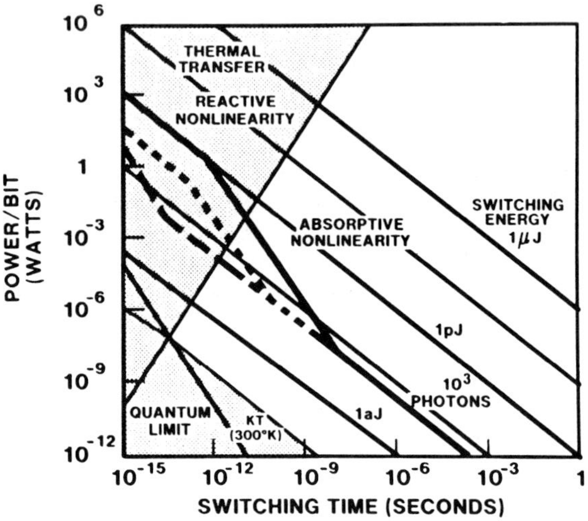

Fig. 7.2-1. Limits for optical switching devices. Fundamental limits to the power and switching time of devices are dictated by basic thermodynamic and quantum mechanics laws as well as practical power considerations. The three heavy line segments indicate limits for bistable optical devices. The frequency scale (at bottom) applies for repetitive switching. $\lambda = 0.85$ μm. Solid line: no resonator. Short dashed line: resonator ($F = 30$). Long dashed line: exciton and resonator ($F = 30$). P. W. Smith (1983, 1984b).

known materials, the fundamental limits for optical devices can be approached very closely in λ^2 cross sections.

In Fig. 7.2-3 Smith compares optical bistable devices with semiconductor electronic and Josephson devices. He concludes that within the 10^{-6} to 10^{-11}s region, one cannot hope to switch with substantially less power than required for semiconductor electronic devices and that appreciably lower switching energies are possible with Josephson technology. "At switching speeds of 10^{-12} through 10^{-14} s, however, bistable optical devices appear to have no competition. This unique capability for subpicosecond switching is one of the most exciting aspects of this new technology." Operation in this region is possible for nonlinear refractive index devices in which the power dissipated can be far less than the power required for switching. Because such short times place the operating point in the thermal transfer region, he concludes that "it does not appear feasible to design a general purpose high-speed digital optical computer. However, for many applications these thermal limits will not present severe problems." He mentions factors that could make optical switches competitive: need for freedom from electromagnetic interference, need for very rapid parallel processing of information

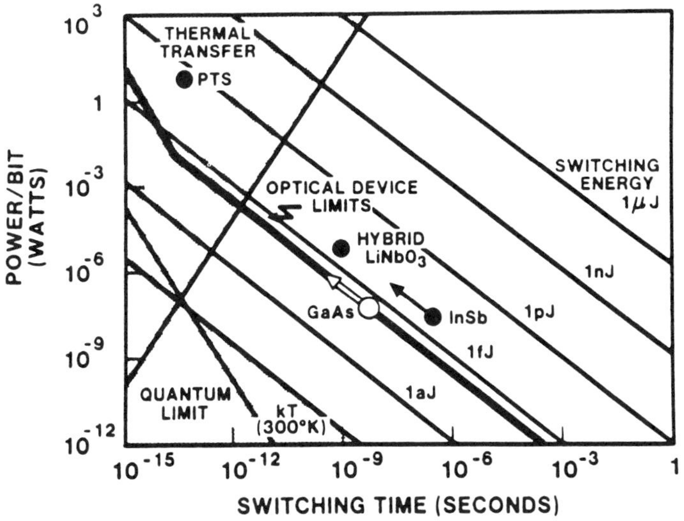

Fig. 7.2-2. Extrapolated current device performance. P. W. Smith (1983).

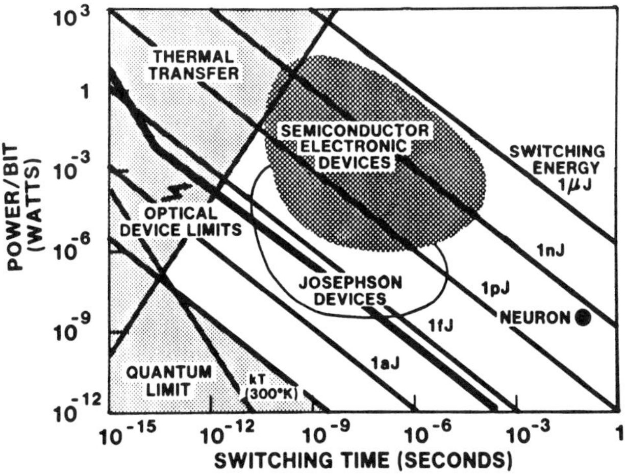

Fig. 7.2-3. Comparison of optical switching device limits with other technologies. P. W. Smith (1983, 1984b).

(which may already be in the form of light), need for very-high-speed multiplexing and demultiplexing to fully exploit the tremendous potential of optical fiber transmission systems.

7.3. NONLINEAR REFRACTIVE INDICES

7.3.1. Comparisons between Materials

In Chapter 3 a number of intrinsic optical bistability experiments were described. In nearly every case, the basic mechanism was an intensity-dependent refractive index. The Kramers-Kronig relations express the relationship between the absorption spectrum $\alpha(\omega) \propto \chi''(\omega)$ of a material and its refractive index $n(\omega) \propto \chi'(\omega)$. Consequently, if the absorption spectrum changes, so does the refractive index as depicted in Fig. 7.3-1. An absorption resonance gives rise to the well-known anomalous dispersion with higher refractive index below the resonance and lower above. If the absorption profile can be made to change by increasing the light intensity, then the refractive index is also intensity-dependent, i.e., non-linear, and the material can be used for dispersive optical bistability. After a comparison summary of various materials in Section 7.3.1, various mechanisms for optical nonlinearities in semiconductors most likely to find application in nonlinear optical signal processing are discussed in succeeding sections.

Band filling, treated in Section 7.3.2, can be envisaged as in Fig. 7.3-2; if light of energy exceeding the bandgap is sufficiently intense to fill all of the lower states in the conduction band faster than they decay, the bandgap effectively shifts up to the photon energy. This shift in band edge changes the absorption spectrum by removing an absorption peak at the band edge, resulting in a reduction of the refractive index below the band edge as in Fig. 7.3-1. This band filling saturation is also known as dynamic Burstein-Moss shifting.

Exciton nonlinear refraction, summarized in Section 7.3.3, can be viewed similarly. Less energy is required to create an electron and hole that are bound to each other to form an exciton, than to create a free electron and free hole. The exciton resonance appears as a sharp absorption peak at an energy slightly below the band edge. Saturation of the exciton absorption, for example by creating so many excitons that they overlap and shorten each other's lifetimes by screening the Coulomb potential responsible for their binding, changes the refractive index as in Figs. 7.3-1 and 3.6-9. The intensity dependence of a two-photon absorption peak arising from biexcitons, i.e. excitonic molecules, gives rise similarly to a nonlinear refractive index but of the opposite sign since the absorption increases with increased light intensity (Section 3.8.4).

Nonlinear refraction from an electron-hole plasma is described in Section 7.3.4. Free electrons and holes in semiconductors behave like mobile charge carriers that can be moved by applied electric (optical) fields. Their effective masses are often less than 10% of that of a free electron, so they move easily and produce large effects in the free-carrier Drude model.

Table 7.3-1 summarizes the results of the bistability experiments of Chapter 3 by listing for each nonlinear medium, the mechanism believed responsible for the nonlinearity, a lower limit $\delta n/I_{\uparrow}$ to n_2, the input power $P_{\uparrow}$ used for switch-up, and the switching times $\tau_{\uparrow}$ and $\tau_{\downarrow}$. From optical bistability experiments one can deduce a lower limit for n_2 by dividing the index change needed, roughly $\lambda/(2L_{NL}\mathcal{F})$ where L_{NL} is the physical length of the nonlinear medium in the

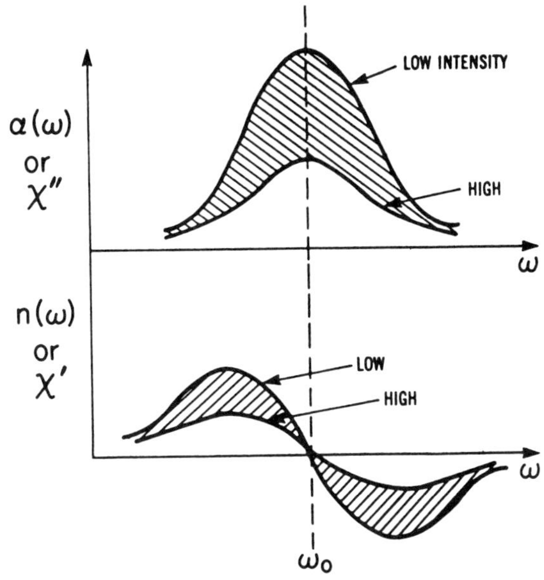

Fig. 7.3-1. Saturation of absorption and refraction for a simple absorption line as a function of frequency. D. A. B. Miller (1983).

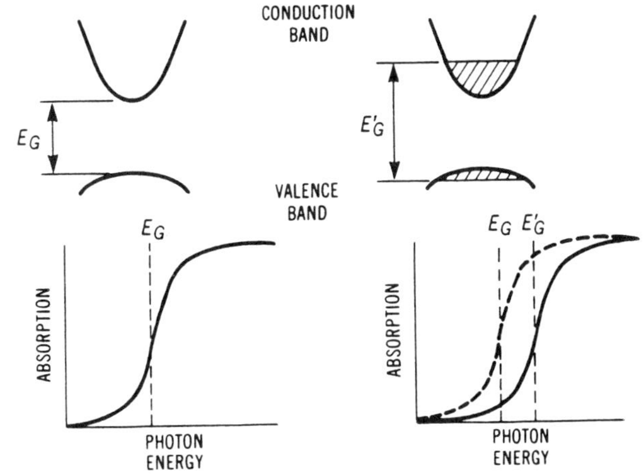

Fig. 7.3-2. Schematic illustration of the shift in optical absorption in a direct-gap semiconductor due to band filling (excitonic effects are omitted for simplicity). D. A. B. Miller (1983).

Table 7.3-1

MECHANISMS OF INTENSITY-DEPENDENT REFRACTIVE INDICES USED FOR OPTICAL BISTABILITY

Medium (Section)	Nonlinear mechanism	$\frac{\delta n}{I_\uparrow} = \frac{\lambda(2L_{NL}\mathscr{F})^{-1}}{I_\uparrow}$ (cm²/kW)	$P_\uparrow$ (mW)	$\tau_\uparrow$	$\tau_\downarrow$
Na (3.2)	hyperfine optical pumping**	$(\pm 10^{-3})$*	13	1-10 μs	10-50 μs
Na (3.2.5a)	2-level saturation**	$(\simeq 10^{-7})$*	60	both <	1 μs
Na:self-focusing (3.9)	optical pumping**	$(+6 \times 10^{-7})$*	150	both ≃	20 μs
Ruby (3.3)	far-away states**	2×10^{-6}	20	3 to	20 ms
Nitrobenzene (3.4)	molecular orientation	$+1.3 \times 10^{-10}$	20 MW/cm²		τ_M=45 ps
CS_2(3.4)	molecular orientation	$+2 \times 10^{-10}$	13 MW/cm²		τ_M=2 ps
Nematic liquid crystal film (3.4)	molecular reorientation	1	≃ 1000		1 s
Te(3.8.1)	free-carrier plasma	$\simeq 10^{-6}$			100 ns
SbSI (3.8.3)	free carriers	50 estimated	2		few s

ZnS, ZnSe (3.5)	thermal	few x 10^{-5}	5		few ms
Color filter (3.5)	thermal	+3 x 10^{-3} (d ≃ 1 mm)	1500		few ms
		+3 x 10^{-4} (d ≃ 25 μm)	18	both	≃ 1 ms
Si (3.5)	thermal	2 x 10^{-4}	500		6 x 10^{4}
Dyes (3.5)	thermal	6 x 10^{-5}	5		several ms
GaAs (3.5)	thermal	+4 x 10^{-4}	100	few ms	≃ 20 ms
GaAs (3.6)	free exciton**	-0.005	100	<200 ps	<40 ns
InSb (3.7)	band filling**	-0.088	8	<1 ns	≃ 500 ns
InSb (3.7)	two-photon	2 x 10^{-4}	100 kW/cm^2		≃ 50 ns
CdS (3.8.2)	bound exciton**	≃ 0.015	0.75		1 ns
CuCl (3.8.4)	biexciton**	10^{-7} (≃ 10^{-5} calc.)	10 MW/cm^2		ps predicted
InAs (3.8.5)	band filling**	-0.04	3	<1 μs	<1 μs
CdHgTe (3.8.6)	plasma & band filling (two photon)**	10^{-5}	500 kW/cm^2	<0.1 μs	≃ 0.4 μs

*density dependent

**resonantly enhanced, i.e., frequency dependent

cavity, by the intensity used in observing bistability. In some cases (Na and GaAs) saturation effects mean n_2 may be much larger than $\delta n/I_\uparrow$. In some cases (Na, ruby, GaAs, InSb, CdS) n_2 is frequency dependent and the observed $\delta n/I_\uparrow$ may be determined by other factors such as α_B or available laser frequencies. In some cases (all thermal; GaAs) δn is very geometry-dependent because of thermal conduction, excitation diffusion, or surface interactions. The nonlinear mechanism has a finite relaxation time τ_M; in most cases n_2 has been deduced from the expansion $n = n_0 + n_2 I$ using measurements made with pulse lengths much longer than τ_M.

As discussed in Section 7.1, optimizing n_2 is not the only consideration for bistability. If background absorption α_B dominates over end-mirror transmissivities (T) then $n_2 n_0^2/\lambda^3 \alpha_B$ should be maximized for a guided-wave bistable device. Of course, since speed is usually important and since n_2 generally decreases with decreasing medium response time τ_M, it is $n_2 n_0^2/\lambda^3 \alpha_B \tau_M$ that should be maximized.

If one compares the values of $\delta n/I_\uparrow \tau_\downarrow$ as a measure of the size of the nonlinearity and the speed, one finds that the only mechanisms with values as large as 10^{-2} to 10 $(\mathrm{cm})^2/(\mathrm{kW}\text{-}\mu\mathrm{s})$ are resonant nonlinearities in semiconductors. For this reason most research directed toward fast practical devices is concentrated upon semiconductors. Consequently we will focus on the measurements and mechanisms in semiconductors; for a concise summary of the physical mechanisms see D. A. B. Miller (1983). Since Table 7.3-1 gives only $\delta n/I_\uparrow$ from bistability measurements, Table 7.3-2 cites a few references on n_2 measurements. The third-order nonlinear susceptibility $\chi^{(3)}$ is related to the nonlinear refractive index n_2' via $n'_2 = 4\pi\chi^{(3)}/n_0$,

$$\chi^{(3)}(\mathrm{esu}) = 3\left[\frac{n_0}{4\pi}\right]^2 n_2(\mathrm{cm^2/kW})\,, \tag{7.3-1}$$

where n_0 is the background refractive index; see Appendix I. The nonlinear refractive index n_{NL} is usually defined as $n_{NL} = n_0 + n_2'|\xi|^2$, when a perturbation expansion is justified, where n_2' is in esu. Alternatively, let $n_{NL} = n_0 + n_2 I$; with I in $\mathrm{kW/cm^2}$, $n_2 = 4\pi \times 10^{10}\, n_2'/n_0 c$; i.e., one sees that n_2 in centimeters squared per kilowatt is numerically roughly $4/n_0$ times n_2' in esu units, and $\chi^{(3)}$ is numerically 3 to 12 times smaller than n_2'. Thus by using these units, order of magnitude comparisons can be made without conversions. In this book we have not distinguished between n_2 and n_2' except by their units.

Chang (1981) has reviewed n_2 materials with an emphasis upon those with submicrosecond response times. His survey of 111 representative optical media is a good starting point in considering nonlinear materials for optical devices. However, the largest n_2 in his tables is only a few times 10^{-4} $\mathrm{cm^2/kW}$ for the intrinsic exciton in CdS at 77 K. Our Tables 7.3-1,2 contain several more promising candidates. For extensive discussions of physical mechanisms for optical refractive index nonlinearities see the reviews by D. A. B. Miller, Miller, and Smith (1981); Jain (1982); Jain and Klein (1982); and Wherrett (1983a); also see Glass (1983).

Even at "low" intensities it is not always true that the refractive index can be expanded in powers of the intensity I. Hill, Parry, and Miller (1982) and A. Miller and Parry (1984b), for example, find $\delta n = -(7.0 \pm 3.5) \times 10^{-3} I^{1/3}$ where I is the incident intensity in watts per centimeter squared; the $I^{1/3}$ dependence is explained by the predominant role of Auger recombination in limiting the saturation of band-edge resonant absorption. They use the bandgap resonant saturation model of D. A. B. Miller, Seaton, Prise, and Smith (1981) [Section 7.3.2] extended to account for the intrinsic Auger recombination process. So even though it is common practice to speak of n_2 or $\chi^{(3)}$ phenomena, one must be careful of situations where $n = n_0 + n_2 I$ is a poor approximation over the range of I's used because of saturation. If bistability or degenerate four-wave-mixing is observed at only one intensity I, defining n_2 as $\delta n/I$ can be misleading.

There are several techniques for measuring n_2 or $\chi^{(3)}$. Perhaps the most popular method at present is optical phase conjugation in the form of degenerate four-wave mixing which, for example, has been applied to the measurement of both small [P. W. Smith, Tomlinson, Eilenberger and Maloney (1981)] and very large [Khan, Bennet and Kruse (1981); D. A. B. Miller, Chemla, Eilenberger, Smith, Gossard, and Wiegmann (1983)] nonlinearities. Wavefront encoding leading to changes in the far-field profile has been used, for example by Weaire, Wherrett, Miller and Smith (1979). Various techniques utilize a change in interference fringes [Milan and Weber (1976)].

7.3.2. Band Filling Nonlinear Refraction (InSb, InAs)

D. A. B. Miller, Seaton, Prise and Smith (1981) have performed measurements of n_2 for InSb at 77 K and found good agreement with their semi-empirical model. The presence of a very large n_2 just below the bandgap in InSb at 4 and 77 K was first reported by D. A. B. Miller, Mozolowski, Miller, and Smith (1978); their far-field beam profiles gave evidence of strong transverse effects with only 10 W/cm^2 intensity incident upon the 7.5-mm-long antireflection-coated sample. They conclude that the nonlinearity is not thermal. Weaire, Wherrett, Miller, and Smith (1979) found good agreement between measured and calculated profiles only if a defocusing mechanism was assumed; see Fig. 7.3-3. Their analysis assumes thin-sample encoding, i.e., that diffraction can be neglected within the sample whose only effect is to encode a radially dependent phase shift such that the beam emerges from the crystal at z = 0 with an amplitude

$$\xi(r,0) = \xi(0,0)\exp\left[-\frac{r^2}{w_0^2} + i\phi_0 \exp\left[-\frac{2r^2}{w_0^2}\right]\right] \tag{7.3-2}$$

where

$$\phi_0 = -\frac{\omega}{c} n_2 I_p \left[\frac{1-e^{-\alpha L}}{\alpha}\right] \tag{7.3-3}$$

and I_p is the peak beam intensity. They then solve the subsequent diffraction problem using an expansion in Gaussian functions. Firth, Seaton, Wright, and Smith (1982) show spatial hysteresis in experimental and computer far-field profiles under bistability conditions; also see Section 2.9.

The agreement between the InSb data and the model of D. A. B. Miller, Seaton, Prise, and Smith (1981) is exhibited in Fig. 7.3-4b. The beam-profile distortion technique described above was used to measure n_2. Four different samples were used so that large enough $n_2 L$ effects with small enough αL effects could be achieved over a 100 cm^{-1} range of detuning. The physics of their semi-empirical model was outlined in Section 3.7; here we summarize the mathematics. They assume every absorbed photon creates a free electron and hole (perhaps by a

Table 7.3-2

SOME REPORTED VALUES OF NONLINEAR REFRACTIVE INDICES

Medium	λ (μm)	Nonlinear Mechanism	$\lvert n_2(\mathrm{cm^2/kW})\rvert$	Response Time (s)	Reference
InSb(77K)	5.4	Band filling	1	0.4×10^{-6}	D.A.B. Miller, Seaton, Prise and Smith (1981)
$Hg_{1-x}Cd_xTe$ $x \simeq 0.2$ (77K)	10.6	Bandgap-resonant intrinsic carrier excitation	0.4		Khan, Bennet and Kruse (1981)
GaAs (10K)	0.82	Free exciton	0.4 deduced from α_{NL}	few x 10^{-9}	Gibbs, McCall, Venkatesan, Passner, Gossard & Wiegmann (1980)
GaAs/ $Al_xGa_{1-x}As$ MQW (300K)	0.84	Free exciton	0.2	20×10^{-9}	Miller, Chemla, Eilenberger, Smith, Gossard, and Wiegmann (1983)
InAs (77K)	3.1	Band filling	0.03	0.33×10^{-6}	Poole and Garmire (1984d,e)
InSb (300K)	10.6	Two photon	2×10^{-4}	50×10^{-9}	Kar, Mathew, Smith, Davis, & Prettl (1983)
$Hg_{1-x}Cd_xTe$	10.6	Intervalence band	3×10^{-5}	10^{-12}	Yuen (1983b)
Various		Thermal effects	10^{-5} to 10^{-4}	10^{-1} to 1	Chang (1981)

CuCl(<10K)	0.38	Biexciton	$\approx 10^{-6}$		Itoh and Katohno (1982); Kuwata, Mita and Nagsawa (1981)
$Hg_{1-x}Cd_xTe$	10.6	Conduction-band non-parabolicity	5×10^{-7} to 5×10^{-6}		Khan, Kruse and Ready (1980)
InAs	10.6	Induced polarization	$+5 \times 10^{-10}$		Wynne (1969)
Ge, Si, GaAs	10.6	Induced polarization	3×10^{-11} to 3×10^{-10}		Wynne (1969)
Ge	10.6	Induced polarization	$(4 \text{ to } 7)10^{-11}$		Watkins, Phipps, and Thomas (1980)
Chlorobenzene, bromobenzene, & 10 other organic compounds	1.32		1.1×10^{-11} 1.4×10^{-11} $< 3 \times 10^{-11}$	20×10^{-12} 13×10^{-12} $<100 \times 10^{-12}$	Witte, Galanti, and Volk (1980)
CS_2, nitrobenzene, etc.		Molecular-orientation Kerr effect	3×10^{-13} to 3×10^{-11}	0.1 to 1×10^{-12}	Chang (1981)
CdS	0.69	Induced polarization	2×10^{-12}		Baranovskii, Borshch, Brodin & Kamuz (1971)
Borosilicate, fused silica, & other glasses	1.06		$(2 \text{ to } 5)\ 10^{-13}$		Milam and Weber (1976)

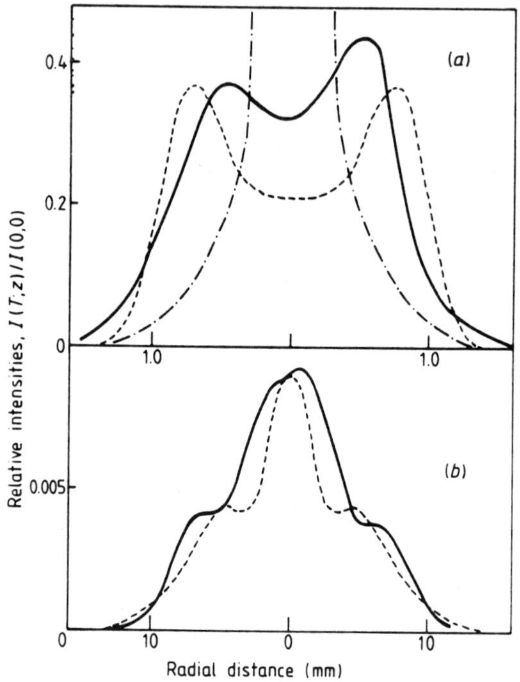

Fig. 7.3-3. Experimental (solid lines) and theoretical (broken lines) intensity profiles (a) in the near field at 7 cm from the sample and (b) in the far field at 189 cm. Data for 130-mW laser beam of 1.67-mm-spot diameter and wavenumber 1886 cm^{-1}. InSb sample 7.5-mm-long, anti-reflection coated on both ends, $N_D - N_A \simeq 3.8 \times 10^{-14} cm^{-3}$ (n-type); temperature $\simeq$ 5 K. Theoretical profiles are shown for the self-defocusing condition (dashed lines) in the near and far field and for the self-focusing condition (dot-dashed line) in the near field to emphasize that the experimental results originate from a defocusing; the magnitude of the on-axis nonlinear phase shift was 3.5 for both simulations. In the far field, the focusing and defocusing results are practically identical. The theoretical and experimental plots are normalized to give the same power levels. Weaire, Wherrett, Miller, and Smith (1979).

phonon-assisted transition for below-bandgap absorption), and they approximate the Fermi-Dirac distributions by Boltzmann distributions. Pumping the InSb with incident radiation of photon energy $\hbar\omega$ and intensity I leads to generated populations of electrons and holes,

$$\delta N = \frac{\alpha(\hbar\omega)I\tau_R}{\hbar\omega} \tag{7.3-4}$$

in the steady state, where τ_R is the interband recombination time. The direct interband absorption coefficient $\alpha_d(\hbar\omega')$ for photon energy $\hbar\omega'$, above the bandgap energy, can be written in the parabolic-band approximation as

$$\alpha_d(\hbar\omega') = \frac{8(2m)^{1/2}e^2}{3\hbar^2 c}\left[\frac{\mu}{m}\right]^{3/2}\frac{1}{n_o}\frac{mP^2}{\hbar^2}\frac{(\hbar\omega'-E_G)^{1/2}}{\hbar\omega'}[1 - f_e(E_c) - f_h(E_v)] \tag{7.3-5}$$

with $\mu = m_c m_v/(m_c + m_v)$ and where μ, m_c, m_v are the reduced, conduction-band and valence-band effective masses, respectively. P is the momentum matrix element defined by Kane (1957):

$$P = -\frac{i\hbar}{m}\langle S|pz|Z\rangle; \tag{7.3-6}$$

and m is the mass of a free electron. They consider only the absorption between the heavy hole and conduction bands and neglect the much weaker light-hole-band contribution. The absorption is weighted according to the occupation probabilities $f_e(E_c)$ and $f_h(E_v)$ of electrons and holes in the conduction and valence bands, respectively, at the energies appropriate for the direct transition at photon energy $\hbar\omega'$,

$$E_c = \frac{m_v}{m_c + m_v}(\hbar\omega' - E_G), \tag{7.3-7}$$

$$E_v = \frac{m_c}{m_c + m_v}(\hbar\omega' - E_G). \tag{7.3-8}$$

Using the Boltzmann approximation and the usual parabolic density of states, one may write the changes δf_e and δf_h in f_e and f_h in the presence of the radiation field as

$$\delta f_e(E_c) = 4\pi^{3/2}\left[\frac{\hbar^2}{2m_c}\right]^{3/2}\frac{\delta N}{(kT)^{3/2}}\exp(-E_c/kT), \tag{7.3-9}$$

$$\delta f_h(E_v) = 4\pi^{3/2}\left[\frac{\hbar^2}{2m_v}\right]^{3/2}\frac{\delta N}{(kT)^{3/2}}\exp(-E_v/kT). \tag{7.3-10}$$

Since $m_v > 10\, m_c$ in direct-bandgap III-V compound semiconductors, δf_h is negligible relative to δf_e. Then

$$\delta\alpha_d(\hbar\omega') \simeq -\frac{16\pi^{3/2}}{3}\frac{e^2\hbar}{m_c}\left[\frac{\mu}{m_c}\right]^{3/2}\frac{mP^2}{\hbar^2}\frac{1}{n_0}\frac{\delta N}{(kT)^{3/2}}\frac{(\hbar\omega'-E_G)^{1/2}}{\hbar\omega'} \times \exp\left[\frac{-\mu(\hbar\omega'-E_G)}{m_c kT}\right]. \tag{7.3-11}$$

From the Kramers-Krönig relations they find the change in refractive index at photon energy $\hbar\omega$ as

$$\delta n(\hbar\omega) = \frac{\hbar c}{\pi}\int_0^\infty \frac{\delta\alpha_d(\hbar\omega')}{(\hbar\omega')^2-(\hbar\omega)^2}\, d(\hbar\omega'). \tag{7.3-12}$$

Then

$$n_2 \equiv \frac{\delta n}{I} = -\frac{8\sqrt{\pi}}{3}\frac{e^2\hbar^2}{m}\frac{\mu}{m_c}\frac{mP^2}{\hbar^2}\frac{1}{n_0 kT}\frac{\alpha(\hbar\omega)\tau_R}{(\hbar\omega)^3}\, J\left[\frac{\mu(\hbar\omega-E_G)}{m_c kT}\right] \tag{7.3-13}$$

where

$$J(a) = \int_0^\infty \frac{x^{1/2}e^{-x}}{x-a}\, dx \tag{7.3-14}$$

is plotted in Fig. 7.3-4a, with the restrictions $|\hbar\omega - E_G| \ll E_G$ and $kT \ll E_G$ to simplify the integral allowing it to be expressed as a function of only one parameter. They emphasize that the Boltzmann (i.e., nondegenerate) approximation leads to a δn proportional to I, but at high intensities the nonlinearity will saturate and can only be described with use of the full Fermi-Dirac statistics. The theoretical values of n_2 are shown in Fig. 7.3-4b for which they used the measured values for $\alpha(\hbar\omega)$, $\mu/m_c \simeq 0.97$, $m_c \simeq 0.014$ m, $m_v \simeq 0.4$ m, $mP^2/\hbar^2 \simeq 11$ eV, $n_0 \simeq 4$, $E_G \simeq 1835\ \text{cm}^{-1}$, and $\tau_R \simeq 400$ ns. It is seen that the theoretical values, based on measured parameters, agree well with the detailed data.

Poole and Garmire (1984c) obtain excellent agreement with measured values of n_2 in InAs [Section 3.8.5] using the above theory modified to include the light-hole-band contribution, a different band-edge expression for linear absorption, and the complete Fermi-Dirac distribution function.

Noting that P^2 and n_0 are substantially constant and $\mu/m_c \simeq 1$ for all III-V compound semiconductors, they predict for the change σ in refractive index for one excited electron-hole pair in unit volume

$$\sigma = \frac{n_2\hbar\omega}{\alpha(\hbar\omega)\tau_R} \simeq 1.7 \times 10^{-17}\frac{1}{T}\frac{1}{(\hbar\omega)^2}\, J\left[\frac{\hbar\omega-E_G}{kT}\right]\ \text{cm}^3. \tag{7.3-15}$$

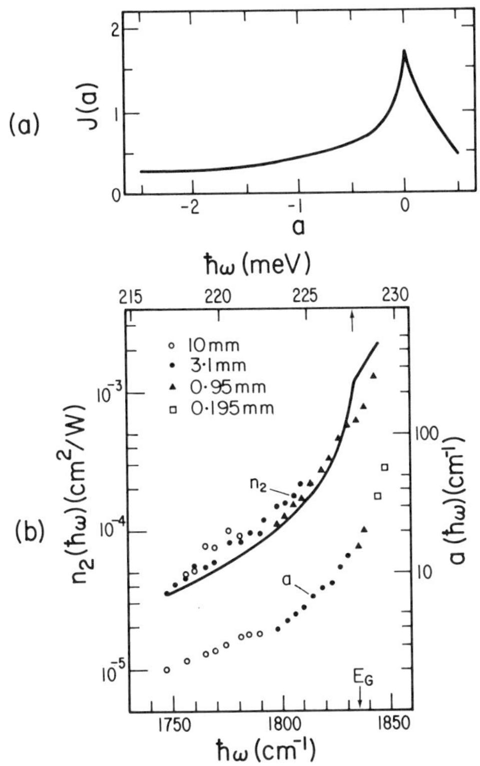

Fig. 7.3-4. (a) The dimensionless resonance function J(a) plotted against dimensionless energy parameter $a = \mu(\hbar\omega - E_G)/m_c kT$. $\mu/m_c \simeq 1$ for all III-V semiconductors. (b) Measured n_2 and absorption coefficient α as a function of photon energy $\hbar\omega$ compared with theory for n_2. The error in α measurement is of the order of the point size. The solid line is calculated with use of the measured α (smoothed for clarity). All the parameters in this theoretical curve are based on measured values. D. A. B. Miller, Seaton, Prise, and Smith (1981).

Thus large effects are still predicted for higher energies, for example, at $\lambda \simeq 1\ \mu m$. They do not treat electron-hole correlation effects.

Wherrett and Higgins (1982a,b) and Higgins and Wherrett (1982) have calculated n_2 by both a band-filling or blocking model and a saturation model consisting of independent two-level absorbers. They note that for device applications one wants a large n_2 and small α. By either model one achieves this by operating at frequencies just beneath a high joint density of states, these states being radiatively coupled. The band edge of a pure semiconductor is not ideal in this respect, having a $\Delta E^{1/2}$ dependence of the density of states, which rises from zero at the edge. They suggest two improvements: (a) Use doped materials which have a finite free-carrier population at zero light intensity, so that the effective band edge is shifted from $\Delta E = 0$ to the position of the Fermi energy, and hence to a region of higher density of states. (b) Apply a magnetic field of the order of Teslas in strength to "sharpen" the band-edge density of states; see Sen (1983).

7.3.3. Exciton-Resonant Nonlinear Refraction (GaAs, CdS)

Coulomb correlation effects change the band-edge density of states from parabolic to a step function and may even lead to a free-exciton resonance. The goal of operating close to a strong absorption line or edge, as in ruby and as discussed at the end of Section 7.3.2, motivated the selection of GaAs as a good candidate for optical bistability. The successful use of the free-exciton resonance in bulk GaAs and in GaAs/AlGaAs multiple-quantum-well (MQW) devices is described in Section 3.6. A simple two-level-transition model of the free-exciton resonance is given there and is used to deduce an n_2:

$$n_{EX}(\lambda) = \frac{\alpha_{EX}\lambda}{4\pi}\,\frac{\Delta}{1 + \Delta^2 + I/I_s}$$

$$n_2 \equiv \left.\frac{dn_{EX}}{dI}\right|_{I=0} = -\frac{\alpha_{EX}\lambda}{4\pi I_s}\,\frac{\Delta}{(1+\Delta^2)^2}$$

from nonlinear absorption data. Although that model was useful in the design of experiments, the many-body treatment of Section 7.3.4 should be a much better description.

D. A. B. Miller, Chemla, Eilenberger, Smith, Gossard, and Tsang (1982) have studied the room-temperature optical nonlinearity in an MQW sample consisting of alternate 10.2-nm GaAs layers and 20.7-nm $Al_{0.28}Ga_{0.72}As$ layers of total thickness $\simeq 2.4$ µm. The low-intensity absorption spectrum is shown in Fig. 3.6-15 and the saturation dependence in Fig. 3.6-16. The temperature dependence of the linewidth Γ_{EX} of the lower energy MQW exciton resonance is shown in Fig. 7.3-5. Up to 150 K, Γ_{EX} is temperature independent and probably arises from fluctuations of layer thickness of the order of one atomic layer. The usual dominant thermal broadening mechanism for exciton resonances at room temperature in 3-D semiconductors is LO phonon absorption, such that

$$\Gamma_{EX} = \Gamma_0 + \Gamma_B\,[\exp(\hbar\omega_{LO}/kT) - 1]^{-1}, \qquad (7.3\text{-}16)$$

where $\hbar\omega_{LO} = k(428\text{ K})$ is the LO phonon energy in GaAs and Γ_B is a measure of the phonon broadening. They obtain a good fit for $\Gamma_0 = 2.0$ meV and $\Gamma_B = 5.5$ meV. The broadening parameter in this fit (5.5 meV per phonon per mode) is, however, less than that calculated and measured for bulk GaAs (> 7 meV), contributing to the relative clarity of the MQW exciton spectrum. The phonon contribution to Γ_{EX} of $\Delta E = \Gamma_{EX} - \Gamma_0 \simeq 1.8$ meV at room temperature corresponds to a mean time τ_0 to phonon absorption of $\simeq$ 0.4 ps using the relation $\Delta E = \hbar/\tau_0$ for a Lorentzian lineshape. Since LO-phonon absorption ionizes the exciton, this model implies that 0.4 ps is also the mean time between the creation of an exciton and its ionization. On the other hand, the recovery time of an absorption change was found to be $\simeq$ 21 ns. Device turn-off time is then determined not by the exciton lifetime but by the much longer recombination time of the carriers that screen the excitons as discussed in Section 3.6.

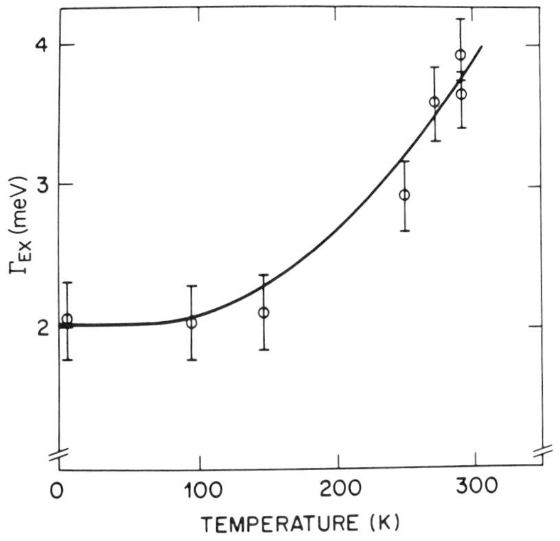

Fig. 7.3-5. Half width at half maximum, Γ_{EX}, of the lower energy (i.e., heavy hole) MQW exciton as a function of temperature. Solid line--theoretical fit of LO phonon broadening model (two adjustable parameters) $\Gamma_{EX}(\text{meV}) = 2 + 5.5/[\exp(\hbar\omega_{LO}/kT) - 1]$ where $\hbar\omega_{LO}/k \simeq 428$ K. Chemla, Miller, Smith, Gossard, and Tsang (1982).

An n_2 of 0.2 cm²/kW has been measured by forward degenerate four-wave mixing for a room-temperature GaAs/GaAlAs structure of sixty-five 96-Å quantum wells near the exciton resonances [D. A. B. Miller, Chemla, Eilenberger, Smith, Gossard, and Wiegmann (1983)]. They observe a diffraction efficiency of $\simeq 10^{-4}$ with $\simeq$ 18 W/cm² average intensity from a mode-locked laser [D. A. B. Miller, Chemla, Smith, Gossard, and Wiegmann (1983)].

D. A. B. Miller, Chemla, Eilenberger, Smith, Gossard, and Tsang (1982) raise the question whether the room-temperature bistability in a GaAs MQW described in Section 3.6.2 arises from excitonic or interband saturation because of the several hundred kW/cm² intensity required. Subsequent studies suggest that transverse diffusion of excitons and carriers results in excitation diameters of 20 to 30 μm even when the input is focused more tightly. In fact for input diameters below 20 μm the switch-up power is almost independent of input diameter. Using

40- to 50-μm minimum diameters, MQW bistability has been seen with an intensity of 3 kW/cm². This intensity for off-resonance bistability is not much more than the on-resonance plane-wave 580 W/cm² minimum saturation intensity found for the exciton by D. A. B. Miller, Chemla, Eilenberger, Smith, Gossard, and Tsang (1982).

Further evidence for transverse diffusion is that the center of the beam switches on (off) about 50 ns before (after) the outer edges of the beam when the input has a 1-μs triangular-shaped temporal dependence; see Gibbs, Jewell, Moloney, Tai, Tarng, Weinberger, Gossard, McCall, Passner, and Wiegmann (1982). Also transverse effects give dramatic changes in the profiles which could probably be used to deduce n_2 values as done for InSb (Section 7.3.2). The intensity profile of the output beam is observed as shown in Fig. 7.3-6a. A microscope objective is placed between the etalon and an aperture. Experimentally, one sees drastic changes in the profile as the output objective is translated along the propagation direction. Since the image plane (i.e., aperture) stays fixed, this corresponds to selecting different object planes. The nonlinearity in GaAs defocuses the beam, so that moving the objective closer to the etalon probes the virtual object planes--as suggested by the dashed beams at the bottom of Fig. 7.3-6a. The experimental profile is mapped out by oscillating a mirror back and forth at a rate that is slow compared with the 40-kHz rate of the input pulses. This sweeps the output across an aperture to obtain transverse beam profiles as shown in Fig. 7.3-6b. Good agreement is found between the data and one-transverse-dimension numerical simulations (Section 2.9) of "good" ring-cavity bistability in which the output is free-space propagated backwards in space corresponding to the movement of the imaging lens. Figure 7.3-7 shows the similarity between the simulations and observed profiles.

7.3.4. Many-Body Theory of Optical Bistability in Semiconductors

Löwenau, Schmitt-Rink, and Haug (1982) have developed a unified microscopic theory of the observed optical bistabilities in narrow-gap semiconductors, such as InSb and GaAs, by calculating the complex, nonlinear dielectric function for arbitrary free-carrier concentrations from an integral equation for the polarization function. The dominant optical nonlinearity in optical bistability in GaAs arises from ionization of the exciton caused by the screening of the Coulomb potential by electronic excitations (Mott transition); see Section 7.3.3. In InSb it arises from filling of the bands with free carriers; see Section 7.3.2. In the lower or "off" state of the bistability a relatively small number of electron-hole (e-h) pairs is excited, while in the upper or "on" state a relatively dense e-h plasma is generated. The dominant correlations in the low-density regime are caused by the attractive e-h Coulomb interaction, which gives rise to the formation of excitons in GaAs. The exciton binding energy in bulk GaAs at low temperatures and in GaAs/AlGaAs superlattice structures even at room temperature is larger than its broadening. (In InSb the broadening exceeds the binding, so the exciton resonance is not resolved.) In the high-density regime the screening of the Coulomb interaction by intraband scattering is the most important process. Their theory consistently combines these low- and high-density processes, but it neglects many-exciton effects expected to be important at intermediate e-h densities.

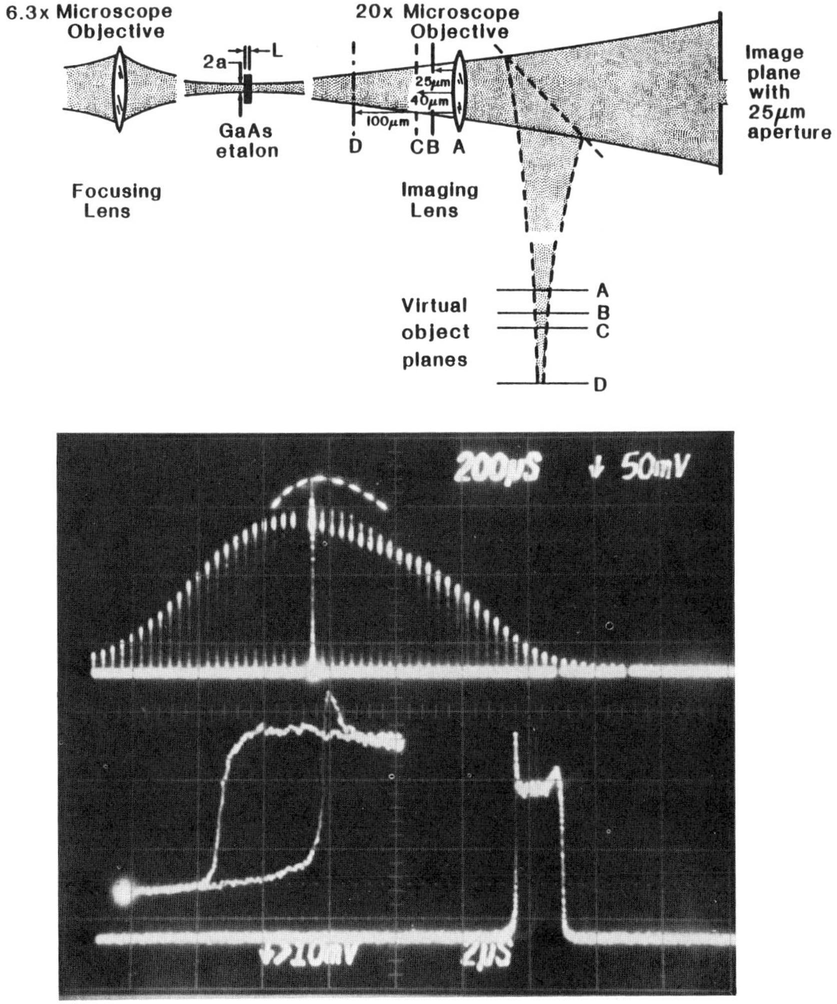

Fig. 7.3-6. (a) Arrangement for observing transverse profiles as a function of longitudinal position. L = 4.9 μm, 2a = 8μm (FWHM), $\mathcal{F}$ = 12, n_0 = 3.5, λ = 0.88 μm. (b) Transverse profiles. Top: each vertical line represents one input pulse of ∿ 4-μs duration. The dashed line is the envelope of the peak of the overshoot in the output intensity. Lower right: time trace of the output for the transverse position intensified at the top (2 μs/cm). Center left: corresponding hysteresis loop. Gibbs, Jewell, Moloney, Tai, Tarng, Weinberger, Gossard, McCall, Passner, and Wiegmann (1983).

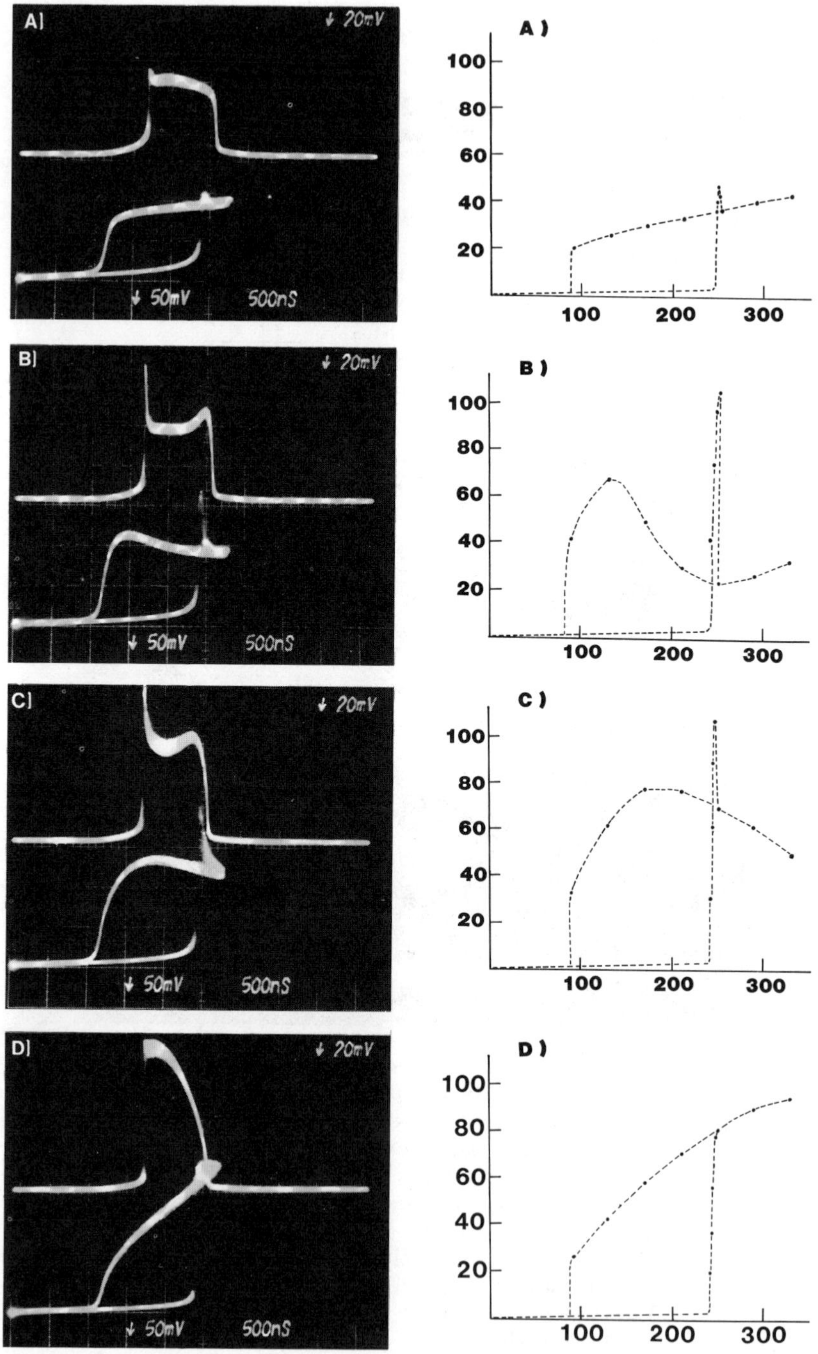
A)
↓ 20mV
↓ 50mV
500nS
B)
↓ 20mV
↓ 50mV
500nS
C)
↓ 20mV
↓ 50mV
500nS
D)
↓ 20mV
↓ 50mV
500nS
A)
100
80
60
40
20
100
200
300
B)
100
80
60
40
20
100
200
300
C)
100
80
60
40
20
100
200
300
D)
100
80
60
40
20
100
200
300

←

Fig. 7.3-7. (A) Output through 25-μm aperture of Fig. 7.3-6a with output of sample imaged on the aperture. Output with 20X objective moved in by (B) 25 μm, (C) 40 μm, and (D) 100 μm. (A') - (D') similar hysteresis loops from 1D computer simulations. Gibbs, Jewell et al. (1983).

The e-h correlation can be described by the polarization function P_{eh}, which also determines the optical dielectric function

$$\varepsilon(\omega) = \varepsilon_\infty - \frac{8\pi e^2}{m^2} \sum_{\mathbf{k},\mathbf{k}'} M_{vc}(\mathbf{k}) M_{vc}^*(\mathbf{k}') \, P_{eh}(\mathbf{k},\mathbf{k}';\omega) \, . \tag{7.3-17}$$

ε_∞ is the background dielectric constant; $M_{vc}(\mathbf{k})$ is the dipole matrix element for direct optical transitions between the valence and the conduction band. Its form is taken from the **k·p** perturbation theory. m and e are are the mass and charge of a free electron, respectively. Since their result is not expressible in a simple formula, for further details see: Haug (1982); Schmitt-Rink, Löwenau, and Haug (1982); Singh, Schmitt-Rink, and Haug (1981); Klingshirn and Haug (1980).

The resulting absorption and refraction spectra $\alpha(\omega)$ and $n(\omega)$, respectively, are shown in Fig. 7.3-8 for GaAs for various free-carrier concentrations N_C. For the unexcited crystal ($N_C = 0$), one sees for a broadening only 5% of the binding energy that the 1s- and 2s-exciton absorption lines are well resolved. With increasing N_C, the ionization continuum shifts to lower energies. The lowest exciton resonance shifts very little because of the compensation between the weakening of the e-h binding and the bandgap reduction. [An exciton blue shift of almost a half width has been reported in a 53Å-well, 56Å-barrier MQW structure at low temperatures: Peyghambarian, Gibbs, Jewell, Antonetti, Migus, Hulin, and Mysyrowicz (1984).] The curve with $N_C = 5\times10^{15}$ cm^{-3} is slightly above the Mott transition, but still a large excitonic resonance is present. The excitonic enhancement would be reduced slightly if the dependence of Γ_{EX} upon N_C were included. Some experimental nonlinear absorption spectra are displayed in Fig. 7.3-9 and are seen to be in good qualitative agreement with Fig. 7.3-8. Nonlinear refraction data are not available; however, the index changes shown in Fig. 7.3-8 easily account for the optical bistability in GaAs (Section 3.6). For an application of this theory to CdS data see Schmitt-Rink, Löwenau, Haug, Bohnert, Kreissl, Kempf, and Klingshirn (1983).

The main difference between InSb and GaAs is the opposite ratio of the damping linewidth to the exciton binding energy, so the exciton resonances are not resolved in InSb at 77 K; see Fig. 7.3-10. The excitonic enhancement still causes large deviations from the square-root dependence of $\alpha(\omega)$ close to the band edge for small N_C. Since real bound e-h pairs never exist, the free-carrier concentrations can in the stationary case be expressed directly in terms of the laser intensity via a simple rate equation [Koch, Schmitt-Rink, and Haug (1981a)]. The resulting relation is shown in Fig. 7.3-11a; saturation effects are clearly seen. The changes of the refractive index and absorption at a fixed frequency with intensity are shown in Fig. 7.3-11b. The calculated and measured index changes are compared in Fig. 7.3-12, showing less agreement than the phenomenological theory of Fig. 7.3-4b.

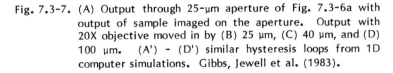

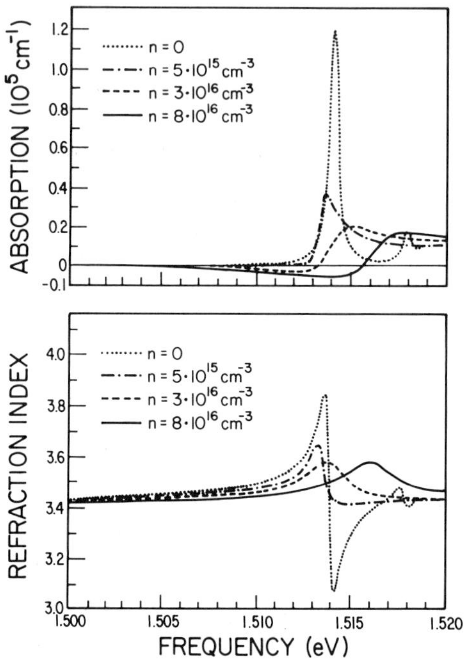

Fig. 7.3-8. Calculated spectra of absorption and refraction for GaAs at various free-carrier densities N_C, labelled as n in the figure. The temperature of the electronic excitation is assumed to be 10 K. Löwenau, Schmitt-Rink, and Haug (1982).

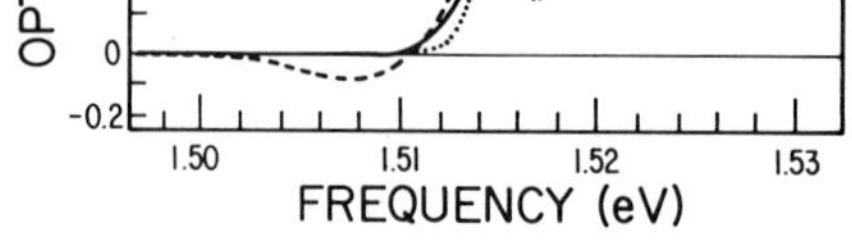

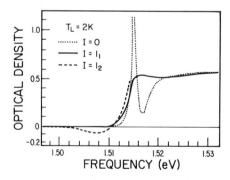

Fig. 7.3-9. Experimental absorption spectra for GaAs at various excitation intensities I. The curves with $I = 0, I_1$ are taken from Shah, Leheny, and Wiegmann (1977), the curve with $I = I_2$ is taken from Hildebrand, Faltenmeier, and Pilkuhn (1976). Löwenau, Schmitt-Rink, and Haug (1982).

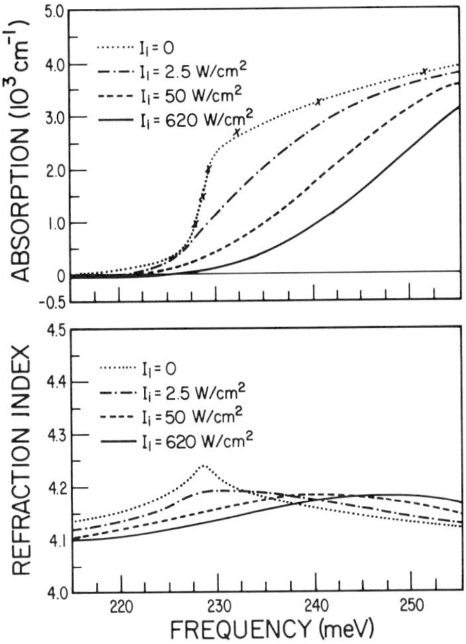

Fig. 7.3-10. Calculated spectra of absorption and refraction for InSb at various internal excitation intensities I_i. The frequency of the pump beam was assumed to be $\omega_0 = 225$ meV. The electronic temperature is T = 77 K. The experimental points of the low-intensity absorption spectrum are taken from Gobeli and Fan (1960). Löwenau, Schmitt-Rink, and Haug (1982).

Steyn-Ross and Gardiner (1983) treat the quantum theory of excitonic optical bistability in the cases of high and low exciton densities, finding qualitative agreement between the high-density calculations and the observed GaAs bistability (Section 3.6). Goll and Haken (1983) also have a many-body theory of optical bistability.

7.3.5. Electron-Hole Plasma Nonlinear Refraction ($Hg_{1-x}Cd_xTe$)

Free-carrier linear optical properties can be summarized in the simplest case of free carriers of effective mass m_c by the standard expression for the complex dielectric constant ε, i.e.,

$$\varepsilon = 1 + \frac{4\pi N_c e^2}{m_c} \frac{1}{-\omega^2 - i\Gamma\omega} . \qquad (7.3\text{-}18)$$

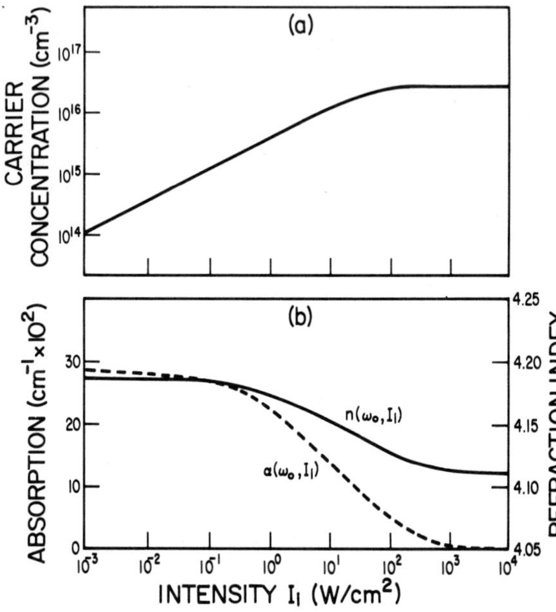

Fig. 7.3-11. (a) Carrier density; (b) absorption $\alpha(\omega_0,I_i)$ and refraction $n(\omega_0,I_i)$ at ω_0 = 225 meV vs the internal excitation intensity I_i for InSb at T = 77K. Löwenau, Schmitt-Rink, and Haug (1982).

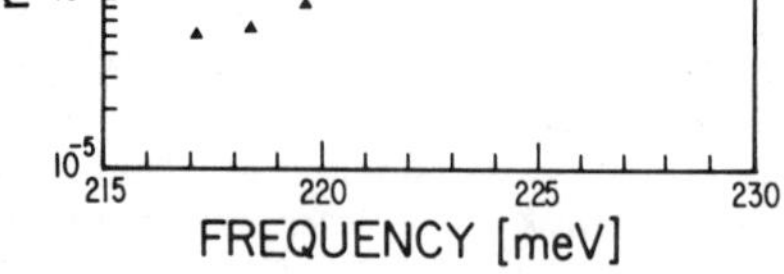

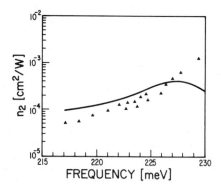

Fig. 7.3-12. Comparison of the calculated changes of the averaged refraction spectrum $-dn/dI$ for intensities I between 30 and 50 W/cm^2 with the measured coefficient n_2, defined by $n(\omega) = n_0(\omega) - n_2(\omega)I$, for InSb at 77 K according to D. A. B. Miller, Seaton, Price, and Smith (1981). Löwenau, Schmitt-Rink, and Haug (1982).

For $\omega^2 \gg \Gamma^2$

$$\delta n = \frac{2\pi e^2 N_C}{n_0 \omega^2 m_C} . \tag{7.3-19}$$

The average number of carriers excited, N_C, is given by

$$N_C = \frac{I}{\hbar\omega} \alpha_{eff} \tau_M \tag{7.3-20}$$

giving

$$n_2 \simeq \frac{\delta n}{I} = \frac{2\pi e^2 \alpha_{eff} \tau_M}{\hbar\omega^3 m_C n_0} = \frac{16\pi^2 \times 10^{10}}{n_0^2 c} \chi_D^{(3)} \tag{7.3-21}$$

in centimeters squared per kilowatt, where $\chi_D^{(3)}$ is the third-order susceptibility in this Drude model.

Elci and Rogovin (1981) perform a density matrix calculation to provide a more precise picture of the frequency dependence of $\chi^{(3)}$, finding

$$\mathrm{Re}\,\chi^{(3)} \simeq \frac{\hbar\omega}{2} \frac{\chi_D^{(3)}}{n_0^4[(\hbar\omega - E_G)^2 + \Gamma]^{1/2}} \tag{7.3-22}$$

where $\Gamma = \Gamma_C = \Gamma_V = \hbar/\tau_M$ and τ_M is the lifetime for electrons and holes.

Jain and Steel (1980) measured a $\chi^{(3)}$ of 5.4×10^{-6} esu at 10.6 μm for their sample and conditions from which they predicted a $\chi^{(3)}$ of over 10^{-2} esu for $Hg_{0.77}Cd_{0.23}Te$ at 180 K. Khan, Bennet, and Kruse (1981) report $\chi^{(3)} \simeq 3 \times 10^{-2}$ esu for $Hg_{0.784}Cd_{0.216}Te$ at 77 K, compared with 8.4×10^{-3} esu calculated from Eq. (7.3-21). For saturation of the HgCdTe $\chi^{(3)}$ by band filling see Yuen and Becla (1983).

7.4. OPTICAL COMPUTING

Goodman (1982, 1984a) has summarized the field of optical data processing as a strong tree with deep roots and new growth. Since the early 1960's the family tree has developed several branches: synthetic aperture radar; pattern recognition using matched filters and using diffraction pattern sampling; acousto-optic signal processing using both space-integrating and time-integrating correlators; discrete optical processing (time-sequential or parallel matrix-vector or matrix-matrix product or systolic filtering); numerical optical processing; and optical interconnections. The field of optical signal processing is a vast one and no attempt will be made to summarize it here. See the Special Issue on Optical Computing of the Proceedings of the IEEE [Caulfield, Horvitz, Tricoles, and Von Winkle (1984] and Alexander (1984), Caulfield, Neff, and Rhodes (1983), Yu (1983), Caulfield (1983), Lee (1981a), Casasent (1978), Cathey (1974).

Goodman (1984b) lists three ways light may be able to penetrate computer technology. The first is through analog optical computing, which is already well established in certain specialized applications: forming images from side-looking radar data, Bragg-cell spectrum analyzers, and a movement afoot toward utilizing Bragg-cell systems to perform computations with digital accuracy. The second is through providing interconnections at various levels in the hierarchy of digital hardware. Machine-to-machine optical fiber interconnects are a reality already and board-to-board and intra-board applications seem certain to follow. At the chip level, serious interconnection problems already exist, but fiber optics seems less attractive than integrated optics or perhaps free-space propagation using holographic optical elements [Goodman, Leonberger, Kung, and Athale (1984)]. "Finally, and undoubtedly furthest from realization, there is the strong possibility that bistable optical devices may someday provide the means for constructing an all-optical computer." [Gibbs, McCall, and Venkatesan (1980); Abraham, Seaton, and Smith (1983); Lohmann (1983); Sawchuck and Strand (1984); Robinson (1984)]. "Such a computer would also make full use of optics as a means for interconnections. However, for such a machine to have maximum possible impact, it is likely that the von Neumann architecture would have to be replaced by other structures having greater inherent parallelism, designed to match the inherent parallelism of optics." [Huang (1984, 1983, 1980)].

What role can optical bistable devices, defined broadly as the family of nonlinear optical devices described in this book, play in optical computing? Practically all of the optical computing to date has employed linear optics. Optics has been used to perform transforms and to form images, but decisions have been made by filters prepared in advance or by an array of detectors and an electronic computer. Input and output have often been the bottlenecks. The important feature of nonlinear devices is that they can make logical decisions [Section 1.2, Sawchuck and Strand (1984), and Lee (1976, 1981b)]. They may be able to determine the correlation between a known and an unknown, greatly reducing the output data. Bistable arrays may be "set" slowly by an electronic processor and then the entire array of information transferred at light speed to another processor or to a logic array. Nonlinear holographic thin-film gratings could be controlled optically to dynamically determine where arrays of optical data are sent. Integrated nonlinear optical switches could be opened or closed for dynamic "rewiring" of a chip. Perhaps a semiconductor etalon can serve as an all-optical or hybrid spatial light modulator [Kingston, Burke, Nichols, and Leonberger (1982); Chemla, Damen, Miller, Gossard, and Wiegmann (1983); D. A. B. Miller, Chemla, Damen, Gossard, Wiegmann, Wood, and Burrus (1984); Efron, Braatz, Little, Schwartz, and Grinberg (1983); Ross, Psaltis, and Anderson (1983)].

We will make no attempt to discuss changes in computer architecture that will be needed to take full advantage of the parallel nature of optical systems. We do note the increasing use of parallelism and dedicated subsystems in electronic computers and choose to interpret that as an evolution toward the massive parallelism possible with light [Waldrop (1984)]. We will make no attempt to summarize the experiments demonstrating some parallel logic operations at $\simeq$ 0.1s time scales using a television camera feeding back electrical signals to the television screen it is recording [Ferrano and Hausler (1980a,b); Lohmann (1977)] or optical feedback to a liquid crystal light valve [Sengupta, Gerlach, and Collins, Jr. (1978); Gerlach, Sengupta, and Collins, Jr. (1980); Chavel, Sawchuck, Strand, Tanguay, Jr., and Soffer (1980); Fatehi, Wasmundt, and Collins, Jr.,(1981, 1984); Athale and Lee (1979a,b; 1981); Efron, Soffer, and Caulfield (1983)]; see the reviews by Collins,

Jr. and Wasmundt (1980), Lee (1981b), and Sawchuck and Strand (1984). We do agree that such studies are increasingly appropriate and should be accelerated in response to recent improvements in nonlinear devices [Robinson (1984)].

Some of the most exciting potential applications are to pattern recognition, artificial intelligence, and associative optical storage where massive parallelism and interconnections would utilize natural advantages of optics. Perhaps optical computers will make their greatest contribution in forming such human-like limited-accuracy real-time decisions. Demonstrations of optical information processing based on the Hopfield (1982) model of neural networks are an especially exciting development [Psaltis and Farhat (1984)].

The study of the physics of bistable systems has been an exciting one; it remains to be seen just how important bistable devices become in optical communications and signal processing.

APPENDIX A

DIFFERENTIAL GAIN WITHOUT POPULATION INVERSION

In Section 2.2 it was shown that a Fabry-Perot cavity containing a nonlinear absorber exhibits bistability for $C > 4$ and that $C = 4$ corresponds to $dY/dX = 0$ or $dI_T/dI_I = \infty$, i.e., infinite ac gain. Consequently, as α_0 is increased from small values, one observes larger and larger differential gain until bistability is reached. It is an interesting aside to see that ac gain can occur, even without an optical cavity, when two-level atoms are subjected to intense coherent radiation.

The sign reversal of the absorption coefficient of a two-level system subjected to powerful monochromatic radiation was predicted to occur without the necessity of population inversion by Rautian and Sobel'man (1962), Mollow (1973), Haroche and Hartmann (1972), and McCall (1974). This ac gain was demonstrated beautifully using an optically pumped rf transition in Cd by Aleksandrov, Bonch-Bruevich, Khodovoi, and Chigir (1973). The effect has been shown in the optical region, first implicitly in the Na optical transistor experiment by Gibbs, McCall, and Venkatesan (1976) and then directly in the Na gain measurements using a weak probe beam at a different frequency by Wu, Ezekiel, Ducloy, and Mollow (1977).

The steady-state solutions of w_0 and v_0, Eq. (2.4-7), as a function of applied field, are shown in Fig. A-1. Note that for all input field strengths, w_0 is always negative, i.e., population inversion never occurs. Also v_0 is always positive, so absorption from the incident field always occurs [Eq. (2.4-4a)]. However, a small modulation on ξ can experience gain:

$$\xi = \xi_0 + \Delta\xi \qquad \text{(A-1)}$$

where $\Delta\xi$ varies slowly compared with $\kappa\xi_0$, γ_L, and γ_T. For then, with $F = \kappa\xi_0(\gamma_L\gamma_T)^{-1/2}$,

$$\frac{\partial\xi_0}{\partial z} + \frac{n_0}{c}\frac{\partial\xi_0}{\partial t} = -\frac{\alpha_0\xi_0}{2}\frac{1}{1+F^2}, \qquad \text{(A-2)}$$

i.e., ξ_0 experiences absorption for all values of F; but

$$\frac{\partial(\Delta\xi)}{\partial z} + \frac{n_0}{c}\frac{\partial(\Delta\xi)}{\partial t} = -\frac{\alpha_0\Delta\xi}{2}\frac{1-F^2}{(1+F^2)^2}. \qquad \text{(A-3)}$$

So for $F > 1$ there is differential field gain in steady state:

$$G \equiv \frac{1}{\Delta\xi}\frac{\partial(\Delta\xi)}{\partial z} = \frac{\alpha_0}{2}\frac{F^2-1}{(1+F^2)^2}, \qquad \text{(A-4)}$$

where

$$\alpha_0 = \frac{8\pi\omega N p^2}{N_0 \hbar c \gamma_T}, \tag{A-5}$$

found from $\partial\xi/\partial z = -\alpha_0\xi/2$ as $\xi \to 0$. The gain depends upon the relaxation times, the light intensity, and the frequency ν of the perturbation (not included in this derivation).

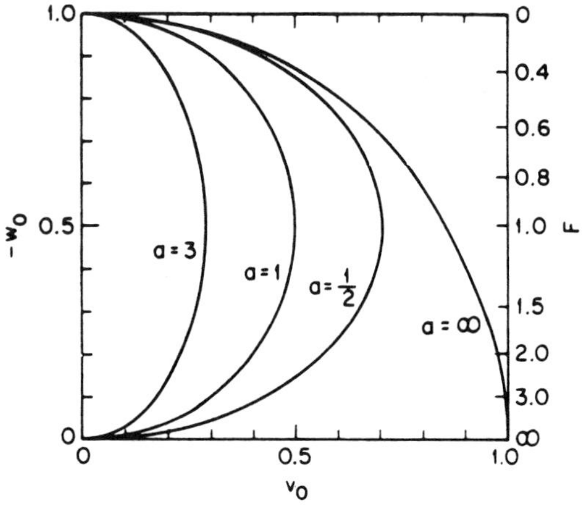

Fig. A-1. The $w < 0$, $v > 0$ quadrant of the Bloch circle $w^2 + v^2 = 1$. The circle represents the motion of an undamped Bloch vector subject to an electromagnetic field that is pulsed with temporal width $\ll T_2$. The ellipses represent the steady-state solutions given by Eq. (2.4-7). Note that a differential increase in F, for $F > 1$, leads to a decrease in v_0, and consequently to instability and differential gain. $a = \gamma_T/\gamma_L$. McCall (1974).

Often optical-absorption lines are inhomogeneously broadened, perhaps owing to a range of Doppler velocity shifts or a variation in static local crystalline-field potentials. In such a case, one may not correctly speak of a single homogeneously broadened absorption line, but instead must include a parameter $\Delta\omega$, denoting the difference between the frequency ω_a of a given homogeneously broadened isochromat and the applied frequency ω, in Bloch's equations. After solving Bloch's equations, the macroscopic polarization envelope is found by summing the contribution from each isochromat. Bloch's equations look the same except for the introduction of a complex field (see Appendix C):

$$\dot{\bar{u}} = +\Delta\omega\bar{v} - w\kappa E_i - \gamma_T\bar{u} \tag{A-6}$$

$$\dot{\bar{v}} = -\Delta\omega\bar{u} - w\kappa E_r - \gamma_T\bar{v} \tag{A-7}$$

$$\dot{w} = \overline{v}\kappa E_r + \overline{u}\kappa E_i - \gamma_L(w + 1); \tag{A-8}$$

and Maxwell's equation is given by Eq. (2.1-8). With $E = E_0 + \Delta E_r$, $\Delta E_i = 0$, and

$$\Delta E_r = \text{Re}\, \Delta\xi\, e^{i\nu t}. \tag{A-9}$$

McCall has calculated the real part of the gain defined by

$$\Delta v(\Delta\omega) = -\gamma_T^{-1}\, \text{Re}\, G(\Delta\omega, \nu)\Delta\xi\, e^{i\nu t} \tag{A-10}$$

as a function of the probe modulation frequency ν for various inhomogeneous broadening (assuming $g(\Delta\omega) \propto \exp[-T^2(\Delta\omega/\gamma_T)^2]$ and relaxation ($a = \gamma_T/\gamma_L$): see Fig. A-2. Gain decreases with inhomogeneous broadening (smaller T) and as ν approaches γ_L. The case T = 0, which never has a zero-frequency instability, does develop regions near $\nu = \kappa E_0$ of gain or instability for relatively large values of F. See Section 2.4.3 for effects of inhomogeneous broadening on bistability. McCall (1974) also investigated transverse mode and degeneracy effects on gain. He also obtained a bistability curve including standing wave effects by expanding the polarization in powers of e^{2ikz} (see Section 2.5.2).

McCall calculated gain for modulation of an on-resonance field; alternatively, one can calculate gain for a probe beam of arbitrary detuning from the driving field and the atomic resonance, e.g., Mollow (1973); see Section 6.2.

The presence of ac field gain does not ensure ac intensity gain:

$$\Delta I_T = [\xi_0 \exp(-\alpha_\xi L) + \Delta\xi \exp(\alpha_{\Delta\xi} L)]^2 - [\xi_0 \exp(-\alpha_\xi L)]^2$$

$$= 2\xi_0 \exp(-\alpha_\xi L)\Delta\xi \exp(\alpha_{\Delta\xi} L)$$

with $\alpha_\xi \equiv \alpha_0 [2(1 + F^2)]^{-1}$ and $\alpha_{\Delta\xi} \equiv \alpha_0(F^2 - 1)[2(1 + F^2)^2]^{-1}$ from Eqs. (A-2 and 3)

$$\Delta I_I \simeq (\xi_0 + \Delta\xi)^2 - \xi_0^2 \simeq 2\xi_0\Delta\xi$$

$$\frac{\Delta I_T}{\Delta I_I} \simeq \exp[(\alpha_{\Delta\xi} - \alpha_\xi)L] \simeq \exp[-\alpha_0 L/(1 + F^2)^2],$$

which is less than unity. Placement of the nonlinear atoms within a Fabry-Perot cavity does ensure ac intensity gain under appropriate conditions.

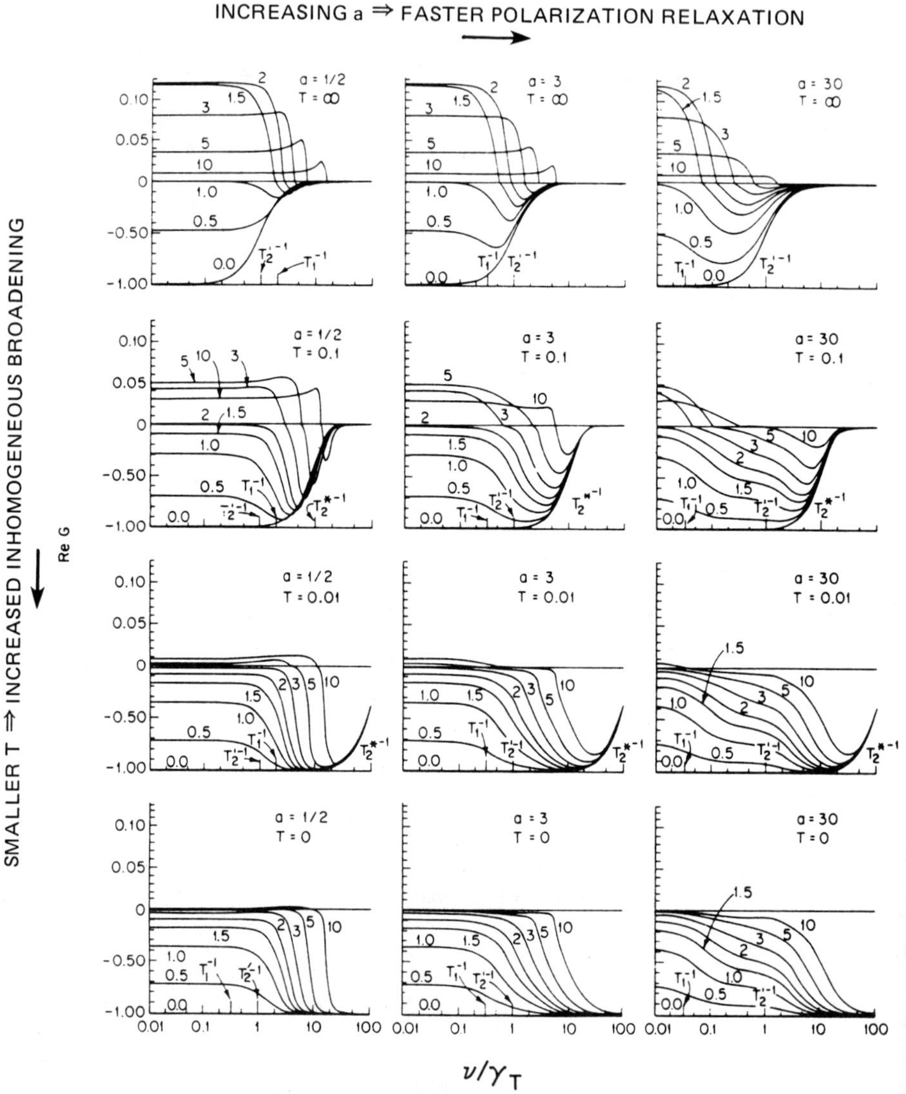

Fig. A-2. Re G as a function of ν and F for various ratios of $a = \gamma_T/\gamma_L$ and various inhomogeneous width parameters T. The numbers 0.0, 0.5, 1.0, ..., 10 represent the value of F for each curve. Note the scale change for Re G > 0. $\gamma_L = T_1^{-1}$; $\gamma_T = T_2^{-1}$. McCall (1974).

APPENDIX B

FABRY-PEROT BOUNDARY CONDITIONS

At the output of the Fabry-Perot cavity at $z = L$ (see Fig. 2.2-1) a choice of phase is available, so we choose boundary conditions

$$E_F(L,t) = \frac{1}{\sqrt{T}} E_T(L,t); \quad E_B(L,t) = \sqrt{R}\, E_F(L,t), \tag{B-1}$$

where $\mathcal{E}_j = E_j\, e^{i(\omega t \pm kz)}$ + c.c. and "+" applies for j = I,F,T and "-" for j = R,B.

At the entrance mirror, phases are important, and we may note that the boundary conditions consist of a matrix M relating the outgoing fields E_R and E_F with the incoming fields E_I and E_B. The mirrors are assumed to be linear, so the matrix relation is linear. Since both mirrors are lossless, the matrix is unitary. Furthermore, if, for example, $E_B = 0$, then the reflectivity R specifies an absolute value of one of the matrix elements. We may therefore write, where the $\sqrt{T}$ and $\sqrt{R}$ magnitudes are obvious from examining each division,

$$\begin{pmatrix} E_F(0) \\ E_R \end{pmatrix} = \begin{pmatrix} \sqrt{T}\, e^{i\alpha} & \sqrt{R}\, e^{i\beta} \\ -\sqrt{R}\, e^{i\gamma} & \sqrt{T}\, e^{i\delta} \end{pmatrix} \begin{pmatrix} E_I \\ E_B(0) \end{pmatrix} \tag{B-2}$$

at $z = 0$, where α, β, γ, and δ are real. The minus sign prefacing one element is anticipatory. Unitarity demands $M^{-1} = M^{\dagger}$, so $\alpha + \delta = \beta + \gamma (\text{mod} 2\pi)$. We now assume that the intensity-independent part of the real part of the refractive index of the intracavity medium has effects all included in k in the $e^{i(\omega t \pm kz)}$ factor. Then in steady state, with all time derivatives zero, E_F and E_B are independent of z. The choice $\alpha = \beta = \gamma = \delta = 0$ then yields $E_I = E_T$, i.e. the mirrors are adjusted for 100% transmission in the "empty" cavity case at frequency ω. The mirrors are then an integral number of one-half wavelengths apart.

If the entrance mirror is now moved a distance less than one-half wavelength, the diagonal elements will not change, since α, β, γ, δ are clearly independent of R, and in the limit $R \to 0$, the mirror movement changes nothing physical. We therefore set $\alpha = \delta = 0$. For us, consequently, the most general entrance mirror boundary condition is

$$\begin{pmatrix} E_F(0) \\ E_R \end{pmatrix} = \begin{pmatrix} \sqrt{T} & \sqrt{R} e^{i\beta} \\ -\sqrt{R}\, e^{-i\beta} & \sqrt{T} \end{pmatrix} \begin{pmatrix} E_I \\ E_B(0) \end{pmatrix}, \tag{B-3}$$

where β is called the detuning parameter.

One may ask how four parameters, α, β, γ, δ with one constraint $\alpha + \delta = \beta + \gamma$ (mod 2π) ended up as one parameter. Implicit in the condition $\alpha = \beta = \gamma = \delta = 0$ are two conventions regarding the phase of E_R and E_I. If we had chosen other conditions, allowing 100% transmission, we would have still found $|E_I| = |E_T|$. Furthermore, the definition of the phase of E_R is not important here.

In general, one has an optical cavity with boundary conditions. The plane-parallel Fabry-Perot cavity is only a special case used for illustrative purposes. This derivation was taken from McCall and Gibbs (1982a).

APPENDIX C

MAXWELL-BLOCH EQUATIONS

The Maxwell equations for a medium without free charges, currents, or magnetization are

$$\boldsymbol{D} = n_0^2\boldsymbol{E} + 4\pi\boldsymbol{P}, \qquad \text{(C-1)}$$

$$\nabla \cdot (\boldsymbol{E} + 4\pi\boldsymbol{P}) = 0, \qquad \text{(C-2)}$$

$$\nabla \times \boldsymbol{E} = -\frac{1}{c}\frac{\partial \boldsymbol{B}}{\partial t}, \qquad \text{(C-3)}$$

$$\nabla \times \boldsymbol{B} = \frac{1}{c}\frac{\partial \boldsymbol{D}}{\partial t}, \qquad \text{(C-4)}$$

in Gaussian units. For a discussion of Maxwell's equations and a comparison of units see Jackson (1962). The background refractive index is n_0. Using vector identities:

$$\nabla \times (\nabla \times \boldsymbol{E}) = -\frac{1}{c}\frac{\partial}{\partial t}\frac{1}{c}\frac{\partial}{\partial t}(n_0^2\boldsymbol{E} + 4\pi\boldsymbol{P}) \qquad \text{(C-5)}$$

$$= \nabla(\nabla \cdot \boldsymbol{E}) - \nabla^2\boldsymbol{E} \qquad \text{(C-6)}$$

$$= -4\pi\nabla(\nabla \cdot \boldsymbol{P}) - \nabla^2\boldsymbol{E} \qquad \text{(C-7)}$$

$$= -\frac{n_0^2}{c^2}\frac{\partial^2\boldsymbol{E}}{\partial t^2} - \frac{4\pi}{c^2}\frac{\partial^2\boldsymbol{P}}{\partial t^2}, \qquad \text{(C-8)}$$

one obtains the second-order wave equation

$$\left[\nabla^2 - \frac{n_0^2}{c^2}\frac{\partial^2}{\partial t^2}\right]\boldsymbol{E} = \frac{4\pi}{c^2}\frac{\partial^2\boldsymbol{P}}{\partial t^2} \qquad \text{(C-9)}$$

assuming $\nabla(\nabla\cdot\boldsymbol{P})$ is zero, i.e., the system is spatially homogeneous.

The second-order wave equation reduces to two first-order equations if the slowly varying envelope approximation (SVEA) is appropriate, namely that the electric field can be written as (with $k \equiv n_0\omega/c$):

$$E(z,t) = \xi(z,t)\, e^{i[\omega t - kz - \phi(z,t)]} + \text{c.c.}$$

$$= 2\xi(z,t)\cos[\omega t - kz - \phi(z,t)] \qquad \text{(C-10)}$$

where ξ and ϕ vary insignificantly in an optical wavelength or period:

$$\frac{\partial \xi}{\partial z} \lambda \ll \xi; \quad \frac{\partial \xi}{\partial t} \ll \omega \xi. \tag{C-11}$$

Denoting $\partial\xi/\partial z$ by ξ', $\partial\xi/\partial t$ by $\dot{\xi}$, and $[\omega t - kz - \phi(z,t)]$ by Ω,

$$E' = [\xi' - i\xi(k + \phi')]\, e^{i\Omega} + \text{c.c.} \tag{C-12}$$

$$E'' = [\xi'' - 2i\xi'(k + \phi') - \xi(k + \phi')^2 - i\xi\phi'']\, e^{i\Omega} + \text{c.c.} \tag{C-13}$$

$$\dot{E} = [\dot{\xi} + i(\omega - \dot{\phi})\xi]\, e^{i\Omega} + \text{c.c.} \tag{C-14}$$

$$\ddot{E} = [\ddot{\xi} + 2i\dot{\xi}(\omega - \dot{\phi}) - \xi(\omega - \dot{\phi})^2 - i\xi\ddot{\phi}]\, e^{i\Omega} + \text{c.c.} \tag{C-15}$$

Consistent with the SVEA, the terms containing ξ'', ϕ'', $\ddot{\xi}$; or $\ddot{\phi}$ are dropped. Substituting into Eq. (C-9) the definition (with N the density of two-level atoms with dipole moment p)

$$P(z,t) = Np[u(z,t) - iv(z,t)]\, e^{i\Omega} + \text{c.c.} \tag{C-16}$$

$$= 2Np(u \cos\Omega + v \sin\Omega)$$

and dropping $\ddot{v}$, $\Omega\dot{v}$, $\ddot{u}$, and $\Omega\dot{u}$ terms compared with $\Omega^2 v$ and $\Omega^2 u$, one finds

$$\xi' + \frac{n_0 \dot{\xi}}{c} = \frac{-2\pi\omega N p v}{n_0 c}, \tag{C-17}$$

and

$$\xi\left[\phi' + \frac{n_0 \dot{\phi}}{c}\right] = \frac{2\pi\omega N p u}{n_0 c}. \tag{C-18}$$

From Eq. (C-16) it is clear that u is the component of the polarization in phase with E and v is the component lagging by 90°.

Alternatively E can be written as

$$E(z,t) = (E_r + iE_i)\exp[i(\omega t - kz)] + \text{c.c.} \tag{C-19}$$

$$= 2E_r \cos(\omega t - kz) - 2E_i \sin(\omega t - kz)$$

and

$$P(z,t) = Np[\bar{u}(z,t) - i\bar{v}(z,t)]\exp[i(\omega t - kz)] + \text{c.c.}$$

$$= 2Np[\bar{u}\cos(\omega t - kz) + \bar{v}\sin(\omega t - kz)].$$

Similarly,

$$\mathcal{E}'' = [(E_r'' + iE_i'') - 2ik(E_r' + iE_i') - k^2(E_r + iE_i)] \exp[i(\omega t - kz)] + c.c. \quad \text{(C-20)}$$

$$\ddot{\mathcal{E}} = \left[(\ddot{E}_r + i\ddot{E}_i) + 2i\omega(\dot{E}_r + i\dot{E}_i) - \omega^2(E_r + iE_i)\right] \exp[i(\omega t - kz)] + c.c. \quad \text{(C-21)}$$

$$E_r' + \frac{n_0\dot{E}_r}{c} = -\frac{2\pi\omega Np\bar{v}}{n_0 c} \quad \text{(C-22)}$$

$$E_i' + \frac{n_0\dot{E}_i}{c} = -\frac{2\pi\omega Np\bar{u}}{n_0 c} \quad \text{(C-23)}$$

where $\bar{u} = u\cos\phi - v\sin\phi$ and $\bar{v} = v\cos\phi + u\sin\phi$ and u, v, and ϕ are the variables used in Eqs. (C-17) and (C-18). Equations (C-22) and C-23) could have been obtained directly from Eqs. (C-17) and (C-18). With

$$E \equiv E_r + iE_i \quad \text{(C-24)}$$

$$E' + \frac{n_0\dot{E}}{c} = \frac{-2\pi\omega Np(\bar{v} + i\bar{u})}{n_0 c} \quad \text{(C-25)}$$

$$E' + \frac{n_0\dot{E}}{c} = -\frac{i2\pi\omega Np(\bar{u} - i\bar{v})}{n_0 c}. \quad \text{(C-26)}$$

Equations (C-17) and (C-18) or, equivalently, Eqs. (C-22) and (C-23) describe the alteration of the slowly varying amplitude of the electromagnetic wave as a result of its interaction with the material system described by u and v. Transverse variations were dropped by replacing ∇^2 by $\partial^2/\partial z^2$; they can be restored, assuming the problem has cylindrical symmetry, by adding

$$\frac{\lambda}{4\pi n_0 r_p^2}\left[\xi\nabla_T^2\phi + 2\frac{\partial\phi}{\partial\rho}\frac{\partial\xi}{\partial\rho}\right] \quad \text{(C-27)}$$

to the left side of Eq. (C-17) and

$$-\frac{\lambda}{4\pi n_0 r_p^2}\left[\nabla_T^2\xi - \xi\left[\frac{\partial\phi}{\partial\rho}\right]^2\right] \quad \text{(C-28)}$$

to the left side of Eq. (C-18) where

$$\nabla_T^2 \equiv \frac{\partial^2}{\partial \rho^2} + \frac{1}{\rho}\frac{\partial}{\partial \rho}. \tag{C-29}$$

The radial parameter ρ can be normalized (i.e., $\rho = r/r_p$) to the half-width at half-maximum radius r_p of the laser beam or to any other convenient radial scale factor.

The equations of motion for the material system interacting with a coherent light beam are found from Schroedinger's equation of motion

$$i\hbar\,\dot{\psi}(\mathbf{r},t) = \mathcal{H}(\mathbf{r},\mathbf{p})\,\psi(\mathbf{r},t) \tag{C-30}$$

where ψ is the wave function. The total Hamiltonian, $\mathcal{H}$, is the sum of the unperturbed Hamiltonian $\mathcal{H}_0$ and the electric dipole interaction energy

$$\mathcal{V} = -e\mathbf{r} \cdot \mathbf{E}. \tag{C-31}$$

For an atom with two levels a and b with energies $(\hbar\omega_a/2)$ and $(-\hbar\omega_a/2)$, $\mathcal{H}_0\psi_a = (+\hbar\omega_a/2)\,\psi_a$ and $\mathcal{H}_0\psi_b = (-\hbar\omega_a/2)\,\psi_b$.

The wave function can be written

$$\psi(r,t) = c_a \exp\left[\frac{-i\omega_a t}{2}\right] u_a(\mathbf{r}) + c_b \exp\left[\frac{+i\omega_a t}{2}\right] u_b(\mathbf{r}). \tag{C-32}$$

If $\boldsymbol{E} = \hat{x}2\xi\cos\omega t$ then with $\kappa \equiv 2p/\hbar$

$$\dot{c}_a = i\kappa\xi\cos\omega t\,\exp(i\omega_a t)\,c_b(t) \tag{C-33}$$

$$\dot{c}_b = i\kappa\xi\cos\omega t\,\exp(-i\omega_a t)\,c_a(t) \tag{C-34}$$

$$p \equiv e\int_{-\infty}^{+\infty} d^3r\,u_a{}^*(\mathbf{r})\;x\;u_b(\mathbf{r}). \tag{C-35}$$

The evolution of the system can be described compactly with the aid of the density matrix

$$\rho = \begin{pmatrix} \rho_{aa} & \rho_{ab} \\ \rho_{ba} & \rho_{bb} \end{pmatrix} = |\psi\rangle\langle\psi|$$

$$= \begin{pmatrix} c_a\exp(-i\omega_a t/2) \\ c_b\exp(+i\omega_a t/2) \end{pmatrix} \begin{pmatrix} c_a^*\exp(+i\omega_a t/2) & c_b^*\exp(-i\omega_a t/2) \end{pmatrix}$$

$$= \begin{pmatrix} |c_a|^2 & c_a c_b^*\exp(-i\omega_a t) \\ c_a^* c_b\exp(i\omega_a t) & |c_b|^2 \end{pmatrix} \tag{C-36}$$

$$\dot{\rho}_{aa} = c_a\dot{c}_a^* + \dot{c}_a c_a^*$$
$$= c_a(-i\kappa\xi \cos\omega t \exp(-i\omega_a t))c_b^*$$
$$+ (i\kappa\xi \cos\omega t \exp(i\omega_a t))c_b c_a^*$$
$$= i\kappa\xi \cos\omega t(\rho_{ba} - \rho_{ab}) \quad \text{(C-37)}$$

$$\dot{\rho}_{ab} = \dot{c}_a c_b^* \exp(-i\omega_a t) + c_a\dot{c}_b^* \exp(-i\omega_a t) - i\omega_a c_a c_b^* \exp(-i\omega_a t)$$
$$= i\kappa\xi \cos\omega t(\rho_{bb} - \rho_{aa}) - i\omega_a\rho_{ab} \,. \quad \text{(C-38)}$$

Notice that Eqs. (C-37) and (C-38) contain terms that oscillate very rapidly at the optical frequency ω. One can remove these rapid oscillations by transforming to a rotating frame:

$$\rho_{ab} = \tilde{\rho}_{ab}\, e^{-i\omega t} \quad \text{(C-39a)}$$

$$\rho_{ba} = \tilde{\rho}_{ba}\, e^{+i\omega t} \quad \text{(C-39b)}$$

$$\rho_{aa} = \tilde{\rho}_{aa}, \quad \rho_{bb} = \tilde{\rho}_{bb} \,. \quad \text{(C-40)}$$

Then

$$\dot{\rho}_{aa} = (i\kappa\xi/2)\,(e^{i\omega t} + e^{-i\omega t})(\tilde{\rho}_{ba}\, e^{i\omega t} - \tilde{\rho}_{ab}\, e^{-i\omega t})$$
$$\simeq \frac{i\kappa\xi}{2}\,(\tilde{\rho}_{ba} - \tilde{\rho}_{ab}) \quad \text{(C-41)}$$

$$\dot{\tilde{\rho}}_{ab} = -i(\omega_a - \omega)\tilde{\rho}_{ab} + \frac{i\kappa\xi}{2}\,(e^{2i\omega t} + 1)(\rho_{bb} - \rho_{aa})$$
$$\simeq -i(\omega_a - \omega)\tilde{\rho}_{ab} + \frac{i\kappa\xi}{2}\,(\rho_{bb} - \rho_{aa})\,, \quad \text{(C-42)}$$

where the neglect of the rapidly oscillating terms $e^{\pm 2i\omega t}$ is called the rotating wave approximation. It is unnecessary if the light is circularly polarized, for then there is no counter-rotating component in the rotating frame. Introducing the phenomenological decay rate $\gamma_L = 1/T_1$ for the population difference $\rho_{aa} - \rho_{bb}$ and $\gamma_T = 1/T_2$ for the polarization components ρ_{ab} and ρ_{ba}:

$$\dot{\tilde{\rho}}_{ab} = -i(\omega_a - \omega)\tilde{\rho}_{ab} + (i\kappa\xi/2)(\rho_{bb} - \rho_{aa}) - \gamma_T\tilde{\rho}_{ab} \quad \text{(C-43)}$$

$$\dot{\rho}_{aa} - \dot{\rho}_{bb} = (i\kappa\xi)(\tilde{\rho}_{ba} - \tilde{\rho}_{ab}) - \gamma_L[(\rho_{aa} - \rho_{bb}) - (\rho_{aa}(\infty) - \rho_{bb}(\infty))].$$

(C-44)

The Bloch equations are obtained by setting

$$u = \tilde{\rho}_{ab} + \tilde{\rho}_{ba} = \langle\sigma_x\rangle = \mathrm{Tr}(\tilde{\rho}\sigma_x) \quad \text{(C-45)}$$

$$v = i(\tilde{\rho}_{ba} - \tilde{\rho}_{ab}) = -\langle\sigma_y\rangle \quad \text{(C-46)}$$

$$w = \rho_{aa} - \rho_{bb} = \langle\sigma_z\rangle\,, \quad \text{(C-47)}$$

where the Pauli spin matrices are

$$\sigma_x = \begin{pmatrix} 0 & 1 \\ 1 & 0 \end{pmatrix}, \quad \sigma_y = \begin{pmatrix} 0 & -i \\ i & 0 \end{pmatrix}, \text{ and } \sigma = \begin{pmatrix} 1 & 0 \\ 0 & -1 \end{pmatrix}. \quad \text{(C-48)}$$

Then Eqs. (C-43) and (C-44) become

$$\dot{u} = v\Delta\omega - \gamma_T u \quad \text{(C-49)}$$

$$\dot{v} = -u\Delta\omega - \kappa\xi w - \gamma_T v \quad \text{(C-50)}$$

$$\dot{w} = \kappa\xi v - \gamma_L[w - w(\infty)]\,, \quad \text{(C-51)}$$

where $\Delta\omega = \omega_a - \omega$ and $w(\infty) \to -1(\rho_{aa} = 0, \rho_{bb} = 1)$ for $\xi = 0$. In setting $E = 2\xi\cos\omega t$, no allowance was made for phase variations ϕ. If $E = 2\xi\cos(\omega t - \phi)$ then $\Delta\omega$ in Eqs. (C-49) and (C-50) goes over to $\Delta\omega + \dot{\phi}$.

If one would like to consider a superposition of $\hat{x}$ and $\hat{y}$ polarizations, the relation between the x dipole moment p_x and the y dipole moment p_y must be obtained. In spherical coordinates $x = r\sin\theta\cos\phi$ and $y = r\sin\theta\sin\phi$ so the relative moments can be obtained by concentrating on the ϕ integration. Let $u_a \propto e^{i\phi}$ and $u_b \propto 1$. Then

$$\int_0^{2\pi} (u_a{}^* x u_b)\,d\phi \propto \int_0^{2\pi} e^{-i\phi}\,\frac{(e^{i\phi} + e^{-i\phi})}{2}\,d\phi = \pi \quad \text{(C-52)}$$

$$\int_0^{2\pi} (u_a{}^* y u_b)\,d\phi \propto \int_0^{2\pi} e^{-i\phi}\,\frac{(e^{i\phi} - e^{-i\phi})}{2i}\,d\phi = -i\pi \quad \text{(C-53)}$$

$$\int_0^{2\pi} (u_b{}^* x u_a)\,d\phi \propto \pi \quad \text{(C-54)}$$

$$\int_0^{2\pi} (u_b{}^* y u_a)\,d\phi \propto +i\pi. \quad \text{(C-55)}$$

So the matrix for **p** = er is

$$\begin{pmatrix} 0 & p(\hat{x} - i\hat{y}) \\ p(\hat{x} + i\hat{y}) & 0 \end{pmatrix} = p(\hat{x}\sigma_x + \hat{y}\sigma_y) . \tag{C-56}$$

$$-e\mathbf{r} \cdot \mathbf{E} = -p(E_x\sigma_x + E_y\sigma_y) = -p \begin{pmatrix} 0 & E_x - iE_y \\ E_x + iE_y & 0 \end{pmatrix} . \tag{C-57}$$

In matrix notation

$$\mathcal{H}_0 = \frac{\hbar\omega_a}{2}\begin{pmatrix} 1 & 0 \\ 0 & -1 \end{pmatrix} = \frac{\hbar\omega_a}{2}\sigma_z , \tag{C-58}$$

and the evolution equation is

$$i\hbar\dot{\rho}(t) = [\mathcal{H}, \rho(t)]. \tag{C-59}$$

Transformation to the rotating frame is by

$$T = e^{+i\sigma_z[\omega t - kz - \phi]/2} = e^{+i\sigma_z\Omega/2} = \begin{pmatrix} e^{+i\Omega/2} & 0 \\ 0 & e^{-i\Omega/2} \end{pmatrix} . \tag{C-60}$$

$$\tilde{\mathcal{H}} = T\mathcal{H}T^{-1}; \quad \tilde{\rho} = T\rho T^{-1}. \tag{C-61}$$

Then $\dot{\tilde{\rho}} = \frac{\partial T}{\partial t}\rho T^{-1} + T\dot{\rho}T^{-1} + T\rho\frac{\partial T}{\partial t} = i\,\frac{(\omega - \dot{\phi})}{2}\,[\sigma_z\tilde{\rho} - \tilde{\rho}\sigma_z] + T\dot{\rho}T^{-1} =$ $\frac{i(\omega - \dot{\phi})}{2}\,[\sigma_z\tilde{\rho} - \tilde{\rho}\sigma_z] + i\hbar[\tilde{\mathcal{H}}, \tilde{\rho}]$. One can define a Hamiltonian so that

$$i\hbar\dot{\tilde{\rho}} = [\tilde{\tilde{\mathcal{H}}}, \tilde{\rho}]; \tag{C-62}$$

i.e. $\tilde{\tilde{\mathcal{H}}} = \tilde{\mathcal{H}} - \hbar(\omega - \dot{\phi})\sigma_z/2$ where

$$\tilde{\tilde{\mathcal{H}}} = \begin{pmatrix} \hbar(\Delta\omega + \dot{\phi})/2 & -p(E_x - iE_y)e^{+i\Omega} \\ -p(E_x + iE_y)e^{-i\Omega} & -\hbar(\Delta\omega + \dot{\phi})/2 \end{pmatrix} . \tag{C-63}$$

If $E_y = 0$ and $E_x = 2\xi\cos\Omega$, then the rotating wave approximation leads to Eq. (C-43) and (C-44). If the light is circularly polarized such that $\boldsymbol{E} = \xi(\hat{x}\cos\Omega + \hat{y}\sin\Omega)$, then

$$\tilde{\tilde{\mathcal{H}}} = \begin{pmatrix} \hbar(\Delta\omega + \dot{\phi})/2 & -p\xi \\ -p\xi & -\hbar(\Delta\omega + \dot{\phi})/2 \end{pmatrix} \tag{C-64}$$

and Eqs. (C-43) and (C-44) result without any approximation. On the other hand a transformation $T' = e^{-i\sigma_z\Omega/2}$ is needed for $\boldsymbol{E} = \xi(\hat{x}\cos\Omega - \hat{y}\sin\Omega)$ to be stationary in the rotating frame, but then one needs $\omega_a \rightarrow -\omega_a$ for $\Delta\omega$ to appear since $\tilde{\mathscr{H}} = \mathscr{H} + \hbar(\omega - \dot{\phi})\sigma_z/2$.

Regardless of the polarization of $\boldsymbol{E}$ one could choose to keep track of two amplitudes

$$\boldsymbol{E} = (E_r + iE_i) \frac{\boldsymbol{E}}{|\boldsymbol{E}|} e^{i(\omega t - kz)} + \text{c.c.} \tag{C-65}$$

rather than one amplitude and a phase

$$\boldsymbol{E} = \xi \frac{\boldsymbol{E}}{|\boldsymbol{E}|} e^{i(\omega t - kz - \phi)} + \text{c.c.} . \tag{C-66}$$

If $\boldsymbol{E}/|\boldsymbol{E}| = \hat{x}$ and $T = e^{i\sigma_z(\omega t - kz)/2}$, then Eq. (C-63) becomes, with $E = E_r + iE_i$ and using Eq. (C-65),

$$\tilde{\mathscr{H}} = \begin{pmatrix} \hbar\Delta\omega/2 & -p[Ee^{i(\omega t - kz)} + E^* e^{-i(\omega t - kz)}]\, e^{i(\omega t - kz)} \\ -p[Ee^{i(\omega t - kz)} + E^* e^{-i(\omega t - kz)}] e^{-i(\omega t - kz)} & -\hbar\Delta\omega/2 \end{pmatrix}$$

$$= \begin{pmatrix} \hbar\Delta\omega/2 & -pE^* \\ -pE & -\hbar\Delta\omega/2 \end{pmatrix} \tag{C-67}$$

in the rotating wave approximation. Then

$$\dot{\bar{u}} = \bar{v}\Delta\omega - \kappa E_i w - u\gamma_T , \tag{C-68}$$

$$\dot{\bar{v}} = -\bar{u}\Delta\omega - \kappa E_r w - v\gamma_T , \tag{C-69}$$

$$\dot{w} = \kappa E_r \bar{v} + \kappa E_i \bar{u} - (w+1)\gamma_L . \tag{C-70}$$

The dipole moment and spontaneous lifetime τ_0 are related by

$$\frac{1}{\tau_0} = \frac{4}{3} \frac{\omega^3(\sqrt{2}p)^2}{\hbar c^3} . \tag{C-71}$$

Consider a three-level system in which the only relaxation is spontaneous emission from upper level a to ground level b and intermediate level c (Slusher and Gibbs (1972)). Then

$$\gamma_L = \frac{1}{T_1} = \frac{1}{2\tau_{ac}} + \frac{1}{\tau_{ab}} \tag{C-72}$$

and

$$\gamma_T = \frac{1}{T_2} = \frac{1}{2\tau_{ac}} + \frac{1}{2\tau_{ab}} \tag{C-73}$$

and Eq. (C-70) is modified to

$$\dot{w} = \kappa E_r \bar{v} + \kappa E_i \bar{u} - \gamma_L (w + x) \tag{C-74}$$

$$\dot{x} = -(2\gamma_T - \gamma_L)(w + x) \tag{C-75}$$

where

$$x = \rho_{aa} + \rho_{bb} = \langle I \rangle \tag{C-76}$$

is the population in the levels a and b coupled by the external coherent field.

The Maxwell-Bloch equations as discussed here are semiclassical in that operator products such as $\xi(t)v(t)$ have consistently been factored in every expectation value:

$$\langle \xi(t)v(t) \rangle = \langle \xi(t) \rangle \langle v(t) \rangle .$$

Then $\xi \equiv \langle \xi(t) \rangle$ is interpreted to be a purely classical electric field. Using this factorization of operator products ignores quantum correlations; see Allen and Eberly (1975). For most bistability purposes this approximation is well justified and a great simplification. Fluctuation phenomena involving a small number of absorbers or of photons require the full quantized field treatment. So does a complete treatment of the spectra of the transmitted and fluorescence light.

For discussions of Maxwell-Bloch equations see Allen and Eberly (1975), Bloch (1946), Brewer (1977), Feynman, Vernon and Hellwarth (1957), McCall and Hahn (1969), Sargent (1977), Slusher and Gibbs (1972), Yariv (1975).

APPENDIX D

FABRY-PEROT CAVITY OPTIMIZATION WITH LINEAR ABSORPTION AND NONLINEAR REFRACTIVE INDEX

In Section 2.4.3 it was shown that far off-resonance from a saturable two-level transition the nonlinear absorption can be negligible when the nonlinear refraction is adequate for bistability. In many practical situations, even though the nonlinear absorption is negligible, there is often a linear background absorption characterized by α_B. D. A. B. Miller (1981a) has given an insightful treatment of this case.

Assume an intensity reflectivity R_F for the input mirror at $z = 0$ and R_B for the output mirror at $z = L$. Then the boundary conditions

$$E_F(0) = (1 - R_F)^{1/2} E_I + (R_F)^{1/2} E_B(0) \qquad \text{(D-1)}$$

$$E_B(0) = E_F(L)\,(R_B)^{1/2} \exp(i\beta/2 - \alpha_B L/2) \qquad \text{(D-2)}$$

$$E_F(L) = E_F(0) \exp(i\beta/2 - \alpha_B L/2) = E_T/\sqrt{T_B} \qquad \text{(D-3)}$$

give

$$\mathscr{T} \equiv \frac{I_T}{I_I} = \frac{(1 - R_B)(1 - R_F)(1 - A)}{(1 - R_\alpha)^2} \, \frac{1}{1 + F \sin^2(\gamma I_{eff} - \delta)}\,, \qquad \text{(D-4)}$$

where $1\text{-}A = \exp(-\alpha_B L)$, $R_\alpha = (R_F R_B)^{1/2} \exp(-\alpha_B L)$, $F = 4R_\alpha/(1 - R_\alpha)^2 = 4\mathscr{F}^2/\pi^2$, and the round-trip phase shift β is given by

$$\frac{\beta}{2} = \gamma I_{eff} - \delta\,, \qquad \text{(D-5)}$$

where $-\delta = \beta_0/2$ of Eq. (2.3-6). Miller's calculation of I_{eff} parallels Marburger and Felber (1978). Assume a nonlinear polarization $n_2|E|^2 E$, e.g., for a Bloch resonance far off resonance (Section 2.4.3) or any material with an index approximated by $n = n_0 + n_2 I$; then

$$\frac{\partial^2 E}{\partial z^2} + k^2 E = \left[ik\alpha_B - \frac{4\pi\omega^2}{c^2} n_2 |E|^2\right] E\,. \qquad \text{(D-6)}$$

If

$$E = [\xi_F \exp(-i\phi_F) \exp(-ikz) + \xi_B \exp(-i\phi_B) \exp(ikz)] \exp(i\omega t)\,, \qquad \text{(D-7)}$$

then

$$\frac{\partial \phi_F}{\partial z} = \frac{2\pi\omega n_2}{n_0 c}[\xi_F^2 + 2\xi_B^2], \qquad \text{(D-8)}$$

$$\frac{\partial \phi_B}{\partial z} = -\frac{2\pi\omega n_2}{n_0 c}[\xi_B^2 + 2\xi_F^2], \qquad \text{(D-9)}$$

$$\frac{\partial \xi_F}{\partial z} = -\frac{\alpha_B}{2}\xi_F, \qquad \text{(D-10)}$$

$$\frac{\partial \xi_B}{\partial z} = +\frac{\alpha_B}{2}\xi_B\,. \qquad \text{(D-11)}$$

I_{eff} is then defined by

$$\phi_F - \phi_B = 2\gamma I_{eff} = \frac{6\pi\omega n_2}{n_0 c}\int_0^L [\xi_B^2(z) + \xi_F^2(z)]dz, \qquad \text{(D-12)}$$

with $\gamma = 12\pi^2\omega n_2 L/n_0^2 c^2$. Also

$$\xi_F^2(z) = \xi_F^2(0)\exp(-\alpha_B z), \qquad \text{(D-13)}$$

$$\xi_B^2(z) = \xi_B^2(0)\exp(+\alpha_B z)\,. \qquad \text{(D-14)}$$

Integrating Eq. (D-12) using Eqs. (D-13,14,2,3),

$$\mathscr{T} = \frac{\alpha_B L(1 - A)(1 - R_B)}{A(1 + R_{B\alpha})}\frac{I_{eff}}{I_I} \qquad \text{(D-15)}$$

with $n_0 c\xi_T^2/4\pi = I_T$ and $R_{B\alpha} = (1 - A)R_B$.

The transmission $\mathscr{T}$ must simultaneously satisfy Eqs. (D-4) and (D-15); the solution could be done graphically (Section 2.6). Miller's analytic solution can be pursued as follows. Let

$$\mathscr{T} = \frac{c_T}{U} \qquad \text{(D-16)}$$

with

$$U \equiv 1 + F\sin^2(\gamma I_{eff} - \delta). \qquad \text{(D-17)}$$

Then

$$\frac{\partial \mathscr{T}}{\partial I_{eff}} = \frac{\partial \mathscr{T}}{\partial U}\frac{\partial U}{\partial I_{eff}} = -\frac{c_T}{U^2}\frac{\partial U}{\partial I_{eff}}\,. \qquad \text{(D-18)}$$

The maximum slope occurs for

$$\frac{\partial^2 \mathcal{T}}{\partial I^2_{eff}} = 0 = \frac{2c_T}{U^3}\left[\frac{\partial U}{\partial I_{eff}}\right]^2 - \frac{c_T}{U^2}\frac{\partial^2 U}{\partial I^2_{eff}} \tag{D-19}$$

$$\frac{\partial U}{\partial I_{eff}} = 2F\gamma \sin(\gamma I_{eff} - \delta)\cos(\gamma I_{eff} - \delta) \tag{D-20}$$

$$\frac{\partial^2 U}{\partial I^2_{eff}} = 2F\gamma^2[\cos^2(\gamma I_{eff} - \delta) - \sin^2(\gamma I_{eff} - \delta)]. \tag{D-21}$$

Let $z = \sin^2(\gamma I_{eff} - \delta)$, then Eq. (D-19) implies

$$\left[\frac{2}{1 + Fz}\right] 4F^2\gamma^2 z(1 - z) = 2F\gamma^2(1 - 2z).$$

Then

$$2Fz^2 - (3F + 2)z + 1 = 0 \tag{D-22}$$

$$z = \sin^2(\gamma I_{eff} - \delta) = \frac{3F + 2 \pm ((3F + 2)^2 - 8F)^{1/2}}{4F}$$

$$= \frac{3F + 2 - ((F + 2)^2 + 8F^2)^{1/2}}{4F} \tag{D-23}$$

where the negative-sign choice keeps $z < 1$ and also $z \to 0$ as $F = 4\mathcal{F}^2/\pi^2 \to \infty$. Then, using a subscript "max" to denote the value for $(\partial\mathcal{T}/\partial I_{eff})$ being a maximum

$$U_{max} = 1 + Fz = \frac{[3(F + 2) - ((F + 2)^2 + 8F^2)^{1/2}]}{4} \tag{D-24}$$

$$\equiv \frac{G(F)}{4}$$

$$\left.\frac{\partial U}{\partial I_{eff}}\right|_{max} = 2F\gamma(z(1 - z))^{1/2}$$

$$= \frac{\gamma[(F + 2)((F + 2)^2 + 8F^2)^{1/2} - (F + 2)^2 - 2F^2]^{1/2}}{\sqrt{2}}$$

$$\equiv \frac{\gamma H(F)}{\sqrt{2}}. \tag{D-25}$$

Then

$$\left.\frac{\partial \mathcal{T}}{\partial I_{eff}}\right|_{max} = \frac{(1 - R_B)(1 - R_F)(1 - A)}{(1 - R_\alpha)^2}\frac{16\gamma}{\sqrt{2}}\frac{H(F)}{[G(F)]^2}. \tag{D-26}$$

Equating $(\partial\mathcal{T}/\partial I_{eff})_{max}$ to the gradient of the "straight line," Eq. (D-15) gives the expression for the critical input intensity for the onset of bistability I_C:

$$I_C = I_I\Big|_{Min} = \frac{\sqrt{2}}{16\pi\beta'}\,\frac{(1 - R_\alpha)^2}{A(1 + R_{B\alpha})(1 - R_F)}\,\frac{[G(F)]^2}{H(F)} \qquad \text{(D-27)}$$

where $\beta' \equiv \gamma/\pi\alpha_B L$, i.e., $I_C \propto \lambda\alpha_B/n_2$ as discussed in Section 7.1. Equating Eqs. (D-4 and 15) with $I_I = I_C$ and U given by Eq. (D-24) yields

$$\gamma I_{eff}\Big|_C = \frac{\sqrt{2}}{4}\,\frac{G(F)}{H(F)}\,. \qquad \text{(D-28)}$$

The critical value δ_C of the cavity mistuning δ is obtained from Eqs. (D-23,28):

$$\delta_C = \frac{\sqrt{2}}{4}\,\frac{G(F)}{H(F)} - \sin^{-1}\left[-\left[\frac{1}{4F}\left(3F + 2 - ((F + 2)^2 + 8F^2))^{1/2}\right)\right]^{1/2}\right]. \qquad \text{(D-29)}$$

The sine itself is always negative at the point of maximum positive slope on the periodic function, and thus the negative square root is taken in the $\sin^{-1}$ term. The quantity δ_C depends only on $F = 4\mathcal{F}^2/\pi^2$. From Eq. (2.1-28)

$$-\frac{\delta_C}{\pi} = \frac{\beta_0'}{2\pi} = \frac{\nu_C - \nu}{FSR}\,, \qquad \text{(D-30)}$$

where the free spectral range (FSR) for a Fabry-Perot cavity is $c/2n_0L$. The values of δ_C/π as a function of $\mathcal{F}$ are shown in Fig. D-1. Eq. (D-27) can be written as

$$I_C = \frac{1}{\beta'}\cdot\frac{1}{\mu} \qquad \text{(D-31)}$$

where β' is the material figure of merit in this plane-wave limit and μ is a cavity figure of merit. Then μ is factored into μ_0, the cavity figure of merit for equal mirror reflectivities, and ρ, which describes how μ is influenced by $R_B \neq R_F$:

$$\mu(R_\alpha, R_{B\alpha}, A) = \rho(R_\alpha, R_{B\alpha}, A)\mu_0(R_\alpha, A), \qquad \text{(D-32)}$$

$$\mu_0(R_\alpha, A) = \frac{16\pi}{\sqrt{2}}\,\frac{A(1 + R_\alpha)(1 - R_\alpha/(1 - A))}{(1 - R_\alpha)^2}\,\frac{H(F)}{[G(F)]^2} \qquad \text{(D-33)}$$

$$\rho(R_\alpha, R_{B\alpha}, A) = \frac{(1 + R_{B\alpha})[1 - R_\alpha^2/R_{B\alpha}(1 - A)]}{(1 + R_\alpha)[1 - R_\alpha/(1 - A)]} . \quad \text{(D-34)}$$

For a given R_α and A, ρ is maximum when $R_{B\alpha} = R_B \exp(-\alpha_B L)$ is maximum which occurs for $R_B = 1$:

$$\rho_{max} = 2 - \frac{A[1 - R_\alpha/(1 - A)]}{1 + R_\alpha} . \quad \text{(D-35)}$$

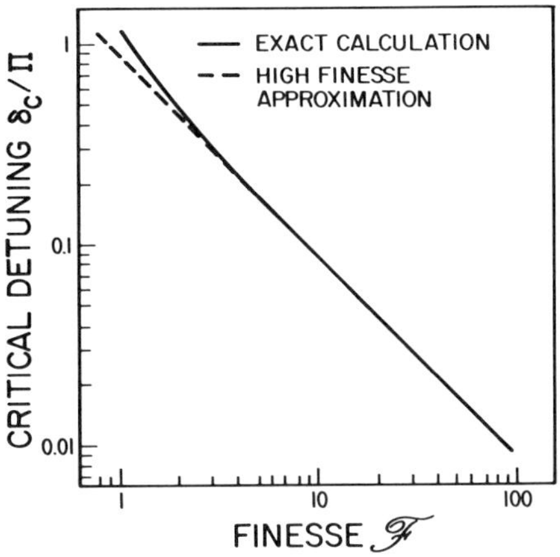

Fig. D-1. Critical detuning of the cavity for the onset of bistability as a function of the cavity finesse. D. A. B. Miller (1981a). © 1981 IEEE

The second term is always positive and usually small (for $R_\alpha \geq 0.5$, $\rho_{max} \geq 1.8$ for all A). Thus, the effect of altering the relative reflectivities of the mirrors can give a reduction in I_c for any given R_α and A of at most a factor of 2. In the limit $R_B = 1$, the cavity transmission vanishes and the reflectivity would be 100% were it not for absorption, which is then crucial to bistable operation.

The dependence of μ_0 upon the absorption per pass A is shown in Fig. D-2. Note that there is no single optimum cavity, but there is an optimum R for a given A. μ_0 can also be expressed as a function of finesse $\mathcal{F} = \pi\sqrt{R_\alpha}/(1 - R_\alpha)$ and $\mathcal{T}_{max} = (1 - R)^2(1 - A)/[1 - R(1 - A)]^2$ from Eq. (D-4) with $R_B = R_F = R$ and $\sin^2(\gamma I_{eff} - \delta) = 0$. Inverting, using $R_\alpha = (1 - A)R$

$$R = 1 + \frac{\pi^2 \mathcal{T}_{max}}{2\mathcal{F}^2} - \frac{\pi(\mathcal{T}_{max})^{1/2}}{\mathcal{F}} \left[1 + \frac{\pi^2 \mathcal{T}_{max}}{4\mathcal{F}^2}\right]^{1/2} , \quad \text{(D-36)}$$

$$A = 1 - \frac{1}{R}\left[1 + \frac{\pi^2}{2\mathcal{F}^2} - \frac{\pi}{\mathcal{F}}\left[1 + \frac{\pi^2}{4\mathcal{F}^2}\right]^{1/2}\right]. \quad \text{(D-37)}$$

Figure D-3 shows that if the optimum $\mathcal{T}_{max}$ is chosen for each finesse $\mathcal{F}$, then μ_0 increases monotonically with increasing $\mathcal{F}$. This is the usual dispersive bistability result: the higher the finesse the lower $I_c\beta'$ needs to be. There are constructional limitations in addition to α_B limitations. If one is limited to a certain $\mathcal{F}$ then the optimum reflectivity can be found by differentiating Eq. (D-32) with fixed R_α; yielding for the optimum value of the equal reflectivities

$$R_{opt} = 1 - A_{opt} = \sqrt{R_\alpha} = \left[1 + \left[\frac{\pi}{2\mathcal{F}}\right]^2\right]^{1/2} - \frac{\pi}{2\mathcal{F}} \quad \text{(D-38)}$$

or

$$(1 - R_{opt}) = 1 - \exp(-\alpha_B L),$$

i.e., for equal reflectivities the input power is minimized for a reflectivity such that the transmission per mirror equals the loss per one-way pass. Equation (D-38) can be used several different ways. If fabrication fixes $\mathcal{F}$, then select R by Eq. (D-38); A is then determined by L and $\alpha_B(\Delta\omega)$. Alternatively, obtaining enough phase shift without too much absorption may suggest a value for A which then determines R_{opt}. In practice, the dependences of α_B and n_2 upon $\Delta\omega$ need to be included in the optimization process as well.

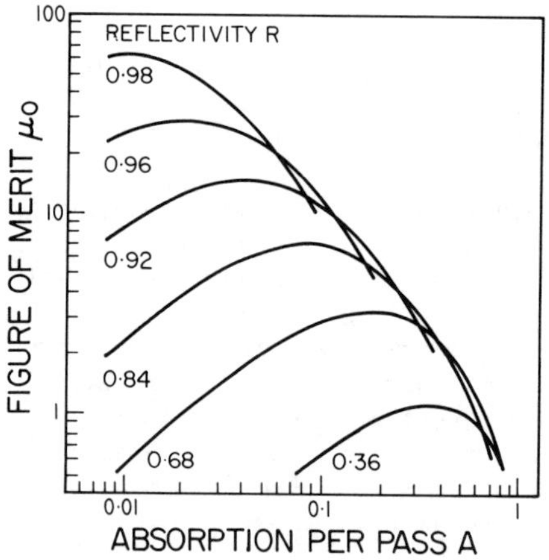

Fig. D-2. Cavity figure of merit μ_0 (equal mirror reflectivities) as a function of mirror reflectivity R and absorption per pass A = 1 - exp($-\alpha_B L$). D. A. B. Miller (1981a). © 1981 IEEE

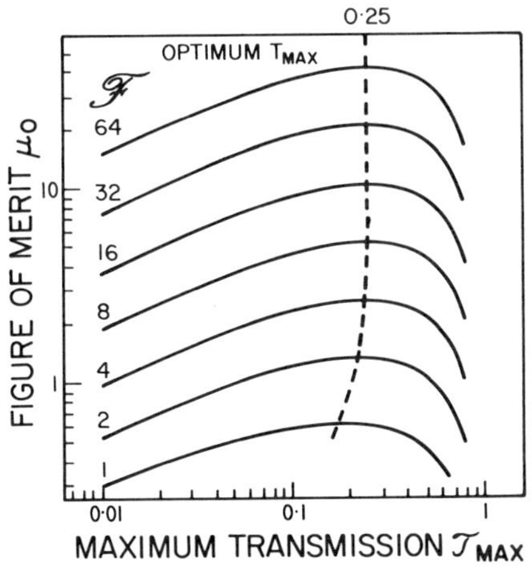

Fig. D-3. Cavity figure of merit μ_0 (equal mirror reflectivities) as a function of peak (fractional) transmission $\mathcal{T}_{max}$ and cavity finesse $\mathcal{F}$. D. A. B. Miller (1981a).
© 1981 IEEE

APPENDIX E

INSTABILITY OF NEGATIVE-SLOPE PORTION OF S-SHAPED CURVE OF I_T VERSUS I_I

Section 6.2 examines the stability of coupled Maxwell-Bloch equations; Eq. (6.2-26) for the eigenvalue of the lowest-order resonant mode shows that all the stationary states lying on the curve $x = x(y)$ with negative slope are unstable. Here the derivation of Goldstone (1984) shows the same for any bistable system with

$$\frac{I_T}{I_I} = \mathscr{T}(\beta) \tag{E-1}$$

and with a single characteristic response time for the control parameter β:

$$\tau \frac{d\beta}{dt} = -(\beta - \beta_0 - \beta_2 I_T) . \tag{E-2}$$

$$= \beta_2 I_I \mathscr{T}(\beta) - \beta + \beta_0 . \tag{E-3}$$

Choose $\beta_0 = 0$ for simplicity. As noted in Section 2.6, bistability requires that the slope of the straight line $\mathscr{T}(\beta) = \beta/(\beta_2 I_I)$ in Fig. 2.6-1 be less than the slope of Eq. (E-1) assuming $\beta_2 > 0$, i.e.,

$$\frac{\mathscr{T}(\beta)}{\beta} < \frac{d\mathscr{T}(\beta)}{d\beta} \quad \text{or} \quad \frac{1}{\beta_2 I_I}\left(1 - \frac{\beta}{\mathscr{T}}\mathscr{T}'\right) < 0 \tag{E-4}$$

where $\mathscr{T}' \equiv d\mathscr{T}/d\beta$ or equivalently

$$1 - \beta_2 I_I \mathscr{T}'(\beta) < 0 . \tag{E-5}$$

Now perturb the control parameter β from its equilibrium value $\overline{\beta}$:

$$\beta = \overline{\beta} + \delta\beta(t) . \tag{E-6}$$

Eq. (E-3) becomes

$$\tau \frac{d\delta\beta}{dt} + [1 - \beta_2 I_I \mathscr{T}'(\beta)]\, \delta\beta = 0 . \tag{E-7}$$

For negative-slope regions Eq. (E-5) applies, and Eq. (E-7) has exponentially growing solutions. Hence, the negative-slope regions are unstable. By identical arguments positive-slope regions are stable. For systems with two or more comparable response times (e.g. cavity round-trip time t_R, cavity decay time τ_C, and one or more material response time τ_M) this analysis is inadequate; see Sections 6.1 through 6.4.

APPENDIX F

QUANTUM POPULATION PULSATION APPROACH TO RESONANCE FLUORESCENCE AND OPTICAL BISTABILITY INSTABILITIES

F.1. INTRODUCTION

The discussion in Section 2.8 of the spectra from an optical bistability cavity treats the effects of cooperative feedback of the spontaneously emitting atoms on the transmitted light. The effects of stimulated emission and multiwave mixing on the spectrum are not considered. In this appendix a theoretical approach is outlined which allows these phenomena to be included. A more complete derivation and discussion of its relevance to optical bistability and other areas in quantum optics is given by Sargent III, Holm, and Zubairy (1985).

The theory is a multimode extension of the Scully and Lamb (1976) theory of the laser and derives the quantized-field version of the population pulsations. For simplicity, only the special case of two modes will be considered here. This allows for the derivation of the resonance fluorescence spectrum and the semiclassical gain/absorption coefficient of a weak field in the presence of a strong field. Hence we can demonstrate how spontaneous emission leads to a sidemode build-up resulting in a cavity instability. The derivation of the spectra of Section 2.8 uses the Heisenberg picture and the quantum regression theorem. This approach is in the Schroedinger picture and so the time dependence of the operators is contained in the density operator ρ. However, we shall use the atomic states of the undressed atom, that is, the eigenstates of the unperturbed Hamiltonian. Since the strong, near-resonant field Stark splits the atomic levels, the states we are using are no longer eigenstates, and hence they are not stationary states. This gives rise to pulsations in the atomic level populations, and hence the term population pulsations.

F.2. THEORY

We wish to describe atomic transitions in a two-level system in the presence of a strong field coupling the two states. Let the upper level be the state a and the lower level the state b. We will identify the field resulting from a spontaneous emission as mode 1 and the strong resonant field as mode 2. All other field modes have been traced over. Hence an atom-field state is described by these three indices. For example, the state $|a,n_1,n_2\rangle$ means that the atom is in the upper state, mode 1 has n_1 photons, and mode 2 has n_2 photons. Initially, of course, $n_1 = 0$, since spontaneous emission arises out of the vacuum. We will allow it to be a perfectly general integer to be able to treat the growth of a cavity sidemode, although we will always consider n_1 to be small such that mode 1 will not saturate. This contrasts with the situation for mode 2, which is a strong, saturating field. Because n_2 is large, we will frequently approximate $n_2+1 \simeq n_2$.

Consider the four atom-field states shown in Fig. F-1. The states are labeled numerically for typographical simplicity. The state $|5\rangle$, for example, is the state $|a,n_1,n_2\rangle$ described above. When a spontaneous emission occurs, an atom in the state $|5\rangle$ goes to the state $|1\rangle = |b,n_1+1,n_2\rangle$, i.e., the atom goes to the ground state and mode 1 acquires an additional photon. However, due to mode 2, the states $|5\rangle$ and $|1\rangle$ are coupled to the states $|2\rangle = |b,n_1,n_2+1\rangle$ and $|4\rangle = |a,n_1+1,n_2-1\rangle$ respectively. Transitions to these states describe the Rabi flopping of the atoms between the upper and lower states due to the strong mode 2. Because we are treating mode 1 to only second-order in the field, these are the only four basis states necessary to solve the two-mode problem.

Our Hamiltonian (in radians/second) is

$$\mathscr{H} = (\omega_a - \omega_2)\sigma_z + \sum_{j=1}^{2} [(\omega_j - \omega_2)a_j^\dagger a_j + (g a_j U_j \sigma^\dagger + \text{adjoint})] \tag{F-1}$$

In this expression a_j is the annihilation operator for the jth field mode, $U_j = U_j(\mathbf{r})$ is the corresponding spatial mode factor, σ and σ_z are the atomic spin-flip and probability-difference operators, ω_a and ω_j are the atomic and field frequencies, and g is the atom-field coupling constant. The rotating wave approximation has been made, and this Hamiltonian is in an interaction picture rotating at the strong field frequency ω_2. We wish to determine the transition rate of the atoms between the states $|5\rangle$ and $|1\rangle$ as a function of frequency, since this yields the resonance fluorescence spectrum. In order to do this, we need to derive the equation of motion for the probability P_{n_1} of having n_1 mode photons. This is given by the trace of the rate of change of the atom-field operator ρ over the atomic states:

$$\dot{\rho}_{n_1 n_2;n_1 n_2} = \dot{\rho}_{an_1 n_2;an_1 n_2} + \dot{\rho}_{bn_1 n_2;bn_1 n_2} \tag{F-2}$$

In the above equation $\rho_{an_1 n_2;an_1 n_2} = \langle an_1 n_2|\rho|an_1 n_2\rangle = \rho_{55}$, and so on. The equations of motion for the probabilities ρ_{55} and ρ_{11} are calculated by projecting the standard density operator equation of motion

$$\dot{\rho} = -i[\mathscr{H},\rho] + \text{relaxation terms} \tag{F-3}$$

onto the appropriate basis states of Fig. F-1:

$$\dot{\rho}_{55} = -\gamma_L \rho_{55} - [iV_{51}\rho_{15} + iV_{52}\rho_{25} + \text{c.c.}] \tag{F-4}$$

$$\dot{\rho}_{11} = +\gamma_L \rho_{55} + [iV_{51}\rho_{15} + iV_{41}\rho_{14} + \text{c.c.}] \tag{F-5}$$

where $V_{51} = gU_1\sqrt{n_1+1}$, $V_{52} = gU_2\sqrt{n_2+1}$, $V_{41} = gU_2\sqrt{n_2}$, and γ_L is the (longitudinal) decay rate of the upper level a to the lower level b. In the $\dot{\rho}_{11}$ equation the approximation of $n_2 \simeq n_2 + 1$ is made, as discussed above. This will be done uniformly throughout, so we will set $V_{41} = V_{52} = V_2$, and for simplicity we will set $V_{51} = V_1$. Substituting Eqs. (F-4,5) into Eq. (F-2), and letting $P_{n_1} = \rho_{n_1 n_2;n_1 n_2}$, we find

$$\dot{P}_{n_1} = \dot{\rho}_{55} + \dot{\rho}_{11}|_{n_1\to(n_1-1)}$$

$$= iV_1^*\rho_{51} - i(V_1^*\rho_{51})|_{n_1\to(n_1-1)} + \text{c.c.} \tag{F-6}$$

This result shows that the rate of change of the probability of the sidemode photon number is simply related to the matrix element connecting the states |5> and |1>, the states between which spontaneous emission occurs. Because ρ_{51} relates an upper state to a lower one, we may think of it as a dipole matrix element. The above equation is similar to the result of the semiclassical slowly varying amplitude and phase approximation that relates an optical field to an induced polarization.

To obtain the dipole element ρ_{51} we again use Eqs. (F-1,3) to determine its equation of motion as well as all of the matrix elements to which it is coupled. We find

$$\dot{\rho}_{51} = -(\gamma_T + i\Delta\omega_1)\rho_{51} + iV_1[\rho_{55}-\rho_{11}] + iV_2[\rho_{54} - \rho_{21}] \tag{F-7}$$

$$\dot{\rho}_{54} = -(\gamma_L + i\Delta\omega_{21})\rho_{54} - i[V_1\rho_{14} - \rho_{51}V_2^* + V_2\rho_{24}] \tag{F-8}$$

$$\dot{\rho}_{21} = \gamma_L\rho_{54} - i\Delta\omega_{21}\rho_{21} + i[\rho_{25}V_1 - V_2^*\rho_{51} + \rho_{24}V_2] \tag{F-9}$$

$$\dot{\rho}_{24} = -(\gamma_T - i\Delta\omega_3)\rho_{24} - iV_2^*[\rho_{54} - \rho_{21}] \tag{F-10}$$

where γ_T is the dipole (transverse) decay rate, $\Delta\omega_{21} = \omega_2 - \omega_1$, $\Delta\omega_n = \omega_a - \omega_n$ and $\omega_3 = 2\omega_2 - \omega_1$. We consider the field amplitudes to vary slowly compared to atomic lifetimes, which are on the order of γ_L^{-1}. This is analogous to the rate equation approximation in laser physics. Thus we solve Eqs. (F-7 to 10) in steady state to obtain ρ_{51} in terms of known quantities. To do this, we recall that the assumption of a weak sidemode field implies that V_1 can only appear to second order. This means that the matrix elements ρ_{55}, ρ_{11}, ρ_{41}, and ρ_{52}, which connect states of the same n_1 number, can be factored into the corresponding semiclassical value determined by the V_2 interaction alone multiplied by the probability of having n_1 or n_1+1 photons. Specifically,

$$\rho_{55} = \rho_{aa}P_{n_1} = \frac{I_2L_2/2}{1 + I_2L_2}P_{n_1} \tag{F-11}$$

$$\rho_{11} = \rho_{bb}P_{n_1+1} = \frac{1 + I_2L_2/2}{1 + I_2L_2}P_{n_1+1} \tag{F-12}$$

$$\rho_{52} = \rho_{ab}P_{n_1} = -\frac{iV_2D_2}{1 + I_2L_2}P_{n_1} = iV_2D_2d_{052} \tag{F-13}$$

$$\rho_{41} = \rho_{ab}P_{n_1+1} = -\frac{iV_2D_2}{1 + I_2L_2}P_{n_1+1} = iV_2D_2d_{041} \tag{F-14}$$

where ρ_{aa}, ρ_{bb}, and ρ_{ab} are the semiclassical atomic density matrix elements found in Chapter 2, $I_2 = 4|V_2|^2/\gamma_T\gamma_L$, D_2 is the ω_2 case of the complex Lorentzians

$$D_n = \frac{1}{\gamma + i(\omega_a - \omega_n)}$$

and

$$L_2 = \frac{\gamma^2}{\gamma^2 + (\omega_a - \omega_2)^2} .$$

Equations (F-7 to F-10) can then be solved for the four matrix elements ρ_{51}, ρ_{54}, ρ_{21}, and ρ_{24}, which connect states of different values of n_1.

We may eliminate the ρ_{54} term appearing in Eq. (F-9) by using the following relationship:

$$\rho_{0110} = \rho_{54} + \rho_{21} . \qquad \text{(F-15)}$$

Substituting this into Eq. (F-9) and solving in steady state,

$$\rho_{21} = \gamma_L D_a \rho_{0110} + iD_a[\rho_{25}V_1 - V_2^*\rho_{51} + \rho_{24}V_2] \qquad \text{(F-16)}$$

where $D_a = 1/(\gamma_L + i\Delta\omega_{21})$. Similarly, we set Eq. (F-8) equal to zero to find

$$\rho_{54} = -iD_a[V_1\rho_{14} - \rho_{51}V_2^* + V_2\rho_{24}] . \qquad \text{(F-17)}$$

The off-diagonal matrix element of the field density operator ρ_{0110} may be solved for in a manner similar to that for the $\dot{P}_{n_1}$ equation. We take the time derivative of Eq. (F-15) and substitute Eqs. (F-8,9):

$$\begin{aligned} \dot{\rho}_{0110} &= \dot{\rho}_{54} + \dot{\rho}_{21} \\ &= -i\Delta\omega_{21}\rho_{0110} - iV_1(\rho_{14} - \rho_{25}) . \end{aligned} \qquad \text{(F-18)}$$

Solving this equation in steady state we find

$$\rho_{0110} = -\frac{iV_1(\rho_{14} - \rho_{25})}{i\Delta\omega_{21}} = \frac{-iV_1V_2^*D_2^*(d_{041} - d_{052})}{i\Delta\omega_{21}} \qquad \text{(F-19)}$$

where the dipole elements ρ_{14} and ρ_{25} are given above. Here $\Delta\omega_{21}$ typically has a size on the order of the atomic decay constants which cause the atomic response to equilibrate rapidly in comparison with the field transients. As $\Delta\omega_{21} \to 0$, this approximation breaks down when the time $1/\Delta\omega_{21}$ is comparable to the times over which the field varies. For a steady state atom-field interaction, as we are considering here, this leads to the delta function spectrum associated with Rayleigh scattering as discussed later. Hence the off-diagonal matrix element ρ_{0110}, which describes the interference between the two modes, is directly connected to the elastic (Rayleigh) portion of the spectrum.

Finally, we solve Eqs. (F-7 and F-10) in steady state. This gives

$$\rho_{51} = iD_1(V_1 d_{051} + V_2 d_{154}) \qquad \text{(F-20)}$$

$$\rho_{24} = -iD_3^* V_2^* d_{154}, \qquad \text{(F-21)}$$

where $d_{051} = \rho_{55} - \rho_{11}$, and $d_{154} = \rho_{54} - \rho_{21}$. We substitute these dipole elements into the population pulsation equations (F-16 and F-17) to find

$$\rho_{54} = -D_a V_2^*[V_1 D_2^* d_{041} + D_1(V_1 d_{051} + V_2 d_{154}) + D_3^* V_2 d_{154}] \qquad \text{(F-22)}$$

$$\rho_{21} = \gamma_L D_a \rho_{0110} + D_a V_2^*[V_1 D_2^* d_{052} + D_1(V_1 d_{051} + V_2 d_{154}) + D_3^* V_2 d_{154}] \ . \qquad \text{(F-23)}$$

Subtracting Eq. (F-23) from (F-22), solving for d_{154}, and substituting Eq. (F-19) for ρ_{0110} we have

$$d_{154} = -\frac{V_1 V_2^*[2D_a d_{051} D_1 + D_a(1 - \gamma_L/i\Delta\omega_{21})d_{041}D_2^* + D_a(1 + \gamma_L/i\Delta\omega_{21})d_{052}D_2^*]}{1 + \frac{\gamma_T}{2} I_2 F(D_1 + D_3^*)} \qquad \text{(F-24)}$$

where the complex dimensionless population pulsation factor F is $\gamma_L/(\gamma_L + i\Delta\omega_{21})$. Substituting Eq. (F-24) into Eq. (F-20) we obtain

$$\rho_{51} = iD_1 V_1 \times \left[d_{051} - \frac{\gamma_T}{2} I_2 F \frac{\left[d_{051}D_1 + D_2^*/2[(1-\gamma_L/i\Delta\omega_{21})d_{041} + (1+\gamma_L/i\Delta\omega_{21})d_{052}]\right]}{1 + \frac{\gamma_T}{2} I_2 F(D_1 + D_3^*)} \right] . \qquad \text{(F-25)}$$

Substituting ρ_{51} into Eq. (F-6) we finally have

$$\dot{P}_{n_1} = -(n_1+1)[A_1 P_{n_1} - (B_1 + \omega/2Q_1)P_{n_1+1}]$$

$$+ n_1[A_1 P_{n_1-1} - (B_1 + \omega/2Q_1)P_{n_1}] + \text{c.c.} \qquad \text{(F-26)}$$

where the coefficients

$$A_1 = \frac{g^2 D_1}{1+I_2 L_2}\left[\frac{I_2 L_2}{2} - \frac{\gamma_T}{2} I_2 F \frac{\left[\frac{I_2 L_2}{2} D_1 - D_2^*(1+\gamma_L/i\Delta\omega_{21})\right]}{1 + \frac{\gamma_T}{2} I_2 F(D_1 + D_3^*)}\right] \qquad (F\text{-}27)$$

$$B_1 = \frac{g^2 D_1}{1+I_2 L_2}\left[1+\frac{I_2 L_2}{2} - \frac{\gamma_T}{2} I_2 F \frac{\left[(1+\frac{I_2 L_2}{2})D_1 + D_2^*(1-\gamma_L/i\Delta\omega_{21})\right]}{1 + \frac{\gamma_T}{2} I_2 F(D_1 + D_3^*)}\right] \qquad (F\text{-}28)$$

and where we include a cavity loss term of ν/Q_1 for the mode 1.

F.3. DISCUSSION

Equation (F-26) has a straightforward physical interpretation. Each term can be understood in terms of the probabilities that the atoms make transitions resulting in the emission or absorption of a mode 1 photon. For example,

$$(B_1 + B_1^* + \frac{\omega_a}{Q_1})(n_1+1)P_{n_1+1} =$$

(absorption rate from an n_1+1 photon field + cavity loss rate)
x (probability of n_1+1 photons).

The A_1 coefficient is proportional to the population of the upper level a and involves spontaneous and stimulated emissions, whereas the B_1 coefficient is proportional to the population of the lower level b and involves absorptions. We are primarily interested in the growth of mode 1, which can be described by the average photon number $\langle n_1\rangle = \Sigma n_1 P_{n_1}$. Using Eq. (F-26), we find the equation of motion of $\langle n_1\rangle$:

$$\frac{d}{dt}\langle n_1\rangle = (A_1 - B_1 - \omega_a/2Q_1)\langle n_1\rangle + A_1 + \text{c.c.} \quad . \qquad (F\text{-}29)$$

The quantity $\langle n_1\rangle$ is essentially the intensity in mode 1, and hence Eq. (F-29) can be considered to be a quantum mechanical version of Beer's law, where $A_1 - B_1$ represents the complex absorption coefficient and A_1 + c.c. the term resulting from spontaneous emission. In free-space no build-up of mode 1 occurs and the time derivative of $\langle n_1\rangle$ denotes the counting rate of a photon counter observing the resonance fluorescence spectrum A_1 + c.c. This result was first obtained by Mollow (1969). In the limit of $I_2 >> 1$, $\gamma_L = 2\gamma_T$, this yields the standard three peaked spectrum. Note that the term proportional to $\gamma_L/i\Delta\omega_{21}$ leads to the delta function spectrum of resonant Rayleigh scattering. As mentioned above, $A_1 - B_1$ yields the semiclassical complex absorption coefficient of a weak field in the presence of a strong field [Mollow (1972) and Sargent III (1978)]. Taking the difference between Eqs. (F-27) and (F-28), we have

$$A_1 - B_1 = \frac{g^2 D_1}{1+I_2 L_2}\left[1 - \frac{(\frac{\gamma_T}{2} I_2 F)[D_1 + D_2^*]}{1 + \frac{\gamma_T}{2} I_2 F(D_1 + D_3^*)}\right] . \tag{F-30}$$

In laser and optical bistability cavities, however, there is a build-up of $\langle n_1 \rangle$. This theory therefore demonstrates how laser and optical bistability instabilities grow from spontaneous emission. To obtain the spectrum in this case, we solve for the steady state value of $\langle n_1 \rangle$ using Eq. (F-29). This yields

$$\langle n_1 \rangle = \frac{A}{B - A + \omega_a/Q_1} \tag{F-31}$$

where $A = A_1 + A_1^*$ and $B = B_1 + B_1^*$. Because of the effect of stimulated emissions and absorptions by the A - B term in Eq. (F-31), the spectrum of the emergent radiation is altered when $A - B \simeq \omega_a/Q_1$ [Holm, Sargent, III, and Stenholm (1984)]. This result is independent of the cooperative emission effects on the spectrum discussed in Section 2.8. For the case when the directions of both mode 1 and mode 2 are collinear, however, the situation is even more complicated since this will give rise to a double sidemode emission. The theory for this case is presented by Sargent III, Holm, and Zubairy (1985), and some results are presented in Holm, Sargent III, and Stenholm (1984).

This appendix was written by D. A. Holm.

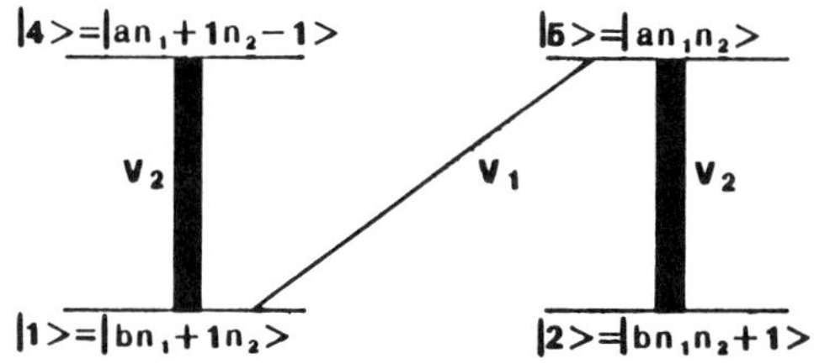

Fig. F-1. Four atom-field states for quantum population-pulsation treatment.

APPENDIX G

FAST-FOURIER-TRANSFORM SOLUTION OF TRANSVERSE EFFECTS

The propagation equation to be solved is Eq. (2.9-13):

$$\frac{\partial E(\vec{\rho},Z,\tau)}{\partial(Z/L_T)} = -\left[\frac{(\alpha_0 L_T/2)(1+i\Delta)}{(1+\Delta^2+|E|^2/E_s^2)} + \frac{i(\ln 2)}{4\pi F}\nabla_T^2\right] E(\vec{\rho},Z,\tau), \qquad \text{(G-1)}$$

where L_T is the cavity round-trip length. Formally integrating with respect to $\zeta \equiv Z/L_T$, one finds, assuming the cavity is filled with homogeneously broadened two-level atoms,

$$E(\vec{\rho},\zeta,\tau) = \exp\left[A\zeta + \int_0^\zeta B(\zeta')d\zeta'\right] E(\vec{\rho},0,\tau), \qquad \text{(G-2)}$$

where $A \equiv -i(\ln 2)\nabla_T^2/4\pi F$ and $B \equiv -(\alpha_0 L_T/2)(i + i\Delta)/(1+\Delta^2+|E|^2/E_s^2)$. This propagation equation is still exact, but it is complicated by the fact that the two terms appearing in square brackets are noncommuting operators. Also the integral over the medium term B is implicit.

In the numerical solution the field is propagated over small intervals in ζ (i.e., $\delta\zeta$). To evaluate the effect of the exact propagator on the field one would need to expand the exponential operator in an infinite series in ζ. Higher terms in the series would involve both operators occurring in multiple commutator brackets, and the algebra would rapidly get out of hand. Alternatively one can seek the best approximation to $\exp[A\zeta + \int_0^\zeta B(\zeta')d\zeta']$, i.e., a representation which matches its power series expansion to as high an order as possible in ζ but retaining its simplicity. The following approximation

$$\exp\left[A\zeta + \int_0^\zeta B(\zeta')d\zeta'\right] \simeq \exp\left[\frac{A\zeta}{2}\right] \exp\left[\int_0^\zeta B(\zeta')d\zeta'\right] \exp\left[\frac{A\zeta}{2}\right] \qquad \text{(G-3)}$$

is accurate to third order in ζ (if $B(\zeta)$ is sufficiently slowly varying). The gain-sheet approximation breaks the full $0 \rightarrow \zeta$ interval in subintervals $\delta\zeta$ and uses this approximation scheme to propagate the solution from sheet to sheet, i.e.,

$$E(\vec{\rho},\zeta_{n+1},\tau) = \exp\left[\frac{A\delta\zeta}{2}\right] \exp\left[\int_{\zeta_n}^{\zeta_{n+1}} B(\zeta')d\zeta'\right] \exp\left[\frac{A\delta\zeta}{2}\right] E(\vec{\rho},\zeta_n,\tau) \quad \text{(G-4)}$$

where $\delta\zeta = \zeta_{n+1} - \zeta_n$. Note both end points are unique in that free-space propagation is over $\delta\zeta/2$ and not over $\delta\zeta$ as with intermediate sheets. The implicit integral can be approximated by a simple quadrature (i.e., error order $\delta\zeta$) or by a higher order iteration. Larger $\delta\zeta$ steps can be taken if the integral is approximated by an iteration scheme of order $(\delta\zeta)^3$. See Lax, Batteh, and Agrawal (1981).

This appendix was written by J.V. Moloney.

APPENDIX H

CRITICAL EXPONENTS IN OPTICAL BISTABILITY TRANSIENTS

$I_T(t)$ is observed when I_I is switched from I_0 to I_1, and the switching time τ is measured as a function of the length L of the nonlinear Fabry-Perot cavity. Let L_C be the length for which τ becomes infinite (critical slowing down). In all the experimental situations (both one- and two-photon dispersive bistability in Rb) studied by Cribier, Giacobino, and Grynberg (1983b), the value of τ agreed with a law

$$\tau \sim (L - L_C)^{-\alpha}$$

where α was found to be 0.5 within experimental uncertainty.

The inverse-square-root law has been derived in Grynberg, Biraben, and Giacobino (1981), Grynberg and Cribier (1983), and Garmire, Marburger, Allen, and Winful (1979). The second reference shows that it corresponds to a general property of a system close to the critical point in the case of a single-beam optical bistability. The law is shown here, following Cribier, Giacobino, and Grynberg (1983b), for the simple case where the two time constants are very different. For example, the nonlinear medium time constant τ_M is often much longer than the cavity decay time τ_C. Assume the phase ϕ of the field evolves according to

$$\tau_M \frac{d\phi}{dt} = g(\phi, L, I_I) \,. \qquad \text{(H-1)}$$

The static solutions for ϕ are given by $g(\phi_0, L, I_0) = 0$ and $g(\phi_1, L, I_1) = 0$. When I_I changes from I_0 to I_1, the system evolves from ϕ_0 to ϕ_1 according to Eq. (H-1). The critical point (ϕ_C, L_C) corresponds to

$$g(\phi_C, L_C, I_0) = 0 \qquad \text{(H-2)}$$

and

$$g'_\phi(\phi_C, L_C, I_0) = 0 \,, \qquad \text{(H-3)}$$

where g'_ϕ denotes $\partial g/\partial\phi$, etc. Close to the critical point, Eq. (H-1) becomes

$$\tau_M \frac{d\phi}{dt} = g'_L (L - L_C) + g''_\phi (\phi - \phi_C)^2/2 \,. \qquad \text{(H-4)}$$

The analysis of this equation shows that the time taken to cross the critical point is of the order of [Grynberg and Cribier (1983)]:

$$\tau \sim \frac{\tau_M}{(g''_\phi \, g'_L)^{1/2}} (L - L_c)^{-1/2} . \tag{H-5}$$

This is a fair estimation of the switching time for points which are close to the critical point. In their experiments $\tau_c \simeq \tau_M \simeq 25$ ns but τ ranged from 2 to 30 μs, i.e., 100 to 1000 times longer than τ_c or τ_M; see Cribier, Giacobino, and Grynberg (1983a,b). Goldstone (1984) shows the inverse-square-root dependence by integrating Eq. (E-2).

For a linear or power-law time dependence for the incident light, Rozanov (1980b) finds a response time proportional to the square root of the rate of increase of the input intensity.

APPENDIX I

RELATIONSHIP BETWEEN n_2 and $\chi^{(3)}$

The relationship between the nonlinear refraction coefficient n_2 and the nonlinear susceptibility $\chi^{(3)}$ depends upon the system of units [Ovadia (1984)]. Begin with cgs units and define the nonlinear refraction coefficient as

$$n_{NL} = n_0 + n_2' \xi^2 , \qquad \text{(I-1)}$$

where ξ is the slowly-varying envelope of the following electric field

$$E(z,t) = \xi(z,t)\, e^{i[kz - \omega t - \phi(z,t)]} + \text{c.c.}$$

For the dielectric constant ε_{NL}, one has

$$\varepsilon_{NL} = n_{NL}^2 = n_0^2 + 2n_0 n_2' \xi^2 = 1 + 4\pi(\chi_L + \chi_{NL}) . \qquad \text{(I-2)}$$

However, $n_0^2 = 1 + 4\pi\chi_L$, and $\chi_{NL} \simeq \chi^{(3)} \langle E{\cdot}E\rangle = 2\chi^{(3)} \xi^2$. Thus, Eq. (I-2) reduces to

$$2n_0 n_0' \xi^2 = 8\pi\chi^{(3)} \xi^2$$

or

$$n_2' \text{ (esu)} = \frac{4\pi}{n_0} \chi^{(3)} \text{ (esu)} . \qquad \text{(I-3)}$$

In MKS units, the nonlinear index of refraction is

$$n_{NL} = n_0 + n_2 I . \qquad \text{(I-4)}$$

$$I(\text{MKS}) = \frac{1}{2} n_0 \varepsilon_0 c \langle E{\cdot}E\rangle = n_0 \varepsilon_0 c(\xi')^2 \qquad \text{(I-5)}$$

where $\varepsilon_0(\text{MKS}) = 36\pi \times 10^9$, ξ' is in volts/m, and I is in W/m² = (newton - m/s)m⁻² = 10³ (dyne-cm/s)cm⁻². The latter units are those of I(esu) given by

$$I(\text{esu}) = \frac{n_0 c}{4\pi} \xi^2 \qquad \text{(I-6)}$$

where ξ is in statvolts/cm and 3×10^4 volts/m = 1 statvolt/cm. The equivalence of Eqs. (I-5) and (I-6) can be shown by a careful substitution of conversion factors. Clearly from Eqs. (I-1) and (I-4),

$$n_2' \xi^2 = n_2 I ; \qquad \text{(I-7)}$$

therefore

$$n_2 = \frac{n_2' \xi^2}{I} = \frac{4\pi}{n_0}\left[\frac{\xi^2}{I}\right]\chi^{(3)}\text{ (esu) .}$$

If n_2 is in cm^2/kW then it is numerically almost equal to $\chi^{(3)}$(esu)--actually about 3 to 10 times larger for semiconductors:

$$I(kW/cm^2) = 10^{-7}\, I(W/m^2) = 10^{-7}\, n_0 \varepsilon_0 c(\xi')^2$$

$$= 10^{-7}\, n_0 \frac{1}{36\pi \times 10^9}(3 \times 10^8)(3 \times 10^4 \xi)^2 = 3n_0\xi^2/4\pi \text{ .} \qquad \text{(I-8)}$$

Then

$$n_2(cm^2/kW) = \frac{1}{3}\left[\frac{4\pi}{n_0}\right]^2 \chi^{(3)}\text{(esu) .} \qquad \text{(I-9)}$$

$$n_2(MKS) = \chi^{(3)}(MKS)/n_0\varepsilon_0 \text{ .}$$

REFERENCES

This list of references does not include all articles in Bowden, Gibbs, and McCall (1984) and Miller, Smith, and Wherrett (1984). An attempt has been made to make this list complete, but the discovery of missing articles every day is proof to the contrary. The author would appreciate copies of missing articles.

[2.5] denotes that the reference in question is cited in Section 2.5, whereas (2.5) denotes that the reference is relevant to Section 2.5 even though it is not cited there.

Abraham, E., and Bullough, R. K. (1980). "Absorptive and dispersive optical bistability," p. 245 in Laser Advances and Applications. Proceedings of the Fourth National Quantum Electronics Conference, B. S. Wherrett, ed. (Wiley, Edinburgh, Scotland).

Abraham, E., Bullough, R. K., and Hassan, S. S. (1979). "Space and time-dependent effects in optical bistability," Opt. Commun. **29**, 109. [2.5.2, 5.1]

Abraham, E., Bullough, R. K., and Hassan, S. S. (1980). "Optical bistability in a Fabry-Perot cavity," J. Opt. Soc. Am. **70**, 589. (2.5.2)

Abraham, E., and Firth, W. J. (1983). "Periodic oscillations and chaos in a Fabry-Perot cavity containing a nonlinearity of finite response time," Opt. Acta **30**, 1541. [6.3]

Abraham, E., Firth, W. J., and Carr, J. (1982). "Self-oscillation and chaos in nonlinear Fabry-Perot resonators with finite response time," Phys. Lett. A **91A**, 47. [6.3]

Abraham, E., and Hassan, S. S. (1980). "Effects of inhomogeneous broadening on optical bistability in a Fabry-Perot cavity," Opt. Commun. **35**, 291. [5.1]

Abraham, E., Hassan, S. S., and Bullough, R. K. (1980). "Dispersive optical bistability in a Fabry-Perot cavity," Opt. Commun. **33**, 93. (2.4.3, 2.5.2)

Abraham, E., Seaton, C. T., and Smith, S. D. (1983). "The optical computer," Sci. Am. **248**, 85. [7.4]

Abraham, E., and Smith, S. D. (1982a). "Optical bistability and related devices," Rep. Prog. Phys. **45**, 815. [1.2, 2.7]

Abraham, E., and Smith, S. D. (1982b). "Nonlinear Fabry-Perot interferometers," J. Phys. E **15**, 33. [1.2]

Abraham, N. B. (1983). "A new focus on laser instabilities and chaos," Laser Focus, p. 73, May. [6.3]

Abraham, N. B., Coleman, M. D., Maeda, M., and Wesson, J. C. (1982). "Single-mode instabilities in high-gain gas lasers," Appl. Phys. B **28**, 169. [6.4]

Abraham, N. B., Gollub, J. P., and Swinney, H. L. (1984). "Testing nonlinear dynamics," Physica **11D**, 252. [6.3]

Abraham, N. B., Lugiato, L. A., and Narducci, L. M., eds. (1985). Feature Issue on Instabilities in Active Optical Media, J. Opt. Soc. Am. B **2**(1). (6.3, 6.4)

Abram, I. (1983). "Nonlinear-optical properties of biexcitons: single-beam propagation," Phys. Rev. B **28**, 4433. (3.8.4)

Abram, I., and Maruani, A. (1982). "Calculation of the nonlinear dielectric function in semiconductors," Phys. Rev. B **26**, 4759. [3.10]

Abram, I., Maruani, A., Chemla, D. S., Bonnouvrier, F., and Batifol, E. (1983). "Theoretical and experimental aspects of the nonlinear spectroscopy of biexcitons in CuCl," Physica **117B** and **118B**, 301. (3.8.4)

Adonts, G. G., Jotyan, G. P., and Kanetsyan, E. G. (1982). "Polarisational optical tristability in the Fabry-Perot interferometer," p. 179 in Proceedings of the XI National Conference on Coherent and Nonlinear Optics, Yerevan, USSR. [3.10]

Agarwal, G. S. (1980). "Bistability in coherent two-photon processes," Opt. Commun. **35**, 149. [3.10]

Agarwal, G. S. (1982a). "Spectrum of single-atom fluctuations in cooperative systems," Phys. Rev. A **26**, 680. [2.8]

Agarwal, G. S. (1982b). "Existence of multistability in systems with complex order parameters," Phys. Rev. A **26**, 888. [3.10]

Agarwal, G. S., and Dattagupta, S. (1982). "Higher-order phase-transitions in systems far from equilibrium--multicritical points in two-mode lasers," Phys. Rev. A **26**, 880. (1.3)

Agarwal, G. S., and Lakshmi, P. A. (1980). "Cooperative effects in optical double resonance," Opt. Commun. **35**, 461.

Agarwal, G. S., Narducci, L. M., Feng, D. H., and Gilmore, R. (1978). "Dynamical approach to steady state and fluctuations in optically bistable systems," p. 281 in Coherence and Quantum Optics IV, L. Mandel and E. Wolf, eds. (Plenum Press, New York). [2.8, 6.5]

Agarwal, G. S., Narducci, L. M., Feng, D. H., and Gilmore, R. (1980). "Fokker-Planck equation approach to optical bistability in the bad-cavity limit," Phys. Rev. A **21**, 1029. [6.5]

Agarwal, G. S., Narducci, L. M., Gilmore, R., and Feng, D. H. (1978). "Optical bistability: a self-consistent analysis of fluctuations and the spectrum of scattered light," Phys. Rev. A **18**, 620. [2.8, 5.6]

Agarwal, G. S., and Shenoy, S. R. (1981a). "Bistability in irradiated Josephson junctions," p. 85 in Optical Bistability, C. M. Bowden, M. Ciftan, and H.R. Robl, eds. (Plenum Press, New York). (6.5)

Agarwal, G. S., and Shenoy, S. R. (1981b). "Observability of hysteresis in first-order equilibrium and nonequilibrium phase transitions," Phys. Rev. A **23**, 2719. (6.5)

Agarwal, G. S., and Singh, S. (1984). "Optical bistability, chaos in the coherent two-photon processes," p. 885 in Coherence and Quantum Optics V, L. Mandel and E. Wolf, eds. (Plenum Press, New York). (6.3)

Agarwal, G. S., and Tewari, S. P. (1980). "Optical bistability with dispersion," Phys. Rev. A **21**, 1638. [2.8]

Agrawal, G. P. (1981a). "Use of bidirectional ring cavity for optically bistable devices," IEEE J. Quantum Electron. **QE-17**, 134. [3.10.3]

Agrawal, G. P. (1981b). "Effect of mode coupling on optical bistability in a bidirectional ring cavity," Appl. Phys. Lett. **38**, 505. Erratum: **39**, 365 (1981). [3.10.3]

Agrawal, G. P. (1981c). "Lasers with three-level absorbers," Phys. Rev. A **24**, 1399. (1.3)

Agrawal, G. P. (1981d). "Optical bistability in a phase-conjugate Fabry-Perot cavity," Opt. Commun. **37**, 366. [3.10.3]

Agrawal, G. P. (1982). "Use of a bidirectional ring cavity for optical bistable devices," IEEE J. Quantum Electron. **QE-18**, 214. [3.10.3]

Agrawal, G. P. (1983a). "Resonant intracavity phase conjugation in two and three-level systems," J. Phys. (Paris) **44**, C2-125. [3.10.3]

Agrawal, G. P. (1983b). "Intracavity resonant degenerate four-wave mixing; bistability in phase conjugation," J. Opt. Soc. Am. **73**, 654. [3.10.3]

Agrawal, G. P., and Carmichael, H. J. (1979). "Optical bistability through nonlinear dispersion and absorption," Phys. Rev. A **19**, 2074. [2.4, 2.5.2]

Agrawal, G. P., and Carmichael, H. J. (1980). "Inhomogeneous broadening and the mean-field approximation for optical bistability in a Fabry-Perot interferometer," Opt. Acta **27**, 651. (2.4.3)

Agrawal, G. P., and Flytzanis, C. (1980a). "Optical bistability through two-photon processes," Bull. Am. Phys. Soc. **25**, 359. [3.10.1]

Agrawal, G. P., and Flytzanis, C. (1980b). "Two-photon double-beam optical bistability," Phys. Rev. Lett. **44**, 1058. [3.10.1]

Agrawal, G. P., and Flytzanis, C. (1981a). "Bistability and hysteresis in phase-conjugated reflectivity," IEEE J. Quantum Electron. **QE-17**, 374. [3.10.3]

Agrawal, G. P., and Flytzanis, C. (1981b). "Theory of two-photon double-beam optical bistability," Phys. Rev. A **24**, 3173. [3.10.1]

Agrawal, G. P., and Flytzanis, C. (1981c). "Active two-beam optical bistability," p. 221 in Optical Bistability, C. M. Bowden, M. Ciftan, and R. H. Robl, eds. (Plenum Press, New York).

Agrawal, G. P., Flytzanis, C., Frey, R., and Pradère, P. (1981). "Bistable reflectivity of phase-conjugated signal through intracavity degenerate four-wave mixing," Appl. Phys. Lett. **38**, 492. [3.10.3]

Akins, R., Athale, R., and Lee, S. H. (1979). "Feedback in analog and digital optical image processing. A review," Proc. SPIE **202**, 71.

Akopyan, R. S., Zel'dovich, B. Ya., and Tabiryan, N. V. (1983). "Nonlinear Fabry-Perot resonator using a photoinduced Freedericksz transition," Sov. Tech. Phys. Lett. **9**, 200.

Aksenov, E. T., Kukhtarev, A. V., Lipovskii, A. A., and Pavlenko, A. V. (1983). "A bistable hybrid optical device using an integrated modulator with an induced dielectric channel," Sov. Phys.-Tech. Phys. **28**, 185. [4.4]

Albert, J., Vincent, D., and Tremblay, R. (1981). "Hybrid bistable optical device using an acoustooptic waveguide modulator," Can. J. Phys. **59**, 1251. [4.4]

Alcock, A. J., and Corkum, P. B. (1979). "Ultra-fast switching of infrared radiation by laser-produced carriers in semiconductors," Can. J. Phys. **57**, 1280. [5.3]

Aleksandrov, D. B., Bonch-Bruevich, A. M., Khodovoi, V. A., and Chigir, N. A. (1973). "Change of absorption line shape and of dispersion of a two-level system in a quasiresonant monochromatic radiation field," JETP Lett. **18**, 58. [A]

Alexander, T. (1984). "Computing with light at lightning speeds," Fortune, p. 82, July 23. [7.4]

Ali, M., Maize, S., Thompson, B. V., and Hermann, J.A. (1983). "Mixed species optical bistability (dynamical behaviour) and a note on the two-photon Fabry-Perot system," paper 35 in Programme of Sixth National Quantum Electronics Conference, University of Sussex, 19-22 September. [3.10.1]

Allen, L., and Eberly, J. H. (1975). Optical Resonance and Two-Level Atoms (John Wiley, New York). [2.4, C]

Allen, S. D., Garmire, E., Marburger, J. H., and Winful, H. (1978). "Transient effects in bistable optical devices," J. Opt. Soc. Am. **68**, 1360. [5.1, 5.6, 6.1]

Anthore, R., Flament, P., Gousebet, G., Rhazi, M., and Weill, M. E. (1982). "Interaction between a laser beam and some liquid media," Appl. Opt. **21**, 2. [6.4]

Antoranz, J. C., Bonilla, L. L., Gea, J., and Velarde, M. G. (1982). "Bistable limit cycles in a model for a laser with a saturable absorber," Phys. Rev. Lett. **49**, 35. [1.3, 6.4]

Arakelian, S. M., Karayan, A. S., and Chilingaryan, Yu. S. (1983). "Dynamics of a nonlinear Fabry-Perot cavity with a nematic liquid crystal," Opt. Spectrosc. **55**, 298. (3.4, 5.1)

Arecchi, F. T. (1979). "Experimental aspects of transition phenomena in quantum optics," p. 357 in Dynamical Critical Phenomena and Related Topics, C. P. Enz, ed. (Springer-Verlag, Berlin).

Arecchi, F. T. (1981). "Transient statistics in optical instabilities," p. 327 in Lasers and Applications, W. O. N. Guimares, C. T. Lin, and A. Mooradian, eds. (Springer-Verlag, Berlin). [2.7]

Arecchi, F. T. (1983). "Turbulence and 1/f spectra in quantum optics," p. 8 in Laser Spectroscopy VI, H. P. Weber and W. Lüthy, eds. (Springer-Verlag, Berlin).

Arecchi, F. T. (1984). "Turbulence and 1/f noise in quantum optics," p. 935 in Coherence and Quantum Optics V, L. Mandel and E. Wolf, eds. (Plenum Press, New York). (6.3)

Arecchi, F. T., Giusfredi, G., Petriella, E., and Salieri, P. (1982). "Low threshold optical bistability with optical pumping," Appl. Phys. B **29**, 79. [3.2]

Arecchi, F. T., and Harrison, R. G., eds. (1985). Instabilities and Chaos in Quantum Optics (Springer-Verlag, Berlin). [6.3]

Arecchi, F. T., Kurmann, J., and Politi, A. (1983). "A new class of optical multistabilities and instabilities induced by atomic coherence," Opt. Commun. **44**, 421. [6.4]

Arecchi, F. T., Lippi, G., Tredicce, J. R., and Abraham, N. B. (1984). "Spontaneous oscillations, generalized multistability, and intermittency route to chaos in a bidirectional CO_2 ring laser," J. Opt. Soc. Am. B **1**, 497. (6.3)

Arecchi, F. T., and Lisi, F. (1982). "Hopping mechanism generating 1/f noise in nonlinear systems," Phys. Rev. Lett. **49**, 94. [6.4] See also two comments on this article by M. R. Beasley, D. D'Humieres, and B. A. Huberman and by R. F. Voss, Phys. Rev. Lett. **50**, 1328-1330 (1983).

Arecchi, F. T., Meucci, R., Puccioni, G., and Tredicce, J. (1982). "Experimental evidence of subharmonic bifurcations, multistability, and turbulence in a Q-switched gas laser," Phys. Rev. Lett. **49**, 1217. [6.4]

Arecchi, F. T., and Politi, A. (1978). "Optical bistability in a resonant two-photon absorber," Lett. Nuovo Cimento **23**, 65. [3.10]

Arecchi, F. T., and Politi, A. (1979). "Generalized Fokker-Planck equation for a nonlinear Brownian motion with fluctuations in the control parameter," Opt. Commun. **29**, 361. (6.5)

Arecchi, F. T., and Salieri, P. (1982). "Optical bistability," Phys. Bull. **33**, 20. [1.2]

Areshev, I. P. (1982). "Optical multistability in the transmission of elliptically polarized wave by a nonlinear interferometer," Proceedings of the XI National Conference on Coherent and Nonlinear Optics, Yerevan, USSR.

Areshev, I. P., Murina, T. A., Rosanov, N. N., and Subashiev, V. K. (1983). "Polarization and amplitude optical multistability in a nonlinear ring cavity," Opt. Commun. **47**, 414. [3.10.2]

Areshev, I. P., and Subashiev, V. K. (1981). "Optical bistability in the case of nonlinear absorption and refraction," Sov. Tech. Phys. Lett. **7**, 283.

Areshev, I. P., and Subashiev, V. K. (1982). "Optical amplitude and polarization multistability in a nonlinear interferometer," Sov. Tech. Phys. Lett. **8**, 588. [3.10.2]

Arimondo, E., Casagrande, F., Lugiato, L. A., and Glorieux, P. (1983). "Repetitive passive Q-switching and bistability in lasers with saturable absorbers," Appl. Phys. B **30**, 57. (1.3)

Arimondo, E., and Dinelli, B. M. (1983). "Optical bistability of a CO_2 laser with intracavity saturable absorber: experiment and model," Opt. Commun. **44**, 277. (1.3)

Arimondo, E., Gozzini, A., Lovitch, L., and Pistelli, E. (1981). "Microwave dispersive bistability in a confocal Fabry-Perot microwave cavity," p. 151 in Optical Bistability, C. M. Bowden, M. Ciftan, and H. R. Robl, eds. (Plenum Press, New York). [2.4, 3.10.7]

Armbruster, D. (1983). "An organizing center for optical bistability and self-pulsing," Z. Phys. B **53**, 157. (6.2, 6.3, 6.4)

Armstrong, Jr., L. (1979). "Bistability effects in cooperative multiphoton ionisation," J. Phys. B **12**, L719. [3.10]

Arrathoon, R., and Hassoun, M. H. (1984). "Optical threshold logic elements for digital computation," Opt. Lett. **10**, 143. (7.4)

Asquini, M. L., and Casagrande, F. (1981). "Optical bistability in a bidirectional ring cavity," Z. Phys. B **44**, 233. [6.2]

Athale, R. A., and Lee, S. H. (1979a). "Development of an optical parallel logic device and a half-adder circuit for digital optical processing," Opt. Eng. **18**, 513. [7.4]

Athale, R. A., and Lee, S. H. (1979b). "Bistability and other nonlinear characteristics of the photoconductor -- twisted nematic liquid crystal device with optical feedback," J. Opt. Soc. Am. **69**, 1422. [7.4]

Athale, R. A., and Lee, S. H. (1981). "Bistability and thresholding by a new photoconductor-twisted nematic liquid crystal device with optical feedback," Appl. Opt. **20**, 1424.

Austin, J. W. (1972). "Laser saturable resonators and criteria for their bistable operation," PhD Dissertation, University of Southern California. [1.4, 2.5, 3.1, 5.2.]

Austin, J. W., and DeShazer, L. G. (1971). "Optical characteristics of a saturable absorber inside a Fabry-Perot interferometer," J. Opt. Soc. Am. **61**, 650. [1.4, 2.5, 3.1, 3.2.]

Auston, D. H. (1977). "Picosecond nonlinear optics," p. 123 in Ultrashort Light Pulses, S. L. Shapiro, ed. (Springer-Verlag, Berlin). (7.3)

Averbukh, I. S., Kovarskii, V. A., and Perel'man, N. F. (1979). "Nonequilibrium first-order type phase transition in an electron-vibration system," Phys. Lett. **74A**, 36.

Averbukh, I. S., Kovarskii, V. A., and Perel'man, N. F. (1980a). "Optical transitions of nuclear spin systems," p. 173 in Optical Bistability, C. M. Bowden, M. Ciftan, and H. R. Robl, eds. (Plenum Press, New York). (3.10.7)

Averbukh, I. S., Kovarskii, V. A., and Perel'man, N. F. (1980b). "Optical multistability and light self-modulation at double-resonance," JETP Lett. **32**, 255. [3.10.2]

Averbukh, I. S., Kovarskii, V. A., and Perel'man, N. F. (1982). "Ac Stark effect induced self-pulsing in two-photon resonant optically bistable systems," Phys. Lett. A **91A**, 401. [6.4]

Azéma, A., Botineau, J., Gires, F., and Saissy, A. (1978). "Guided-wave measurement of the 1.06-μm two-photon absorption coefficient in GaAs epitaxial layers," J. Appl. Phys. **49**, 24. [5.3]

Azéma, A., Botineau, J., Gires, F., Saissy, A., and Vanneste, C. (1975). "Interaction between free carrier and optical field along the surface of a semiconductor material," Opt. Commun. **15**, 80. [5.3]

Azéma, A., Botineau, J., Gires, F., Saissy, A., and Vanneste, C. (1976a). "Optical determination of free carrier parameters in an epitaxial GaAs layer," Appl. Phys. **9**, 47. [5.3]

Azéma, A., Botineau, J., Gires, F., Saissy, A., and Vanneste, C. (1976b). "Propagation d'une onde infrarouge guidée en présence de porteurs libres," Rev. Phys. Appl. **11**, 239. [5.3]

Baker, H. C., and Armstrong, Jr., L. (1981). "Bistability in multiphoton ionization with recombination," Opt. Lett. **6**, 357. [3.10]

Bakiev, A. M., Dneprovskii, V. S., Kovaliuk, Z. D., and Stadnik, V. A. (1983a). "Optical bistability in GaSe," Doklady Akademii Nauk SSSR **271**, 611. Translation: Sov. Phys.-Dokl. (USA) **28**, 579 (1983). [3.8.7]

Bakiev, A. M., Dneprovskii, V. S., Kovalyuk, Z. D., and Stadnik, V. A. (1983b). "Optical bistability related to excitons in an uncooled semiconductor," JETP Lett. **38**, 596. [3.8.7]

Ballagh, R. J., Cooper, J., Hamilton, M. W., Sandle, W. J., and Warrington, D. M. (1981). "Optical bistability in a Gaussian cavity mode," Opt. Commun. **37**, 143. [2.9, 3.2]

Ballagh, R. J., Cooper, J., and Sandle, W. J. (1981). "Effective two-state behavior of a collisionally perturbed Zeeman degenerate atomic transition and its application to optical bistability," J. Phys. B **14**, 3881. [3.2]

Band, Y. B. (1983). "Optical bistability in thin slabs of nonlinear media," Proc. SPIE **380** (Los Alamos Conference on Optics '83), 389.

Band, Y. B. (1984). "Optical bistability in nonlinear media: an exact method of calculation," J. Appl. Phys. **56**, 656.

Baranovskii, I. V., Borshch, A. A., Brodin, M. S., and Kamuz, A. M. (1971). "Self-focusing of ruby laser radiation in CdS crystals," Sov. Phys. JETP **33**, 861. [7.3]

Barbarino, S., Gozzini, A., Maccarrone, F., Longo, I., and Stampacchia, R. (1982). "Critical slowing-down in microwave absorptive bistability," Nuovo Cimento B **71B**, 183. [3.10.7, 5.6]

Bar-Joseph, I., and Silberberg, Y. (1983). "The mechanism of instabilities in an optical cavity," Opt. Commun. **48**, 53. [6.3]

Bar-Joseph, I., and Silberberg, Y. (1984). "Self-oscillations of counterpropagating waves in a two-level medium," J. Opt. Soc. Am. B **1**, 498. (6.4)

Basov, N. G. (1965). "Semiconductor lasers," Science **149**, 821. [1.2]

Basov, N. G. (1966). "Theory of pulsating conditions for lasers," IEEE J. Quantum Electron. **QE-2**, 542. (1.2, 1.3)

Basov, N. G. (1968). "0-1-dynamics of injection lasers," IEEE J. Quantum Electron. **QE-4**, 855. (1.2, 1.3)

Basov, N. G., Culver, W. H., and Shah, B. (1972). "Applications of lasers to computers," p. 1649 in Laser Handbook, Vol. 2, F. T. Arecchi and E. O. Schulz-Dubois, eds. (American Elsevier, New York). [1.2, 1.3]

Basov, N. G., and Morozov, V. N. (1970). "Contribution to the theory of the dynamics of injection lasers," Sov. Phys. JETP **30**, 338. (1.2, 1.3)

Basov, N. G., Morozov, V. N., Nikitin, V. V., and Semenov, A. S. (1968). "Investigation of GaAs laser radiation pulsations," Sov. Phys. Semicond. **1**, 1305. [6.1] (1.3)

Batovrin, V. K., and Novokreshchenov, V. K. (1978). "Hysteresis effect in second-harmonic generation in a laser," Sov. Tech. Phys. Lett. **4**, 325. (1.3)

Bazhenov, V. Y., Bogatov, A. P., Eliseev, P. G., Okhotnikov, O. G., Pak, G. T., Rakhval'skii, M. P., Soskin, M. S., Taranenko, V. B., and Khairetdinov, K. A. (1981). "Bistable operation and spectral tuning of an injection laser with an external dispersive resonator," Sov. J. Quantum Electron. **11**, 510. (1.3)

Bazhenov, V. Yu., Bogatov, A. P., Eliseev, P. G., Okhotnikov, O. G., Pak, G. T., Rakhvalsky, M. P., Soskin, M. S., Taranenko, V. B., and Khairetdinov, K. A. (1982). "Bistable operation and spectral tuning of injection laser with external dispersive cavity," IEE Proc. **129**, Part I, 77. (1.3)

Bel'skii, A. M., Nesterenko, T. M., and Khapalyuk, A. P. (1978). "Structure of the radiation field of a Fabry-Perot resonator filled with an active medium with quadratic inhomogeneities," Zhurnal Prikladnoi Spektroskopii **28**, 229. (1.3)

Belyaletdinov, I. F., Zolotov, E. M., and Prokhorov, A. M. (1979). "Optically activated thin-film switch," Sov. Tech. Phys. Lett. (USA) **5**, 12. (4.4)

Benza, V., and Lugiato, L. A. (1979a). "Dressed mode description of optical bistability," Z. Physik B **35**, 383. [6.2]

Benza, V., and Lugiato, L. A. (1979b). "Analytical treatment of the transient in absorptive optical bistability," Lett. Nuovo Cimento **26**, 405. [5.6]

Benza, V., and Lugiato, L. A. (1981a). "Semiclassical and quantum statistical dressed mode description of optical bistability," p. 9 in Optical Bistability, C. M. Bowden, M. Ciftan, and H. R. Robl, eds. (Plenum Press, New York). [6.2]

Benza, V., and Lugiato, L. A. (1981b). "Analytical treatment of self-pulsing in absorptive optical bistability," p. 760 in Proceedings of the International Conference on Lasers '80, C. B. Collins, ed. (STS Press, McLean, Virginia). (6.2)

Benza, V. and Lugiato, L. A. (1982). "Dressed mode description of optical bistability. II. Analytical treatment of self-pulsing," Z. Phys. B **47**, 79. [6.2]

Benza, V., Lugiato, L. A., and Meystre, P. (1980). "Analytical description of self-pulsing in absorptive optical bistability," Opt. Commun. **33**, 113. [6.2]

Berger, M. (1984). "Bistable diodes handle switching in TDM optical system," Electronics, May 31, p. 65. (1.3)

Berreman, D. W. (1983). "Numerical modelling of twisted nematic devices," Philos. Trans. Roy. Soc. London A **309**, 203.

Bischofberger, T., and Shen, Y. R. (1978a). "Transient behavior of a nonlinear Fabry-Perot," Appl. Phys. Lett. **32**, 156. [2.9, 3.2, 3.4, 5.1]

Bischofberger, T., and Shen, Y. R. (1978b). "Nonlinear behavior of a Fabry-Perot interferometer filled with a Kerr liquid," J. Opt. Soc. Am. **68**, 642. [2.9, 3.2, 3.4, 5.1]

Bischofberger, T., and Shen, Y. R. (1979a). "Nonlinear Fabry-Perot filled with CS_2 and nitrobenzene," Opt. Lett. **4**, 40. [2.9, 3.2, 3.4]

Bischofberger, T., and Shen, Y. R. (1979b). "Theoretical and experimental study of the dynamic behavior of a nonlinear Fabry-Perot interferometer," Phys. Rev. A **19**, 1169. [2.9, 3.2, 3.4, 5.1]

Bistrov, V. P., Gnedoy, S. A., Shipilov, K. F., and Shmonov, T. A. (1982). "Operating regimes of hybrid optical feedback circuits," p. 188 in Proceedings of the XI National Conference on Coherent and Nonlinear Optics, Yerevan, USSR. [6.3]

Bjorkholm, J. E. (1967). "Arrangement for passive transmission pulsing of a Q-switched laser," U.S. Patent 3,500,241, filed Oct. 23, 1967, granted Mar. 10, 1970. [1.4, 4.1, 6.1]

Bjorkholm, J. E., and Ashkin, A. (1974). "CW self-focusing and self-trapping of light in sodium vapor," Phys. Rev. Lett. **32**, 129. [3.2, 3.9.1]

Bjorkholm, J. E., Smith, P. W., and Tomlinson, W. J. (1982). "Optical bistability based on self-focusing: an approximate analysis," IEEE J. Quantum Electron. **QE-18**, 2016. [3.9.1]

Bjorkholm, J. E., Smith, P. W., Tomlinson, W. J., and Kaplan, A. E. (1982). "Optical bistability based on self-focusing," Opt. Lett. **6**, 345. [3.9.1, 6.1]

Bjorkholm, J. E., Smith, P. W., Tomlinson, W. J., Pearson, D. B., Maloney, P. J., and Kaplan, A. E. (1981). "Optical bistability based on self-focusing," IEEE J. Quantum Electron. **QE-17**, 118. [3.9.1]

Bloch, F. (1946). "Nuclear induction," Phys. Rev. **70**, 460. [C]

Bloom, D. M., Mollenauer, L. F., Lin, C., Taylor, D. W., and DelGaudio, A. M. (1979). "Direct demonstration of distortionless picosecond-pulse propagation in kilometer-length optical fibers," Opt. Lett. **4**, 297. [1.2]

Bloom, D. M., Shank, C. V., Fork, R. L., and Teschke, O. (1978). "Sub-picosecond optical gating and wavefront conjugation," p. 96 in Picosecond Phenomena, C. V. Shank, E. P. Ippen, S. L. Shapiro, eds. (Springer-Verlag, Berlin). [5.8]

Blow, K. J., and Doran, N. J. (1983). "Soliton dynamics in a synchronously pumped ring cavity," in Colloquium on Non-linear Optical Waveguides (IEE, London, England).

Boardman, A. D., and Egan, P. (1983). "Nonlinear surface polaritons in a slab," in Colloquium on Non-linear Optical Waveguides (IEE, London, England).

Bogatov, A. P. Eliseev, P. G., Okhotnikov, O. G., and Pak, G. T. (1978). "Hysteresis of the output radiation power of cw ALGaAs heterolasers," Sov. J. Quantum Electron. **8**, 1408. [1.3]

Boggess, T. F., Moss, S. C., Boyd, I. W., and Smirl, A. L. (1984). "Silicon optical Zener switch," p. 270 in Conference on Lasers and Electro-Optics, Technical Digest (IEEE, New York). [5.2]

Bohnert, K., Kalt, H., and Klingshirn, C. (1983). "Intrinsic absorptive optical bistability in CdS," Appl. Phys. Lett. **43**, 1088. [2.10, 3.8]

Boiko, B. B., Dzhilavdari, I. Z., and Petrov, N. S. (1975). "Reflection of plane light from a nonlinear transparent isotropic medium," J. Appl. Spectroscopy **23**, 1511. [3.10.4]

Boiko, B. B., Insarova, N. I., Olefir, G. I., and Petrov, N. S. (1983). "Thermal variation in the refractive index of a thin absorbing layer in conditions of strong optical irradiation," J. Appl. Spectrosc. **39**, 946.

Boiko, B. B., and Petrov, N. S. (1982). "Hysteresis in the reflection of light at the boundary with a nonlinear medium in the presence of a transition layer," J. Appl. Spectrosc. **37**, 1386.

Boiko, B. B., and Petrov, N. S. (1982). "Phenomenological theory of reflection and transmission of light at a boundary with a nonlinear medium," J. Appl. Spectrosc. **36**, 342.

Boiko, B. B., and Petrov, N. S. (1982). "Hysteresis in reflection of light from the boundary of the nonlinear medium with the transition layer," p. 197 in Proceedings of the XI National Conference on Coherent and Nonlinear Optics, Yerevan, USSR. (3.10.4)

Bol'shov, L. A., Kiselev, V. P., and Reshetin, V. P. (1983). "Appearance of stochastic regimes of light pulse reflection by a nonlinear Fabry-Perot interferometer," Sov. J. Quantum Electron. **13**, 921.

Bomberger, W. D., Findakly, T., and Chen, B. (1982). "Integrated optical logic devices," Proc. SPIE **317**, 23. (3.10.5, 4.4)

Bonchbruevich, A. M., Przibelskii, S. G., and Chigir, N. A. (1978). "Amplification of weak light in 2-level system without population-inversion," Vest. Mosk. Univ. Ser. Fiz. Mat. I. Astron. **19**, 35.

Bonifacio, R., ed. (1982). Dissipative Systems in Quantum Optics: Resonance Fluorescence, Optical Bistability, Superfluorescence (Springer-Verlag, Berlin).

Bonifacio, R., Gronchi, M., and Lugiato, L. A. (1978). "Photon statistics of a bistable absorber," Phys. Rev. A **18**, 2266. [1.3, 6.5]

Bonifacio, R., Gronchi, M, and Lugiato, L. A. (1979a). "Dispersive bistability in homogeneously broadened systems," Nuovo Cimento **53B**, 311. [2.1, 2.4]

Bonifacio, R., Gronchi, M., and Lugiato, L. A. (1979b). "Self-pulsing in bistable absorption," Opt. Commun. **30**, 129. [6.2]

Bonifacio, R., Gronchi, M., and Lugiato, L. A. (1981). "Instabilities in optical bistability; transform from cw to pulsed," p. 31 in Optical Bistability, C. M. Bowden, M. Ciftan, and H. R. Robl, eds. (Plenum Press, New York). [6.2]

Bonifacio, R., and Lugiato, L. A. (1976). "Cooperative effects and bistability for resonance fluorescence," Opt. Commun. **19**, 172. [2.4, 2.8, 3.10, 5.1, 5.6]

Bonifacio, R., and Lugiato, L. A. (1978a). "Optical bistability and cooperative effects in resonance fluorescence," Phys. Rev. A **18**, 1129. [2.4, 2.5.2, 2.8, 5.1, 5.6, 6.5]

Bonifacio, R., and Lugiato, L. A. (1978b). "Cooperative effects in optical bistability and resonance fluorescence," p. 249 in Coherence and Quantum Optics IV, L. Mandel and E. Wolf, eds. (Plenum Press, New York). [2.4, 2.8, 6.5]

Bonifacio, R., and Lugiato, L. A. (1978c). "Bistable absorption in a ring cavity," Lett. Nuovo Cimento **21**, 505. [2.4]

Bonifacio, R., and Lugiato, L. A. (1978d). "Mean field model for absorptive and dispersive bistability with inhomogeneous broadening," Lett. Nuovo Cimento **21**, 517. [2.1, 2.4]

Bonifacio, R., and Lugiato, L. A. (1978e). "Photon statistics and spectrum of transmitted light in optical bistability," Phys. Rev. Lett. **40**, 1023. [2.8, 6.5]

Bonifacio, R., and Lugiato, L. A. (1978f). "Instabilities for a coherently driven absorber in a ring cavity," Lett. Nuovo Cimento **21**, 510. [6.2]

Bonifacio, R., Lugiato, L. A., Farina, J. D., and Narducci, L. M. (1981). "Long time evolution for a one-dimensional Fokker-Planck process: application to absorptive optical bistability," IEEE J. Quantum Electron. **QE-17**, 357. [6.5]

Bonifacio, R., Lugiato, L. A., and Gronchi, M. (1979). "Theory of optical bistability," p. 426 in Laser Spectroscopy IV, H. Walther and K. W. Rothe, eds. (Springer-Verlag, Berlin). [2.4]

Bonifacio, R., and Meystre, P. (1978). "Transient response in optical bistability," Opt. Commun. **27**, 147. [5.1, 5.6]

Bonifacio, R., and Meystre, P. (1979). "Critical slowing down in optical bistability," Opt. Commun. **29**, 131. [5.6]

Borodulin, V. I., Konyaev, V. P., Novikova, E. R., Tager, A. A., Tregub, D. P., and Elenkrig, B. B. (1983). "Experimental investigation of inhomogeneously pumped injection lasers" Sov. J. Quantum Electron. **13**, 398. (1.3)

Borshch, A. A., Brodin, M. S., Volkov, V. I., and Kukhtarev, N. V. (1981). "Optical bistability and hysteresis in a reversed wave in the course of degenerate six-photon interaction in cadmium sulfide," Sov. J. Quantum Electron. **11**, 777. [3.8.2, 3.10.3]

Borshch, A., Brodin, M., Volkov, V., and Kukhtarev, N. (1982). "Optical bistability and hysteresis in phase conjugation by degenerate six-photon mixing," Opt. Commun. **41**, 213. [3.10.3]

Borshch, A., Brodin, M., Volkov, V., Kukhtarev, N., and Starkov, V. (1982). "Intrinsic optical hysteresis in phase conjugation by CdS crystals," p. 186 in Proceedings of the XI National Conference on Coherent and Nonlinear Optics, Yerevan, USSR. [3.8.2]

Borshch, A., Brodin, M., Volkov, V., Kukhtarev, N., and Starkov, V. (1984). "Optical hysteresis and bistability in phase conjugation by degenerate six-photon mixing," J. Opt. Soc. Am. A **1**, 40. [3.10.3]

Bosacchi, B., and Narducci, L. M. (1982). "Optical bistability in the resonant frustrated-total-reflection optical cavity." J. Opt. Soc. Am. **72**, 1761. [3.10.4, 3.10.5]

Bosacchi, B., and Narducci, L. M. (1983). "Optical bistability in the frustrated-total-reflection optical cavity," Opt. Lett. **8**, 324. [3.10.4, 3.10.5]

Bösiger, P., Brun, E., and Meier, D. (1978). "Ruby NMR laser: A phenomenon of spontaneous self-organization of a nuclear spin system," Phys. Rev. A **18**, 671.

Bösiger, P., Brun, E., and Meier, D. (1979). "Collective ordering phenomena and instabilities of the ^{27}Al nuclear spin system in ruby," Phys. Rev. A **20**, 1073. [3.10.7]

Bösiger, P., Brun, E., and Meier, D. (1981). "Bistability and phase transitions of nuclear spin systems," p. 173 in Optical Bistability, C. M. Bowden, M. Ciftan, and H. R. Robl, eds. (Plenum Press, New York). [3.10.7]

Botineau, J., Gires, F., and Vanneste, C. (1974). "Modulation d'une onde infrarouge guidée par porteurs libres," C. R. Acad. Sci. **278**, 171. [5.3]

Bourdet, G. (1981). Private communication, ENSTA, Palaiseau, France. [4.5]

Bowden, C. M. (1981). "Optical bistability based upon atomic correlation in a small volume," p. 405 in Optical Bistability, C. M. Bowden, M. Ciftan, and H. R. Robl, eds. (Plenum Press, New York). [3.10.6]

Bowden, C. M., Ciftan, M., and Robl, H. R., eds. (1981). Optical Bistability (Plenum Press, New York). [1.1]

Bowden, C. M., Gibbs, H. M., and McCall, S., eds. (1984). Optical Bistability 2 (Plenum Press, New York). [1.1, 6.5]

Bowden, C. M., and Sung, C. C. (1979). "First- and second-order phase transitions in the Dicke model: relation to optical bistability," Phys. Rev. A **19**, 2392. [3.10.6]

Bowden, C. M., and Sung, C. C. (1980). "Cooperative optical bistability in a small volume without mirrors," J. Opt. Soc. Am **70**, 589. [3.10.6] (2.10)

Bowden, C. M., and Sung, C. C. (1981). "Cooperative transients in inter-atomic correlation in the presence of an externally applied coherent field: relation to intrinsic mirrorless optical bistability," Proc. SPIE **317**, 276. [3.10.6]

Boyd, G. D., Cheng, J., and Thurston, R. N. (1982). "A multiplexible bistable nematic liquid crystal display using thermal erasure," Appl. Phys. Lett. **40**, 936.

Bradley, D. J., and Stallard, W. A. (1983). "Bistability and slow oscillation in an external cavity semiconductor diode laser," in Colloquium on Non-linear Optical Waveguides (IEE, London, England).

Brand, H., Graham, R., and Schenzle, A. (1980). "Wigner distribution of bistable steady states in optical subharmonic generation," Opt. Commun. **32**, 359. [6.5]

Brand, H., and Schenzle, A. (1978). "A soluble stochastic model for first-order type nonequilibrium phase transitions," Phys. Lett. **68A**, 427. [6.5]

Brand, H., Schenzle, A., and Schröder, G. (1982). "Lower and upper bounds for the eigenvalues of the Fokker-Planck equation in detailed balance," Phys. Rev. A **25**, 2324. (6.5)

Brewer, R. G. (1977). "Coherent optical spectroscopy," p. 341 in Frontiers in Laser Spectroscopy, Vol. 1, R. Balian, S. Haroche, and S. Liberman, eds. (North-Holland, Amsterdam). [C]

Broggi, G., and Lugiato, L. A. (1984). "Transient noise-induced optical bistability," Phys. Rev. A **29**, 2949.

Brun, E., Derighetti, B., Holzner, R., and Meier, D. (1983). "The NMR-laser: a nonlinear solid state system showing chaos," Helv. Phys. Acta **56**, 825. [6.3]

Brunner, W., Fischer, R., and Paul, H. (1984). "Chaos and order in the spectral behavior of multimode lasers," J. Opt. Soc. Am. B **1**, 487. (1.3)

Brunner, W., and Paul, H. (1980). "Theory of high-frequency modulation of Ar-ion laser radiation due to stable two-mode oscillation," Opt. Commun. **35**, 421. (1.3)

Brunner, W., and Paul, H. (1982). "Regular and chaotic behaviour of multimode gas lasers," Appl. Phys. B **28**, 168. [6.4]

Buckingham, A. D. (1956). "Birefringence resulting from the application of an intense beam of light to an isotropic medium," Proc. Phys. Soc. **B69**, 344. [5.8]

Bullough, R. K., and Hassan, S. S. (1983). "Optical extinction theorem in the nonlinear theory of optical multistability," Proc. SPIE **369**, 257.

Bullough, R. K., Hassan, S. S., and Tewari, S. P. (1983). "Refractive index theory of optical bistability," p. 229 in Quantum Electronics and Electro-Optics, P. L. Knight, ed. (John Wiley, New York).

Bulsara, A. R., Schieve, W. C., and Gragg, R. F. (1978). "Phase transitions induced by white noise in bistable optical systems," Phys. Lett. **68A**, 294. [6.5]

Burt, M. G. (1983). "Enhanced bandgap resonant non-linear susceptibility in quantum-well heterostructures," Electronics Lett. **19**, 132.

Carenco, A., and Menigaux, L. (1980). "Optical bistability using a directional coupler and a detector monolithically integrated in GaAs," Appl. Phys. Lett. **37**, 880. [4.4]

Carenco, A., Menigaux, L., Delpech, Ph., Denuit, B., and Salvi, M. (1980). "Monolithic integration of a detector and a directional coupler switch in GaAs-bistable operation," p. 252 in Proceedings of the Sixth European Conference on Optical Communication (IEE, London, England). (4.4)

Carmichael, H. J. (1980). "The mean-field approximation and validity of a truncated Bloch hierarchy in absorptive bistability," Opt. Acta **27**, 147. [2.5.2]

Carmichael, H. J. (1981), "The fluorescent spectrum and collective effects in absorptive bistability," Z. Phys. B **42**, 183. [2.8]

Carmichael, H. J. (1983a). "Many-mode stability analysis in the mean-field limit for absorptive bistability," Phys. Rev. A **28**, 480.

Carmichael, H. J. (1983b). "Chaos in non-linear optical-systems," p. 64 in Laser Physics, J. D. Harvey and D. F. Walls, eds. (Springer-Verlag, Berlin). (6.3, 6.4)

Carmichael, H. J. (1984). "Optical bistability and multimode instabilities," Phys. Rev. Lett. **52**, 1292. [6.3]

Carmichael, H. J. (1984). "Optical bistability and multimode instabilities in a standing-wave cavity," talk WT2, Optical Society of America 1984 Annual Meeting, Technical Program. (6.3)

Carmichael, H. J., and Agrawal, G. P. (1980). "Steady-state formulation of optical bistability for a Doppler-broadened medium in a Fabry-Perot," Opt. Commun. **34**, 293. [2.5.2]

Carmichael, H. J., and Agrawal, G. P. (1981). "Absorptive and dispersive bistability for a Doppler-broadened medium in a Fabry-Perot: steady-state description," p. 23 in Optical Bistability, C. M. Bowden, M. Ciftan, and H. R. Robl, eds. (Plenum Press, New York). [2.5.2]

Carmichael, H. J., and Agrawal, G. P. (1981). "Doppler-broadened medium in a Fabry-Perot: steady-state description," p. 237 in Optical Bistability, C. M. Bowden, M. Ciftan, and H. R. Robl, eds. (Plenum Press, New York).

Carmichael, H. J., Asquini, L. and Lugiato, L. A. (1984). In preparation.

Carmichael, H. J., and Hermann, J. A. (1980). "Analytic description of optical bistability including spatial effects," Z. Phys. B **38**, 365. [2.5]

Carmichael, H. J., Savage, C. M., and Walls, D. F. (1983). "From optical tristability to chaos," Phys. Rev. Lett. **50**, 163. [6.4]

Carmichael, H. J., Savage, C. M., and Walls, D. F. (1984). "Multistability, self-oscillation and chaos in nonlinear optics," p. 957 in Coherence and Quantum Optics V, L. Mandel and E. Wolf. eds. (Plenum Press, New York). (6.3)

Carmichael, H. J., Snapp, R. R., and Schieve, W. C. (1982). "Oscillatory instabilities leading to 'optical turbulence' in a bistable ring cavity," Phys. Rev. A **26**, 3408. [5.1, 6.3]

Carmichael, H. J., and Walls, D. F. (1977). "Hysteresis in the spectrum for cooperative resonance fluorescence," J. Phys. B **10**, L685. [2.8]

Carmichael, H. J., Walls, D. F., Drummond, P. D., and Hassan, S. S. (1983). "Quantum theory of optical bistability. III. Atomic-fluorescence in a low-Q cavity," Phys. Rev. A **27**, 3112. [2.8] (6.5)

Carney, J. K., and Fonstad, C. G. (1981). "Double-heterojunction laser diodes with multiply segmented contacts," Appl. Phys. Lett. **38**, 303.

Carr, J. (1984). "One-dimensional approximations for a quadratic Ikeda map," Phys. Lett. A **104A**, 59. (6.3)

Carter, G. M., Chen, Y. J., and Tripathy, S. K. (1983). "Intensity-dependent index of refraction in multilayers of polydiacetylene," Appl. Phys. Lett. **43**, 891. [7.2]

Casagrande, F., Eschenazi, E., and Lugiato, L. A. (1984). "Fluctuation theory in quantum-optical systems. II. Rigorous adiabatic elimination of the hierarchy method," Phys. Rev. A **29**, 239.

Casagrande, F., and Lugiato, L. A. (1978). "On the first-order phase transition in the laser with saturable absorber," Nuovo Cimento **48B**, 287. [6.5]

Casagrande, F., and Lugiato, L. A. (1980). "Antibunching in absorptive optical bistability," Nuovo Cimento **55B**, 173. [6.5]

Casagrande, F., and Lugiato, L. A. (1982). "Quantum statistical treatment of open systems, laser dynamics and optical bistability," p. 53 in Quantum Optics. Proceedings of the South African Summer School in Theoretical Physics, C. A. Engelbrecht, ed. (Springer-Verlag, Berlin).

Casagrande, F., Lugiato, L. A., and Asquini, M. L. (1980). "Instabilities for a coherently driven absorber in a Fabry-Perot cavity," Opt. Commun. **32**, 492. [6.2]

Casasent, D., ed. (1978). Optical Data Processing (Springer-Verlag, Berlin). [7.4]

Cathey, W. T. (1974). Optical Information Processing and Holography (John Wiley, New York). [7.4]

Caulfield, H. J., ed. (1983). 10th International Optical Computing Conference (IEEE Computer Society Press, Silver Spring, MD). [7.4]

Caulfield, H. J., Horovitz, S., Tricoles, G. P., and Von Winkle, W. A., eds. (1984). "Special issue on optical computing," Proc. IEEE **72**, 755-979. [7.4]

Caulfield, H. J., Neff, J. A., and Rhodes, W. T. (1983). "Optical computing: the coming revolution in optical signal processing," Laser Focus, November, p. 100. [7.4]

Cecchi, S., Giusfredi, G., Petriella, E., and Salieri, P. (1982). "Observation of optical tristability in sodium vapors," Phys. Rev. Lett. **49**, 1928. [3.10]

Chang, C. L., and Tsai, C. S. (1983). "Electro-optic analog-to-digital converter using channel waveguide Fabry-Perot modulator array," Appl. Phys. Lett. **43**, 22. [3.10.5]

Chang, Joon-Sung, and Lim, Ki-Soo (1981). "Optical bistability using a $LiNbO_3$ electro-optic modulator," J. Korean Phys. Soc. **14**, 153. (4.5)

Chang, T. Y. (1981). "Fast self-induced refractive index changes in optical media: a survey," Opt. Eng. **20**, 220. [7.3]

Changjun, L., and Stegeman, G. (1983). "Bistability in prism-waveguide coupling," J. Opt. Soc. Am. **73**, 1960. [3.10.5]

Chase, L. L., Peyghambarian, N., Grynberg, G., and Mysyrowicz, A. (1979). "Direct creation of a high density of biexcitons at k=0 in CuCl," Opt. Commun. **28**, 189. [3.8]

Chavel, P., Sawchuck, A. A., Strand, T. C., Tanguay, Jr., A. R., and Soffer, B. H. (1980). "Optical logic with variable-grating-mode liquid crystal devices," Opt. Lett. **5**, 398. [7.4]

Chemla, D. S. (1980). "Non-linear optical properties of condensed matter," Rep. Prog. Phys. **43**, 1191. (7.3)

Chemla, D. S., Damen, T. C., Miller, D. A. B., Gossard, A. C., and Wiegmann, W. (1983). "Electroabsorption by Stark effect on room-temperature excitons in GaAs/GaAlAs multiple quantum well structures," Appl. Phys. Lett. **42**, 864. [7.4]

Chemla, D. S., and Maruani, A. (1982). "Nonlinear optical effects associated with excitonic-molecules in large gap semiconductors," Progress Quantum Electronics **8**, 1. (3.8.2, 3.8.4, 7.3)

Chemla, D. S., Miller, D. A. B., Smith, P. W., Gossard, A. C., and Tsang, W. T. (1982). "Large optical nonlinearities in room-temperature GaAs-GaAlAs multiple quantum well structures," in Conference on Lasers and Electro-Optics Technical Digest (IEEE, New York). [3.6]

Chemla, D. S., Miller, D. A. B., Smith, P. W., Gossard, A. C., and Wiegmann, W. (1984). "Room temperature excitonic nonlinear absorption and refraction in GaAs/AlGaAs multiple quantum well structures," IEEE J. Quantum Electron. **QE-20**, 265. (3.6, 7.3)

Chen, L., Li, C. F., and Hong, J. (1984). "Ikeda instability and degree of stability in optical bistability," p. 103 in Conference Digest of the United States-Japan Seminar on Coherence, Incoherence, and Chaos in Quantum Electronics, Nara, M. D. Levenson, and T. Shimizu, coordinators. (6.3)

Chen, Y. C., and Liu, J. M. (1985). "Polarization bistability in semiconductor lasers," Appl. Phys. Lett. **46**, 16. (1.3)

Cheng, J., Thurston, R. N., Boyd, G. D., and Meyer, R. B. (1982). "A nematic liquid crystal storage display based on bistable boundary layer configurations," Appl. Phys. Lett. **40**, 1007.

Cheung, M. M., Durbin, S. D., and Shen, Y. R. (1983). "Optical bistability and self-oscillation of a nonlinear Fabry-Perot interferometer filled with a nematic liquid crystal film," Opt. Lett. **8**, 39. [3.4, 6.1]

Chikama, T. (1984). "Noise-induced enhancement of the optical bistability," J. Phys. Soc. Jpn. **53**, 1252. (6.5)

Cho, Y., and Umeda, T. (1984). Chaos in laser oscillations with delayed feedback: numerical analysis and observation using semiconductor lasers," J. Opt. Soc. Am. B **1**, 497. (6.3)

Chrostowski, J. (1982). "Noisy bifurcations in acousto-optic bistability," Phys. Rev. A **26**, 3023. [6.3]

Chrostowski, J., and Delisle, C. (1979a). "Asymmetrical Fabry-Perot interferometer with feedback," Appl. Opt. **18**, 2358. [4.5]

Chrostowski, J., and Delisle, C. (1979b). "Bistable piezoelectric Fabry-Perot-interferometer," Can. J. Phys. **57**, 1376. [4.5]

Chrostowski, J., and Delisle, C. (1982). "Bistable optical switching based on Bragg diffraction," Opt. Commun. **41**, 71. [4.5]

Chrostowski, J., Delisle, C., and Tremblay, R. (1982). "Acousto-optic bistable generator," J. Opt. Soc. Am. **72**, 1770. [6.3]

Chrostowski, J., Delisle, C., and Tremblay, R. (1983). "Optical oscillations in an acoustooptic bistable device," Can. J. Phys. **61**, 188. [6.5]

Chrostowski, J., Delisle, C., Vallee, R., Carrier, D., and Boulay, L. (1982). "Multistable nonlinear Fabry-Perot interferometer," Can. J. Phys. **60**, 1303. [4.5]

Chrostowski, J., Vallee, R., and Delisle, C. (1983). "Self-pulsing and chaos in acoustooptic bistability," Can. J. Phys. **61**, 1143. [6.3]

Chrostowski, J., and Zardecki, A. (1979). "Noise transformation by optically bistable system," Opt. Commun. **29**, 230. [6.5]

Chrostowski, J., Zardecki, A., and Delisle, C. (1981). "Time-dependent fluctuations and phase hysteresis in dispersive bistability," Phys. Rev. A **24**, 345. [6.5]

Ciftan, M. (1980). "Hysteresis as a cooperative phenomenon," p. 311 in Proceedings of the Southeast Conference '80 (IEEE, New York).

Cohen-Tannoudji, C. (1976). "Atoms in strong resonant fields," in Frontiers in Laser Spectroscopy, R. Balian, S. Haroche, and S. Liberman, eds. (North-Holland, Amsterdam). [2.8]

Collins, Jr., S. A., ed. (1980a). Special Issue on Feedback in Optics, Opt. Eng. **19**, 441-487. [1.2]

Collins, Jr., S. A. (1980b). "Feedback in optics," Opt. Eng. **19**, 441. [4.1]

Collins, Jr., S. A. (1983). "Optical pulse generator using a Hughes liquid crystal light valve," in the Proceedings of the NASA Conference on Optical Information Processing for Aerospace Applications. (7.4)

Collins, Jr., S. A., Fatehi, M. T., and Wasmundt, K. C. (1980). "Optical logic gates using a Hughes liquid crystal light valve," Proc. SPIE **232**, 168. (7.4)

Collins, Jr., S. A., Gerlach, U. H., and Zakman, Z. M. (1979). "Optical feedback for generating arrays of bistable elements," Proc. SPIE **185**, 36.

Collins, Jr., S. A., and Wasmundt, K. C. (1980). "Optical feedback and bistability: a review," Opt. Eng. **19**, 478. [7.4]

Cooperman, G. D., and Winful, H. G. (1981). "Self-pulsing and chaos in a bistable distributed-feedback structure," J. Opt. Soc. Am. **71**, 1634. [6.4]

Craig, A. E., Olson, G. A., and Sarid, D. (1983). "Experimental observation of the long-range surface-plasmon polariton," Opt. Lett. **8**, 380. [3.10.5]

Craig, D., Mathew, J. G. H., Kar, A. K., Miller, A., Smith, S. D. (1984). "Nonlinear refraction and nonlinear absorption in InAs," p. 279 in Optical Bistability 2, C. M. Bowden, H. M. Gibbs and S. L. McCall, eds., (Plenum Press, 1984). [7.3.1]

Craig, D., Mathew, J. G. H., and Miller, A. (1984). "Optical switching in CdHgTe," Conference on Infrared Physics 3, Zurich, July.

Craig, D., Miller, A., Mathew, J. G. H., and Kar, A. K. (1985). "Fast optical switching and bistability in room temperature CdHgTe at 10.6 μm," Infrared Phys. **25**, 289. [3.8.6]

Cresser, J. D., and Meystre, P. (1981). "The role of phases in the transient dynamics of nonlinear interferometers," p. 265 in Optical Bistability, C. M. Bowden, M. Ciftan, and H. R. Robl, eds. (Plenum Press, New York).

Cribier, S., Giacobino, E., and Grynberg, G. (1983a). "Quantitative investigation of critical slowing down in all-optical bistability," Opt. Commun. **47**, 170. [5.6, H]

Cribier, S., Giacobino, E., and Grynberg, G. (1983b). "Transients in optical bistability near the critical point," p. 319 in Laser Spectroscopy VI, H. P. Weber and W. Lüthy, eds. (Springer-Verlag, Berlin). [5.6, H]

Cronin-Golomb, M., White, J. O., Fischer, B., and Yariv, A. (1982). "Exact solution of a nonlinear model of four-wave mixing and phase conjugation," Opt. Lett. **7**, 313. [3.10.3]

Cross, P. S., Schmidt, R. V., and Thornton, R. L. (1977). "Optically-controlled two channel integrated-optical switch," TUB1 in Topical Meeting on Integrated and Guided Wave Optics (Optical Society of America, Washington, DC). (4.4)

Cross, P. S., Schmidt, R. V., Thornton, R. L., and Smith, P. W. (1978). "Optically controlled two channel integrated-optical switch," IEEE J. Quantum Electron. **QE-14**, 577. [4.4]

Crutchfield, J. P., and Huberman, B. A. (1980). "Fluctuations and the onset of chaos," Phys. Lett. **77A**, 407. [6.3]

Crutchfield, J., Nauenberg, M., and Rudnick, J. (1981). "Scaling for external noise at the onset of chaos," Phys. Rev. Lett. **46**, 933. [6.3]

Dagenais, M. (1982). "Low power optical bistability using bound excitons," J. Opt. Soc. Am. **72**, 1835. [3.8.2]

Dagenais, M. (1983). "Low-power optical saturation of bound excitons with giant oscillator strength, " Appl. Phys. Lett. **43**, 742. [3.8.2,7.3.1]

Dagenais, M. (1984). "Optical hysteresis in fast transient experiments near the band gap of cadmium sulfide," Appl. Phys. Lett. **45**, 1267 (1984). (3.8.2)

Dagenais, M., and Sharfin, W. F. (1984). "Cavityless optical bistability due to light-induced absorption in cadmium sulfide," Appl. Phys. Lett. **45**, 210. [2.10]

Dagenais, M., and Sharfin, W. F. (1985). "Picojoule, subnanosecond, all-optical switching using bound excitons in CdS," Appl. Phys. Lett. **46**, 230. [3.8.2]

Dagenais, M., Sharfin, W. F., and Winful, H. G. (1984). "Low power optical bistability in cadmium sulfide platelets," J. Opt. Soc. Am. B **1**, 476. [2.10]

Dagenais, M., and Winful, H. (1982). "Low-power saturation of bound excitons in cadmium sulfide platelets," J. Opt. Soc. Am. **72**, 1782. [3.8.2]

Dagenais, M., and Winful, H. (1983a). "Low-power saturation of bound excitons and observation of optical bistability," p. 108 in Conference on Lasers and Electro-Optics Technical Digest (IEEE, New York). [3.8.2]

Dagenais, M., and Winful, H. G. (1983b). "Bound-exciton optical nonlinearity, transverse bistability, and self-pulsation in cadmium sulfide," J. Opt. Soc. Am. **73**, 1863. (3.8.2)

Dagenais, M., and Winful, H. (1984a). "Low power optical bistability near bound excitons in cadmium sulfide," p. 267 in Optical Bistability 2, C. M. Bowden, H. M. Gibbs, and S. L. McCall, eds. (Plenum Press, New York). [3.8.2, 6.1]

Dagenais, M., and Winful, H. G. (1984b). "Low power transverse optical bistability near bound excitons in cadmium sulfide," Appl. Phys. Lett. **44**, 574. [3.8.2]

Damen, T. C., and Duguay, M. A. (1980). "An opto-electronic regenerative pulser," Electron. Lett. **16**, 166. [6.3]

Damen, T. C., Duguay, M. A., Wiesenfeld, J. M., Stone, J., and Burrus, C. A. (1980). "Picosecond pulses from an optically pumped GaAs laser," p. 38 in Picosecond Phenomena, R. Hochstrasser, W. Kaiser, and C. V. Shank, eds. (Springer-Verlag, Berlin). [3.6]

Davis, P., and Ikeda, K. (1984). "T^3 in a model of a nonlinear optical resonator," Phys. Lett. A **100A**, 455. (6.3)

DeFonzo, A. P., Lee, C. H., and Mak, P. S. (1979). "Optically controllable millimeter wave phase shifter," Appl. Phys. Lett. **35**, 575. [5.3]

Delfino, G. (1982). "Optical hybrid bistability related to feedback configurations: an analysis," Appl. Opt. **21**, 4377. (4.5)

Dembinski, S. T., Kossakowski, A., Lugiato, L. A., and Mandel, P. (1978a). "Semiclassical and quantum theory of the bistability in lasers containing saturable absorbers II," Phys. Rev. A **18**, 1145. (1.3)

Dembinski, S. T., Kossakowski, A., Wolniewicz, L., Lugiato, L. A., and Mandel, P. (1978). "Correlation functions for a laser with saturable absorber near threshold," Z. Phys. B **32**, 107. (6.5)

Derstine, M. W., Gibbs, H. M., Hopf, F. A., and Kaplan, D. L. (1982). "Bifurcation gap in a hybrid optically bistable system," Phys. Rev. A **26**, 3720. [6.3]

Derstine, M. W., Gibbs, H. M., Hopf, F. A., and Kaplan, D. L. (1983). "Alternate paths to chaos in optical bistability," Phys. Rev. A **27**, 3200. [6.3]

Derstine, M. W., Gibbs, H. M., Hopf, F. A., and Rushford, M. C. (1982). "Possible observation of optical chaos in an all-optical bistable device," J. Opt. Soc. Am. **72**, 1753. [6.1]

Derstine, M. W., Gibbs, H. M., Hopf, F. A., and Sanders, L. D. (1984). "Distinguishing chaos from noise in an optically bistable system," J. Opt. Soc. Am. B **1** 464. [6.3]

Derstine, M. W., Hopf, F. A., Kaplan, D. L., and Gibbs, H. M. (1982). "Bifurcation gap in a hybrid bistable device," J. Opt. Soc. Am. **72**, 1770. [6.3]

Dingle, R. (1975). "Confined carrier quantum states in ultrathin semiconductor heterostructures," p. 21 in Festkörperprobleme XV (Advances in Solid State Physics), H. J. Quiesser, ed. (Pergamon-Vieweg, Braunschweig). [3.6]

Dlodlo, T. S. (1981). "Laser with a saturable absorber--dispersive effects," Physica **111C**, 353. (1.3)

Dlodlo, T. S. (1981). "Bistable mirrors for self-pulsing lasers," Phys. Lett. **84A**, 107. (1.3)

Dlodlo, T. S. (1983). "Multistable laser through a nonlinear mirror," Phys. Rev. A **27**, 1435. (1.3)

Donnelly, J. P., DeMeo, N. L., Ferrante, G. A., Nichols, K. B., and O'Donnell, F. J. (1984). "Optical guided-wave gallium arsenide monolithic interferometer," Appl. Phys. Lett. **45**, 360. (3.10.5)

Dorsel, A., McCullen, J. D., Meystre, P., Vignes, E., and Walther, H. (1983). "Optical bistability and mirror confinement induced by radiation pressure," Phys. Rev. Lett. **51**, 1550. [3.10.7]

Dorsel, A., McCullen, J. D., Meystre, P., Vignes, E., and Walther, H. (1984). "Radiation pressure induced optical bistability and mirror confinement," p. 353 in Optical Bistability 2, C. M. Bowden, H. M. Gibbs, and S. L. McCall, eds., (Plenum Press, New York). [3.10.7]

Dove, B. L., and Russell, T. B. (1984). Proceedings of the AGARD Workshop on "Digital Optical Circuit Technology," Schliersee, West Germany, Sept.

Doyle, W. M., and Gerber, W. D. (1968). "Frequency discrimination characteristics of an elliptically polarized dual-polarization gas laser," IEEE J. Quantum Electron. **QE-4**, 870. [1.3]

Drummond, P. D. (1979). "Nonequilibrium transitions in quantum optical systems," PhD Dissertation, University of Waikato, New Zealand. [2.4, 2.8, 2.9]

Drummond, P. D. (1980). "Optical bistability in radially varying beams," Bull. Am. Phys. Soc. **25**, 565. [2.9]

Drummond, P. D. (1981a). "Optical bistability in a radially varying mode," IEEE J. Quantum Electron. **QE-17**, 301. [2.9]

Drummond, P. D. (1981b). "Applications of the generalized P-representation to optical bistability," p. 481 in Optical Bistability, C. M. Bowden, M. Ciftan, and H. R. Robl, eds. (Plenum Press, New York).

Drummond, P. D. (1982). "Adiabatic results for populations in optical bistability," Opt. Commun. **40**, 224. [5.1, 6.5]

Drummond, P. D., and Carmichael, H. J. (1978). "Volterra cycles and the cooperative fluorescence critical point," Opt. Commun. **27**, 160. [3.10]

Drummond, P. D., and Gardiner, C. W. (1980). "Generalised P-representations in quantum optics," J. Phys. A **13**, 2353. [6.5]

Drummond, P. D., Gardiner, C. W., and Walls, D. F. (1981). "Quasiprobability methods for nonlinear chemical and optical systems," Phys. Rev. A **24**, 914. [6.5]

Drummond, P. D., McNeil, K. J., and Walls, D. F. (1979). "Bistability and photon antibunching in sub/second harmonic generation," Opt. Commun. **28**, 255. [3.10.7, 6.5]

Drummond, P. D., McNeil, K. J., and Walls, D. F. (1980a). "Statistical properties of an incoherently driven nonlinear interferometer," Phys. Rev. A **22**, 1672. [3.10.7, 6.5]

Drummond, P. D., McNeil, K. J., and Walls, D. F. (1980b). "Non-equilibrium transitions in sub/second harmonic generation. I. Semiclassical theory." Opt. Acta **27**, 321. [6.5]

Drummond, P. D., McNeil, K. J., and Walls, D. F. (1980c). "Non-equilibrium transitions in sub/second harmonic generation. II. Quantum theory," Opt. Acta, **28** 211. (6.5)

Drummond, P. D., and Walls, D. F. (1980a). "Transmitted spectrum of dispersive optical bistability," J. Opt. Soc. Am. **70**, 589. [2.8]

Drummond, P. D., and Walls, D. F. (1980b). "Quantum theory of optical bistability I: nonlinear polarizability model," J. Phys. A **13**, 725. [2.8, 6.5]

Drummond, P. D., and Walls, D. F. (1981). "Quantum theory of optical bistability. II. Atomic fluorescence in a high-Q cavity," Phys. Rev. A **23**, 2563. [6.5]

Duguay, M. A. (1976). "The ultrafast optical Kerr shutter," p. 161 in Progress in Optics XIV, E. Wolf, ed. (North-Holland, Amsterdam). [5.8]

Duguay, M. A., Damen, T. C., Stone, J., Wiesenfeld, J. M., and Burrus, C. A. (1980). "Picosecond pulses from an optically pumped ribbon-whisker laser," Appl. Phys. Lett. **37**, 370 (received April 11, 1980). [3.6]

Duguay, M. A., and Hansen, J. W. (1969). "An ultrafast light gate," Appl. Phys. Lett. **15**, 192. [5.3, 5.8]

Durbin, S. D., Arakelian, S. M., Cheung, M. M., and Shen, Y. R. (1982). "Nonlinear optical effects in a nematic liquid crystal resulting from laser-induced refractive indices," Appl. Phys. B **28**, 145. [3.4]

Durbin, S. D., Arakelian, S. M., Cheung, M. M., and Shen, Y. R. (1983). "Highly nonlinear optical effects in liquid-crystals," J. Phys. (Paris) **44**, C2-161. (3.4)

Durham, T. (1983). "Shedding a light on the optical device potential," Computing **11**, 26. (1.2)

Dutta, N. K., Agrawal, G. P., and Focht, M. W. (1984). "Bistability in coupled cavity semiconductor lasers," Appl. Phys. Lett. **44**, 30. (1.3)

Dziura, T. G., and Hall, D. G. (1983). "Bistable operation of two semiconductor lasers in an external cavity: rate-equation analysis," IEEE J. Quantum Electron. **QE-19**, 441. (1.3)

Dziura, T. G., and Hall, D. G. (1985). "Semiclassical theory of bistable semiconductor lasers including radial mode variation," Phys. Rev. A **31**, 1551. (1.3, 2.9)

Eckmann, J.-P. (1981). "Roads to turbulence in dissipative dynamical systems," Rev. Mod. Phys. **53**, 643. (6.3)

Efron, U., Braatz, P. O., Little, M. J., Schwartz, R. N., and Grinberg, J. (1983). "Silicon liquid crystal light valves: status and issues," Opt. Eng. **22**, 682. [7.4]

Efron, U., Soffer, B. H., and Caulfield, H. J. (1983). "The applications of silicon liquid crystal light valves to optical data processing: a review," p. 105 in Proceedings of the NASA Conference on Optical Information Processing for Aerospace Applications. [7.4]

Eichler, H. J. (1983). "Optical multistability in silicon observed with a cw laser at 1.06 μm," Opt. Commun. **45**, 62. [3.5.2]

Eichler, H. J., Heritage, J. P., and Beisser, F. A. (1981). "Optical tuning of a silicon Fabry-Perot-interferometer by a pulsed 1.06 μm laser," IEEE J. Quantum Electron. **QE-17**, 2351. [5.2]

Eichler, H. J., Massmann, F., and Zaki, C. (1982). "Modulation and compression of Nd:YAG laser pulses by self-tuning of a silicon cavity," Opt. Commun. **40**, 302. [5.2]

Eichler, H. J., Massmann, F., Zaki, C., and Heritage, J. (1982). "Self-modulation of Nd-YAG laser pulses in silicon and optical multistability," Appl. Phys. B **28**, 13. [3.5, 5.2]

Elci, A., and Rogovin, D. (1981). "Four-wave mixing and phase conjugation near the band edge," Phys. Rev. B **24**, 5796. [7.3]

Elci, A., Scully, M. O., and O'Hare, J. M. (1983). "Electron-hole-pair generation and hysteresis in semiconductors," Phys. Rev. B **27**, 4779. (2.10, 3.7, 7.3.5)

Elgin, J. N., and Sarkar, S. (1984). "Exact results for photon statistics of an optical bistability type system," Opt. Acta **31**, 131. (6.5)

Emel'yanov, V.I., and Zokhdi, Z. (1980). "Bistability and hysteresis of static polarization induced by laser illumination of crystals," Sov. J. Quantum Electron. **10**, 869. [3.10.6]

Emshary, C. A., Al-Saidi, I. A., Harrison, R. G., and Firth, W. J. (1983). "Observation of optical bistability in an all-optical passive ring cavity containing a molecular gas," paper 13 in Programme of Sixth National Quantum Electronics Conference, University of Sussex, 19-22 September. [3.2.5b]

Engelbrecht, C. A., ed. (1981). Quantum Optics. Proceedings of the South African Summer School in Theoretical Physics (Springer-Verlag, Berlin).

Englund, J. C., Schieve, W. C., Zurek, W., and Gragg, R. F. (1981). "Fluctuations and transitions in the absorptive optical bistability," p. 315 in Optical Bistability, C. M. Bowden, M. Ciftan, and H. R. Robl, eds. (Plenum Press, New York).

Englund, J. C., Snapp, R. R., and Schieve, W. C. (1983). "Fluctuations, instabilities, and chaos in the laser-driven nonlinear ring cavity," in Progress in Optics, Vol. 21, E. Wolf, ed. (North-Holland, Amsterdam). [6.3]

Épshtein, E. M. (1978). "Optical thermal breakdown of a semiconducting plate," Sov. Phys. Tech. Phys. **23**, 983. [2.10]

Erneux, T., and Mandel, P. (1983). "Temporal aspects of absorptive optical bistability," Phys. Rev. A **28**, 896. [5.1]

Erneux, T., and Mandel, P. (1984). "Stationary, harmonic, and pulsed operations of an optically bistable laser with saturable absorber. II," Phys. Rev. A **30**, 1902. (6.3, 6.4).

Farhat, N., and Psaltis, D. (1984). "New approach to optical information processing based on the Hopfield model," talk ThP6, Optical Society of America 1984 Annual Meeting, Technical Program. (7.4)

Farina, J. D., Narducci, L. M., Yuan, J. M., and Lugiato, L. A. (1980). "Time evolution of an optically bistable system--local relaxation and tunneling," Opt. Eng. **19**, 469. [6.5]

Farina, J. D., Narducci, L. M., Yuan, J. M., and Lugiato, L. A. (1981). "Short- and long-time transient evolution in absorptive optical bistability," p. 337 in Optical Bistability, C. M. Bowden, M. Ciftan, and H. R. Robl, eds. (Plenum Press, New York).

Farmer, J. D. (1982). "Chaotic attractors of an infinite-dimensional dynamical system," Physica **4D**, 366. [6.3]

Fasano, J. J., Deck, R. T., and Sarid, D. (1982). "Enhanced nonlinear-propagation constant of a long-range surface-plasma wave," J. Opt. Soc. Am. **72**, 1747. [3.10.5]

Fatehi, M. T., Collins, Jr., S. A., and Wasmundt, K. C. (1981). "The optical computer goes digital," Optical Spectra, p. 39, January.

Fatehi, M. T., Wasmundt, K. C., and Collins, Jr., S. A. (1981). "Optical logic gates using liquid crystal light valve: Implementation and application example," Appl. Opt. **20**, 2250. [7.4]

Fatehi, M. T., Wasmundt, K. C., and Collins, Jr., S. A. (1984). "Optical flip-flops and sequential logic circuits using a liquid crystal light valve," Appl. Opt. **23**, 2163. [7.4]

Fazekas, P. (1981). "Laser-induced phase transition in amorphous $GeSe_2$ films," Phil. Mag. B **44**, 435. [6.1]

Fehenback, G. W., Schäfer, W., Treusch, J., and Ulbrich, R. G. (1982). "Transient optical spectra of a dense exciton gas in a direct-gap semiconductor," Phys. Rev. Lett. **49**, 1281.

Feigenbaum, M. J. (1980). "Universal behavior in nonlinear systems," Los Alamos Science **1**, 4. [6.3]

Feinberg, J. (1982). "Asymmetric self-defocusing of an optical beam from the photorefractive effect," J. Opt. Soc. Am. **72**, 46. [3.10.3]

Felber, F. S., and Marburger, J. H. (1976). "Theory of nonresonant multistable optical devices," Appl. Phys. Lett. **28**, 731. [2.6]

Feldman, A. (1978a). "Bistable optical systems based on Pockels cells," J. Opt. Soc. Am. **68**, 1448. [4.5]

Feldman, A. (1978b). "Ultralinear bistable electrooptic polarization modulator," Appl. Phys. Lett. **33**, 243. [4.5, 5.4]

Feldman, A. (1979a). "Bistable optical systems based on a Pockels cell," Opt. Lett. **4**, 115. [4.5]

Feldman, A. (1979b). "Steady-state analysis of a modified bistable optical system," J. Opt. Soc. Am. **69**, 1421. [4.5]

Ferrano, G., and Häusler, G. (1980a). "A bistable TV-optical feedback system," Opt. Commun. **32**, 375.

Ferrano, G., and Häusler, G. (1980b). "TV optical feedback systems," Opt. Eng. **19**, 442. [7.4]

Feynman, R. P., Vernon, Jr., F. L., and Hellwarth, R. W. (1957). "Geometrical representation of the Schrödinger equation for solving maser problems," J. Appl. Phys. **28**, 49. [C]

Firth, W. J. (1981). "Stability of nonlinear Fabry-Perot resonators," Opt. Commun. **39**, 343. [6.3]

Firth, W. J. (1983). "Dynamical instabilities in nonlinear resonators," paper 9 in Programme of Sixth National Quantum Electronics Conference, University of Sussex, 19-22 September.

Firth, W. J., and Abraham, E. (1984). "Physical interpretation of the route to chaos in nonlinear resonators," J. Opt. Soc. Am. B **1**, 489. (6.3)

Firth, W. J., Abraham, E., and Wright, E. M. (1982). "Ikeda oscillation and chaos in folded Fabry-Perot resonators," Appl. Phys. B **28**, 170. [6.3]

Firth, W. J., Abraham, E., Wright, E. M., Galbraith, I., and Wherrett, B. S. (1984). "Diffusion, diffraction and reflection in semiconductor optical bistability devices," Proc. Royal Society Meeting on Optical Bistability, Dynamic Nonlinearity and Photonic Logic, London. [2.9]

Firth, W. J., Seaton, C. T., Wright, E. M., and Smith, S. D. (1982). "Spatial hysteresis in optical bistability," Appl. Phys. B **28**, 131. [2.9, 3.7, 7.3]

Firth, W. J., and Wright, E. M. (1982a). "Theory of Gaussian-beam optical bistability," Opt. Commun. **40**, 233. [2.9]

Firth, W. J., and Wright, E. M. (1982b). "Oscillation and chaos in a Fabry-Perot bistable cavity with Gaussian input beam," Phys. Lett. A **92A**, 211. [6.3]

Firth, W. J., Wright, E. M., and Cummins, E. J. D. (1984). "Connection between Ikeda instability and phase conjugation," p. 111 in Optical Bistability 2, C. M. Bowden, H. M. Gibbs, and S. L. McCall, eds. (Plenum Press, New York). [6.3]

Fisher, R. A., ed. (1983). Optical Phase Conjugation (Academic Press, New York). [5.8]

Fleck, Jr., F. A. (1968). "Evolution of a Q-switched laser pulse from noise," Appl. Phys. Lett. **13**, 365. [2.5]

Flytzanis, C., Agrawal, G. P., Tang, C. L. (1981). "Critical behavior in phase conjugation," p. 317 in Lasers and Applications, W. O. N. Guimaraes, C.-T. Lin, and A. Mooradian, eds. (Springer-Verlag, Berlin). [3.10.3]

Flytzanis, C., and Tang, C. L. (1980). "Light-induced critical behavior in the four-wave interaction in nonlinear systems," Phys. Rev. Lett. **45**, 441. [3.10.3]

Fork, R. L. (1982). "The physics of optical switching," Phys. Rev. A **26**, 2049. [7.2]

Fork, R. L., Tomlinson, W. J., and Heilos, L. J. (1966). "Hysteresis in an He-Ne laser," Appl. Phys. Lett. **8**, 162. [1.3]

Fujimoto, J. G., Weiner, A. M., and Ippen, E. P. (1984). "Generation and measurement of optical pulses as short as 16 fs," Appl. Phys. Lett. **44**, 832. [5.2]

Gabrielse, G., Dehmelt, H., and Kells, W. (1985). "Observation of a relativistic, bistable hysteresis in the cyclotron motion of a single electron," Phys. Rev. Lett. **54**, 537. [3.10.7]

Galarneau, P., Chin, S. L., Xing-Xiao, M., Farkas, G., and Yergeau, F. (1983). "Single pulse detection of self-focusing in a polyatomic molecular gas using the bistability characteristics," Proc. SPIE **380**, 392 (Los Alamos Conference on Optics '83). (3.9)

Gao, J. Y., Narducci, L. M., Sadiky, H., Squicciarini, M., and Yuan, J. M. (1984). "Higher-order bifurcations in a bistable system with delay," Phys. Rev. A **30**, 901. [6.3]

Gao, J. Y., Narducci, L. M., Schulman, L. S., Squicciarini, M., and Yuan, J. M. (1983). "Route to chaos in a hybrid bistable system with delay," Phys. Rev. A **28**, 2910. [6.3]

Gao, J. Y., Yuan, J. M., and Narducci, L. M. (1983). "Instabilities and chaotic behavior in a hybrid bistable system with a short delay," Opt. Commun. **44**, 201. [6.3]

Gardiner, C. W. (1981). "Aspects of classical and quantum theory of stochastic bistable systems," p. 53 in Stochastic Nonlinear Systems in Physics, Chemistry, and Biology. Proceedings of the Workshop, L. Arnold and R. Lefever, eds. (Springer-Verlag, Berlin). (6.5)

Gardiner, K., and Vaughan, J. M. (1983). "Resonant modes in a dispersive cavity," Proc. SPIE **369**, 253. [3.2.3]

Garmire, E. (1979). "Progress in bistable optical devices," Proc. SPIE **176**, 12. [1.2]

Garmire, E. (1981a). "Signal processing with a nonlinear Fabry-Perot," Proc. SPIE **269**, 69. [1.2]

Garmire E. (1981b). "Introduction to bistable optical devices for optical signal processing," Proc. SPIE **218**, 27. [1.2, 3.10.5] (4.3)

Garmire, E., Allen, S. D., and Marburger, J. (1978). "Bistable optical devices for integrated optics and fiber optics applications," Proc. SPIE **139**, 174. (1.2, 4.3)

Garmire, E., Allen, S. D., and Marburger, J. (1979). "Bistable optical devices for integrated optics and fiber optics applications," Opt. Eng. **18**, 194.

Garmire, E., Allen, S. D., Marburger, J., and Verber, C. M. (1978). "Multimode integrated optical bistable switch," Opt. Lett. **3**, 69. [4.4]

Garmire, E., Goldstone, J. A., and Hendrix, J. (1981). "Ultrashort pulse detection with nonlinear Fabry-Perots," IEEE J. Quantum Electron. **QE-17**, 118. (5.2)

Garmire, E., Marburger, J. H., and Allen, S. D. (1978). "Incoherent mirrorless bistable optical devices," Appl. Phys. Lett. **32**, 320. [4.3]

Garmire, E., Marburger, J. H., Allen, S. D., and Winful, H. G. (1979). "Transient response of hybrid bistable optical devices," Appl. Phys. Lett. **34**, 374. [4.5, 5.1, 5.6, H]

Gea-Banacloche, J., Chow, W. W., and Scully, M. O. (1983). "Laser cavity dumping using optical bistability," Opt. Commun. **46**, 43. (1.3)

Gerlach, U. H., Sengupta, U. K., and Collins, Jr., S. A. (1980). "Single-spatial light modulator bistable optical matrix device using optical feedback," Opt. Eng. **19**, 452. [7.4]

Germey, K., Schutte, F. J., Tiebel, R., and Worlitzer, K. (1982). "Quantum statistics of multimode optical bistability with trilinear interaction," Ann. Phys. (Germany) **39**, 170.

Giacobino, E., Cribier, S., Grynberg, G., and Biraben, F. (1982). "Study of transients in intrinsic optical bistability," Appl. Phys. B **29**, 170. [5.6]

Giacobino, E., Devaud, M., Biraben, F., and Grynberg, G. (1980). "Doppler-free two-photon dispersion and optical bistability in rubidium vapor," Phys. Rev. Lett. **45**, 434. [3.10]

Giacobino, E., and Grynberg, G. (1984). "Switching in all-optical bistability," International Conference on Lasers '84, San Francisco.

Gibbs, H. M. (1982). "Optical bistability," 1982 Yearbook of Science and Technology (McGraw-Hill, New York).

Gibbs, H. M. (1982). "Optical logic--What is it? Where does it fit in?" p. F4.2 in National Telesystems Conference. NTC '82. Systems for the Eighties (IEEE, New York). [1.2]

Gibbs, H. M. (1983a). "Progress toward all-optical logic using semiconductors," p. 92 in Conference on Lasers and Electro-Optics Technical Digest (IEEE, New York). [1.2]

Gibbs, H. M. (1983b). "Optical bistability for logic," Bull. Am. Phys. Soc. **28**, 1316. [1.2]

Gibbs, H. M. (1984). "Physics of optical bistability," talk ThF1, Optical Society of America 1984 Annual Meeting, Technical Program.

Gibbs, H. M. (1984). "Optical bistability and nonlinear optical signal processing," International Conference on Lasers '84, San Francisco.

Gibbs, H. M., Churchill, G. G., and Salamo, G. J. (1974). "Faraday rotation under cw saturation and self-induced transparency conditions," Opt. Commun. **12**, 396. [3.2]

Gibbs, H. M., Derstine, M. W., Hopf, F. A., Jewell, J. L., Kaplan, D. L., Moloney, J. V., Shoemaker, R. L., Tai, K., Tarng, S. S., Watson, E. A., Gossard, A. C., McCall, S. L., Passner, A., Venkatesan, T. N. C., and Wiegmann, W. (1981). "Transient phenomena in optical bistability," p. 265 in Lasers '81, C. B. Collins, ed. (STS Press, McLean, Virginia). [2.9]

Gibbs, H. M., Derstine, M. W., Hopf, F. A., Kaplan, D. L., Rushford, M., C., Shoemaker, R. L., Weinberger, D. A., and Wing, W. H. (1982). "Optical bistability instabilities: regenerative pulsations, periodic oscillations, and optical chaos," talk FL2 in Conference on Lasers and Electro-Optics Technical Digest (IEEE, New York). [3.9.1, 6.1, 6.3]

Gibbs, H. M., Gossard, A. C., McCall, S. L., Passner, A., Wiegmann, W., and Venkatesan, T. N. C. (1979). "Saturation of the free exciton resonance in GaAs," Solid State Commun. **30**, 271. [3.6]

Gibbs, H. M., Hopf, F. A., Kaplan, D. L., Derstine, M. W., and Shoemaker, R. L. (1981). "Periodic oscillations and chaos in optical bistability; possible guided-wave all-optical square-wave oscillators," Proc. SPIE **317**, 297. [6.3]

Gibbs, H. M., Hopf, F. A., Kaplan, D. L., and Shoemaker, R. L. (1981). "Observation of chaos in optical bistability," Phys. Rev. Lett. **46**, 474; J. Opt. Soc. Am. **71**, 367. [4.5, 6.3]

Gibbs, H. M., Jewell, J. L., Moloney, J. V., Rushford, M. C., Tai, K., Tarng, S. S., Weinberger, D. A., Gossard, A. C., McCall, S. L., Passner, A., and Wiegmann, W. (1982). "Room-temperature optical bistability and self-defocusing in semiconductor etalons," Appl. Phys. B **29**, 171. [2.9]

Gibbs, H. M., Jewell, J. L., Moloney, J. V., Tai, K., Tarng, S. S., Weinberger, D. A., Gossard, A. C., McCall, S. L., Passner, A., and Wiegmann, W. (1983). "Optical bistability, regenerative pulsations, and transverse effects in room-temperature GaAs-AlGaAs superlattice etalons," J. Phys. (Paris) **44**, C2-195. [3.6, 7.3]

Gibbs, H. M., Jewell, J. L., Moloney, J. V., Tarng, S. S., Tai, K., Watson, E. A., Gossard, A. C., McCall, S. L., Passner, A., Venkatesan, T. N. C., and Wiegmann, W. (1982). "Switching of a GaAs bistable etalon: external switching on and off, regenerative pulsations, transverse effects and lasing," Proc. SPIE **321**, 67. [2.9, 5.5]

Gibbs, H. M., Jewell, J. L., Tarng, S. S., Gossard, A. C., and Wiegmann, W. (1981). "Regenerative pulsations in an optical bistable GaAs etalon," IEEE J. Quantum Electron. **QE-17**(12), 42. [6.1]

Gibbs, H. M., McCall, S. L., Gossard, A. C., Passner, A., Wiegmann, W., and Venkatesan, T. N. C. (1980). "Controlling light with light: optical bistability and optical modulation," p. 441 in Laser Spectroscopy IV, H. Walther and K. W. Rothe, eds. (Springer-Verlag, Berlin). [3.5, 3.6]

Gibbs, H. M., McCall, S. L., Passner, A., Gossard, A. C., Wiegmann, W., and Venkatesan, T. N. C. (1979). "Progress toward practical optical bistable devices," CLEA 1979; IEEE J. Quantum Electron. **QE-15**, 108D (received January 5). [3.5, 3.6, 3.7, 5.5]

Gibbs, H. M., McCall, S. L., and Venkatesan, T. N. C. (1975). "Amplifying characteristics of a cavity-enclosed nonlinear medium," U.S. Patents 4,012,699, filed March 28, 1975, granted March 15, 1977; and 4,121,167, filed October 27, 1976, granted October 17, 1978. [3.2, 3.3, 3.5, 3.6, 3.7]

Gibbs, H. M., McCall, S. L., and Venkatesan, T. N. C. (1976). "Differential gain and bistability using a sodium-filled Fabry-Perot interferometer," Phys. Rev. Lett **36**, 1135. [1.2, 2.3, 3.2, 3.5, 5.3, A, C]

Gibbs, H. M., McCall, S. L., and Venkatesan, T. N. C. (1977). "Optical bistability and differential gain," p. 111 in Coherence in Spectroscopy and Modern Physics, F. T. Arecchi, R. Bonifacio, and M. O. Scully, eds. (Plenum Press, New York). [2.3, 3.2]

Gibbs, H. M., McCall, S. L., and Venkatesan, T. N. C. (1979). "Optical bistability," Optics News **5**, 6. [1.1]

Gibbs, H. M., McCall, S. L., and Venkatesan, T. N. C. (1980). "Optical bistable devices: the basic components of all-optical systems?" Opt. Eng. **19**, 463. [1.2, 3.6, 7.1, 7.4]

Gibbs, H. M., McCall, S. L., and Venkatesan, T. N. C. (1981). "Optical bistable devices: the basic components of all-optical systems?" Proc. SPIE **269**, 75. (3.6)

Gibbs, H. M., McCall, S. L., Venkatesan, T. N. C., Gossard, A. C., Passner, A., and Wiegmann, W. (1979). "Optical bistability in semiconductors," Appl. Phys. Lett. **35**, 451 (received June 15; received March 15 by Phys. Rev. Lett.). [3.5, 3.6, 5.4]

Gibbs, H. M., McCall, S. L., Venkatesan, T. N. C., Passner, A., Gossard, A. C., and Wiegmann, W. (1980a). "Optical bistability and modulation in semiconductor etalons," p. MB5-1 in Integrated and Guided-Wave Optics, Technical Digest (IEEE, New York). (3.6)

Gibbs, H. M., McCall, S. L., Venkatesan, T. N. C., Passner, A., Gossard, A. C., and Wiegmann, W. (1980b). "Nonlinear optical susceptibilities of semiconductors and optical bistability," p. 9 in Proceedings of the Topical Conference on Basic Optical Properties of Materials (U.S. Government Printing Office, Washington). (3.6)

Gibbs, H. M., McCall, S. L., Venkatesan, T. N. C., Passner, A., Gossard, A. C., and Wiegmann, W. (1980c). "Optical bistability in a GaAs etalon," p. 109 in Optical Bistability, C. M. Bowden, M. Ciftan, and H. R. Robl, eds. (Plenum Press, New York). (3.6)

Gibbs, H. M., McCall, S. L., Venkatesan, T. N. C., Passner, A., Gossard, A. C., and Wiegmann, W. (1981). "Optical bistability and optical nonlinearities in GaAs, Proceedings of the International Conference on Excited States and Multiresonant Nonlinear Optical Processes in Solids, Aussois, France, March 18-20 (Les Editions de Physique, Orsay, France). [3.6]

Gibbs, H. M., Olbright, G. R., Peyghambarian, N., Haug, H. (1985). "Kinks in increasing absorption optical bistability," paper TuM5 in Conference on Lasers and Electro-Optics Technical Digest (IEEE, New York). [2.10]

Gibbs, H. M., Olbright, G. R., Peyghambarian, N., Schmidt, H. E., Koch, S. W., and Haug, H. (1985). "Kinks: longitudinal excitation discontinuities in increasing absorption bistability," Phys. Rev. A, to be published. (2.10)

Gibbs, H. M., and Slusher, R. E. (1971). "Optical pulse compression by focusing in a resonant absorber," Appl. Phys. Lett. **18**, 505. [5.2]

Gibbs, H. M., Tarng, S. S., Jewell, J. L., Tai, K., Weinberger, D. A., Gossard, A. C., McCall, S. L., Passner, A., and Wiegmann, W. (1982). "Optical bistability in semiconductor etalons," Workshop on Opto-Electronics, La Jolla, California, March 23. [3.6]

Gibbs, H. M., Tarng, S. S., Jewell, J. L., Weinberger, D. A., Tai, K., Gossard, A. C., McCall, S. L., Passner, A., and Wiegmann, W. (1982a). "Room-temperature excitonic optical bistability in GaAs etalons," postdeadline paper FL6 in Conference on Lasers and Electro-Optics Technical Digest (IEEE, New York). [3.6, 6.1]

Gibbs, H. M., Tarng, S. S., Jewell, J. L., Weinberger, D. A., Tai, K., Gossard, A. C., McCall, S. L., Passner, A., and Wiegmann, W. (1982b). "Room-temperature excitonic optical bistability in a GaAs-GaAlAs superlattice etalon," Appl. Phys. Lett. **41**, 221. [3.6]

Gibbs, H. M., Venkatesan, T. N. C., McCall, S. L., Passner, A., Gossard, A. C., and Wiegmann, W. (1979). "Optical modulation by optical tuning of a cavity," Appl. Phys. Lett. **34**, 511. [1.2, 5.3]

Giglio, M. (1982). "Transition to chaotic behavior via period doubling bifurcations," Appl. Phys. B **28**, 165. [6.3]

Gilmore, R., and Narducci, L. M. (1978). "Relation between the equilibrium and nonequilibrium critical properties of the Dicke model," Phys. Rev. A **17**, 1747.

Gioggia, R. S., and Abraham, N. B. (1984). "Anomalous mode pulling, instabilities, and chaos in a single-mode, standing-wave 3.39-μm He-Ne laser," Phys. Rev. A **29**, 1304. [6.3]

Glas, P., Klehr, A., and Müller, R. (1982). "Transient and stationary properties in bistable operation of a GaAs laser coupled to an external resonator," Opt. Commun. **44**, 196. (1.3)

Glas, P., and Müller, R. (1982). "Bistable operation of a GaAs-AlGaAs diode-laser coupled to an external resonator of narrow spectral bandwidth," Opt. Quantum Electron. **14**, 375. (1.3)

Glas, P., Müller, R., and Klehr, A. (1983). "Bistability, self-sustained oscillations, and irregular operation of a GaAs laser coupled to an external resonator," Opt. Commun. **47**, 297. (1.3)

Glass, A. M. (1983). "Material limitations in optical signal processing," p. 106 in Conference on Lasers and Electro-Optics, Technical Digest (IEEE, New York). [7.3] (7.4)

Glass, A. M., and Negran, T. J. (1974). "Optical gating and logic with pyroelectric crystals," Appl. Phys. Lett. **24**, 81. [5.3]

Gobeli, G. W., and Fan, H. Y. (1960). "Infrared absorption and valence band in indium antimonide," Phys. Rev. **119**, 613.

Goldobin, I. S., Kurnosov, V. D., Luk'yanov, V. N., Pak, G. T., Semenov, A. T., Shelkov, N. V., Yakubovich, S. D., and Yashumov, I. V. (1981). "Bistable cw injection heterolaser," Sov. J. Quantum Electron. **11**, 526. (1.3)

Goldstone, J. A., (1984). "Optical bistability," in Laser Handbook, Volume 4, M. Stitch and M. Bass, eds. (North-Holland, Amsterdam). [1.1, 6.1, E, H] (5.6, 6)

Goldstone, J. A., and Garmire, E. M. (1981a). "On the dynamic response of nonlinear Fabry-Perot interferometers," IEEE J. Quantum Electron. **QE-17**, 366. [5.1, 5.7, 6.1]

Goldstone, J. A., and Garmire, E. (1981b). "Regenerative oscillation in the nonlinear Fabry-Perot," J. Opt. Soc. Am. **71**, 1633. [6.1]

Goldstone, J. A., and Garmire, E. M. (1983a). "Self-pulsing in a nonlinear Fabry-Perot with a long medium-response time," p. 90 in Conference on Lasers and Electro-Optics Technical Digest (IEEE, New York). [6.1]

Goldstone, J. A., and Garmire, E. M. (1983b). "Regenerative oscillation in the nonlinear Fabry-Perot interferometer," IEEE J. Quantum Electron. **QE-19**, 208. [6.1]

Goldstone, J. A., and Garmire, E. M. (1984a). "Macroscopic manifestations of microscopic optical bistability," paper TUBB3, IQEC '84 [2.10] and; J. Opt. Soc. Am. B **1**, 466. (2.10)

Goldstone, J. A., and Garmire, E. (1984b). "Intrinsic optical bistability in nonlinear media," Phys. Rev. Lett. **53**, 910 [2.10]

Goldstone, J. A., Garmire, E., and Ho, P.-T. (1980). "Overshoot switching in bistable optical devices," p. MB6 in Integrated and Guided-Wave Optics Technical Digest (IEEE, New York). (5.6)

Goldstone, J. A., Ho, P.-T., and Garmire, E. (1980). "Overshoot switching, alternate switching, and subharmonic generation in bistable optical devices," Appl. Phys. Lett. **37**, 126. [5.6]

Goldstone, J. A., Ho, P.-T., and Garmire, E. (1981). "Transient phenomena in bistable optical devices," p. 187 in Optical Bistability, C. M. Bowden, M. Ciftan, and H. R. Robl, eds. (Plenum Press, New York).

Golik, L. L., Grigor'yants, A. V., Elinson, M. I., and Balkarei, Yu. I. (1983). "Hysteresis and nonlinear thermooptic waves in a semiconductor Fabry-Perot interferometer," Opt. Commun. **46**, 51. [3.5.2]

Goll, J., and Haken, H. (1980). "Theory of optical bistability of excitons," Phys. Status Solidi B **101**, 489. (3.6, 7.3)

Goll, J., and Haken, H. (1983). "Saturation of interband transitions in semiconductors and the effect of optical bistability," Phys. Rev. A **28**, 910. [7.3]

Gollub, J. P., and Benson, S. V. (1980). "Many routes to turbulent convection," J. Fluid Mech. **100**, 449. [6.3]

Gomi, M., Miyazawa, Y., Uchino, K., Abe, M., and Nomura, S. (1982). "Optical stabilizer using a bistable optical device with a PMN electrostrictor," Appl. Opt. **21**, 2616 (1982). [4.5]

Gomi, M., Uchino, K., Abe, M., and Nomura, S. (1981). "Bistable optical device using the electrostrictive displacement transducer," Jpn. J. Appl. Phys. **20**, L375. [4.5]

Goodman, J. W. (1982). "Architectural development of optical data processing systems," J. Electrical and Electronics Engineering, Australia **2** (3), 139. [7.4]

Goodman, J. W. (1984a). "The optical data processing family tree," Opt. News, May/June, p. 25. [7.4]

Goodman, J. W. (1984b). "Optical computing: how far can light waves penetrate computer technology?" J. Opt. Soc. Am. B **1**, 464. [7.4]

Goodman, J. W., Leonberger, F. J., Kung, S.-Y., and Athale, R. A. (1984). "Optical interconnections for VLSI systems," Proc. IEEE **72**, 850. [7.4]

Gossard, A. C. (1979). "GaAs/AlGaAs layered films," Thin Solid Films **57**, 3. (3.6)

Gouesbet, G., Rhazi, M., and Weill, M. E. (1983). "New heartbeat phenomenon, and the concept of 2-D optical turbulence," Appl. Opt. **22**, 304. [6.4]

Gozzini, A., Maccarrone, F., and Longo, I. (1982a). "Absorptive optical bistability at microwave frequencies," Il Nuovo Cimento **1D**, 489. [2.4, 3.10.7]

Gozzini, A., Maccarrone, F., and Longo, I. (1982b). "Multistability at microwave frequencies," Nuovo Cimento B **72B**, 220. [2.4,3.10.7]

Gragg, R. F., Schieve, W. C., and Bulsara, A. R. (1979). "Stochastic-master-equation description of an optical bistability: a comparison with the microscopic theory," Phys. Rev. A **19**, 2052. [6.5]

Graham, R., and Schenzle, A. (1981). "Dispersive optical bistability with fluctuations," Phys. Rev. A **23**, 1302. [6.5]

Graham, R., and Schenzle, A. (1981). "Dispersive optical bistability with fluctuations," p. 293 in Optical Bistability, C. M. Bowden, M. Ciftan, and H. R. Robl, eds. (Plenum Press, New York).

Grant, D. E., Drummond, P. D., and Kimble, H. J. (1982). "Evolution of hysteresis in absorptive bistability with two-level atoms," J. Opt. Soc. Am. **72**, 1760. [3.2.5b]

Grant, D. E., and Kimble, H. J. (1982). "Optical bistability for two-level atoms in a standing-wave cavity," Opt. Lett. **7**, 353. Erratum **8**, 66 (1983). [3.2.5b]

Grant, D. E., and Kimble, H. J. (1983). "Transient response in absorptive bistability," Opt. Commun. **44**, 415. [3.2.5b]

Grassberger, P., and Procaccia, I. (1983). "Estimation of the Kolmogorov entropy from a chaotic signal," Phys. Rev. A **28**, 2591. [6.3]

Gray, R. W., and Casperson, L. W. (1978). "Optooptic modulation based on gain saturation," IEEE J. Quantum Electron. **QE-14**, 893. [5.3,6.3]

Greene, R. L., and Bajaj, K. K. (1983). "Binding energies of Wannier excitons in GaAs-$Ga_{1-x}Al_xAs$ quantum well structures," Solid State Commun. **45**, 831. [3.6]

Greene, W. P., Gibbs, H. M., Passner, A., McCall, S. L., and Venkatesan, T. N. C. (1980). "An undergraduate optical bistability experiment," Bull. Am. Phys. Soc. **25**, 23. (4.3)

Greene, W. P., Gibbs, H. M., Passner, A., McCall, S. L., and Venkatesan, T. N. C. (1980). "Optical bistability: an undergraduate experiment," Opt. News **6**(2), 16. [4.3]

Grigor'yants, A. V., Golik, L. L., Rzhanov, Yu. A., Elinson, M. I., and Balkarei, Yu. I. (1984). "Switching waves in a multistable Fabry-Perot interferometer with a thermooptic nonlinearity," Sov. J. Quantum Electron. **14**, 714. (3.5.2)

Grischkowsky, D. (1974). "Optical pulse compression," Appl. Phys. Lett. **25**, 566. [5.2]

Grischkowky, D. (1977). "A light-controlled light modulator," Appl. Phys. Lett. **31**, 437. [5.3]

Grischkowsky, D. (1978). "Nonlinear Fabry-Perot interferometer with subnanosecond response times," J. Opt. Soc. Am. **68**, 641. [3.4]

Gronchi, M., Benza, V., Lugiato, L. A., Meystre, P., and Sargent III, M. (1981). "Analysis of self-pulsing in absorptive optical bistability," Phys. Rev. A **24**, 1419. [6.2]

Gronchi, M., and Lugiato, L. A. (1978). "Fokker-Planck equation for optical bistability," Lett. Nuovo Cimento **23**, 593. (6.5)

Gronchi, M., and Lugiato, L. A. (1980). "Exact solution for dispersive optical bistability with inhomogeneous broadening," Opt. Lett. **5**, 108. [2.1, 2.4]

Grothe, H., and Proebster, W. (1983). "Monolithic integration of InGaAsP/InP LED and transistor--a light-coupled bistable electro-optical device," Electron. Lett. **19**, 194. [4.5]

Grynberg, G., Biraben, F., and Giacobino, E. (1981). "Transients in optical bistability," Appl. Phys. B **26**, 155. [5.6, H]

Grynberg, G., and Cribier, S. (1983). "Critical exponents in dispersive optical bistability," J. Phys. Lett. **44**, L449. [H]

Grynberg, G., Devaud, M., Flytzanis, C., and Cagnac, B. (1980). "Doppler-free two-photon dispersion," J. Phys.(Paris) **41**, 931. [3.10.1]

Gugliemi, R., and Carenco, A. (1981). "Optical bistability in $LiNbO_3$ using monolithically integrated directional coupler and detector," p. 61 in First European Conference on Integrated Optics (IEE, London). [4.4]

Gurevich, S. A., Portnoi, E. L., Raikh, M. E., Ryvkin, B. S., Timofeev, F. N., and Fronts, K. (1982). "Bistable operation of a semiconductor laser," Sov. Tech. Phys. Lett. **8**, 381. (1.3)

Haag, G., Munz, M., Marowsky, G. (1981). "Bistability in the operation of an organic dye laser," IEEE J. Quantum Electron. **QE-17**, 349. (1.3)

Hajtó, J., and Apai, P. (1980). "Investigation of laser induced light absorption oscillation," J. Non-Crystalline Solids **35-36**, 1085. [6.1]

Hajtó, J., and Janossy, I. (1983). "Optical bistability observed in amorphous semiconductor films," Philos. Mag. B **47**, 347. [2.10, 6.1]

Hajtó J., Janossy, J., and Firth, A. (1983). "Explanation of the laser-induced oscillatory phenomenon in amorphous semiconductor films," Philos. Mag. B **48**, 311. [6.1]

Hajtó J., Janossy, J., and Forgacs, G. (1982). "Laser-induced optical anisotropy in self-supporting amorphous $GeSe_2$ films," J. Phys. C **15**, 6293. [2.10]

Hajtó, J., Zentai, G., and Somogyi, I. K. (1977). "Light-induced transmittance oscillation in $GeSe_2$ thin films," Solid State Commun. **23**, 401. [6.1]

Haken, H., ed. (1981). Chaos and Order in Nature (Springer-Verlag, Berlin, 1981).

Hall, D. G., and Dziura, T. G. (1984). "Transverse effects in the bistable operation of lasers containing saturable absorbers," Opt. Commun. **49**, 146. [2.9] (1.3)

Hambenne, J. B., and Sargent III, M. (1975). "Physical interpretation of bistable unidirectional ring-laser operation," IEEE J. Quantum Electron. **QE-11**, 90. (1.3)

Hamilton, M. W., Ballagh, R. J., and Sandle, W. J. (1982). "Polarization switching in a ring cavity with resonantly driven J = 1/2 to J = 1/2 atoms," Z. Phys. B **49**, 263. [3.10.2]

Hanamura, E. (1973). "Giant two-photon absorption due to excitonic molecule," Solid State Commun. **12**, 951. [3.8]

Hanamura, E. (1981a). "Optical bistable system responding in pico-second," Solid State Commun. **38**, 939. [3.8, 3.10]

Hanamura, E. (1981b). "Dynamics of coherent excitations. Optical bistable system responding in pico-second," p. 51 in Proceedings of the International Conference on Excited States and Multiresonant Nonlinear Optical Processes in Solids, Aussois, France, March 18-20 (Les Editions de Physique, Orsay, France). [3.8, 3.10]

Hanamura, E. (1982). "Optical bistability in semiconductors," Solid State Phys. **17**, 70.

Hanggi, P., Bulsara, A. R., and Janda, R. (1980). "Spectrum and dynamic-response function of transmitted light in the absorptive optical bistability," Phys. Rev. A **22**, 671. [6.5]

Hanggi, P., Grabert, H., Talkner, P., and Thomas, H. (1984). "Bistable systems: master equation versus Fokker-Planck modeling," Phys. Rev. A **29**, 371. [6.5]

Hanggi, P., Marchesoni, F., and Grigolini, P. (1984). "Bistable flow driven by coloured Gaussian noise: a critical study," Z. Phys. B **56**, 333. [6.5]

Hanggi, P., Mroczkowski, T. J., Moss, F., and McClintock, P. V. E. (1985). "Bistability driven by colored noise: theory and experiment," Phys. Rev. A, May. (6.5)

Hanggi, P., and Riseborough, P. (1983). "Activation rates in bistable systems in the presence of correlated noise," Phys. Rev. A **27**, 3379. (6.5)

Hano, M., and Kayano, H. (1979). "Magneto-optic bistable device using magnetic thin film waveguide," Trans. Inst. Electron. Commun. Eng. Jpn. **E62**, 792.

Harder, Ch., Lau, K. Y., and Yariv, A. (1981). "Bistability and pulsations in cw semiconductor lasers with a controlled amount of saturable absorption," Appl. Phys. Lett. **39**, 382. [1.3,6.1]

Harder, Ch., Lau, K. Y., and Yariv, A. (1982a). "Bistability and negative resistance in semiconductor lasers," Appl. Phys. Lett. **40**, 124. [1.3]

Harder, Ch., Lau, K. Y., and Yariv, A. (1982b). "Experimental observation of phase transition, bistability and critical slowing down in lasers," Appl. Phys. B **28**, 139. [1.3, 5.6]

Harder, Ch., Lau, K. Y., and Yariv, A. (1982c). "Bistability and pulsations in semiconductor lasers with inhomogeneous current injection," IEEE J. Quantum Electron. **QE-18**, 1351. [1.3, 6.1]

Harder, Ch., Lau, K. Y., and Yariv, A. (1982d). "Static and dynamic characteristics of bistable semiconductor injection lasers," p. FC1 in Topical Meeting on Integrated and Guided-Wave Optics (Optical Society of America, Washington, DC). (1.3)

Harder, Ch., Smith, J. S., Lau, K. Y., and Yariv, A. (1983). "Passive mode locking of buried heterostructure lasers with nonuniform current injection," Appl. Phys. Lett. **42**, 772. [6.1]

Haroche, S., and Hartmann, F. (1972). "Theory of saturated-absorption line shapes," Phys. Rev. A **6**, 1280. [A]

Harrison, R. G., Al-Saidi, I. A., Cummins, E. J. D., and Firth, W. J. (1985). "Evidence fór optical bistability in millimeter gas cells," Appl. Phys. Lett. **46**, 532.

Harrison, R. G., Emshary, C., Al-Saidi, I., Firth, W. J., and Abraham, E. (1983). "Optical bistability and Ikeda instabilities in an all optical passive ring cavity," postdeadline paper THU30, Conference on Lasers and Electro-Optics, Technical Digest (IEEE, New York). (6.3)

Harrison, R. G., Firth, W. J., and Al-Saidi, I. A. (1984a). "Observation of bifurcation to chaos in an all-optical Fabry-Perot resonator," Phys. Rev. Lett. **53**, 258. [6.3]

Harrison, R. G., Firth, W. J., and Al-Saidi, I. A. (1984b). "Observation of period-doubling to chaos in an all-optical Fabry-Perot resonator," J. Opt. Soc. Am. B **1**, 488. (6.3)

Harrison, R. G., Firth, W. J., Emshary, C. A., and Al-Saidi, I. A. (1983). "Observation of period doubling in an all-optical resonator containing NH_3 gas," Phys. Rev. Lett. **51**, 562. [6.3]

Harrison, R. G., Firth, W. J., Emshary, C. A., and Al-Saidi, I. A. (1984). "Observation of optical hysteresis in an all-optical passive ring cavity containing molecular gas," Appl. Phys. Lett. **44**, 716. [3.2, 5b, 6.3]

Harrison, R. G., Firth, W. J., Emshary, C. A., Al-Saidi, I. A., and Abraham, E. (1983). "Observation of Ikeda instabilities in an all-optical resonator containing two level nonlinear media," paper 12 in Programme of Sixth National Quantum Electronics Conference, University of Sussex, 19-22 September. (6.3)

Hasegawa, H., Nakagomi, T., Mabuchi, M., and Kondo, K. (1980). "Nonequilibrium thermodynamics of lasing and bistable optical systems," J. Stat. Phys. **23**, 281. (1.3, 6.5)

Hassan, S. S., and Bullough, R. K. (1981). "The driven Dicke model and its macroscopic extension: bistability or bifurcation?" p. 367 in Optical Bistability, C. M. Bowden, M. Ciftan, and H. R. Robl, eds. (Plenum Press, New York).

Hassan, S. S., Bullough, R. K., Puri, R. R., and Lawande, S. V. (1980). "Intensity fluctuations in a driven Dicke model," Physica **103A**, 213. [3.10]

Hassan, S. S., Drummond, P. D., and Walls, D. F. (1978). "Dispersive optical bistability in a ring cavity," Opt. Commun. **27**, 480. [2.4]

Hassan, S. S., Tewari, S. P., and Abraham, E. (1982). "Standing wave treatment applied to 3-level bistable systems," Opt. Commun. **40**, 461. [3.10]

Hassan, S. S., and Walls, D. F. (1978). "Photon statistics and absorption spectrum in cooperative resonance fluorescence," J. Phys. A **11**, L87. (2.8)

Haug, H. (1982). "Nonlinear optical phenomena and bistability in semiconductors," p. 149 in Festkörperprobleme, XXII, (Advances in Solid State Physics), J. Treusch, ed. (Vieweg, Braunschweig). [3.8, 7.3]

Haug, H., Koch, S. W., März, R., and Schmitt-Rink, S. (1981a). "Optical nonlinearity and instability in semiconductors due to biexciton formation," p. 15 in Proceedings of the International Conference on Excited States and Multiresonant Nonlinear Optical Processes in Solids, Aussois, France, March 18-20 (Les Editions de Physique, Orsay, France). [3.8]

Haug, H., Koch, S. W., März, R., and Schmitt-Rink, S. (1981b). "Optical nonlinearity and bistability in semiconductors due to biexciton formation," J. Lumin. **24/25**, 621. [3.8.4]

Haug, H., Koch, S. W., Neumann, R., and Schmidt, H. E. (1982). "Optical bistability due to a two-photon absorption resonance with fluctuations," Z. Phys. B **49**, 79. [3.8] (6.5)

Haug, H., März, R., and Schmitt-Rink, S. (1980). "Dielectric function for semiconductors with a high exciton concentration," Phys. Lett. **77A**, 287. [3.8]

Haus, H. A., Ippen, E. P., Lattes, A., and Leonberger, F. J. (1982). "Optical exclusive OR gate," Appl. Phys. B **28**, 283. [3.10.5]

Haus, J. W., Bowden, C. M., and Sung, C. C. (1984). "Steady-state and pulse behavior of transmitted light through dispersive nonlinear media in a ring cavity," J. Opt. Soc. Am. B **1**, 742. (6.3)

Häusler, G., and Seckmeyer, G. (1984). "Chaos and cooperation in pictorial feedback systems," p. 362 in Optics in Modern Science and Technology, Conference Digest of the 13th Congress of the International Commission for Optics, Sapporo.

Heffernan, D. M. (1983). "Single-parameter characterization of bistability in double contact injection lasers," Phys. Lett. A **94A**, 106. (1.3)

Heffernan, D. M. (1984). "Long switching times in absorptive bistable systems," Phys. Lett. A **104A**, 169.

Hegarty, J., and Jackson, K. A. (1984). "High-speed modulation and switching with gain in a GaAsAl traveling-wave optical amplifier," Appl. Phys. Lett. **45**, 1314. (5.8)

Held, G.A., Jeffries, C., and Haller, E. E. (1984). "Observation of chaotic behavior in an electron-hole plasma in Ge," Phys. Rev. Lett. **52**, 1037. [6.3]

Helleman, R. H. G., ed. (1980). Nonlinear Dynamics (New York Academy of Sciences, NY). [6.3]

Hendow, S. T., Dunn, R. W., Chow, W. W., and Small, J. G. (1982). "Observation of bistable behavior in the polarization of a laser," Opt. Lett. **7**, 356. (1.3)

Hendow, S. T., and Sargent III, M. (1982a). "The role of population pulsations in single-mode laser instabilities," Opt. Commun. **40**, 385. [6.2]

Hendow, S. T., and Sargent III, M. (1982b). "Effects of detuning on single-mode laser instabilities," Opt. Commun. **43**, 59. [6.2]

Henneberger, F., and May, V. (1982). "Nonlinear optical response and retarded photon self-energy. Comment on Nonlinear optical response and optical bistability due to excitonic molecules," Phys. Status Solidi (b) **113**, K147. [3.8.4]

Henneberger, F., and Rossmann, H. (1984). "Resonatorless optical bistability based on increasing nonlinear absorption," Phys. Status Solidi (b) **121**, 685. [2.10]

Hermann, J. A. (1980). "Spatial effects in optical bistability," Opt. Acta **27**, 159. [2.5.2]

Hermann, J. A. (1981). "Stability conditions for two-photon optical bistability in a ring cavity," Opt. Commun. **37**, 431. [3.10.1]

Hermann, J. A. (1982a). "Tristability and oscillating instabilities in a dispersive two-beam system," Opt. Commun. **44**, 62. (3.10, 6.2, 6.3, 6.4)

Hermann, J. A. (1982b). "Instabilities in coherently driven two-photon absorption within a ring cavity," Phys. Lett. **90A**, 178. [6.2]

Hermann, J. A. (1984). "Beam propagation and optical power limiting with nonlinear media," J. Opt. Soc. Am. B **1**, 729.

Hermann, J. A., and Carmichael, H. J. (1982). "Stark effect in dispersive optical bistability," Opt. Lett. **7**, 207. [2.4]

Hermann, J. A., and Carmichael, H. J. (1984). "Analytic solutions for two-photon absorption in counterpropagating beams in the cubic approximation," Z. Phys. B **54**, 263.

Hermann, J. A., and Elgin, J. N. (1981). "The mechanisms for optical bistability in a two-photon resonant system," Phys. Lett. **86A**, 461. [3.10.1, 5.3]

Hermann, J. A., Elgin, J. N., and Knight, P. L. (1982). "The role of the interactive Kerr effect in two-photon optical multistability," Z. Phys. B **45**, 255. [3.10.1]

Hermann, J. A., and Grigg, M. E. (1984). "Optical power limiting with nonlinear media," Opt. Commun. **49**, 367. [5.2]

Hermann, J. A., and Thompson, B. V. (1980). "Two-photon optical bistability with spatial effects," Phys. Lett. A **79**, 153. (3.10.1)

Hermann, J. A., and Thompson, B. V. (1981). "Analytic description of multiphoton optical bistability in a ring cavity," p. 199 in Optical Bistability, C. M. Bowden, M. Ciftan, and H. R. Robl, eds. (Plenum Press, New York).

Hermann, J. A., and Thompson, B. V. (1982). "Dispersive two-photon optical multistability," Opt. Lett. **7**, 301. [3.10.1]

Hermann, J. A., and Walls, D. F. (1982). "Theory of two-photon optical tristability," Phys. Rev. A **26**, 2085. [3.10.1]

Hermann, J.-P., and Smith, P. W. (1980). "Non-linear Fabry-Perot containing the polydiacetylene PTS," J. Opt. Soc. Am. **70**, 656. (3.10.4)

Herzog, U. (1983). "Photon statistics in absorptive multi-photon optical bistability," Opt. Acta **30**, 1781. (3.10.1, 6.1)

Hey, J. D., and Hopf, F. A. (1982). "Non-linear optics," p. 211 in Quantum Optics. Proceedings of the South African Summer School in Theoretical Physics, C. A. Engelbrecht, ed. (Springer-Verlag, Berlin).

Higgins, H. A., and Wherrett, B. S. (1982). "The intensity dependence of interband Landau absorption," Opt. Commun. **41**, 350. [7.3.2]

Hildebrand, O., Faltenmeier, B., and Pilkuhn, M. H. (1976). "Direct determination of reduced band gap and chemical potential in an electron-hole plasma in high-purity GaAs," Solid State Commun. **19**, 841.

Hill, J. R., Parry, G., and Miller, A. (1982). "Non-linear refractive index changes in CdHgTe at 175 K with 10.6 μm radiation," Opt. Commun. **43**, 151. [7.3.1]

Hillman, L. W., Boyd, R. W., and Stroud, Jr., C. R. (1982). "Natural modes for the analysis of optical bistability and laser instability," Opt. Lett. **7**, 426. [6.4]

Hinov, H. P., Kukleva, E., and Derzhanski, A. I. (1983). "Electro-optical behavior of a novel strong-weak and weak-weak anchored large-pitch cholesteric," Mol. Cryst. Liq. Cryst. (GB) **98**, 109.

Hioe, F. T., and Singh, S. (1981). "Correlations, transients, bistability, and phase-transition analogy in two-mode lasers," Phys. Rev. A **24**, 2050. (1.3)

Hochstrasser, R. M., Kaiser, W., and Shank, C. V. (1980). Picosecond Phenomena II (Springer-Verlag, New York). [1.2]

Hoff, F. (1984). "Optical bistable circuits," Slaboproudy Obz. **44**, 145.

Holm, D. A., Sargent III, M., and Stenholm, S. (1984). "Effects of stimulated emission and phase conjugation on resonance fluorescence spectra," talk TuO10, Optical Society of America 1984 Annual Meeting, Technical Program. [F]

Holmes, L. (1983). "Optical bistability--the promise of lightwave logic and switching," Electron. Power **29**, 581.

Hönerlage, B., Bigot, J. Y., and Levy, R. (1984). "Optical bistability in CuCl," p. 253 in Optical Bistability 2, C. M. Bowden, H. M. Gibbs, and S. L. McCall, eds. (Plenum Press, New York). [3.8, 3.10]

Hong-Jun, Z., Jian-Hua, D., Jun-Hui, Y., and Cun-Xiu, G. (1981). "A bistable liquid crystal electro-optic modulator," Opt. Commun. **38**, 21. [4.5]

Hong-Jun, Z., Jian-Hua, D., Peng-Ye, W., and Chao-Ding, J. (1983). "Chaos in liquid crystal hybrid optical bistable devices," p. 322 in Laser Spectroscopy VI, H. P. Weber and W. Lüthy, eds. (Springer-Verlag, Berlin).

Hopf, F. A. (1982). "Chaos and periodicity in optical bistability," J. Opt. Soc. Am. **72**, 1753. [6.3]

Hopf, F. A. (1983). "Bifurcations to chaos in optical bistability," J. Phys. (Paris) **44**, C2-193. (6.3)

Hopf, F. A. (1984). "Chaos in optical bistability," p. 81 in Conference Digest of the United States-Japan Seminar on Coherence, Incoherence, and Chaos in Quantum Electronics, Nara, M. D. Levenson, and T. Shimizu, coordinators. (6.3)

Hopf, F. A., and Bowden, C. M. (1984). "Heuristic model for fluctuations in mirrorless optical bistability," talk WM6, Optical Society of America 1984 Annual Meeting, Technical Program. (3.10.6, 6.5)

Hopf, F. A., Bowden, C. M., and Louisell, W. (1984). "Mirrorless optical bistability with the use of the local-field correction," Phys. Rev. A **29**, 2591. [3.10.6]

Hopf, F. A., Derstine, M. W., Gibbs, H. M., and Rushford, M. C. (1984). "Chaos in optics," p. 67 in Optical Bistability 2, C. M. Bowden, H. M. Gibbs, and S. L. McCall, eds. (Plenum Press, New York). [6.1, 6.3]

Hopf, F. A., Kaplan, D. L., Gibbs, H. M., and Shoemaker, R. L. (1982). "Bifurcations to chaos in optical bistability," Phys. Rev. A **25**, 2172. [6.3]

Hopf, F. A., Kaplan, D. L., Shoemaker, R. L., and Gibbs, H. M. (1981). "Path to turbulence of an optical bistable system," J. Opt. Soc. Am. **71**, 1634. [6.3]

Hopf, F. A., and Meystre, P. (1979). "Numerical studies of the switching of a bistable optical memory," Opt. Commun. **29**, 235. [5.1, 5.6, 6.5]

Hopf, F. A., and Meystre, P. (1980). "Phase-switching of a dispersive non-linear interferometer," Opt. Commun. **33**, 225. [5.7]

Hopf, F. A., Meystre, P., Drummond, P., and Walls, D. F. (1979). "Anomalous switching in dispersive optical bistability," Opt. Commun. **31**, 245. [5.7]

Hopf, F. A., and Shakir, S. A. (1981). "Frequency switching in dispersive optical bistability," p. 281 in Optical Bistability, C. M. Bowden, M. Ciftan, and H. R. Robl, eds. (Plenum Press, New York).

Hopfield, J. J. (1982). "Neural networks and physical systems with emergent collective computational abilities," Proc. Natl. Acad. Sci. **79**, 2554. [7.4]

Huang, A. (1980). "Design for an optical general purpose digital computer," Proc. SPIE **232**, 119. [7.4]

Huang, A. (1983). "Parallel algorithms for optical digital computers," p. 13 in 10th International Optical Computing Conference (IEEE Computer Society Press, Silver Spring, Maryland). [7.4]

Huang, A. (1984). "Architectural considerations involved in the design of an optical digital computer," Proc. IEEE **72**, 780. [7.4]

Huang, A. (1984). "Computer architecture based on optical bistable devices," talk ThF3, Optical Society of America 1984 Annual Meeting, Technical Program. (7.4)

Hutcheson, L. D., ed. (1985). Integrated Optical Circuits and Components: Design and Applications. (Marcel Dekker, New York). (3.10.5)

Ikeda, K. (1979). "Multiple-valued stationary state and its instability of the transmitted light by a ring cavity system," Opt. Commun. **30**, 257. [2.9, 6.3]

Ikeda, K. (1983). "Optical turbulence: chaos in optical bistability," J. Phys. (Paris) **44**, C2-183. (6.3)

Ikeda, K. (1984a). "Chaos and optical bistability: bifurcation structure," p. 875 in Coherence and Quantum Optics V, L. Mandel and E. Wolf, eds. (Plenum Press, New York). (6.3)

Ikeda, K. (1984b). "Optical chaos due to a competition between multiple oscillations," J. Opt. Soc. Am. B **1**, 487. (6.3)

Ikeda, K., and Akimoto, O. (1981). "Chaos in a bistable optical system," U.S.-Japan Workshop, "Nonequilibrium statistical physics problems in fusion plasmas--stochasticity and chaos," Kyoto, November.

Ikeda, K., and Akimoto, O. (1982a). "Instability leading to periodic and chaotic self-pulsations in a bistable optical cavity," Phys. Rev. Lett. **48**, 617. [2.9, 6.4]

Ikeda, K., and Akimoto, O. (1982b). "Successive bifurcations and dynamical multistability in a bistable optical system: a detailed study of the transition to chaos," Appl. Phys. B **28**, 170. (6.3)

Ikeda, K., and Akimoto, O. (1982c). "Chaos in a bistable optical system," p. 147 in US-Japan Joint Institute for Fusion Theory. Workshop on Nonequilibrium Statistical Physics Problems in Fusion Plasmas--Stochasticity and Chaos (Institute of Plasma Physics, Nagoya, Japan).

Ikeda, K., Daido, H., and Akimoto, O. (1980). "Optical turbulence: chaotic behavior of transmitted light from a ring cavity," Phys. Rev. Lett. **45**, 709. [6.3]

Ikeda, K., Kondo, K., and Akimoto, O. (1982). "Successive higher-harmonic bifurcations in systems with delayed feedback," Phys. Rev. Lett. **49**, 1467. [6.3]

Ikeda, K., and Mizuno, M. (1984). "Frustrated instabilities in nonlinear optical resonators," Phys. Rev. Lett. **53**, 1340. (6.3)

Ikeda, K., Mizuno, M., and Davis, P. (1984). "Competition mechanisms leading to optical chaos," p. 87 in Conference Digest of the United States-Japan Seminar on Coherence, Incoherence, and Chaos in Quantum Electronics, Nara, M. D. Levenson, and T. Shimizu, coordinators. (6.3)

Inaba, H. (1981). "Laser operation with output feedback pumping and first-order phase transition analogy," Phys. Lett. A **86A**, 452. (1.3)

Inaba, H. (1984). "Optically bistable operation of laser by output-feedback pumping and loss-modulation schemes with first-order phase-transition analogies," J. Opt. Soc. Am. B **1**, 467. (1.3)

Ippen, E. P., Eilenberger, D. J., and Dixon, R. W. (1980a). "Picosecond pulse generation with diode lasers," p. 21 in Picosecond Phenomena II, R. Hochstrasser, W. Kaiser, and C. V. Shank, eds. (Springer-Verlag, New York). [6.1]

Ippen, E. P., Eilenberger, D. J., and Dixon, R. W. (1980b). "Picosecond pulse generation by passive mode locking of diode lasers," Appl. Phys. Lett. **37**, 267. [6.1]

Ippen, E. P., and Shank, C. V. (1975). "Picosecond response of a high-repetition-rate CS_2 optical Kerr gate," Appl. Phys. Lett. **26**, 92. [5.8]

Ippen, E. P., and Shank, C. V. (1978). "Sub-picosecond spectroscopy," Physics Today **31(5)**, 41. [5.8]

Ishibashi, T., Tarucha, S., and Okamoto, H. (1982). "Exciton associated optical-absorption spectra of AlAs/GaAs super-lattices at 300 K," Institute of Physics Conference Series **1982**, 587.

Ishihara, S. (1982). "Optical computing," Oyo Buturi (Japan) **51**, 570.

Ito, H., and Inaba, H. (1980). "Bistable optical devices," J. Inst. Electron. Commun. Eng. Jpn. **63**, 1025.

Ito, H., Ogawa, Y., and Inaba, H. (1979a). "Integrated Mach-Zehnder type bistable optical device and its application to optical multivibrators," p. 11.3 in Proceedings of the Optical Communication Conference (Philips Res. Labs., Eindhoven, Netherlands). (4.5)

Ito, H., Ogawa, Y., and Inaba, H. (1979b). "Integrated bistable optical device using Mach-Zehnder interferometric optical waveguide," Electron. Lett. **15**, 283. [4.5]

Ito, H., Ogawa, Y., and Inaba, H. (1980). "Integrated astable optical multivibrator using Mach-Zehnder interferometric optical switches," Electron. Lett. **16**, 593. [4.5]

Ito, H., Ogawa, Y., and Inaba, H. (1981). "Analysis and experiments on integrated optical multivibrators using electrooptically controlled bistable optical devices," IEEE J. Quantum Electron. **QE-17**, 325. [4.5, 6.1]

Ito, H., Ogawa, Y., Makita, K., and Inaba, H. (1979). "Integrated bistable optical multivibrator using electrooptically controlled directional coupler switches," Electron. Lett. **15**, 791. [4.5]

Itoh, T., and Katohno, T. (1982). "Broadening of excitonic molecule state under the giant two-photon excitation in CuCl," J. Phys. Soc. Japan **51**, 707. [3.8.4]

Jackson, J. D. (1962). Classical Electrodynamics (Wiley, New York). [C]

Jain, K., and Pratt, Jr., G. W. (1976). "Optical transistor," Appl. Phys. Lett. **28**, 719. [3.10, 5.4]

Jain, R. K. (1982). "Degenerate 4-wave mixing in semiconductors application to phase conjugation and to picosecond-resolved studies of transient carrier dynamics," Opt. Eng. **21**, 199. [7.3]

Jain, R. K., and Klein, M. B. (1979). "Degenerate four-wave mixing near the band gap of semiconductors," Appl. Phys. Lett. **35**, 454.

Jain, R. K., and Klein, M. B. (1982). "Degenerate four-wave mixing in semiconductors," Chapter 10 in Optical Phase Conjugation, R. A. Fisher, ed. (Academic Press, New York). [7.3]

Jain, R. K., and Lind, R. C. (1983). "Degenerate 4-mixing in semiconductor-doped glasses," J. Opt. Soc. Am. **73**, 647. (7.3)

Jain, R. K., and Steel, D. G. (1980). "Degenerate four-wave mixing of 10.6 μm radiation in $Hg_{1-x}Cd_xTe$," Appl. Phys. Lett. **37**, 1. [7.3]

Jain, R. K., and Steel, D. G. (1982). "Large optical nonlinearities and cw degenerate four-wave mixing in HgCdTe," Opt. Commun. **43**, 72. [6.1]

Jameson, R. S., Hickernell, R. K., and Sarid, D. (1984). "Experimental results of logic operations with control beams in InSb bistable etalon," talk ThF7, Optical Society of America 1984 Annual Meeting, Technical Program. (3.7)

Jamison, S. A., and Nurmikko, A. V. (1978). "Generation of picosecond pulses of variable duration at 10.6 μm," Appl. Phys. Lett. **33**, 598. [5.3]

Jaques, A., and Glorieux, P. (1982). "Observation of bistability in a CO_2 laser exhibiting passive Q-switching," Opt. Commun. **40**, 455. (1.3)

Jensen, S. M. (1979). "An analysis of bistable optical devices," p. 23.4 in Proceedings of the Optical Communication Conference (Philips Res. Labs., Eindhoven, Netherlands). (5.6)

Jensen, S. M. (1980a). "High-speed optical logic devices," Proc. SPIE **218**, 33. (3.10.5, 5.5, 5.6, 7.2)

Jensen, S. M. (1980b). "The non-linear coherent coupler: a new optical logic device," p. MB4 in Integrated and Guided Wave Optics, Technical Digest (IEEE, New York). [3.10.5]

Jensen, S. M. (1982a). "Transient response of nonlinear ring resonators using GaAs with photon energy below band gap," Proc. SPIE **321**, 61. [3.6] (3.10.5)

Jensen, S. M. (1982b). "The nonlinear coherent coupler," IEEE J. Quantum Electron. **QE-18**, 1580. [3.10.5]

Jensen, S. M. (1982c). "The nonlinear coherent coupler," IEEE Trans. Microwave Theory Tech. **MTT-30**, 1568. (3.10.5)

Jerominek, H., Pomerleau, J. Y. D., Tremblay, R., and Delisle, C. (1984). "An integrated acousto-optic bistable device," Opt. Commun. **51**, 6. [4.5]

Jewell, J. L. (1981). "Regenerative pulsations and thermal effects in an optical bistable GaAs etalon," M.S. dissertation, Optical Sciences Center, University of Arizona. [6.1]

Jewell, J. L. (1984). "Fabrication, investigation and optimization of GaAs optical bistable devices and logic gates," Ph.D. dissertation, Optical Sciences, University of Arizona. (3.6)

Jewell, J. L., Gibbs, H. M., Gossard, A. C., Passner, A., and Wiegmann, W. (1983). "Fabrication of GaAs bistable optical devices," Materials Lett. **1**, 148. [3.6]

Jewell, J. L., Gibbs, H. M., Gossard, A. C., and Wiegmann, W. (1982). "Fabrication of high-finesse thin GaAs etalons," J. Opt. Soc. Am. **72**, 1770. [3.6]

Jewell, J. L., Gibbs, H. M., Tarng, S. S., Gossard, A. C., and Wiegmann, W. (1982). "Regenerative pulsations from an intrinsic bistable optical device," Appl. Phys. Lett. **40**, 291. [6.1]

Jewell, J. L., Lee, Y. H., Warren, M., Gibbs, H. M., Peyghambarian, N., Gossard, A. C., and Wiegmann, W. (1984). "Low-energy fast optical logic gates in a room-temperature GaAs étalon," talk ThF6, Optical Society of America 1984 Annual Meeting, Technical Program. (3.6, 5.8)

Jewell, J. L., Lee, Y. H., Warren, M., Gibbs, H. M., Peyghambarian, N., Gossard, A. C., and Wiegmann, W. (1985). "3-pJ 82-MHz optical logic gates in a room-temperature GaAs-AlGaAs multiple-quantum-well etalon," Appl. Phys. Lett. **46**, 918. (1.2, 3.6, 5.8)

Jewell, J. L., Lee, Y. H., Warren, M., Rushford, M. C., Gibbs, H. M., Peyghambarian, N., Gossard, A. C., and Wiegmann, W. (1984). "Optical logic in GaAs Fabry-Perot etalons," p. 156 in Optics in Modern Science and Technology, Conference Digest of the 13th Congress of the International Commission for Optics, Sapporo. (1.2, 3.6, 5.8)

Jewell, J. L., Ovadia, S., Peyghambarian, N., Tarng, S. S., Gibbs, H. M., Gossard, A. C., and Wiegmann, W. (1984). "Room-temperature excitonic optical bistability in bulk GaAs," p. 120 in Conference on Lasers and Electro-Optics, Technical Digest (IEEE, New York). (3.6)

Jewell, J. L., Rushford, M. C., and Gibbs, H. M. (1984). "Use of a single nonlinear Fabry-Perot etalon as optical logic gates," Appl. Phys. Lett. **44**, 172. [1.2]

Jewell, J. L., Rushford, M. C., Gibbs, H. M., and Peyghambarian, N. (1983). "Nonlinear-etalon optical logic gates," J. Opt. Soc. Am. **73**, 1969. (1.2, 3.6)

Jewell, J. L., Rushford, M. C., Gibbs, H. M., and Peyghambarian, N. (1984). "Single-elaton optical logic gates," p. 184 in Conference on Lasers and Electro-Optics, Technical Digest (IEEE, New York).

Jewell, J. L., Tarng, S. S., Gibbs, H. M., Tai, K., Weinberger, D. A., and Ovadia, S. (1984). "Advances in GaAs bistable devices," p. 223 in Optical Bistability 2, C. M. Bowden, H. M. Gibbs, and S. L. McCall, eds. (Plenum Press, New York). [3.6]

Johnson, A. M., and Auston, D. H. (1975). "Microwave switching by picosecond photoconductivity," IEEE J. Quantum Electron. **QE-11**, 283. [5.3]

Johnson, E. J., Riseberg, L. A., Lempicki, A., and Samelson, H. (1975). "Completely optical coincidence logic employing a dye laser," Appl. Phys. Lett. **26**, 444.

Johnson, L. M., and Pratt, Jr., G. W. (1978). "Proposed generation of picosecond second-harmonic pulses in Te," Appl. Phys. Lett. **33**, 1002. [3.10, 5.2]

Kadanoff, L. P. (1983). "Roads to chaos," Physics Today, p. 46, December. [6.3]

Kagan, A. G., and Khanin, Y. I. (1983). "Problems in steady-state theory of a multimode laser with a selective saturable absorber," Sov. J. Quantum Electron. **13**, 88. (1.3)

Kane, E. O. (1957). "Band structure of indium antimonide," J. Phys. Chem. Solids **1**, 249. [7.3.2]

Kaplan, A. E. (1976). "Hysteresis reflection and refraction by a nonlinear boundary--a new class of effects in nonlinear optics," JETP Lett. **24**, 114. [3.10.4]

Kaplan, A. E. (1977). "Theory of hysteresis reflection and refraction of light by a boundary of a nonlinear medium," Sov. Phys. JETP **45**, 896. [3.10.4]

Kaplan, A. E. (1980). "Bistable reflection of light from the boundary of an 'artificial' nonlinear medium," J. Opt. Soc. Am. **70**, 658. [3.10.4]

Kaplan, A. E. (1981a). "Optical bistability due to mutual self-action of counter-propagated light beams," IEEE J. Quantum Electron. **QE-17**, 118. [3.9.2, 3.10.3]

Kaplan, A. E. (1981b). "Optical bistability that is due to mutual self-action of counterpropagating beams of light," Opt. Lett. **6**, 360. [3.9.1, 3.10.3]

Kaplan, A. E. (1981c). "Bistable reflection of light by an electro-optically driven interface," Appl. Phys. Lett. **38**, 67. [3.10.4]

Kaplan, A. E. (1981d). "Conditions of excitation of new waves (LITW) at nonlinear interface and diagram of wave states of the system," IEEE J. Quantum Electron. **QE-17**, 336. [3.10.4]

Kaplan, A. E. (1981e). "Theory of plane wave reflection and refraction by the nonlinear interface," p. 447 in Optical Bistability, C. M. Bowden, M. Ciftan, and H. R. Robl, eds. (Plenum Press, New York).

Kaplan, A. E. (1982a). "Enhancement of laser gyro by nonreciprocal feedback," J. Opt. Soc. Am. **72**, 1748. [3.10.3]

Kaplan, A. E. (1982b). "Bistable cyclotron resonance based on relativistic nonlinearity," Appl. Phys. B **28**, 166. [3.10.7]

Kaplan, A. E. (1982c). "Hysteresis in cyclotron resonance based on weak relativistic-mass effects of the electron," Phys. Rev. Lett. **48**, 138.

Kaplan, A. E. (1983). "Optoelectronic enhancement of the Sagnac effect in a ring resonator and related effect of directional bistability," Appl. Phys. Lett. **42**, 479. [3.10.3]

Kaplan, A. E. (1984). "Hysteretic (bistable) cyclotron resonance in semiconductors," Phys. Rev. B **29**, 820. [3.10.7]

Kaplan, A. E., and Elci, A. (1983). "Cyclotron optical bistability in semiconductors," p. 108 in Conference on Lasers and Electro-Optics, Technical Digest (IEEE, New York).

Kaplan, A. E., and Meystre, P. (1981). "Enhancement of the Sagnac effect due to nonlinearly induced nonreciprocity," Opt. Lett. **6**, 59. [3.10.3, 6.4]

Kaplan, A. E., and Meystre, P. (1982). "Directionally asymmetrical bistability in a symmetrically pumped nonlinear ring interferometer," Opt. Commun. **40**, 229. [3.10.3]

Kaplan, A. E., Smith, P. W., and Tomlinson, W. J. (1981). "Switching of reflection of light at nonlinear interfaces," Proc. SPIE **317**, 305. (3.10.4)

Kaplan, D. L., Gibbs, H. M., Hopf, F. A., Derstine, M. W., and Shoemaker, R. L. (1983). "Periodic oscillations and chaos in optical bistability; possible guided-wave all-optical square-wave oscillators," Opt. Eng. **22**, 161. [6.3]

Kar, A. K., Mathew, J. G. H., and Smith, S. D. (1983). "Optical bistability in InSb at room temperature with two-photon excitation," p. 90 in Conference on Lasers and Electro-Optics, Technical Digest (IEEE, New York). (3.7)

Kar, A. K., Mathew, J. G. H., Smith, S. D., Davis, B., and Prettl, W. (1983). "Optical bistability in InSb at room temperature with two-photon excitation," Appl. Phys. Lett. **42**, 334. [3.7, 3.10.1, 7.3]

Karpushko, F. V., Kireev, A. S., Morozov, I. A., Sinitsyn, G. V., and Strizhenok, N. V. (1977). "Spectral characteristics of nonlinear interferometers in a strong field," Zh. Prikl. Spektros. **26**, 269; J. Appl. Spectrosc. USSR **26**, 204 (1977). [3.5]

Karpushko, F. V., and Sinitsyn, G. V. (1978). "An optical logic element for integrated optics in a nonlinear semiconductor interferometer," Zh. Prikl. Spektros. **29**, 820 [J. Appl. Spectrosc. USSR **29**, 1323 (1978)].[3.5, 3.8, 5.6]

Karpushko, F. V., and Sinitsyn, G. V. (1979). "Switching of a laser emission spectrum by an external optical signal," Sov. J. Quantum Electron. **9**, 520. [3.5, 5.3]

Karpushko, F. V., and Sinitsyn, G. V. (1982). "The anomalous nonlinearity and optical bistability in thin-film interference structures," Appl. Phys. B **28**, 137. [3.5]

Kastal'skii, A. A. (1973). "Optical and electrical effects in circuits containing a Fabry-Perot resonator," Sov. Phys. Semicond. **7**, 635. [2.3, 4.1]

Kawaguchi, H. (1981). "Bistable operation of semiconductor lasers by optical injection," Electron. Lett. **17**, 741. [1.3]

Kawaguchi, H. (1982a). "Bistability and differential gain in semiconductor lasers," Japan J. Appl. Phys. **21**, 371. [1.3]

Kawaguchi, H. (1982b). "Optical input and output characteristics for bistable semiconductor lasers," Appl. Phys. Lett. **41**, 702. [1.3]

Kawaguchi, H. (1982c). "Optical bistable-switching operation in semiconductor lasers with inhomogeneous excitation," IEE Proc. **129**, Part I, 141. (1.3)

Kawaguchi, H. (1984). "Optical bistability and chaos in a semiconductor laser with a saturable absorber," Appl. Phys. Lett. **45**, 1264. (1.3)

Kawaguchi, H., and Iwane, G. (1981). "Bistable operation in semiconductor lasers with inhomogeneous excitation," Electronics Lett. **17**, 167. [1.3.]

Kawaguchi, H., and Otsuka, K. (1984). "A new class of instabilities in a diode laser with an external cavity," Appl. Phys. Lett. **45**, 934. [6.3]

Kazantsev, A. P., Rautian, S. G., and Surdutovich, G. I. (1968). "Theory of a gas laser with nonlinear absorption," Sov. Phys. JETP **27**, 756. (1.3)

Kazberuk, A. V., Karpushko, F. V., and Sinitsyn, G. V. (1978). "Nonlinear thin-film semiconductor interferometer for optical pulse shaping," Sov. Tech. Phys. Lett. **4**, 544. [5.2]

Keyes, R. W. (1975). "Physical limits in digital electronics," Proc. IEEE **63**, 740. [7.2]

Keyes, R. W. (1981). "Fundamental limits in digital information processing," Proc. IEEE **69**, 267. [7.2]

Khadzhi, P. I., Moskalenko, S. A., Rotaru, A. Kh., Belkin, S. N., and Kiseleva, E. S. (1982). "Bistability in a system of coherent excitons, photons, and biexcitons," Sov. Phys.-Solid State **24**, 928. (3.8.4)

Khan, M. A., Bennet, R. L., and Kruse, P. W. (1981). "Bandgap-resonant optical phase conjugation in n-type $Hg_{1-x}Cd_xTe$ at 10.6 μm," Opt. Lett. **6**, 560. [7.3]

Khan, M. A., Kruse, P. W., and Ready, J. F. (1980). "Optical phase conjugation in $Hg_{1-x}Cd_xTe$," Opt. Lett. **5**, 261. [7.3]

Khan, M. A., Kruse, P. W., Wood, R. A., and Park, Y. K. (1983). "Optical phase conjugation and nonlinear Fabry-Perot effects in epitaxial layers of $Hg_{1-x}Cd_xTe$, p. 176 in Conference on Lasers and Electro-Optics, Technical Digest (IEEE, New York). [3.8.6]

Khanin, Ya. I., and Matorin, I. I. (1981). "The effect of hard mode locking in a dye laser," Opt. Quantum Electron. **13**, 439. (1.3)

Khoo, I. C. (1982). "Optical bistability in nematic films utilizing self-focusing of light," Appl. Phys. Lett. **41**, 909. [2.9, 3.4, 3.9.2]

Khoo, I. C., and Hou, J. Y. (1982). "Theory of optical switching and bistability at a nonlinear (liquid-crystal-dielectric) interface," J. Opt. Soc. Am. **72**, 1761. [3.4, 3.10.4]

Khoo, I. C., Hou, J. Y., Normandin, R., and So, V. C. Y. (1983). "Theory and experiment on optical bistability in a Fabry-Perot interferometer with an intracavity nematic liquid-crystal film," Phys. Rev. A **27**, 3251. [3.4]

Khoo, I. C., Hou, J. Y., Shepard, S., and Yan, P. Y. (1983). "Nonlinear Fabry-Perot characteristics of a thin film near the TIR state," p. 92 in Conference on Lasers and Electro-Optics, Technical Digest (IEEE, New York). [3.5]

Khoo, I. C., and Liu, T. H. (1984). "Cavityless optical switching elements using nonlinear self-phase modulation," talk WT1, Optical Society of America 1984 Annual Meeting, Technical Program. (3.9.2)

Khoo, I. C., Liu, T. H., and Normandin, R. (1984). "Transverse self-phase modulation bistability in a nonlinear thin film -- theory and experiment," International Conference on Lasers '84, San Francisco. (3.9.2)

Khoo, I. C., Normandin, R., and So, V. C. Y. (1982a). "Nonlinear Fabry-Perot action and optical bistability using a nematic liquid-crystal film," J. Opt. Soc. Am. **72**, 1761. [3.4]

Khoo, I. C., Normandin, R., and So, V. C. Y. (1982b). "Optical bistability using a nematic liquid crystal film in a Fabry-Perot cavity," J. Appl. Phys. **53**, 7599. [3.4]

Khoo, I. C., and Shepard, S. (1982). "Optical bistability utilizing the self-focusing of light in a nematic liquid-crystal film," J. Opt. Soc. Am. **72**, 1761. [3.4, 3.9.2]

Khoo, I. C., Shepard, S., Nadar, S., and Zhuang, S. L. (1982). "Quantitative theory and experiments on optical imaging and switching properties of nematic liquid crystals," Appl. Phys. B **28**, 141. [3.4]

Khoo, I.C., Yan, P. Y., Liu, T. H., Shepard, S., and Hou, J. Y. (1984a). "Theory and experiment on optical transverse intensity bistability in the transmission through a nonlinear thin (nematic liquid crystal) film," Phys. Rev. A **29**, 2756. [2.9.3, 3.9.2]

Khoo, I. C., Yan, P. Y., Liu, T. H., Shepard, S., and Hou, J. Y. (1984b). "Theory and experiment on transverse intensity bistability in the transmission of a Gaussian laser beam through a nonlinear thin film," J. Opt. Soc. Am B **1**, 477. [2.9.3, 3.9.2]

Kimble, H.J. and Grant, D. E. (1982). "Transient response in absorptive optical bistability," J. Opt. Soc. Am. 72, 1760. [5.6]

Kimble, H. J., Grant, D. E., Rosenberger, A. T., and Drummond, P. D. (1983). "Optical bistability with 2-level atoms," Lecture Notes in Physics **182**, p. 14 in Laser Physics, J. D. Harvey and D. F. Wals, eds. (Springer-Verlag Berlin). [3.2.5]

Kimble, H. J., Rosenberger, A. T., and Drummond, P. D. (1984). "Optical bistability with two-level atoms," p. 1 in Optical Bistability 2, C. M. Bowden, H. M. Gibbs, S. L. McCall, eds. (Plenum Press, New York). [3.2.5b]

Kimble, H. J., Rosenberger, A. T., and Orozco, L. A. (1984). "Evolution of hysteresis and onset of dynamical instability in optical bistability with two-level atoms," International Conference on Lasers '84, San Francisco.

Kingston, R. H., Burke, B. E., Nichols, K. B., and Leonberger, F. J. (1982). "Spatial light modulation using electro-absorption in a GaAs charge-coupled device," Appl. Phys. Lett. **41**, 413. [7.4]

Kingston, R. H., and Leonberger, F. J. (1984). "Fourier transformation using an electroabsorptive CCD spatial light modulator," p. 44 in Conference on Lasers and Electro-Optics, Technical Digest (IEEE, New York). (7.4)

Kitano, M., Yabuzaki, T., and Ogawa, T. (1981a). "Optical tristability," Phys. Rev. Lett. **46**, 926. [3.10.2,6.4]

Kitano, M., Yabuzaki, T., and Ogawa, T. (1981b). "Self-sustained spin precession in an optical tristable system," Phys. Rev. A **24**, 3156. [3.10.2, 6.4]

Kitano, M., Yabuzaki, T., and Ogawa, T. (1983). "Chaos and period-doubling bifurcations in a simple acoustic system," Phys. Rev. Lett. **50**, 713. [6.3]

Kitano, M., Yabuzaki, T., and Ogawa, T. (1984). "Symmetry-recovering crises of chaos in polarization-related optical bistability," Phys. Rev. A **29**, 1288. [6.4]

Kitayama, K., Kimura, Y., and Seikai, S. (1985). "Fiber-optic logic gate," Appl. Phys. Lett. **46**, 317. (3.10.5, 7.4)

Kitayama, K., and Wang, S. (1983). "Optical pulse compression by nonlinear coupling," Appl. Phys. Lett. **43**, 17. [5.2, 3.10.5]

Klein, M. B. (1980). "Stark cell devices for CO_2 lasers," Proc. SPIE **227**, 141. [4.5]

Klein, M. B., Lackner, A. M., Myer, G. B., and Jain, R. K. (1981). "Optical noise-limiting and multistability using EO liquid crystal devices." IEEE J. Quantum Electron. **QE-17**, 162. (3.4)

Klingshirn, C., Bohnert, K., and Kalt, H. (1984). "Optically non-linear and bistable behavior of direct gap semiconductors," AGARD Workshop on "Digital Optical Circuit Technology," Schliersee, West Germany, Sept. [3.8.2]

Klingshirn, C., and Haug, H. (1980). "Optical properties of highly excited direct gap semiconductors," Phys. Rep. **70**, 315. [7.3]

Kobayashi, T., Kothari, N. C., and Uchiki, H. (1984). "Nondegenerate two-photon optical bistability in a Fabry-Perot cavity filled with large permanent dipole molecules," Phys. Rev. A **29**, 2727. [3.10.1]

Kobayashi, T., Sueta, T., Cho, Y., and Matsuo, Y. (1972). "High-repetition-rate optical pulse generator using a Fabry-Perot electro-optic modulator," Appl. Phys. Lett. **21**, 341. [4.1]

Koch, S. W., and Haug, H. (1981). "Two-photon generation of excitonic molecules and optical bistability," Phys. Rev. Lett. **46**, 450. [3.8, 3.10.1]

Koch, S. W., Schmidt, H. E., and Haug, H. (1984). "Optical bistability due to induced absorption: propagation dynamics of excitation profiles," Appl. Phys. Lett. **45**, 932. [2.10]

Koch, S. W., Schmitt-Rink, S., and Haug, H. (1981a). "Theory of optical nonlinearities in InSb," Phys. Status Solidi (B) **106**, 135. [7.3.4]

Koch, S. W., Schmitt-Rink, S., and Haug, H. (1981b). "Calculation of the intensity-dependent changes of the index of refraction in GaAs," Solid State Commun. **38**, 1023.

Kogelnik, H. (1981). "Limits in integrated optics," Proc. IEEE **69**, 232. [7.2]

Kompanets, I. N., Parfenov, A. V., and Popov, Y. M. (1981a). "Bistable properties of spatial light modulator with internal feedback," Opt. Commun. **36**, 415. (4.5, 7.4)

Kompanets, I. N., Parfenov, A. V., and Popov, Y. M. (1981b). "Multistability in optical transmittance of spatial light modulators with internal feedback," Opt. Commun. **36**, 417. (4.5, 7.4)

Kondo, K., Mabuchi, M., and Hasegawa, H. (1980). "On the stationary distribution of photons in optical bistability," Opt. Commun. **32**, 136. (2.7, 6.5)

Kosonocky, W. F. (1964). "Optical system for performing digital logic," U.S. Patent 34331437, filed May 25, 1964, granted March 4, 1969. [1.2, 1.3]

Koster, A., Martinot, P. Pardo, F. Laval, S., Paraire, N., and Neviere, M. (1984). "Bistable optical devices by guided wave excitation," p. 122 in Conference on Lasers and Electro-Optics, Technical Digest (IEEE, New York). [3.10.5]

Kothari, N. C., and Kobayashi, T. (1984). "Single beam two-photon optical bistability in a submicron-size Fabry-Perot cavity," IEEE J. Quantum Electron. **QE-20**, 418. [3.10.1]

Kovarskii, V. A., Perel'man, N. F., and Averbukh, I. Sh. (1982). "Optical bistability on the basis of multiphoton resonance," page 190 in Proceedings of the XI National Conference on Coherent and Nonlinear Optics, Yerevan, USSR. [3.10.2]

Kowarschik, R., and Zimmermann, A. (1982). "Waveguide resonators with distributed Bragg reflectors," Opt. Acta **29**, 455. [3.10.5]

Krauyalene, I., Rimeika, R., and Chiplis, D. (1984). "Electro-optic bistability in an integrated-optics interferometer with external feedback," Sov. Phys. Tech. Phys., Oct.

Kreissl, A., Bohnert, K., Lyssenko, V. G., and Klingshirn, C. (1982). "Experimental investigation on the complex dielectric function of CdS in the exciton and in the plasma limit," Phys. Stat. Solidi b **114**, 537. (3.8.2, 7.3.4)

Krivoshchekov, G. V., Kurbatov, P. F., Smirnov, V. C., and Tumalkin, A. M. (1982). "Bistability and polarization of the radiation emitted by a xenon laser as a result of the $5d[3/2]_1^0$-$6p[3/2]_1$ transition in the ^{136}Xe isotope in a weak magnetic field," Sov. J. Quantum Electron. **12**, 551. (1.3)

Kuhlke, D. (1982). "Bistability and self-sustained intensity oscillations in a ring laser with optical backscattering--an example of a system far from thermal equilibrium," Acta Phys. Pol. A **A61**, 547 (1982).

Kukhtarev, N. V. (1982). "Optical bistability with distributed feedback in cholesteric liquid-crystals," Ukrainskii Fizisheskii Zh. **27**, 291.

Kukhtarev, N. V., Pavlik, B. D., and Semenets, T. I. (1984). "Optical hysteresis in wavefront reversal by traveling holographic gratings," Sov. J. Quantum Electron. **14**, 282. (3.10.3)

Kukhtarev, N. V., and Semenets, T. I. (1981a). "Diffraction bleaching and bistability in wavefront reversal in resonant media," Sov. J. Quantum Electron. **11**, 1216. [3.10.3]

Kukhtarev, N. V., and Semenets, T. I. (1981b). "Optical bistability and hysteresis in four-wave inversion of a wavefront in ferroelectrics," Sov. Phys. Tech. Phys. **26**, 1159 (1981). [3.10.3]

Kukhtarev, N. V., and Starkov, V. N. (1981). "Optical bistability in optical phase conjugation in electrooptic crystals with a diffuse nonlinearity," Sov. Tech. Phys. Lett. **7**, 296. [3.10.3]

Kurnosov, V. D., Luk'yanov, V. N., Sapozhnikov, S. M., Semenov, A. T., and Tambiev, Yu. A. (1976). "Investigation of transient stimulated emission from optically coupled lasers," Sov. J. Quantum Electron. **6**, 982. (1.3)

Kurnosov, V. D., and Semenov, A. T. (1974). "Injection laser with two strongly coupled resonators," Sov. J. Quantum Electron. **4**, 17.

Kus, M., and Wodkiewicz, K. (1982). "Amplitude fluctuations in optical bistability," P. 202 in Proceedings of the XI National Conference on Coherent and Nonlinear Optics, Yerevan, USSR. (6.5)

Kus, M., Wodkiewicz, K., and Gallas, J. A. C. (1983). "Absorptive optical bistability with laser-amplitude fluctuations," Phys. Rev. A **28**, 314. (6.5)

Kuwata, M., Mita, T., and Nagasawa, N. (1981). "Renormalization of polaritons due to the formation of excitonic molecules in CuCl; an experimental aspect," Solid State Commun. **40**, 911. [7.3]

Kwong, S.-K., Cronin-Golomb, M., and Yariv, A. (1984). "Optical bistability and hysteresis with a photorefractive self-pumped phase conjugate mirror," Appl. Phys. Lett. **45**, 1016. [3.10.3]

Lamb, Jr., W. E. (1964). "Theory of an optical maser," Phys. Rev. **134**, A1429. (Received 13 January 1964). [1.3]

Lange, W., Mitschke, F., Deserno, R., and Mlynek, J. (1984). "Magnetically induced relaxation oscillations in a sodium-filled Fabry-Perot resonator," J. Opt. Soc. Am. B **1**, 468. [6.4]

Lasher, G. J. (1964). "Analysis of a proposed bistable injection laser," Solid-State Electron. **7**, 707. (Received 30 March 1964). [1.3]

Lattes, A., Haus, H. A., Leonberger, F. J., and Ippen, E. P. (1983). "An ultrafast all-optical gate," IEEE J. Quantum Electron. **QE-19**, 1718. [3.10.5]

Lau, K. Y., Harder, Ch., and Yariv, A. (1982a). "Dynamical switching characteristics of a bistable injection laser," Appl. Phys. Lett. **40**, 198. [1.3, 5.6]

Lau, K. Y., Harder, Ch., and Yariv, A. (1982b). "Interaction of a bistable injection laser with an external optical cavity," Appl. Phys. Lett. **40**, 369. [1.3]

Lau, K. Y., and Yariv, A. (1984). "Bistable optical electrical/microwave switching using optically coupled monolithically integrated GaAlAs translasers," Appl. Phys. Lett. **45**, 719. (1.3)

Laulicht, I. (1981). "Self-oscillations in a hybrid bistable optical-device," Opt. Quantum Electron. **13**, 295. (6.1)

Laval, S. (1984). "Optical bistability," Rev. Phys. Appl. (France) **19**, 77. [1.2]

Lavallard, P., Bichard, R., and Benoît à la Guillaume, C. (1977). "Saturation of optical-absorption by a nonequilibrium electron-hole plasma in InSb crystals," Phys. Rev. B **16**, 2804.

Lavallard, P., Duong, P. H., and Itoh, T. (1983). "Nonlinear optical transmission of CdSe," Physica **117B** and **118B**, 401. [2.10]

Lawandy, N. M. (1984). "Coherent resonance line broadening effects in absorptive optical bistability," Opt. Commun. **51**, 294.

Lawandy, N. M., Plant, D. V., and MacFarlane, D. L. (1985). "Optical bistability in a dissipative thermally expanding etalon," IEEE J. Quantum Electron. **QE-21**, 108.

Lawandy, N. M., and Rabinovich, W. S. (1984). "Absorptive bistability in a three-level system interacting with two fields," IEEE J. Quantum Electron. **QE-20**, 458. [3.10.2]

Lawandy, N. M., and Willner, R. (1983). "Saturable interferometer theory of optical bistability," IEEE J. Quantum Electron. **QE-19**, 1499. (2.4.1)

Lax, M. (1963). "Formal theory of quantum fluctuations from a driven state," Phys. Rev. **129**, 2342. [2.8]

Lax, M., Batteh, J. H., and Agrawal, G. P. (1981). "Channeling of intense electromagnetic beams," J. Appl. Phys. **52**, 109. [G]

Lebedev, S. A., Kopranekov, V. N., and Goncharova, L. S. (1981). "Optical switch using organic materials," Sov. Phys. Tech. Phys. **26**, 1013. (3.5.2, 7.3)

LeBerre, M., Ressayre, E., and Tallet, A. (1982). "Self-focusing and spatial ringing of intense cw light propagating through a strong absorbing medium," Phys. Rev. A **25**, 1604. [5.2]

LeBerre, M., Ressayre, E., and Tallet, A. (1984). "Analytical description of diffraction-free-encoding optical bistability and instabilities," talk WM1, Optical Society of America 1984 Annual Meeting, Technical Program. (3.9.2, 6.3)

LeBerre, M., Ressayre, E., and Tallet, A. (1984). "Resonant self-focusing of a cw intense light beam," Phys. Rev. A **29**, 2669. (3.9.2)

LeBerre, M., Ressayre, E., Tallet, A., Gibbs, H. M., Peyghambarian, N., Rushford, M. C., and Tai, K. (1983). "The physics of the formation of spatial rings," p. 571 in Proceedings of the International Conference on Lasers '83 (IEEE, New York).

LeBerre, M., Ressayre, E., Tallet, A., and Mattar, F. P. (1985). "Quasi-trapping of Gaussian beams in two-level systems," J. Opt. Soc. Am. B, to be published. [3.9.1]

LeBerre, M., Ressayre, E., Tallet, A., Tai, K., Gibbs, H. M., Rushford, M. C., and Peyghambarian, N. (1984). "Continuous-wave off-resonance rings and continuous-wave on-resonance enhancement," J. Opt. Soc. Am. B **1**, 591. [3.9.2, 5.2]

LeBerre, M., Ressayre, E., Tallet, A., Tai, K., Hopf, F. A., Gibbs, H. M., and Moloney, J. V. (1984). "Optical bistability and instabilities via diffraction-free-encoding and a single feedback mirror," postdeadline paper PD-A5, International Quantum Electronics Conference '84, Anaheim, CA. [2.9, 3.9.2, 6.3]

LeBerre, M., Ressayre, E., Tallet, A., Tai, K., and Gibbs, M. (1985). "Transverse optical bistabilities and instabilities," IEEE J. Quantum Electron., to be published.

Lee, C. H., ed. (1984). Picosecond Optoelectronic Devices (Academic Press, New York). [5.8]

Lee, C. H., Mak, P. S., and DeFonzo, A. P. (1980). "Optical control of millimeter-wave propagation in dielectric waveguides," IEEE J. Quantum Electron. **QE-16**, 277. [5.3]

Lee, C. S., and Osada, H. (1985). "Observation of optical bistability due to resonator configuration transition," Opt. Lett. **10**, 232. (1.3)

Lee, S. H. (1976). "Nonlinear optical processing," p. 255 in Optical Information Processing, Yu. E. Nesterikhin, G. W. Stroke, and W. E. Kock, eds. (Plenum Press, New York). [7.4]

Lee, S. H., ed. (1981a). Optical Information Processing (Springer-Verlag, Berlin). [7.4]

Lee, S. H. (1981b). "Nonlinear optical processing," p. 261 in S. H. Lee, Optical Information Processing (Springer-Verlag, Berlin). [7.4]

Levin, K. H., and Tang, C. L. (1979). "Optical switching and bistability in tunable lasers," Appl. Phys. Lett. **34**, 376. [1.3]

Levy, R., Bigot, J. Y., Hönerlage, B., Tomasini, F., and Grun, J. B. (1983). "Optical bistability due to biexcitons in CuCl," Solid State Commun. **48**, 705. [3.8.4]

Li, C. F. (1979). "Hybrid optical bistable device using a nonlinear feedback," Acta Physica Sinica **2**, 99. (In Chinese) (4.5)

Li, C. F. (1980). "Bistable optical devices with nonlinear feedback," Wuli (China) **9**, 99.

Li, C. F. (1981). "Nonlinear feedback bistable optical devices," Chin. Phys. **1**, 149. [4.5]

Li, C. F. (1982). "Hybrid bistable optical devices," in Conference on Lasers and Electro-Optics, Technical Digest (IEEE, New York). [4.5]

Li, C. F., and Ji, J. R. (1981). "Demonstration of optical bistability using a Michelson interferometer," IEEE J. Quantum Electron. **QE-17**, 1317. [4.5]

Li, Chun-Fei, and Xu, Jing-Chun (1982). "Scanning Fabry-Perot interferometer bistable optical device," Chin. Phys. **2**, 794. (4.5)

Li, F. L. (1982). "Study on two-photon optical multistability," Laser J. **9**, 496. (3.10.1)

Li, F. L. (1983a). "Two-photon optical multistability," IEEE J. Quantum Electron. **QE-19**, 887. [3.10.1, 3.10.2]

Li, F. L. (1983b). "Two-photon optical multistability theory," Chin. J. Phys. **3**, 577. [3.10.1, 3.10.2, 3.10.3]

Li, F. L., Hermann, J. A., and Elgin, J. N. (1982). "Effects of two-photon optical bistability upon phase conjugation," Opt. Commun. **40**, 446. [3.10.3] (3.10.1)

Li, Y., Lin, S., Jin, F., and Wang, Y. (1984). "Bistability of optical guided-wave beam deflector," Chin. J. Phys. **4**, 418. (4.5)

Li, Y.-G., and Zhang, H.-J. (1983). "An analysis on characteristics of the hybrid optical bistability," Acta Phys. Sin. **32**, 301. Chin. J. Phys. (USA).

Li, Yong-Gui, Zhang, Hong-Jun, Yang, Jun-Hui, and Gao, Gun-Xiu (1982). "Bistable characteristics of hybrid nonlinear Fabry-Perot optical bistability," Acta Phys. Sin. **31**, 446.

Liao, C., and Stegeman, G. I. (1984). "Nonlinear prism coupler," Appl. Phys. Lett. **44**, 164. [3.10.5]

Liao, P. F., and Bjorklund, G. C. (1976). "Polarization rotation induced by resonant two-photon dispersion," Phys. Rev. Lett. **36**, 584. [5.3.]

Liao, P. F., and Bjorklund, G. C. (1977). "Polarization rotation effects in atomic sodium vapor," Phys. Rev. A **15**, 2009. [5.3]

Lin, D. Y., and Dienes, A. (1981). "Passive mode locking and optical bistability of an argon-ion laser," Appl. Opt. **20**, 3825. (1.3)

Lisitsa, M. P., Stolyarenko, A. V., and Terekhova, S. F. (1983). "On excitation generation in laser pulse radiation," Ukr. Fiz. Zh. **28**, 835.

Lisitsyn, V. N., and Chebotaev, V. P. (1968). "Hysteresis and 'hard' excitation in a gas laser," JETP Lett. **7**, 1; Sov. Phys. JETP **27**, 227. [1.3]

Litfin, G. (1982). "Optically bistability due to orientation bleaching in color center crystals," Appl. Phys. B **28**, 134. [3.10.7]

Lohmann, A. W. (1977). "Suggestions for hybrid image processing," Opt. Commun. **22**, 165. [7.4]

Lohmann, A. W. (1983). "Chances for optical computing," p. 1 in 10th International Optical Computing Conference (IEEE Computer Society Press, Silver Spring, Maryland). [7.4]

Löwenau, J. P., Schmitt-Rink, S., and Haug, H. (1982). "Many-body theory of optical bistability in semiconductors," Phys. Rev. Lett. **49**, 1511. [3.6, 3.7. 7.3]

Loy, M. M. T. (1975). "A dispersive modulator," Appl. Phys. Lett. **26**, 98. [5.2]

Lubkin, G. B. (1981). "Period-doubling route to chaos shows universality," Physics Today, **17**, March. [6.3]

Lugiato, L. A. (1978). "Instabilities in the laser with injected signal and laser-phase transition analogy," Lett. Nuovo Cimento **23**, 609. [1.3.]

Lugiato, L. A. (1979). "Spectrum of transmitted and fluorescent light in absorptive optical bistability," Nuovo Cimento **50B**, 89. [2.8]

Lugiato, L. A. (1980a). "On the fluorescent spectrum in absorptive optical bistability," Lett. Nuovo Cimento **29**, 375. [2.8]

Lugiato, L. A. (1980b). "Self pulsing in dispersive optical bistability," Opt. Commun. **33**, 108. [6.2]

Lugiato, L. A. (1981). "Many-mode quantum statistical theory of optical bistability," Z. Phys. B **41**, 85. (6.5)

Lugiato, L. A. (1983a). "Optical bistability," Contemporary Physics **24**, 333. [2.7]

Lugiato, L. A. (1983b). "Dressed-mode theory of optical bistability," Phys. Rev. A **28**, 483. [6.2]

Lugiato, L. A. (1984). "Theory of optical bistability," p. 69 in Progress in Optics, Vol. XXI, E. Wolf, ed. (North-Holland, Amsterdam). [1.2, 2]

Lugiato, L. A., Asquini, M. L., and Narducci, L. M. (1982). "The relation between the Bonifacio-Lugiato and Ikeda instabilities in optical bistability," Opt. Commun. **41**, 450. [6.3]

Lugiato, L. A., Benza, V., Farina, J. D., Narducci, L. M., and Seibert, E. J. (1981). "Self-pulsing in absorptive optical bistability," J. Opt. Soc. Am. **71**, 1633.

Lugiato, L. A., Benza, V., and Narducci, M. (1982). "Optical bistability, self-pulsing and higher-order bifurcations," p. 120 in Evolution of Order and Chaos, H. Haken, ed. (Springer, Berlin).

Lugiato, L. A., Benza, V., Narducci, L. M., and Farina, J. D. (1981). "Optical bistability, instabilities and higher order bifurcations," Opt. Commun. **39**, 405. [6.2]

Lugiato, L. A., Benza, V., Narducci, L. M., and Farina, J. D. (1982). "Higher order instabilities in absorptive optical bistability," Appl. Phys. B **28**, 164. [6.2]

Lugiato, L. A., Benza, V., Narducci, L. M., and Farina, J. D. (1983). "Dressed mode description of optical bistability. III. Transient behaviour and higher order bifurcations," Z. Phys. B **49**, 351. [6.2]

Lugiato, L. A., and Bonifacio, R. (1978). "Mean field theory of optical bistability and resonance fluorescence," p. 85 in Coherence in Spectroscopy and Modern Physics, F. T. Arecchi, R. Bonifacio, and M. O. Scully, eds. (Plenum Press, New York). [2.4, 2.7]

Lugiato, L. A., Casagrande, F., and Pizzuto, L. (1982). "Fluctuation theory in quantum-optical systems," Phys. Rev. A **26**, 3438. [6.5]

Lugiato, L. A., Farina, J. D., and Narducci, L. M. (1980). "Quantum-statistical treatment of the transient in absorptive optical bistability--local relaxation," Phys. Rev. A **22**, 253. [6.5]

Lugiato, L. A., and Horowicz, R. J. (1985). "Thermal and parametric noise in dispersive optical bistability. I: General formalism," J. Opt. Soc. Am. B **2**(5).

Lugiato, L. A., Horowicz, R. J., Strini, G., and Narducci, L. M. (1984a). "Instabilities in passive and active optical systems with a Gaussian transverse intensity profile," Phys. Rev. A **30**, 1366. [6.4]

Lugiato, L. A., Horowicz, R. J., Strini, G., and Narducci, L. M. (1984b). "Transverse effects and noise in optical instabilities," Proc. Royal Society Meeting on Optical Bistability, Dynamic Nonlinearity and Photonic Logic, London. [6.4]

Lugiato, L. A., Mandel, P., Dembinski, S., and Kossakowski, A. (1978). "Semiclassical and quantum theories of bistability in lasers containing saturable absorbers," Phys. Rev. A **18**, 238. (1.3)

Lugiato, L. A., Mandel, P., and Narducci, L. M. (1983). "Adiabatic elimination of nonlinear dynamical systems," Phys. Rev. A **29**, 1438.

Lugiato, L. A., and Milani, M. (1983). "Transverse effects and self-pulsing in optical bistability," Z. Phys. B **50**, 171. [6.2]

Lugiato, L. A. and Milani, M. (1984). "Effects of the radial variation of the electric field on some instabilities in optical bistability and lasers," p. 397 in Optical Bistability 2, C. M. Bowden, H. M. Gibbs, and S. L. McCall, eds. (Plenum Press, New York). [2.9, 6.2]

Lugiato, L. A., Milani, M., and Meystre, P. (1982). "Analytical description of anomalous switching in dispersive optical bistability," Opt. Commun. **40**, 307. [5.7]

Lugiato, L. A., and Narducci, L. M. (1984). "Self-pulsing, breathing and chaos in optical bistability and in a laser with injected signal," p. 941 in Coherence and Quantum Optics V. L. Mandel and E. Wolf, eds. (Plenum Press, New York). [6.4] (6.3)

Lugiato, L. A., Narducci, L. M., Bandy, D. K., and Pennise, C. A. (1982a). "Self-pulsing and chaos in a mean-field model of optical bistability," Opt. Commun. **43**, 281. [6.4]

Lugiato, L. A., Narducci, L. M., Bandy, D. K., and Pennise, C. A. (1982b). "Self-pulsing and chaos in a mean-field model of optical bistability," J. Opt. Soc. Am. **72**, 1753. [6.4]

Lugiato, L. A., and Strini, G. (1982a). "On the squeezing obtainable in parametric oscillators and bistable absorption," Opt. Commun. **41**, 67. [6.5]

Lugiato, L. A., and Strini, G. (1982b). "On nonclassical effects in two-photon optical bistability and two-photon laser," Opt. Commun. **41**, 374. [6.5]

Lugiato, L. A., and Strini, G. (1982c). "Fluctuations in multiphoton optical bistability," Opt. Commun. **41**, 447. [6.5]

Lugovoi, V. N. (1969). "On stimulated Raman emission in an optical resonator," Sov. Phys. JETP **29**, 374. [3.10.7]

Lugovoi, V. N. (1977). "On the theory of a nonlinear optical resonator," Opt. Acta **24**, 743. [3.10.7]

Lugovoi, V. N. (1978). "Inverse Cerenkov effect and new phenomena in non-linear optics," Opt. Acta **25**, 337. [3.10.7]

Lugovoi, V. N. (1979a). "Bistability and hysteresis phenomena in an optical parametric oscillator," Phys. Stat. Solidi (b) **94**, 79. [3.10.7]

Lugovoi, V. N. (1979b). "On the mutual quenching of modes in stimulated Raman emission," Phys. Lett. **69A**, 402. [3.10.7]

Lugovoi, V. N. (1979c). "Nonlinear optical resonator containing a medium with a quadratic susceptibility," Sov. Tech. Phys. Lett. **5**, 202. [3.10.7]

Lugovoi, V. N. (1979d). "Nonlinear optical resonators (excited by external radiation) (review)," Sov. J. Quantum Electron. **9**, 1207. [1.2, 3.10.7]

Lugovoi, V. N. (1979e). " The mutual mode quenching in optical second harmonic generation," Sov. Phys. JETP **49**, 985. [3.10.7]

Lugovoi, V. N. (1981). "Bistability and self-pulsing in Cerenkov-type optical parametric interactions," IEEE J. Quantum Electron. **QE-17**, 384. [3.10.7]

Lugovoi, V. N. (1983). "Nonlinear optical phenomena in a multiple interferometer," J. Opt. Soc. Am. **73**, 473. [3.10.7]

Lukosz, W. (1984). "Grating and prism input couplers as bistable integrated optical elements," talk WT5, Optical Society of America 1984 Annual Meeting, Technical Program. (3.10.5)

Lukosz, W. (1984). "New radiation pressure-induced mechano-optical bistability," talk WT6, Optical Society of America 1984 Annual Meeting, Technical Program. (3.10.7)

Lytel, R. (1984). "Optical multistability in collinear degenerate four-wave mixing," J. Opt. Soc. Am. B **1**, 91. [3.10.3]

Macgillivray, W. R. (1983). "Optical bistability and non absorption resonance in atomic sodium," p. 41 in Laser Physics, J. D. Harvey and D. F. Walls, eds. (Springer-Verlag Berlin). [3.2.3]

MacKenzie, H. A., Al-Attar, H. A., and Wherrett, B. S. (1983). "Nonlinear third order susceptibility measurement in indium antimonide by nondegenerate four wave mixing, paper 72 in Programme of Sixth National Quantum Electronics Conference, University of Sussex, 19-22 September. (7.3)

Maeda, M., and Abraham, N. B. (1982). "Measurements of mode-splitting self-pulsing in a single-mode, Fabry-Perot laser," Phys. Rev. A **26**, 3395. [6.4]

Maier, A. A. (1982). "Optical transistors and bistable devices utilizing nonlinear transmission of light in systems with unidirectional coupled waves," Sov. J. Quantum Electron. **12**, 1490. [3.10.5]

Maier, A. A. (1984). "Self-switching of light in a directional coupler," Sov. J. Quantum Electron. **11**, 101. (3.10.5)

Mandel, P., and Erneux, T. (1982a). "Switching times in absorptive optical bistability," Opt. Commun. **42**, 362. [5.1]

Mandel, P., and Erneux, T. (1982b). "Dynamics of nascent hysteresis in optical bistability," Opt. Commun. **44**, 55. [5.1]

Mandel, P., and Erneux, T. (1984). "Stationary, harmonic and pulsed operations of an optically bistable laser with saturable absorber," Phys. Rev. A. **30**, 1893. (6.3,6.4)

Mandel, P., and Fang, F.-K. (1981). "Lasers with two-photon saturable absorbers," Phys. Lett. A **83**, 59. (1.3)

Mandel, P., and Kapral, R. (1983). "Subharmonic and chaotic bifurcation structure in optical bistability," Opt. Commun. **47**, 151. [6.3]

Mandel, L., Roy, R., and Singh, S. (1981). "Optical bistability effects in a dye ring laser," p. 127 in Optical Bistability, C. M. Bowden, M. Ciftan, and H. R. Robl, eds. (Plenum Press, New York). [3.10.2, 6.5] (1.3)

Mandel, L., and Wolf, E., eds. (1984). Coherence and Quantum Optics V (Plenum Press, New York). [6.3, 6.5]

Marburger, J., Allen, S. D., Garmire, E., Levenson, M., and Winful, H. (1978). "Optical bistability without mirrors," J. Opt. Soc. Am. **68**, 642. [4.3, 4.5, 5.6]

Marburger, J. H., and Felber, F. S. (1978). "Theory of a lossless nonlinear Fabry-Perot interferometer," Phys. Rev. A **17**, 335. [2.3, 2.9, E]

Marburger, J., and Garmire, E. (1979). "Bistable optical devices: an overview," p. 395 in Fiber Optics. Advances in Research and Development, B. Bendow and S. S. Mitra, eds. (Plenum Press, New York). [1.2]

Markarov, V. A., Pershin, S. M., Podshivalov, A. A., Zadoian, R. S., and Zheludev, N. I. (1983). "Amplitude and polarization instability of picosecond light pulses exciting a semiconductor optical resonator," Opt. Lett. **8**, 557. (5.2)

Martin-Pereda, J. A. (1984). "Optical nonlinearities between glass and liquid crystal," talk M-4, European Conference on Optics, Optical Systems and Applications. [5.7]

Martin-Pereda, J. A., and Muriel, M. A. (1982). "Instabilities in hybrid liquid crystal optical bistable devices," Appl. Phys. B **28**, 138. [5.7]

Martin-Pereda, J. A., and Muriel, M. A. (1983a). "Laser pulse shaping with liquid crystals," Proc. SPIE **369**, 148. (5.2)

Martin-Pereda, J. A., and Muriel, M. A. (1983b). "Optical memory with twisted nematic liquid crystal (TNLC) devices," Proc. SPIE **369**, 284. (4.5)

Martin-Pereda, J. A., and Muriel, M. A. (1983c). "Instabilities in hybrid optical bistable devices with nonlinear feedback," p. 80 in Conference on Lasers and Electro-Optics, Technical Digest (IEEE, New York). [6.1]

Martin-Pereda, J. A., and Muriel, M. A. (1984). "Light level to electrical frequency conversion with hybrid optical bistable devices," talk J-5, European Conference on Optical Systems and Applications.

Martin-Pereda, J. A., Muriel, M. A., and Otón, J. M. (1981). "Biestabilidad óptica: un nuevo camino para el láser," Mundo Electronico **110**, 139.

Martinot, P., Koster, A., Laval, S., and Carvalho, W. (1982). "Optical bistability from surface plasmon excitation," Appl. Phys. B **29**, 172. [3.10.5]

Martinot, P., Laval, S., and Koster, A. (1984). "Optical bistability from surface plasmon excitation through a nonlinear medium," J. Phys. (Paris) **45**, 597. [3.10.5]

März, R., Schmitt-Rink, S., and Haug, H. (1980). "Optical properties of dense exciton-biexciton systems," Z. Phys. B **40**, 9. [3.8]

Masumoto, Y., and Shionoya, S. (1981). "High intensity effect on two-photon resonant, coherent Raman scattering via excitonic molecules in CuCl and CuBr," Solid State Commun. **38**, 865. [3.8]

Mathew, J. G. H., Craig, D., and Miller, A. (1985). "Optical switching in a CdHgTe étalon at room temperature," Appl. Phys. Lett. **46**, 128. (3.8.6)

Mathew, J. G. H., Kar, A. J., Heckenberg, N. R., and Galbraith, I. (1985). "Time resolved self-defocusing in InSb at room temperature," IEEE J. Quantum Electron. **QE-21**, 94. (2.9, 7.3)

Mathew, J. G. H., Kar, A. K., Heckenberg, N. R., and Smith, S. D. (1983). "Large optical nonlinearities in InSb at room temperature," paper 68 in Programme of Sixth National Quantum Electronics Conference, University of Sussex, 19-22 September. (3.7,7.3)

Mattar, F. P. (1978). "Transverse effect associated with the nonlinear interaction of counter-propagating coherent optical pulses in resonant media," p. 475

in Proceedings of the Eleventh Congress of the International Commission for Optics ICO-11 (Instituto de Optica "Daza de Valdes," CSIC, Sociedad Espanola de Optica, Madrid, Spain). (2.5.2)

Mattar, F. P. (1981). "Effects of propagation, transverse mode coupling and diffraction on nonlinear light pulse evolution," p. 503 in Optical Bistability, C. M. Bowden, M. Ciftan, and H. R. Robl, eds. (Plenum Press, New York). [2.9]

Mattar, F. P., and Eberly, J. H. (1979). "An efficient algorithm for the study of nonlinear resonant propagation of two concomitant optical pulses interacting with a three-level atomic system," p. 61 in Laser-Induced Processes in Molecules, K. L. Kompa and S. D. Smith, eds. (Springer-Verlag, Berlin). [5.3]

Mattar, F. P., Moretti, G., and Franceour, R. E. (1981). "Transient counter-beam propagation in a nonlinear Fabry-Perot cavity," Computer Phys. Commun. **23**, 1. [2.9]

Mattar, F. P., and Newstein, M. C. (1977). "Transverse effects associated with propagation of coherent optical pulses in resonant media," IEEE J. Quantum Electron. **13**, 507.

Mattar, F. P., and Newstein, M. C. (1980). "Adaptive stretching and rezoning as effective computational techniques for 2-level paraxial Maxwell-Bloch simulation," Computer Phys. Commun. **20**, 139.

May, R. M. (1976). "Simple mathematical models with very complicated dynamics," Nature **261**, 459. [6.3]

Mayer, G., and Gires, F. (1964), "Action d'une onde lumineuse intense sur l'indice de réfraction des liquides," Acad. Sci. Paris **258**, 2039. [5.8]

Mayr, M., Risken, H., and Vollmer, H. D. (1981). "Periodic and chaotic breathing of pulses in a ring laser," Opt. Commun. **36**, 480. (6.3)

McCall, S. L. (1974). "Instabilities in continuous-wave light propagation in absorbing media," Phys. Rev. A **9**, 1515. [1.4, 2.5, 3.2, 6.2, C]

McCall, S. L. (1978). "Instability and regenerative pulsation phenomena in Fabry-Perot nonlinear optic media devices," Appl. Phys. Lett. **32**, 284. [1.2, 4.5, 6.1]

McCall, S. L., and Gibbs, H. M. (1978). "Optical bistability via thermal effects in a glass filter," J. Opt. Soc. Am. **68**, 1378. [1.2, 3.5]

McCall, S. L., and Gibbs, H. M. (1979). "Optical bistable devices: recent experimental results," International Conference on Lasers '79, Orlando, Florida, December 17-21.

McCall, S. L., and Gibbs, H. M. (1980). "Standing wave effects in optical bistability," Opt. Commun. **33**, 335. [2.5]

McCall, S. L., and Gibbs, H. M. (1981). "Conditions and limitations in intrinsic optical bistability," p. 1 in Optical Bistability, C. M. Bowden, M. Ciftan, and H. R. Robl, eds. (Plenum Press, New York). [7.2]

McCall, S. L., and Gibbs, H. M. (1982a). "Optical bistability," p. 93 in Dissipative Systems in Quantum Optics, R. Bonifacio, ed. (Springer-Verlag, Berlin). [A]

McCall, S. L., and Gibbs, H. M. (1982b). "Fundamental limits of optical data processing," Applications of Opto-Electronics, La Jolla Institute, March. [7.2]

McCall, S. L., Gibbs, H. M., Churchill, G. G., and Venkatesan, T. N. C. (1975), "Optical transistor and bistability," Bull. Am. Phys. Soc. **20**, 636. [1, 1.4, 3.2, 5.3]

McCall, S. L., Gibbs, H. M., Greene, W. P., and Passner, A. (1978). "Thermal optical bistability," unpublished. [3.5]

McCall, S. L., Gibbs, H. M., Hopf, F. A., Kaplan, D. L., and Ovadia, S. (1982). "Fluctuations in optical bistability: experiments with shot noise," Appl. Phys. B **28**, 99. [6.5]

McCall, S. L., Gibbs, H. M., Kaplan, D. L., and Hopf, F. A. (1981). "Experiments with bistable optical devices with shot noise," International Conference on Lasers '81, New Orleans, talk L.4. [6.5]

McCall, S. L., Gibbs, H. M., and Venkatesan, T. N. C. (1975), "Optical transistor and bistability," J. Opt. Soc. Am. **65**, 1184. [1.4, 3.2]

McCall, S. L., and Hahn, E. L. (1969). "Self-induced transparency," Phys. Rev. **183**, 457. [2.4, C]

McCall, S. L., Ovadia, S., Gibbs, H. M., Hopf, F. A., and Kaplan, D. L. (1984). "Statistical fluctuations in optical bistability induced by shot noise," p. 187 in Optical Bistability 2, C. M. Bowden, H. M. Gibbs, and S. L. McCall, eds. (Plenum Press, New York). [6.5]

McCall, S. L., Ovadia, S., Gibbs, H. M., Hopf, F. A., and Kaplan, D. L. (1985). "Statistical fluctuations in optical bistability induced by shot noise," IEEE J. Quantum Electron., to be published. [6.5]

McCullen, J. D., Meystre, P., and Wright, E. M. (1984). "Mirror confinement and control through radiation pressure," Opt. Lett. **9**, 193. [3.10.7]

McLaughlin, D. W., Moloney, J. V., and Newell, A. C. (1983). "Solitary waves as fixed points of infinite-dimensional maps in an optical bistable ring cavity," Phys. Rev. Lett. **51**, 75. [2.9]

Meier, D., Holzner, R., Derighetti, B., and Brun, E. (1982). "Bistability and chaos in NMR systems," p. 146 in Evolution of Order and Chaos, H. Haken, ed. (Springer, Berlin). [3.10.7, 6.4]

Menigaux, L., and Carenco, A. (1981). "Monolithic integration in GaAs of an electro-optic bistable device," in 11th European Solid State Device Research Conference ESSDERC 81 and the 6th Symposium on Solid State Device Technology SSSDT 81 (European Physical Society, Petit-Lancy, Switzerland).

Meyer, R. B., and Thurston, R. N. (1983). "Discovery of dc switching of a bistable boundary layer liquid crystal display," Appl. Phys. Lett. **43**, 342.

Meystre, P. (1978). "On the use of the mean-field theory in optical bistability," Opt. Commun. **26**, 277. [2.5.2]

Meystre, P. (1979). "Comment on 'Space and time-dependent effects in optical bistability,' by A. Abraham, R. K. Bullough, and S. S. Hassan," Opt. Commun. **30**, 262.

Migus, A., Antonetti, A., Hulin, D., Mysyrowicz, A., Gibbs, H. M., Peyghambarian, N., and Jewell, J. L. (1985), "One-picosecond optical NOR gate at room temperature with a GaAs-AlGaAs multiple-quantum-well nonlinear Fabry-Perot etalon," Appl. Phys. Lett. **46**, 70. [1.2, 3.6, 5.8]

Mihalache, D., Mazilu, D., and Totia, H. (1985). "Bistable states of s-polarized nonlinear waves guided by an asymmetric three layer dielectric structure," Physica Scripta **31**. (3.10.5)

Mihalache, D., Nazmitdinov, R. G., and Fedyanin, V. K. (1984). "p-polarized nonlinear surface waves in symmetric layered structures," Physica Scripta **29**, 269. (3.10.5)

Mihalache, D., and Totia, H. (1984). "s-polarized nonlinear surface and guided waves in an asymmetric layered structure," Rev. Roum. Phys. **29**, 365. (3.10.5)

Milan, D., and Weber, M. J. (1976). "Measurement of nonlinear refractive-index coefficients using time resolved interferometry: Application to optical materials for high-power neodymium lasers," J. Appl. Phys. **47**, 2497. [7.3]

Milani, M., Bonifacio, R., and Scully, M. (1979). "A Model for bistability in a coherently driven Josephson junction," Lett. Nuovo Cimento **26**, 353.

Miller, A. (1983). "Fast optical nonlinearities in semiconductors," paper 14 in Programme of Sixth National Quantum Electronics Conference, University of Sussex, 19-22, September. (7.3)

Miller, A., Craig, D., Parry, G., Mathew, J. G. H., and Kar, A. K. (1984). "Optical bistability in $Cd_xHg_{1-x}Te$," AGARD Workshop on "Digital Optical Circuit Technology," Schliersee, West Germany, Sept. [3.8.6]

Miller, A., Hill, J. R., and Parry, G. (1982). "Low intensity nonlinear refraction in cadmium mercury telluride at 10.6 μm," Appl. Phys. B **29**, 173. (7.3)

Miller, A., Mathew, J. G. H., Craig, D., and Parry, G. (1984). "Band-gap resonant Fabry-Perot nonlinearities and two-photon-induced optical switching in CdHgTe," J. Opt. Soc. Am. B **1**, 475. (3.8.6)

Miller, A., and Miller, D. A. B. (1982). "Dynamic nonlinear optics in semiconductors," Appl. Phys. B **28**, 92. (7.3)

Miller, A., Miller, D. A. B., and Smith, S. D. (1981). "Dynamic non-linear optical processes in semiconductors," Adv. Phys. **30**, 697. [7.3]

Miller, A., and Parry, G. (1984a). "Fast and sensitive nonlinear processes: bistability in CdHgTe," Phil. Trans. Roy. Soc. London A **313**, 277. [3.8.6]

Miller, A., and Parry, G. (1984b). "Optical bistability in semiconductors with density dependent carrier lifetimes," Optics and Quantum Electron. **16**, 339. [7.3.1]

Miller, A., Parry, G., and Daley, R. (1984). "Low-power nonlinear Fabry-Perot reflection in CdHgTe at 10 μm," IEEE J. Quantum Electron. **QE-20**, 710. [3.8.6, 6.1]

Miller, A., Smith, S. D., and Wherrett, B., eds. (1984). Proceedings of the Royal Society International Conference on Optical Bistability, Dynamical Nonlinearities and Photonic Logic, London, March 21-22, Phil. Trans. Roy. Soc. London A **313**, 187-451. [1.1]

Miller, D. A. B. (1981a). "Refractive Fabry-Perot bistability with linear absorption: theory of operation and cavity optimization," IEEE J. Quantum Electron. **QE-17**, 306. [7.1]

Miller, D. A. B. (1981b). See p. 563 in Optical Bistability, C. M. Bowden, M. Ciftan, and H. R. Robl, eds. (Plenum Press, New York). [7.1]

Miller, D. A. B. (1982a). "Bistable optical devices: physics and operating characteristics," Laser Focus **18(4)**, 79. [1.2]

Miller, D. A. B. (1982b). "Optical bistability," Cesk. Cas. Fyz. Sekce A **32**, 582.

Miller, D. A. B. (1983). "Dynamic nonlinear optics in semiconductors: physics and applications," Laser Focus **19(7)**, 61. [7.3]

Miller, D. A. B. (1984a). "Multiple quantum well nonlinearities for optical processing applications," p. 120 in Conference on Lasers and Electro-Optics, Technical Digest (IEEE, New York). (7.3)

Miller, D. A. B. (1984b). "Optical bistability and differential gain resulting from absorption increasing with excitation." J. Opt. Soc. Am. B **1**, 857. [2.10]

Miller, D. A. B., Chemla, D. S., Damen, T. C., Gossard, A. C., Wiegmann, W., Wood, T. H., and Burrus, C. A. (1984a). "Novel hybrid optically bistable switch: the quantum well self-electro-optic effect device," Appl. Phys. Lett. **45**, 13. [2.10, 4.5, 7.4]

Miller, D. A. B., Chemla, D. S., Damen, T. C., Gossard, A. C., Wiegmann, W., Wood, T. H., and Burrus, C. A. (1984b). "Band-edge electroabsorption in quantum well structures: the quantum-confined Stark effect," Phys. Rev. Lett. **53**, 2173. [4.5]

Miller, D. A. B., Chemla, D. S., Damen, T. C., Gossard, A. C., Wiegmann, W., Wood, T. H., and Burrus, C. A. (1984c). "Novel linear optical modulator and an optical level shifter using a quantum-well self-electro-optic effect device," talk FG4, Optical Society of America 1984 Annual Meeting, Technical Program. (4.5)

Miller, D. A. B., Chemla, D. S., Damen, T. C., Wood, T. H., Burrus, C. A., Gossard, A. C., and Wiegmann, W. (1984). "Optical-level shifter and self-linearized optical modulator using a quantum-well self-electro-optic effect device," Opt. Lett. **9**, 567. (4.5)

Miller, D. A. B., Chemla, D. S., Damen, T. C., Wood, T. H., Burrus, C. A., Gossard, A. C., and Wiegmann, W. (1984). "Very-low-energy mirrorless optically bistable switching using a quantum-well self-electro-optic effect device," talk ThF4, Optical Society of America 1984 Annual Meeting, Technical Program. (4.5)

Miller, D. A. B., Chemla, D. S., Eilenberger, D. J., Smith, P. W., Gossard, A. C., and Tsang, W. T. (1982). "Large room-temperature optical nonlinearity in $GaAs/Ga_{1-x}Al_xAs$ multiple quantum well structures," Appl. Phys. Lett. **41**, 679. [3.6, 7.3]

Miller, D. A. B., Chemla, D. S., Eilenberger, D. J., Smith, P. W., Gossard, A. C., and Wiegmann, W. (1983). "Degenerate four-wave mixing in room-temperature GaAs/GaAlAs multiple quantum well structures," Appl. Phys. Lett. **42**, 925. [3.6, 7.3]

Miller, D. A. B., Chemla, D. S., Smith, P. W., and Gossard, A. C. (1983). "Resonant room temperature nonlinear optical processes in GaAs/GaAlAs multiple quantum well structures," p. 136 in Conference on Lasers and Electro-Optics, Technical Digest (IEEE, New York). (3.6, 7.3)

Miller, D. A. B., Chemla, D. S., Smith, P. W., Gossard, A. C., and Tsang, W. T. (1982a). "Room-temperature saturation characteristics of GaAs-GaAlAs multiple quantum well structures and of bulk GaAs," Appl. Phys. B **28**, 96. [3.6]

Miller, D. A. B., Chemla, D. S., Smith, P. W., Gossard, A. C., and Tsang, W. T. (1982b). "Large nonlinearities in room-temperature GaAs structures," J. Opt. Soc. Am. **72**, 1783. [3.6]

Miller, D. A. B., Chemla, D. S., Smith, P. W., Gossard, A. C., and Wiegmann, W. (1983). "Nonlinear optics with a diode-laser light source," Opt. Lett. **8**, 477 [3.6.2, 7.3]

Miller, D. A. B., Gossard, A. C., and Wiegmann, W. (1984). "Optical bistability from increasing absorption," J. Opt. Soc. Am. B **1**, 477 (2.10) and Opt. Lett. **9**, 162 [2.10].

Miller, D. A. B., Harrison, R. G., Johnston, A. M., Seaton, C. T., and Smith, S. D. (1980). "Degenerate four-wave mixing in InSb at 5 K," Opt. Commun. **32**, 478.

Miller, D. A. B., Mozolowski, M. H., Miller, A., and Smith, S. D. (1978). "Nonlinear optical effects in InSb with a cw CO laser," Opt. Commun. **27**, 133. [3.7, 7.3]

Miller, D. A. B., Seaton, C. T., Prise, M. E., and Smith, S. D. (1981). "Band-gap-resonant nonlinear refraction in III-V semiconductors," Phys. Rev. Lett. **47**, 197. [3.7, 7.3]

Miller, D. A. B., Seaton, C. T., and Smith, S. D. (1981). "Optical bistability and transphasor action in semiconductors," Proc. SPIE **236**, 435. [1.2]

Miller, D. A. B., and Smith, S. D. (1979). "Two beam optical signal amplification and bistability in InSb," Opt. Commun. **31**, 101 (received June 20). [3.7, 5.4]

Miller, D. A. B., Smith, S. D., and Johnston, A. (1979). "Optical bistability and signal amplification in a semiconductor crystal: applications of new low-power nonlinear effects in InSb," Appl. Phys. Lett. **35**, 658 (received June 15). [3.7]

Miller, D. A. B., Smith, S. D., and Johnston, A. (1980). "Optical bistability and 'transistor' action in a semiconductor crystal," p. 241 in Proceedings of the Fourth National Quantum Electronics Conference, B. S. Wherrett, ed. (Wiley, Edinburgh, Scotland).

Miller, D. A. B., Smith, S. D., and Seaton, C. T. (1981a). "Optical bistability and multi-stability in the semiconductor InSb," p. 115 in Optical Bistability, C. M. Bowden, M. Ciftan, and H. R. Robl, eds. (Plenum Press, New York). [3.7, 5.4]

Miller, D. A. B., Smith, S. D., and Seaton, C. T. (1981b). "Optical bistability in semiconductors," IEEE J. Quantum Electron. **QE-17**, 312. [3.7]

Miller, D. A. B., Smith, S. D., and Wherrett, B. S. (1980). "The microscopic mechanism of third-order optical nonlinearity in InSb," Opt. Commun. **35**, 221. [3.7]

Miller, D. A. B., and Wood, T. H. (1984). "Quantum well optical modulators and SEED all-optical switches," Opt. News **10**(6), 19. [4.5]

Miller, R. C., Kleinman, D. A., Tsang, W. T., and Gossard, A. C. (1981). "Observation of the excited level of excitons in GaAs quantum wells," Phys. Rev. B **24**, 1134. [3.6]

Milonni, P. W., Ackerhalt, J. R., and Galbraith, H. W. (1983). "Chaos and nonlinear optics: a chaotic Raman attractor," Phys. Rev. A **28**, 887. (6.3)

Mitschke, F., Deserno, R., Mlynek, J., and Lange, W. (1983). "Transients in all-optical bistability using transverse optical pumping: observation of critical slowing down," Opt. Commun. **46**, 135. [5.6]

Mitschke, F., and Fluggen, N. (1984). "Chaotic behaviour of a hybrid optical bistable system without a time delay," Appl. Phys. B **35**, 59. (6.4)

Mitschke, F., Mlynek, J., and Lange, W. (1983). "Observation of magnetically induced optical self-pulsing in a Fabry-Perot resonator," Phys. Rev. Lett. **50**, 1660. [6.3]

Mitsuhashi, Y. (1981). "Saturable Fabry-Perot filter for nonlinear optical image processing," Opt. Lett. **6**, 111. (7.4)

Mlynek, J., Köster, E., Kolbe, J., and Lange, W. (1984). "Phase conjugation with the use of transverse optical pumping," J. Opt. Soc. Am. B **1**, 532. [3.10.3]

Mlynek, J., Mitschke, F., Deserno, R., and Lange, W. (1982). "Optical bistability by transverse optical pumping," Appl. Phys. B **28**, 135. [3.10.2]

Mlynek, J., Mitschke, F., Deserno, R., and Lange, W. (1984). "Optical bistability from three-level atoms with the use of a coherent nonlinear mechanism," Phys. Rev. A **29**, 1297. [3.10.2]

Mollow, B. R. (1969). "Power spectrum of light scattered by two-level systems," Phys. Rev. **188**, 1969. [2.8]

Mollow, B. R. (1972). "Stimulated emission and absorption near resonance for driven systems," Phys. Rev. A **5**, 2217. [F]

Mollow, B. R. (1973). "Propagation of intense coherent light waves in resonant media," Phys. Rev. A **7**, 1319. [C]

Moloney, J. V. (1982a). "Intermittancy in the transition to optical chaos," J. Opt. Soc. Am. **72**, 1753. [6.3]

Moloney, J. V. (1982b). "Bistable behaviour of a detuned Fabry-Perot etalon with a Gaussian input spatial profile under self-focusing and defocusing conditions," Opt. Acta **29**, 1503. [2.9]

Moloney, J. V. (1984a). "Evolution of two-dimensional transverse solitary waves and solitons in an optical bistable resonator," J. Opt. Soc. Am. B **1**, 467. (2.9)

Moloney, J. V. (1984b). "Self-focusing-induced optical turbulence," Phys. Rev. Lett. **53**, 556. [6.3]

Moloney, J. V. (1984c). "Coexistent attractors and new periodic cycles in a bistable ring cavity," Opt. Commun. **48**, 435. [6.3]

Moloney, J. V., Belic, M. R., and Gibbs, H. M. (1982). "Calculation of transverse effects in optical bistability using fast Fourier transform techniques," Opt. Commun. **41**, 379. [2.9]

Moloney, J. V., and Gibbs, H. M. (1982a). "The role of diffractive coupling and self-focusing or defocusing in the dynamical switching of a bistable optical cavity," Phys. Rev. Lett. **48**, 1607. [2.9]

Moloney, J. V., and Gibbs, H. M. (1982b). "The dynamical switching of a bistable optical ring cavity with Gaussian input beam profile," Appl. Phys. B **28**, 100. [2.9]

Moloney, J. V., Gibbs, H. M., Jewell, J. L., Tai, K., Watson, E. A., Gossard, A. C., and Wiegmann, W. (1982). "Transverse effects in optical bistability: computations and observations," p. 156 in Conference on Lasers and Electro-Optics, Technical Digest (IEEE, New York). [2.9]

Moloney, J. V., Hammel, S., and Jones, C. (1984). "Global dynamics of Ikeda's plane-wave map," J. Opt. Soc. Am. B **1**, 499. [6.3]

Moloney, J. V., and Hopf, F. A. (1981). "Transverse effects and chaos in optical bistability," J. Opt. Soc. Am. **71**, 1634. [6.3]

Moloney, J. V., Hopf, F. A., and Gibbs, H. M. (1982a). "Effects of transverse beam variation on bifurcations in an intrinsic bistable ring cavity," Phys. Rev. A **25**, 3442. [6.3]

Moloney, J. V., Hopf, F. A., Gibbs, H. M. (1982b). "Novel bifurcation sequences in a ring bistable cavity with an input Gaussian spatial profile," Appl. Phys. B **28**, 98. [6.3]

Moloney, J. V., Newell, A.C., and McLaughlin, D. W. (1984). "Transverse solitary waves in a dispersive ring bistable cavity containing a saturable nonlinearity," p. 407 in Optical Bistability 2, C. M. Bowden, H. M. Gibbs, and S. L. McCall, eds. (Plenum Press, New York). [2.9]

Moloney, J. V., Sargent III, M., and Gibbs, H. M. (1982). "Transverse effects in absorptive optical bistability," Opt. Commun. **44**, 289. [2.9]

Molter-Orr, L. A., Haus, H. A., and Leonberger, F. J. (1983). "20 GHz optical waveguide sampler," IEEE J. Quantum Electron. **QE-19**, 1877. [4.5]

Moore, S. M. (1984). "The effect of a positive correlation time on optical bistability," Nuovo Cimento B **79B**, 125.

Moore, S. M. (1984). "Amplitude and phase fluctuations in a noisy cavity," Lett. Nuovo Cimento **40**, 348. (6.5)

Mourou, G., Stancampiano, C. V., Antonetti, A., and Orszag, A. (1981). "Picosecond microwave pulses generated with a subpicosecond laser-driven semiconductor switch," Appl. Phys. Lett. **39**, 295. [5.3]

Muller, R. (1980a). "Bistable operation and spatial hole burning in multimode lasers containing saturable absorbers," Z. Phys. B **40**, 257. (1.3)

Muller, R. (1980b). "Bistable operation of multimode lasers containing saturable absorbers," Opt. Commun. **33**, 326. (1.3)

Muller, R., and Glas, P. (1984). "Bistable operation of a tunable laser with and without absorber coupled to an external resonator," Z. Phys. B **54**, 271. (1.3)

Murina, T. A., and Rozanov, N. N. (1981a). "Operation of hybrid optically bistable devices," Sov. J. Quantum Electron. **11**, 711. [6.3]

Murina, T. A., and Rozanov, N. N. (1981b). "Operating regimes of bistable electrooptic devices using Fabry-Perot resonators," Sov. Tech. Phys. Lett. **7**, 374. [6.3]

Murina, T. A., and Rozanov, N. N. (1984). "Fluctuations in hybrid bistable optical devices," Sov. Phys. Tech. Phys. **29**, 100. (6.5)

Nakai, T., Ogasawara, N., and Ito, R. (1983). "Optical bistability in a semiconductor laser amplifier," Jpn. J. Appl. Phys. **22**, L310. (1.3)

Nakajima, H., and Frey, R. (1985). "Observation of bistable reflectivity of a phase-conjugated signal through intracavity nearly degenerate four-wave mixing," Phys. Rev. Lett. **54**, 1798. (3.10.3)

Nakatsuka, H., Asaka, S., Itoh, H., Ikeda, K., and Matsuoka, M. (1983). "Observation of bifurcation to chaos in an all-optical bistable system," Phys. Rev. Lett. **50**, 109. [6.3]

Nakatsuka, H., Asaka, S., Itoh, H., and Matsuoka, M. (1983). "Observation of periodic and chaotic instabilities in an all optical bistable system," J. Phys. (Paris) **44**, C2-205. [6.3]

Nakazawa, M., Tokuda, M., and Uchida, N. (1981). "Self-sustained intensity oscillation of a laser diode introduced by a delayed electrical feedback using an optical fiber and an electrical amplifier," Appl. Phys. Lett. **39**, 379. [6.3]

Narducci, L. M., Bandy, D. K., Pennise, C. A., and Lugiato, L. A. (1983). "Asymptotic relations and period doubling bifurcations in a mean-field model of optical bistability," Opt. Commun. **44**, 207. [6.4]

Narducci, L. M., Feng, D. H., Gilmore, R., and Agarwal, G. S. (1978). "Transient and steady-state behavior of collective atomic systems driven by a classical field," Phys. Rev. A **18**, 1571. [3.10.6]

Narducci, L. M., Gao, J. Y., and Yuan, J. M. (1982). "Instability and chaotic behavior in the Okada-Takizawa bistable system with a delayed feedback," J. Opt. Soc. Am. **72**, 1753. [6.3]

Narducci, L. M., Gao, J. Y., Yuan, J. M., and Squicciarini, M. F., Jr. (1983). "Route to chaos in a bistable system with delay, "J. Opt. Soc. Am **73**, 1960.

Narducci, L. M., Gilmore, R., Feng, D. H., and Agarwal, G. S. (1978). "Quantum analysis of optical bistability and spectrum of fluctuations," Opt. Lett. **2**, 88. [2.8, 5.6, 6.5]

Narducci, L. M., Gilmore, R., Feng, D. H., and Agarwal, G. S. (1979). "Absorption spectrum of optically bistable systems," Phys. Rev. A **20**, 545. [2.8]

Nathan, M. I., Marinace, J. C., Rutz, R. F., Michel, A. E., and Lasher, G. J. (1965). "GaAs injection laser with novel mode control and switching properties," J. Appl. Phys. **36**, 473. [1.3]

Nemenov, M. I., Ryvkin, B. S., and Stepanova, M. N. (1983). "Optical bistability due to the Franz-Keldysh effect with incoherent unpolarized light," Sov. Tech. Phys. Lett. **9**, 260. (4.5)

Nemenov, M. I., Ryvkin, B. S., and Stepanova, M. N. (1984). "Optical memory cell and Franz-Keldysh-effect optical amplifier," Sov. Tech. Phys. Lett. **10**, 199. (4.5)

Neuhauser, W., and Toschek, P. E. (1979). "Bistability of a mode-locked laser induced by optical nutation," Opt. Commun. **28**, 233. [1.3]

Neumann, R., Koch, S. W., Schmidt, H. E., and Haug, H. (1984). "Deterministic chaos and noise in optical bistability," Z. Phys. B **55**, 155. [6.4]

Neviere, M., Vincent, P., Paraire, N., and Reinisch, R. (1984). "Optical bistability enhanced by guided wave resonance," p. 440 in Optics in Modern Science and Technology, Conference Digest of the 13th Congress of the International Commission for Optics, Sapporo. (3.10.5)

Neyer, A., and Voges, E. (1982a). "Dynamics of electrooptic bistable devices with delayed feedback," IEEE J. Quantum Electron. **QE-18**, 2009. [6.3]

Neyer, A., and Voges, E. (1982b). "Hybrid electro-optical multivibrator operating by finite feedback delay," Electron. Lett. **18**, 59. [6.3]

Nguyen, H. X., and Zimmerman, R. (1984). "Nonlinear exciton transmission in CdS," Phys. Stat. Solidi (b) **124**, 191. (3.8.2)

Nikolaus, B., and Grischkowsky, D. (1983a). "12X pulse compression using optical fibers," Appl. Phys. Lett. **42**, 1. [5.2]

Nikolaus, B., and Grischkowsky, D. (1983b). "90-fs tunable optical pulses obtained by two-stage pulse compression," Appl. Phys. Lett. **43**, 228. [5.2]

Nishiyama, Y. (1980a). "Long-time behavior of optical bistability," Phys. Rev. A **21**, 1618. [5.1]

Nishiyama, Y. (1980b). "Optical bistability by phase switching," Phys. Rev. A **22**, 2723. [5.7]

Nomura, S., and Uchino, K. (1982). "Electrostrictive effect in $Pb(Mg_{1/3} \cdot Nb_{2/3})O_3$ type materials," Ferroelectrics **41**, 117. [4.5]

Normandin, R., and Stegeman, G. I. (1979). "Signal processing with nonlinear integrated optics," Jpn. J. Appl. Phys. **19-1**, 465. [3.10.5]

Normandin, R., and Stegeman, G. I. (1980). "Picosecond signal processing with planar, nonlinear integrated optics," Appl. Phys. Lett. **36**, 253. [3.10.5]

Novotny, G. V. (1963). "One GaAs laser is quenched by another," Electronics, July 26, p. 57. (1.2)

Odagiri, Y., Komatsu, K., and Suzuki, S. (1984). "Bistable laser-diode memory for optical time-division switching applications," p. 184 in Conference on Lasers and Electro-Optics, Technical Digest (IEEE, New York).

Ogawa, Y., Ito, H., and Inaba, H. (1981). "New bistable optical device using semiconductor laser diode," Jpn. J. Appl. Phys. **20**, L646. [4.5]

Ogawa, Y., Ito, H., and Inaba, H. (1982). "Bistable optical device using a light emitting diode," Appl. Opt. **21**, 1878.

Okada, M. (1979). "Light modulation by an electrooptic device with feedback," Opt. Commun. **28**, 300. [4.5]

Okada, M. (1980a). "Optical regenerative oscillation and monostable pulse generation in hybrid bistable optical devices," Opt. Commun. **34**, 153. [4.5, 6.1]

Okada, M. (1980b). "Reduction of the switching energy in electrooptic bistable devices," Opt. Commun. **35**, 31. [4.5]

Okada, M. (1980c). "Optical regenerative oscillation and monostable pulse generation in electro-optic bistable devices," p. 246 in Proceedings of the Sixth European Conference on Optical Communication (IEE, London, England).

Okada, M., and Takizawa, K. (1979a). "Optical multistability in the mirrorless electrooptic device with feedback," IEEE J. Quantum Electron. **QE-15**, 82. [4.5]

Okada, M., and Takizawa, K. (1979b). "Multi-functional electrooptic devices with feedback," Jpn. J. Appl. Phys. **18**, 133. [4.5]

Okada, M., and Takizawa, K. (1979c). "Electrooptic nonlinear devices with two feedback signals," IEEE J. Quantum Electron. **QE-15**, 1170. [4.5]

Okada, M., and Takizawa, K. (1980). "Optical regenerative oscillation and monostable pulse generation in electrooptic bistable devices," IEEE J. Quantum Electron. **QE-16**, 770. [4.5, 6.1]

Okada, M., and Takizawa, K. (1981a). "Instability and transient responses of an electrooptic bistable device," IEEE J. Quantum Electron. **QE-17**, 517. [4.5, 6.1]

Okada, M., and Takizawa, K. (1981b). "Instability of an electrooptic bistable device with a delayed feedback," IEEE J. Quantum Electron. **QE-17**, 2135. [6.3]

Okuda, M., and Onaka, K. (1977a). "Effect of bleaching property and optical loss on bistability of saturable optical resonators with distributed Bragg-reflectors," Jpn. J. Appl. Phys. **16**, 303. [2.5, 3.10.5]

Okuda, M., and Onaka, K. (1977b). "Bistability of optical resonator with distributed Bragg-reflectors by using the Kerr effect," Jpn. J. Appl. Phys. **16**, 769. [3.10.5]

Okuda, M., and Onaka, K. (1978). "Response of an optical resonator with distributed Bragg-reflectors to light pulses," Jpn. J. Appl. Phys. **17**, 1105. [3.10.5]

Okuda, M., Onaka, K., and Sakai, T. (1978). "Transient behavior of a saturable optical resonator with distributed Bragg-reflectors," Jpn. J. Appl. Phys. **17**, 2123. [3.10.5]

Okuda, M., Toyota, M., and Onaka, K. (1976). "Saturable optical resonators with distributed Bragg-reflectors," Opt. Commun. **19**, 138. [2.5, 3.10.5]

Okumura, K., Ogawa, Y., Ito, H., and Inaba, H. (1983a). "Self-excited oscillation and monostable operation of a bistable light emitting diode (BILED)," Trans. Inst. Electron. Commun. Eng. Jpn. Part C **J66C**, 401. (1.3)

Okumura, K., Ogawa, Y., Ito, H., and Inaba, H. (1983b). "Studies on multi-functionalization of bistable laser diodes (BILD's) and bistable light emitting diodes (BILED's)," Trans. Inst. Electron. Commun. Eng. Jpn. Part C **J66C**, 393. (1.3)

Okumura, K., Ogawa, Y., Ito, H., and Inaba, H. (1984). "Optical logic inverter and AND elements using laser or light-emitting diodes and photodetectors in a bistable system," Opt. Lett. **9**, 519. (3.10.5, 4.5)

Okumura, K., Ogawa, Y., Ito, H., and Inaba, H. (1984). "Optical bistability and monolithic logic functions based on bistable-laser light-emitting diodes," J. Opt. Soc. Am. B **1**, 465. (1.3)

Olbright, G. R., Gibbs, H. M., Macleod, H. A., Peyghambarian, N., and Tai, K., (1984). "Low-power microsecond optical bistability in ZnS interference filters," talk ThF5, Optical Society of America '84 Annual Meeting, Technical Program. [3.5.1]

Olbright, G. R., Peyghambarian, N., Gibbs, H. M., Macleod, H. A., and Van Milligen, F. (1984). "Microsecond room-temperature optical bistability and crosstalk studies in ZnS and ZnSe interference filters with visible light and milliwatt powers," Appl. Phys. Lett. **45**, 1031. [3.5.1]

Olsson, A., Erskine, D. J., Xu, Z. Y., Schremer, A., and Tang, C. L. (1982). "Nonlinear luminescence and time-resolved diffusion profiles of photoexcited carriers in semiconductors," Appl. Phys. Lett. **41**, 659. [7.2]

Olsson, N. A., and Tsang, W. T. (1983). "Wideband frequency-shift keying with a spectrally bistable cleaved-coupled-cavity semiconductor laser," Electron. Lett. **19**, 808. (1.3)

Olsson, N. A., Tsang, W. T., Logan, R. A., Kaminow, I. P., and Ko, J.-S. (1984). "Spectral bistability in coupled cavity semiconductor lasers," Appl. Phys. Lett. **44**, 375. (1.3)

Ong, H. L. (1985). "External field enhanced optical bistability in nematic liquid crystals," Appl. Phys. Lett. **46**, 822. (3.4)

Orenstein, M., Speiser, S., and Katriel, J. (1984). "An eikonal approximation for nonlinear resonators exhibiting bistability," Opt. Commun. **48**, 367. (5.1)

Orozco, L. A., Rosenberger, A. T., and Kimble, H. J. (1983). "Evolution of hysteresis in dispersive bistability with two-level atoms," J. Opt. Soc. Am. **73**, 1961. [3.2.5b]

Orozco, L. A., Rosenberger, A. T., and Kimble, H. J. (1984). "Intrinsic dynamic instability in optical bistability with two-level atoms," Phys. Rev. Lett. **53**, 2547. [3.2.5b, 6.4]

Orozco, L. A., Rosenberger, A. T., and Kimble, H. J. (1984). "Observation of the single-mode instability in optical bistability," talk WT4, Optical Society of America 1984 Annual Meeting, Technical Program. (6.4)

Otsuka, K. (1982). "Proposal and analysis of laser-amplifier-based optical bistable circuit," IEEE J. Quantum Electron. **QE-18**, 10. (1.3)

Otsuka, K., and Iwamura, H. (1983). "Theory of optical multistability and chaos in a resonant-type semiconductor laser amplifier," Phys. Rev. A **28**, 3153 and **30**, 650. [6.3].

Otsuka, K., and Kawaguchi, H. (1984a). "Period doubling bifurcations in detuned lasers with injected signals," Phys. Rev. A **29**, 2953. [6.3] (6.4)

Otsuka, K., and Kawaguchi, H. (1984b), "New route to optical turbulence in detuned lasers with a compound cavity," Phys. Rev. A **30**, 1575. [6.3]

Otsuka, K., and Kobayashi, S. (1983). "Optical bistability and nonlinear resonance in a resonant-type semiconductor laser amplifier," Electron. Lett. **19**, 262. (1.3)

Oudar, J. L., and Kuszelewicz, R. (1984). "Demonstration of optical bistability with intensity-coupled high gain lasers," Appl. Phys. Lett. **45**, 831. (1.3)

Ovadia, S. (1984). "Fundamental limitations on the operation of optical bistable devices in a ring cavity and in GaAs etalons," PhD dissertation, Optical Sciences Center, University of Arizona. [6.5, I]

Ovadia, S., Gibbs, H. M., Peyghambarian, N., Sarid, D., and Jewell, J. L. (1985). "Evidence that room-temperature GaAs optical bistability is excitonic," Opt. Eng., to be published. [3.6]

Ovadia, S., Gibbs, H. M., Peyghambarian, N., Sarid, D., Jewell, J. L., Gossard, A. C., and Wiegmann, W. (1984). "Evidence that room-temperature GaAs optical bistability is excitonic," talk WM5, Optical Society of America 1984 Annual Meeting, Technical Program. (3.6)

Ovadia, S., and Sargent III, M. (1983). "Two-photon laser and optical bistability instabilities," J. Opt. Soc. Am. **73**, 1969.

Ovadia, S., and Sargent III, M. (1984). "Two-photon laser and optical bistability sidemode instabilities," Opt. Commun. **49**, 447. [6.2]

Paoli, T. L. (1979). "Saturable absorption effects in the self-pulsing (AlGa)As junction laser," Appl. Phys. Lett. **34**, 652. (1.3)

Papanek, J., and Strba, A. (1981). "Multistable optical device based on nematic liquid crystals," Cesk. Cas. Fyz. Sekce A **31**, 146.

Papuchon, M., Schnapper, A., and Puech, C. (1979). "Development and applications of a bistable optical device from an integrated two-arm interferometer," Rev. Tech. Thomson-CSF **11**, 871. (in French) [4.5]

Paraire, N., Vincent, P., Neviere, M., and Koster, A. (1984). "Gratings in nonlinear optics and optical bistability," p. 442 in Optics in Modern Science and Technology, Conference Digest of the 13th Congress of the International Commission for Optics, Sapporo. (3.10.5)

Parigger, C., Zoller, P., and Walls, D. F. (1983). "Effect of Stark shift on two photon optical tristability," Opt. Commun. **44**, 213. [3.10]

Park, Y. K. (1984). "$Hg_{1-x}Cd_xTe$ nonlinear Fabry-Perot filters for optical limiting at 10.6 μm," p. 178 in Conference on Lasers and Electro-Optics, Technical Digest (IEEE, New York). [5.2]

Parry, G., Miller, A., Hill, J. R., and Daley, R. (1983). "All-optical modulation, gain and logic operation in a cadmium mercury telluride etalon," paper 10 in Programme of Sixth National Quantum Electronics Conference, University of Sussex, 19-22 September. (7.3)

Passner, A., Gibbs, H. M., Gossard, A. C., McCall, S. L., Venkatesan, T. N. C., and Wiegmann, W. (1980). "Ultrashort laser: lasing in MBE GaAs layer with perpendicular-to-film optical excitation and emission," IEEE J. Quantum Electron. **QE-16**, 1283 (received January 26, 1979). [3.6]

Paulus, P., Wedding, B., Gasch, A., and Jager, D. (1984). "Bistability and solitons observed in a nonlinear ring resonator," Phys. Lett. A **102A**, 89.

Pepper, D. M., and Klein, M. B. (1979). "Observation of mirrorless optical bistability and optical limiting using Stark tunable gases," IEEE J. Quantum Electron. **QE-15**, 1362. [2.10, 4.5]

Pereda, J. A. M., Muriel, M. A., and Oton, J. M. (1981). "Optical bistability--a new path for lasers," Mundo Electron. **110**, 139.

Perel'man, N. F., Kovarskii, V. A., and Averbukh, I. S. (1981). "Stark instability and cooperative threshold phenomena in double optical resonance," Sov. Phys. JETP **53**, 39. [6.4]

Petersen, P. M. (1984). "Period-doublings and chaos in electro-optic bistable devices with short delay times in the feedback," talk M-3, European Conference on Optics, Optical Systems and Applications.

Petersen, P. M., Ravn, J. N., and Skettrup, T. (1984). "Bifurcations in electrooptic bistable devices with short delay times in the feedback," IEEE J. Quantum Electron. **QE-20**, 690. [6.3]

Peyghambarian, N. (1982). "A study of the energy and momentum distribution of excitonic particles at high densities in CuCl," PhD dissertation, Indiana University. [3.8.4]

Peyghambarian, N. (1983). "Optical bistability: a novel approach to optical signal processing and communications, Proceedings of Optical Information Processing for Aerospace Applications II, NASA Langley Research Center, August 30-31, 1983; NASA publication 230. [1.2]

Peyghambarian, N., Chase, L. L., and Mysyrowicz, A. (1982). "Biexciton resonance linewidth in CuCl: collision broadening or not?" Opt. Commun. **42**, 51. [3.8]

Peyghambarian, N., and Gibbs, H. M. (1984). "Optical nonlinearity and bistability in semiconductors," Proceedings of the International School on Nonlinear Phenomena in Solids, September 21-29, 1984, Varna, Bulgaria, to be published in Nonlinear Phenomena in Solids, A. F. Vavrek, ed. (World Scientific Publishing, Singapore).

Peyghambarian, N., and Gibbs, H. M. (1985). "Optical nonlinearity, bistability, and signal processing in semiconductors," J. Opt. Soc. Am. **2**(7).

Peyghambarian, N., and Gibbs, H. M. (1985). "Optical bistability for optical signal processing and computing," Opt. Eng. **24**, 68. (7.4)

Peyghambarian, N., Gibbs, H. M., Hulin, D., Antonetti, A., Migus, A., and Jewell, J. L. (1984). "Evidence for exciton-exciton interaction in a GaAs-AlGaAs multiple quantum well from dynamic studies with subpicosecond resolution," talk ThJ7, Optical Society of America 1984 Annual Meeting, Technical Program. (3.6)

Peyghambarian, N., Gibbs, H. M., Jewell, J. L., Antonetti, A., Migus, A., Hulin, D., and Mysyrowicz, A. (1984). "Blue shift of the exciton resonance due to exciton-exciton interactions in a multiple-quantum-well structure," Phys. Rev. Lett. **53**, 2433. [3.6]

Peyghambarian, N., Gibbs, H. M., Rushford, M. C., Sarid, D. and Weinberger, D. A. (1983). "Experimental and theoretical investigations of the biexciton optical nonlinearity and bistability in CuCl," J. Opt. Soc. Am. **73**, 1863. [3.10] (3.8.4)

Peyghambarian, N., Gibbs, H. M., Rushford, M. C., and Weinberger, D. A. (1983). "Observation of biexcitonic optical bistability and optical limiting in CuCl," Phys. Rev. Lett. **51**, 1692. [3.8.4]

Peyghambarian, N., Gibbs, H. M., Rushford, M. C., Weinberger, D. A., and Sarid, D. (1983). "Optical bistability using the biexciton two-photon resonance in CuCl," J. Opt. Soc. Am. **73**, 1385. [3.8.4]

Peyghambarian, N., Sarid, D., Gibbs, H. M., Chase, L. L., and Mysyrowicz, A. (1984). "Collision broadening model for the biexciton resonance in CuCl," Opt. Commun. **49**, 125. [3.8.4]

Pineau, P. (1980). "1st steps for the optical transistor," Recherche **11**, 974. (1.2)

Pomeau, Y., and Manneville, P. (1980). "Intermittent transition to turbulence in dissipative dynamical systems," Commun. Math. Phys. **74**, 189. [6.3]

Poole, C. D. and Garmire, E. (1984a). "Nonlinear refraction at the band gap in InAs," J. Opt. Soc. Am. B **1**, 475. [3.8.5]

Poole, C. D., and Garmire, E. (1984b). "Reflective optical bistability at 3 mW in InAs," p. 122 in Conference on Lasers and Electro-Optics, Technical Digest (IEEE, New York). [3.8.5]

Poole, C. D., and Garmire, E. (1984c). "Optical bistability at the band gap in InAs," Appl. Phys. Lett. **44**, 363. [3.8.5, 7.3.2]

Poole, C. D., and Garmire, E. (1984d). "Nonlinear refraction at the absorption edge in InAs," Opt. Lett. **9**, 356. [3.8.5, 7.3.1]

Poole, C. D., and Garmire, E. (1984e). "Nonlinear refraction and nonlinear absorption in InAs," p. 279 in Optical Bistability 2, C. M. Bowden, H. M. Gibbs, and S. L. McCall, eds. (Plenum Press, New York). [7.3.1]

Popescu, I. M., Podoleanu, A. G., and Sterian, P. E. (1983). "Magneto-optical bistable device," Rev. Phys. Appl. (Paris) **18**, 313. (In French) (4.5)

Popescu, I. M., Sterian, P. E., and Stefanescu, E. N. (1982). "Amplification capabilities of the opto-electronics devices realised on the basis of the optical bistability effect," Stud. Cercet. Fiz. **34**, 905.

Poston, T., Walls, D. F., and Zoller, P. (1982). "Multiple bifurcations in coherent N-photon processes," Opt. Acta **29**, 1691. (3.10.1, 3.10.2)

Prakash, J. (1982). "Photo-induced polarization and photo-induced depolarization of impurity-vacancy dipoles," Indian J. Pure Appl. Phys. **20**, 434.

Prokhorov, A. M., Semenov, O. G., Shipilov, K. F., and Shmaonov, T. A. (1979). "Nonlinear effect of powerful laser radiation field on optical cavity filled with Kerr dielectric," Bull. Acad. Sci. USSR, Phys. Ser. **43**, 120. (3.4)

Psaltis, D., and Farhat, N. (1984). "A new approach to optical information processing based on the Hopfield model," p. 24 in Optics in Modern Science and Technology, Conference Digest of the 13th Congress of the International Commission for Optics, Sapporo. [7.4]

Puls, J., and Henneberger, F. (1984). "Saturation of the band-edge absorption of CdS_xSe_{1-x} mixed crystals at room temperature," Phys. Stat. Solidi (b) **121**, K187. (3.8.2)

Puri, R. R., Bullough, R. K., and Hassan, S. S. (1982). "Non-trivial quantum model of optical bistability," Appl. Phys. B **29**, 174. (6.5)

Puri, R. R., Hildred, G. P., Hassan, S. S., and Bullough, R. K. (1983). "Strictly quantum optical bistability," paper 31 in Programme of Sixth National Quantum Electronics Conference, University of Sussex, 19-22 September.

Puri, R. R., Lawande, S. V., and Hassan, S. S. (1980). "Dispersion in the driven Dicke model," Opt. Commun. **35**, 179. [3.10.6]

Quint, D. W., Johnson, L. M., Jain, K., and Pratt, Jr., G. W. (1979). "Optical modulation of second-harmonic generation in tellurium," J. Appl. Phys. **50**, 3073. [2.9.2e, 3.10.7, 5.4]

Rautian, S. G., and Sobel'man, I. I. (1962), "Line shape and dispersion in the vicinity of an absorption band, as affected by induced transitions," Sov. Phys. JETP **14**, 328. [C]

Reid, Margaret, McNeil, K. J., and Walls, D. F. (1981). "Unified approach to multiphoton lasers and multiphoton bistability," Phys. Rev. A **24**, 2029. [6.5]

Reid, M. D., and Walls, D. F. (1983). "Quantum fluctuations in the two-photon laser," Phys. Rev. A **28**, 332. (6.5)

Reynaud, S. (1983). "Resonance fluorescence--the dressed atom approach," Ann. Phys. Fr. **8**, 315. (In French) (2.8)

Reynolds, D. C., and Collins, T. C. (1981). Excitons: Their Properties and Uses (Academic Press, New York). [3.6]

Risken, H., and Nummedal, K. (1968). "Self-pulsing in lasers," J. Appl. Phys. **39**, 4662. [6.2]

Robinson, A. L. (1984). "Multiple quantum wells for optical logic," Science **225**, 822. [3.6, 4.5, 7.4]

Rosenberger, A. T., Orozco, L. A., and Kimble, H. J. (1983a). "Anomalous behavior on the upper branch in optical bistability," J. Opt. Soc. Am. **73**, 1961. [3.2.5b]

Rosenberger, A. T., Orozco, L. A., and Kimble, H. J. (1983b). "Observation of absorptive bistability with 2-level atoms in a ring cavity," Phys. Rev. A **28**, 2569. [3.2.5b]

Rosenberger, A. T., Orozco, L. A., and Kimble, H. J. (1984a). "Observation of the single-mode instability in optical bistability," postdeadline paper PD-A4, International Quantum Electronics Conference '84. Anaheim, CA. [6.4]

Rosenberger, A. T., Orozco, L. A., and Kimble, H. J. (1984b). "Optical bistability: steady-state and transient behavior," p. 62 in Fluctuations and Sensitivity in Nonequilibrium Systems, W. Horsthemke and D. Kondepudi, eds. (Springer-Verlag, Berlin). [3.2.5b, 5.6]

Rosenberger, A. T., Orozco, L. A., and Kimble, H. J. (1984c). "What is the best measure of cavity loss in optical bistability," talk WT3, Optical Society of America 1984 Annual Meeting, Technical Program.

Ross, W. E., Psaltis, D., and Anderson, R. H. (1983). "Two-dimensional magneto-optic spatial light modulator for signal processing," Opt. Eng. **22**, 485. [7.4]

Rossman, H., Henneberger, F., and Voigt, J. (1983). "Memory effect in the excitonic transmission of CdS," Phys. Status Solidi B **115**, K63. [2.10]

Roy, R., and Mandel, L. (1980). "Optical bistability and first order phase transition in a ring dye laser," Opt. Commun. **34**, 133. [3.10.2, 6.5]

Roy, R., and Zubairy, M. S. (1980a). "Beyond the mean-field theory of dispersive optical bistability," Phys. Rev. A **21**, 274. [2.4]

Roy, R., and Zubairy, M. S. (1980b). "Analytic solutions of the optical bistability equations for a standing wave cavity," Opt. Commun. **32**, 163. [2.5.2]

Rozanov, N. N. (1980a). "Allowance for finite beam size in hybrid bistable optical devices," Sov. Phys.-Tech. Phys. **25**, 835. [2.9, 4.2]

Rozanov, N. N. (1980b). "Time-varying operation of hybrid bistable optical devices," Sov. Tech. Phys. Lett. **6**, 77. [2.9, 6.3, H]

Rozanov, N. N. (1981a). "Hysteresis in the self-effects of coherent radiation in media with a resonant nonlinearity," Sov. Tech. Phys. Lett. **7**, 150. [3.10.6]

Rozanov, N. N. (1981b). "Hysteresis phenomena in distributed optical systems," Sov. Phys. JETP **53**, 47. [2.9, 2.10]

Rozanov, N. N. (1982). "Hysteresis and stochastic phenomena in nonlinear optical systems," Bull. Acad. Sci. USSR, Phys. Ser. **46**(10), 29. (2.9,6.3)

Rozanov, N. N. (1983). "Effects of fluctuations in wide-aperture bistable optical systems," Opt. Spectrosc. **55**, 125. (6.5)

Rozanov, N. N., and Semenov, V. E. (1980). "Hysteresis variations of the beam profile in a nonlinear Fabry-Perot interferometer," Opt. Spectrosc. (USSR) **48**, 59. [2.9]

Rozanov, N. N., and Semenov, V. E. (1981). "The kinetics of the hysteresis change of the beam profile in non-linear interferometers," Opt. Commun. **38**, 435. [2.9]

Rozanov, N. N., and Semenov, V. E. (1983). "Light-pulse tailoring with phase control," Sov. Tech. Phys. Lett. **9**, 655.

Rozanov, N. N., Semenov, V. E., and Khodova, G. V. (1982a). "Transverse structure of a field in nonlinear bistable interferometers. I. Switching waves and steady-state profiles," Sov. J. Quantum Electron. **12**, 193. [2.9]

Rozanov, N. N., Semenov, V. E., and Khodova, G. V. (1982b). "Transverse field structure in nonlinear bistable interferometers. II. Time-dependent regimes," Sov. J. Quantum Electron. **12**, 198. [2.9]

Ruelle, D., and Takens, F. (1971). "On the nature of turbulence," Commun. Math. Phys. **20**, 167. [6.3]

Ruschin, S., and Bauer, S. H. (1979). "Bistability, hysteresis and critical behavior of a CO_2 laser, with SF_6 intracavity as a saturable absorber," Chem. Phys. Lett. **66**, 100. [1.3]

Ruschin, S., and Bauer, S. H. (1981). "Bistability of a CO_2 laser with SF_6 intracavity as an absorber: transient effects," Appl. Phys. **24**, 45. [1.3]

Rushford, M. C., Gibbs, H. M., Jewell, J. L. Peyghambarian, N., Weinberger, D.A., and Li, C. F. (1984). "Observation of thermal optical bistability, crosstalk, regenerative pulsations and external switch-off in a simple dye-filled etalon, p. 345 in Optical Bistability 2, C. M. Bowden, H. M. Gibbs, and S. L. McCall, eds. (Plenum Press, New York). [3.5]

Rushford, M. C., Weinberger, D. A., Gibbs, H. M., Li, C. F., and Peyghambarian, N. (1983). "Room-temperature thermal optical bistability in thin-film interference filters and dye-filled etalons," Topical Meeting on Optical Bistability, Technical Digest (Optical Society of America, Washington). [3.5.1]

Ryvkin, B. S. (1979), "Oscillations in the current and the light intensity in laser irradiation of a plane-parallel photoconducting plate," Sov. Tech. Phys. Lett. **5**, 239. [4.5]

Ryvkin, B. S. (1981). "Falling current-voltage characteristic and optical bistability of a resonator photocell in the Franz-Keldysh effect," Sov. Phys. Semicond. **15**, 796.

Ryvkin, B. S., and Stepanova, M. N. (1982). "Bistable optical characteristics of a resonator photocell with two-step optical transitions," Sov. Tech. Phys. Lett. **8**, 413. [4.5]

Sandle, W. J. (1980). "Experiments on optical bistability," p. 225 in Laser Physics, D. F. Walls, and J. D. Harvey, eds. (Academic Press, Sydney). [3.2]

Sandle, W. J., Ballagh, R. J., and Gallagher, A. (1981). "Optical bistability experiments and mean field theories," p. 93 in Optical Bistability, C. M. Bowden, M. Ciftan, and H. R. Robl, eds. (Plenum Press, New York). [2.9]

Sandle, W. J., and Gallagher, A. (1980). "Observation of dispersive and absorptive optical bistability in a two-level system," J. Opt. Soc. Am. **70**, 656. (3.2.5)

Sandle, W. J., and Gallagher, A. (1981). "Optical bistability by an atomic vapor in a focusing Fabry-Perot cavity," Phys. Rev. A **24**, 2017; reported at an unofficial session on optical bistability at the Fourth International Conference on Laser Spectroscopy, Rottach-Egern, June 1979. [3.2]

Sargent III, M. (1977). "Laser theory," in Frontiers in Laser Spectroscopy, Vol. 1, R. Balian, S. Haroche, and S. Liberman, eds. (North-Holland, Amsterdam). [C]

Sargent III, M. (1978). "Spectroscopic techniques based on Lamb's laser theory," Phys. Rep. **43C**, 223. [F]

Sargent III, M. (1980). "Standing-wave optical bistability and instability," Sov. J. Quantum Electron. **10**, 1247. [6.2]

Sargent III, M., Holm, D. A., and Zubairy, M. S. (1985). "Quantum theory of multiwave mixing. I. General formalism," Phys. Rev. A, to be published. [F]

Sargent III, M. Scully, M. O., and Lamb, Jr., W. E. (1974). Laser Physics (Addison-Wesley, London). [1.3]

Sargent III, M., Zubairy, M. S., and DeMartini, F. (1983). "Quantum theory of laser and optical bistability instabilities," Opt. Lett. **8**, 76. [2.8, 6.2, 6.4]

Sarid, D. (1981a). "The nonlinear propagation constant of a surface plasmon," Appl. Phys. Lett. **39**, 889. [3.10.5]

Sarid, D. (1981b). "Analysis of bistability in a ring-channel waveguide," Opt. Lett. **6**, 552. [3.10.5]

Sarid, D. (1981c). "Long-range surface-plasma waves on very thin metal films," Phys. Rev. Lett. **47**, 1927. [3.10.5]

Sarid, D. (1981d). "Intrinsic bistable guided-wave devices: theory and applications," Proc. SPIE **317**, 132. [3.10.5]

Sarid, D., Deck, R. T., Craig, A., Hickernell, R., Jameson, R., and Fasano, J. J. (1982). "Enhancement of optical fields by long-range surface-plasma waves," J. Opt. Soc. Am. **72**, 1720. [3.10.5]

Sarid, D., Deck, R. T., and Fasano, J. J. (1982). "Enhanced nonlinearity of the propagation constant of a long-range surface-plasma wave," J. Opt. Soc. Am. **72**, 1345. [3.10.5]

Sarid, D., and Jameson, R. S. (1984). "Optimization of the two-signal optical NAND (2-son) gate," International Conference on Lasers '84, San Francisco. (3.10.5)

Sarid, D., Jameson, R. S., and Hickernell, R. K. (1984). "Optical bistability on reflection with an InSb etalon controlled by a guided wave," Opt. Lett. **9**, 159. [3.7]

Sarid, D., and Peyghambarian, N. (1984). "Optical bistability and spatial dispersion in exciton-polariton systems," talk WM4, Optical Society of America 1984 Annual Meeting, Technical Program. (3.8.4)

Sarid, D., Peyghambarian, N., and Gibbs, H. M. (1983a). "Analysis of biexcitonic optical bistability in CuCl in the presence of collision broadening," Phys. Rev. B **28**, 1184. [3.8.4]

Sarid, D., Peyghambarian, N., and Gibbs, H. M. (1983b). "Comments on the local-field effect in the biexciton system in CuCl," J. Opt. Soc. Am. **73**, 1385. [3.10.6] (3.8.4)

Sarid, D., and Sargent III, M. (1982). "Tunable nonlinear directional coupler," J. Opt. Soc. Am. **72**, 835.

Sarid, D., and Stegeman, G. I. (1981). "Optimization of the effects of power dependent refractive indices in optical waveguides," J. Appl. Phys. **52**, 5439. [3.10.5]

Sasaki, A., Matsuda, K.-I., Kimura, Y., and Fujita, S. (1982). "High-current InGaAsP-InP phototransistors and some monolithic optical devices," IEEE Trans. Electron Devices **ED-29**, 1382. [4.5]

Sasaki, A., Taneya, M., Yano, H., and Fujita, S. (1984). "Optoelectronic integrated device with light amplification and optical bistability," IEEE Trans. Electron Devices **ED-31**, 805. [4.5]

Satchell, J. S., Parigger, C., and Sandle, W. J. (1983). "Self-pulsing in two-cavity optical bistability," Opt. Commun. **47**, 230. [6.1]

Savage, C. M., Carmichael, H. J., and Walls, D. F. (1982). "Optical multistability and self-oscillations in three level systems," Opt. Commun. **42**, 211. [3.10.2]

Sawchuck, A. A., and Strand, T. C. (1984). "Digital optical computing," Proc. IEEE **72**, 758. [7.4]

Schenzle, A. (1981). "Optical bistability and fluctuations, " p. 275 in LASERS '81, C. B. Collins, ed. (STS Press McLean, VA). [2.9]

Schenzle, A. (1982). "Nonlinear optical phenomena and fluctuations," p. 103 in Quantum Optics. Proceedings of the South African Summer School in Theoretical Physics, C. A. Engelbrecht, ed. (Springer-Verlag, Berlin).

Schenzle, A., and Brand, H. (1978). "Fluctuations in optical bistable systems," Opt. Commun. **27**, 485. [6.5]

Schenzle, A., and Brand, H. (1979a). "Dynamic behaviour of optical bistability," Opt. Commun. **31**, 401. [6.5]

Schenzle, A., and Brand, H. (1979b). "Multiplicative stochastic processes in statistical physics," Phys. Rev. A **20**, 1628. [6.5]

Schenzle, A., and Thel, T. (1984). "Optical bistability with finite bandwidth noise," p. 179 in Optical Bistability 2, C. M. Bowden, H. M. Gibbs, and S. L. McCall, eds. (Plenum, New York). [6.5]

Schmidt, H. E., Haug, H., and Koch, S. W. (1984). "Theoretical explanation of the absorptive optical bistability in semiconductors due to band-gap shrinkage," Appl. Phys. Lett. **44**, 787. [2.10]

Schmidt, H. E., Koch, S. W., and Haug, H. (1983). "Simulations of the dynamics of optical bistability with fluctuations," Z. Phys. B **51**, 85. (6.5)

Schmieder, G., Kempf, K., Bohnert, K., Kobbe, G., Lyssenko, V. G., Kreissl, A, and and Klingshirn, C. (1981). "Renormalization with increasing excitation of the dielectric function in II-VI semiconductors," J. Lumin. **24/25**, 613. (7.8.2, 7.3.4)

Schmitt-Rink, S., and Haug, H. (1981). "Nonlinear energy transport of radiation in dielectrics due to virtual biexciton formation," Phys. Status Solidi B **108**, 377. [3.8.4]

Schmitt-Rink, S., and Haug, H. (1982). "Comment on 'Nonlinear optical response and optical bistability due to excitonic molecules'," Phys. Status Solidi B **113**, K143. (3.8.4.)

Schmitt-Rink, S., Löwenau, J., and Haug, H. (1982). "Theory of absorption and refraction of direct-gap semiconductors with arbitrary free-carrier concentrations," Z. Phys. B **47**, 13. [7.3]

Schmitt-Rink, S., Löwenau, J., Haug, H., Bohnert, K., Kreissl, A., Kempf, K., and Klingshirn, C. (1983). "Theoretical and experimental studies of the transition from the exciton to the plasma phase," Physica **117B** and **118B**, 339. [7.3.4] (3.8.2)

Schnapper, A., Papuchon, M., and Puech, C. (1979). "Optical bistability using an integrated two arm interferometer," Opt. Commun. **29**, 364. [4.5]

Schnapper, A., Papuchon, M., and Puech, C. (1981). "Remotely controlled integrated directional coupler switch," IEEE J. Quantum Electron. **QE-17**, 332. [4.4, 5.5]

Schulz, W. E., MacGillivray, W. R., and Standage, M. C. (1983). "The interaction between single-mode laser radiation and sodium vapour in a Fabry-Perot etalon and a ring optical cavity," Opt. Commun. **45**, 67. [3.2.3]

Schutte, F.-J., Germey, K., Tiebel, R., and Worlitzer, K. (1983). "Optical bistability in multimode systems with second- and third-order dispersive nonlinearity," Opt. Acta **30**, 465.

Schwendimann, P. (1979). "Optical bistability in dispersive and absorptive media," J. Phys. A **12**, L39. (2.4.3)

Scott, J. F., Sargent III, M., and Cantrell, C. D. (1975). "Laser-phase transition analogy: application to first-order transitions," Opt. Commun. **15**, 13. [1.3]

Scully, M. O., and Lamb, Jr., W. E. (1967). "Quantum theory of an optical maser. I. General theory," Phys. Rev. **159**, 209. [F]

Sczaniecki, L. (1981). "Bistability of multi-photon lasers," p. 808 in Proceedings of the International Conference on Lasers '80, C. B. Collins, ed. (STS Press, McLean, Virginia). (1.3)

Seaton, C. T., and Smith, S. D. (1983). "Optical clarity," Electronics **56**, 8.

Seaton, C. T., Smith, S. D., Tooley, F. A. P., Prise, M. E., and Taghizadeh, M. R. (1983). "The realisation of an InSb bistable device as an optical AND gate and its use to measure carrier recombination times," Appl. Phys. Lett. **42**, 131. [5.5]

Seaton, C. T., Tooley, F. A. P., Taghizadeh, M. R., Firth, W. J., and Smith, S. D. (1983). "Optical logic operations in InSb bistable devices," p. 92 in Conference on Lasers and Electro-Optics, Technical Digest (IEEE, New York). (1.2,3.7)

Seidel, H. (1969). "Bistable optical circuit using saturable absorber within a resonant cavity," U.S. Patent 3,610,731, filed May 19, 1969, granted October 5, 1971. [1.4, 3.1, 3.2]

Seko, A. (1980). "All-optical parallel logic operation using fiber laser plate for digital image processing," Appl. Phys. Lett. **37**, 260. (7.4)

Selloni, A., Quattropani, A., Schwendimann, P., and Baltes, H. P. (1981). "Switching in optical bistability by the anharmonic oscillator model," Opt. Acta **28**, 125. (5.2, 5.6, 5.7)

Selloni, A., and Schwendimann, P. (1979). "Anharmonic oscillator model for dispersive optical bistability," Opt. Acta **26**, 1541. (2.4.3)

Sen, P. K. (1983). "Non-linear refraction in magnetized III-V-semiconductors," J. Phys. C **16**, 2603. [7.3.2]

Sengupta, U. K., Gerlach, U. H., and Collins, S. A. (1978). "Bistable optical spatial device using direct optical feedback," Opt. Lett. **3**, 199. [7.4]

Senitzky, I. R., and Genossar, J. (1981). "Cooperation in an 'optical-bistability' system," p. 353 in Optical Bistability, C. M. Bowden, M. Ciftan, and H. R. Robl, eds. (Plenum Press, New York).

Shah, J., Leheny, R. F., and Wiegmann, W. (1977). "Low-temperature absorption spectrum in GaAs in the presence of optical pumping," Phys. Rev. B **16**, 1577. (3.6)

Shank, C. V., Fork, R. L., Greene, B. I., Reinhart, F. K., and Logan, R. A. (1981). "Picosecond nonequilibrium carrier transport in GaAs," Appl. Phys. Lett. **38**, 104. [3.6]

Shank, C. V., Fork, R. L., Leheny, R. F., and Shah, J. (1979). "Dynamics of photoexcited GaAs band-edge absorption with subpicosecond resolution," Phys. Rev. Lett. **42**, 112. [3.6]

Shank, C. V., Ippen, E. P., and Shapiro, S. L., eds. (1978). Picosecond Phenomena (Springer-Verlag, Berlin). [1.2]

Sharfin, W., and Dagenais, M. (1984). "Low power optical bistability near the band gap in cadmium sulfide," International Conference on Lasers '84, San Francisco. [3.8.2]

Sharfin, W. F., and Dagenais, M. (1985). "Room-temperature optical bistability in InGaAsP/InP amplifiers and implications for passive devices," Appl. Phys. Lett. **46**, 819. (1.3)

Shen, Y. R. (1982). "Recent advances in optical bistability," Nature **299**, 779. [1.2]

Shen, Y. R. (1984). "Optical nonlinearity and bistability in liquid crystals," Phil. Trans. Roy. Soc. London A **313**, 187. [2.10]

Shen, Y. R. (1984). The Principles of Nonlinear Optics (John Wiley, New York).

Shenoy, S. R. (1983). "Analogue of optical bistability in driven Josephson-junctions," p. 238 in Lecture Notes in Physics, Vol. 184 (Springer-Verlag, Berlin). [3.10.7]

Shenoy, S. R., and Agarwal, G. S. (1980). "Radiation-induced bistability in Josephson junctions," Phys. Rev. Lett. **44**, 1525. Erratum: **45**, 401. [3.10.7]

Shenoy, S. R., and Agarwal, G. S. (1981). "Nonequilibrium phase transition in a radiation-driven Josephson junction," Phys. Rev. B **23** 1977.

Shore, K. A. (1982). "Semiconductor laser bistable operation with an adjustable trigger," Opt. Quantum Electron. **14**, 321. (1.3)

Shraiman, B., Wayne, C. E., and Martin, P. C. (1981). "A scaling theory for noisy period-doubling transitions to chaos," Phys. Rev. Lett. **46**, 935. [6.3]

Silberberg, Y., and Bar-Joseph, I. (1982). "Instabilities, self-oscillation, and chaos in a simple nonlinear optical interaction," Phys. Rev. Lett. **48**, 1541. [3.10.3, 6.4]

Silberberg, Y., and Bar-Joseph, I. (1984). "Optical instabilities in a nonlinear Kerr medium," J. Opt. Soc. Am. B **1**, 662. (6.3, 6.4)

Silberberg, Y., and Bar-Joseph, I. (1984). "The physical mechanism of optical instabilities," p. 61 in Optical Bistability 2, C. M. Bowden, H. M. Gibbs, and S. L. McCall, eds. (Plenum Press, New York). (6.3, 6.4)

Silberberg, Y., Smith, P. W., Eilenberger, D. J., and Miller, D. A. B. (1984). "Passive mode locking of diode lasers," talk ThJ4, Optical Society of America 1984 Annual Meeting, Technical Program. [6.1]

Silberberg, Y., Smith, P. W., Miller, D. A. B., Tell, B., Gossard, A. C., and Wiegmann, W. (1985). "Fast nonlinear optical response from proton-bombarded multiple quantum well structures," Appl. Phys. Lett. **46**, 701. (3.6.2)

Singh, H. B., Schmitt-Rink, S., and Haug, H. (1981). "Optical dielectric function for highly excited semiconductors in the moment-conserving approximation," Phys. Rev. B **24**, 7304. [7.3]

Singh, S., and Agarwal, G. S. (1983). "Chaos in coherent two-photon processes in a ring cavity," Opt. Commun. **47**, 73. [6.4]

Siva Kumari, A. S., and Shrivastava, K. N. (1982). "Nonequilibrium steady states of ferromagnetics illuminated by microwaves," Physica **114B**, 336. [3.10.7]

Slusher, R. E., and Gibbs, H. M. (1972). "Self-induced transparency in atomic-rubidium," Phys. Rev. A **5**, 1634. [C]

Smith, P. W. (1978). "Interferometer apparatus using electro-optic material with feedback," U.S. Patent 4,196,306; filed May 3, 1978; granted April 1, 1980. [4.2]

Smith, P. W. (1980). "Hybrid bistable optical devices," Opt. Eng. **19**, 456. [4.5]

Smith, P. W., ed. (1981). Special Issue on Optical Bistability, IEEE J. Quantum Electron. **QE-17**, 300-386. [1.2]

Smith, P. W. (1982). "On the physical limits of digital optical switching and logic elements," Bell Syst. Tech. J. **61**, 1975. [7.2]

Smith, P. W. (1983). "Physical limits for optical switching and logic," Electro/83 Professional Program Session Record 11 "Probing the Limits of Optical Processing and Transmission." [7.2]

Smith, P. W. (1984a). "All-optical switching and logic: potential and limitations," p. 184 in Conference on Lasers and Electro-Optics, Technical Digest (IEEE, New York).

Smith, P. W. (1984b). "Applications of all-optical switching and logic." Phil. Trans. Roy. Soc. London A **313**, 349. [7.2]

Smith, P. W. (1984c). "Digital optical signal processing," International Conference on Lasers '84, San Francisco.

Smith, P. W. (1984d). "Device and signal processing aspects of optical bistability," talk THF2, Optical Society of America 1984 Annual Meeting, Technical Program.

Smith, P. W., Ashkin, A., Bjorkholm, J. E., and Eilenberger, D. J. (1984). "Studies of self-focusing bistable devices using liquid suspensions of dielectric particles," Opt. Lett. **10**, 131. [3.9.1, 6.1]

Smith, P. W., Hermann, J.-P., Tomlinson, W. J., and Maloney, P. J. (1979a). "Optical bistability at a nonlinear interface," J. Opt. Soc. Am. **69**, 1421. [3.10.4]

Smith, P. W., Hermann, J.-P., Tomlinson, W. J., and Maloney, P. J. (1979b). "Optical bistability at a nonlinear interface," Appl. Phys. Lett. **35**, 846. [3.10.4]

Smith, P. W., Kaminow, I. P., Maloney, P. J., and Stulz, L. W. (1977). "Integrated electrooptic nonlinear Fabry-Perot devices," p. TUB2 in Proceedings of the Topical Meeting on Integrated and Guided Wave Optics (Optical Society of America, Washington, DC). [4.4]

Smith, P. W., Kaminow, I. P., Maloney, P. J., and Stulz, L. W. (1978). "Integrated bistable optical devices," Appl. Phys. Lett. **33**, 24. [4.4]

Smith, P. W., Kaminow, I. P., Maloney, P. J., and Stulz, L. W. (1979). "Self-contained integrated bistable optical devices," Appl. Phys. Lett. **34**, 62. [4.4]

Smith, P. W., Maloney, P. J., and Ashkin, A. (1982). "Use of a liquid suspension of dielectric spheres as an artificial Kerr medium," Opt. Lett. **7**, 347. [3.9.1, 3.10.4]

Smith, P. W., and Miller, D.A.B. (1982). "Optical bistability," Laser Focus **18**(1), 77.

Smith, P. W., and Tomlinson, W. J. (1981a). "Bistable optical devices promise subpicosecond switching," IEEE Spectrum **18**, 26. [1.2, 3.9.1, 5.8, 7.2]

Smith, P. W., and Tomlinson, W. J. (1981b). "Optical properties of nonlinear interfaces," p. 463 in Optical Bistability, C. M. Bowden, M. Ciftan, and H. R. Robl, eds. (Plenum Press, New York). (3.10.4)

Smith, P. W., and Tomlinson, W. J. (1984a). "Nonlinear optical interfaces," p. 369 in Optical Bistability 2, C. M. Bowden, H. M. Gibbs, and S. L. McCall, eds. (Plenum Press, New York). [3.10.4]

Smith, P. W., and Tomlinson, W. J. (1984b). "Nonlinear optical interfaces: switching behavior," IEEE J. Quantum Electron. **QE-20**, 30. [3.10.4]

Smith, P. W., Tomlinson, W. J., Eilenberger, D. J., and Maloney, P. J. (1981). "Measurement of electronic optical Kerr coefficients," Opt. Lett. **6**, 581. [7.3]

Smith, P. W., Tomlinson, W. J., Maloney, P. J., and Hermann, J.-P. (1980). "Optical bistability and switching at a nonlinear interface," J. Opt. Soc. Am. **70**, 657. (3.10.4)

Smith, P. W., Tomlinson, W. J., Maloney, P. J., and Hermann, J.-P. (1981). "Experimental studies of a nonlinear interface," IEEE J. Quantum Electron. **QE-17**, 340. [3.10.4]

Smith, P. W., Tomlinson, W. J., Maloney, P. J., and Kaplan, A. E. (1982). "Bistability at an electro-optic interface," Opt. Lett. **7**, 57. [3.10.4]

Smith, P. W., and Turner, E. H. (1977a). "Bistable Fabry-Perot resonator," J. Opt. Soc. Am. **67**, 250. [4.2]

Smith, P. W., and Turner, E. H. (1977b). "A bistable Fabry-Perot resonator," Appl. Phys. Lett. **30**, 280. [2.3, 4.2, 4.4]

Smith, P. W., and Turner, E. H. (1977c). "Bistable Fabry-Perot devices," IEEE J. Quantum Electron. **QE-13**, 42. [4.2]

Smith, P. W., and Turner, E. H. (1978). "Nonlinear optical apparatus using Fabry-Perot resonators," U.S. Patent 4,221,472; filed August 28, 1978; granted September 9, 1980. [4.5]

Smith, P. W., Turner, E. H., and Maloney, P. J. (1978). "Electro-optic nonlinear Fabry-Perot devices," IEEE J. Quantum Electron. **QE-14**, 207. [1.2, 3.10.5, 4.5, 5.4]

Smith, P. W., Turner, E. H., and Mumford, B. B. (1978). "Nonlinear electro-optic Fabry-Perot devices using reflected-light feedback," Opt. Lett. **2**, 55. [3.5, 4.5]

Smith, S. D. (1980). "Optical bistability, towards the all-optical computer," Europhys. News **11**, 3.

Smith, S. D. (1981a). "Optical bistability in semiconductors," p. 307 in Lasers and Applications, W. O. N. Guimaraes, C.-T. Lin, and A. Mooradian, eds. (Springer-Verlag, Berlin). [1.2, 3.7]

Smith, S. D. (1981b). "Optically bistable devices," in Colloquium on 'Advances in Optical and Quantum Electronics' (IEE, London).

Smith, S. D. (1982). "Giant nonlinearities, optical bistability and the optical transistor in narrow-gap semiconductors," p. 113 in Physics of Narrow Gap Semiconductors. Proceedings of the 4th International Conference, E. Gornik, H. Heinrich, and L. Palmetshofer, eds. (Springer-Verlag, Berlin). [1.2] (7.3)

Smith, S. D. (1984). "Nonlinear optics in semiconductors and applications to bistable switches, logic gates, and other optical circuit elements," p. 258 in Conference on Lasers and Electro-Optics, Technical Digest (IEEE, New York). (1.2, 3.7)

Smith, S. D. (1984). "Progress for optically bistable switching elements in optical computing," talk E-4, 1984 European Conference on Optics, Optical Systems and Applications.

Smith, S. D., and Abraham, A. (1983). "Optical bistability" in F. T. Arecchi, F. Strumia and H. Walther, eds., Advances in Laser Spectroscopy (Plenum Press, New York).

Smith, S. D., Mathew, J. G. H., Taghizadeh, M. R., Walker, A. C., Wherrett, B. S., and Hendry, A. (1984). "Room temperature visible wavelength optical bistability in ZnSe interference filters," Opt. Commun. **51**, 357. [3.5]

Smith, S. D., and Miller, D. A. B. (1980). "Computing at the speed of light," New Scientist, February 21, p. 554. [1.2]

Smith, S. D., and Miller, D. A. B. (1980). "Infrared optical bistability and transphasor action," p. TH20 in Infrared Lasers. Topical Meeting on Infrared Lasers (Optical Society of America, Washington, D.C.)

Smith, S. D., and Miller, D. A. B. (1981). "Optical bistability and transphasor action using semiconductor materials," p. 597 in Proceedings of the International Conference on the Physics of Semiconductors, Kyoto, Japan, S. Tanaka and Y. Toyozawa, eds. (J. Phys. Soc. Jpn. **49**, Suppl. A). [3.7]

Smith, S.D., Miller, D.A.B., and Wherrett, B. S. (1980). "Ultra-fast bistable switching and optical transistor action in semiconductor non-linear interferometers," p. 425 in the Proceedings of the Second International Symposium on Ultrafast Phenomena in Spectroscopy, Oct. 30-Nov. 5, Reinhardsbrunn, G.D.R.

Smith, S. D., Seaton, C. T., and Prise, M. E. (1982). "Optically bistable switching of a semiconductor resonator with a 30 ps pulse," Appl. Phys. B **28**, 132. [5.5]

Smith, S. D., Seaton, C. T., Prise, M. E., Firth, W. J., Tooley, F., and Pidgeon, C. R. (1983). "Optically bistable switching of a semiconductor resonator with a 30 ps pulse," Physica B and C **117-118 B**, 413. (5.5)

Smith, S. D., and Walker, A. (1984). "Optical bistability strides toward realization as a new technology; field is fruitful for laser and materials research," Laser Focus, p. 18, August. [1.2]

Smith, S. D., and Walker, A. C. (1984). "Optical bistability and its application to computing," p. 448 in Optics in Modern Science and Technology, Conference Digest of the 13th Congress of the International Commission for Optics, Sapporo. (7.4)

Smith, S. D., and Wherrett, B. S. (1983). "Optical bistability in semiconductors," p. 1 in Laser Physics, J. D. Harvey and D. F. Walls, eds. (Springer-Verlag, Berlin). [1.2] (3.7, 7.3)

Snapp, R. R., Carmichael, H. J., and Schieve, W. C. (1981). "The path to 'turbulence': optical bistability and universality in the ring cavity," Opt. Commun. **40**, 68. [6.3]

Soffer, B. H., Boswell, D., Lackner, A. M., Chavel, P. Sawchuck, A.A., Strand, T. C., and Tanguay, Jr., A. R. (1980). "Optical computing with variable grating mode liquid crystal devices," Proc. SPIE **232**, 128. (7.4)

Sohler, W. (1980). "Optical bistable device as electrooptic multivibrator," Appl. Phys. Lett. **36**, 351. [4.5, 6.1]

Sohler, W. (1984). "Non-linear integrated optics," talk E-5, European Conference on Optics, Optical Systems and Applications 1984.

Soileau, M. J., Williams, W. E., and Van Stryland, E. W. (1983). "Optical power limiter with picosecond response time," IEEE J. Quantum Electron. **QE-19**, 731. [5.2]

Song, J.-W., Lee, H.-Y., Shin, S.-Y., and Kwon, Y.-S. (1983). "Periodic window, period doubling, and chaos in a liquid crystal bistable optical device," Appl. Phys. Lett. **43**, 14. [6.3]

Song, Jae-Wong, Shin, Sang-Yung, and Kwon, Young-Se (1984). "Optical bistability, regenerative oscillations, and monostable pulse generation in a liquid crystal bistable optical device," Appl. Opt. **23**, 1521.

Song, J.-W., Yoon, T.-H., and Shin, S.-Y. (1984). "Chaotic sequence and a simple model of a return map in a driven nonlinear oscillator using a light-emitting diode," J. Opt. Soc. Am. B **1**, 488. (6.3)

Spencer, M. B., and Lamb, Jr., W. E. (1972a). "Laser with a transmitting window," Phys. Rev. A **5**, 884. [1.3, 6.1]

Spencer, M. B., and Lamb, Jr., W. E. (1972b). "Theory of two coupled lasers," Phys. Rev. A **5**, 893. [1.3, 6.1]

Spiller, E. (1971). "Saturable resonator for visible light," J. Opt. Soc. Am. **61**, 669. [1.4, 2.5, 3.1, 3.2, 5.2]

Spiller, E. (1972). "Saturable optical resonator," J. Appl. Phys. **43**, 1673. [1.4, 2.5, 3.1, 3.2, 5.2]

Stallard, W. A., and Bradley, D. J. (1983). "Bistability and slow oscillation in an external cavity semiconductor laser," Appl. Phys. Lett. **42**, 858. (1.3)

Staupendahl, G., and Schindler, K. (1980). "A new optical-optical modulator (OOM) in the infrared region," p. 437 in The Proceedings of the Second International Symposium on Ultrafast Phenomena in Spectroscopy, Oct. 30-Nov. 5, Reinhardsbrunn, G.D.R.

Staupendahl, G., and Schindler, K. (1982). "Optical tuning of a tellurium cavity: optical modulation and bistability in the infrared region at room temperature," Optical and Quantum Electron. **14**, 157. [3.8, 5.2, 5.3]

Stefanescu, E. N., Popescu, I. M., and Sterian, P. E. (1984). "The semiclassical approach of the optical bistability," Rev. Roum. Phys. **29**, 183.

Stegeman, G. I. (1982a). "Guided wave approaches to optical bistability," IEEE J. Quantum Electron. **QE-18**, 1610. [3.10.5]

Stegeman, G. I. (1982b). "Comparison of guided wave approaches to optical bistability," Appl. Phys. Lett. **41**, 214. [3.10.5]

Stegeman, G. I. (1982c). "Comparison of guided-wave approaches to optical bistability," J. Opt. Soc. Am. **72**, 1769. [3.10.5]

Stegeman, G. I. (1982d). "Guided wave approaches to optical bistability," IEEE Trans. Microwave Theory Tech. **MMT-30**, 1598. [3.10.5]

Stegeman, G. I. (1983). "High-speed signal processing with nonlinear integrated optics," J. Opt. Commun. **4**, 20. [3.10.5]

Stegeman, G. I., Burke, J. J., and Hall, D. G. (1982). "Nonlinear optics of long range surface plasmons," Appl. Phys. Lett. **41**, 906. [3.10.5]

Stegeman, G. I., and Normandin, R. (1981). "Picosecond signal processing with planar nonlinear integrated optics," Proc. SPIE **317**, 291. [3.10.5]

Stegeman, G. I., and Seaton, C. T. (1984). "Nonlinear surface plasmons guided by thin metal films," Opt. Lett. **9**, 235. (3.10.5)

Stegeman, G. I., Seaton, C. T., Chilwell, J., and Smith, S. D. (1984). "Nonlinear waves guided by thin films," Appl. Phys. Lett. **44**, 830. (3.10.5)

Stegeman, G. I., Seaton, C. T., Shoemaker, R. L., and Valera, J. D. (1984). "Nonlinear guided waves," talk FO2, Optical Society of America 1984 Annual Meeting, Technical Program. (3.10.5)

Stegeman, G. I., Seaton, C. T., and Valera, J. D. (1984). "Nonlinear guided waves: new opportunities for optical bistability," International Conference on Lasers '84, San Francisco. (3.10.5)

Steÿn-Ross, M. L., and Gardiner, C. W. (1983). "Quantum theory of excitonic optical bistability," Phys. Rev. A **27**, 310. [7.3]

Stone, J., Burrus, C. A., and Campbell, J. C. (1980). "Laser action in photopumped GaAs ribbon whiskers," J. Appl. Phys. **51**, 3038 (received December 26, 1979). [3.6]

Street, R. T. (1983). "Switching on to the optical system (optical computer)," Comput. Wkly., No. 864, p. 26.

Sung, C. C., and Bowden, C. M. (1984a). "Effects of mirror reflectivity in excitonic optical bistability," J. Opt. Soc. Am. B **1**, 395. (3.8.4)

Sung, C. C., and Bowden, C. M. (1984b). "Optical bistability of a dense exciton-biexciton system: CuCl," Phys. Rev. A **29**, 1957. (3.8.4)

Sung, C. C., Bowden, C. M., Haus, J. W., and Chiu, W. K. (1984c). "Stationary properties of dispersive optical bistability in CuCl," Phys. Rev. A **30**, 1873. (3.8.4)

Sung, C. C., Haus, J. W., Bowden, C. M. (1984). "Transient behavior of excitonic semiconductor optical bistability," J. Opt. Soc. Am. B **1**, 476. (3.8.4)

Swinney, H. L. (1983). "Observations of order and chaos in nonlinear systems," Physica **7D**, 3. [6.3]

Swinney, H. L., and Gollub, J. P. (1978). "The transition to turbulence," Physics Today, August, p. 41. [6.3]

Swinney. H. L., and Gollub, J. P. (1981). Hydrodynamic Instabilities and The Transition to Turbulence (Springer-Verlag, New York). [6.3]

Szöke, A. (1972). "Bistable optical device," U.S. Patent 3,813,605, filed November 7, 1972, granted May 28, 1974. [1.4, 4.1]

Szöke, A., Daneu, V., Goldhar, J., and Kurnit, N. A. (1969). "Bistable optical element and its applications," Appl. Phys. Lett. **15**, 376 (received September 2, 1969). [1.4, 2.2, 2.4, 2.5, 3.1, 3.2, 5.2, 6.1]

Taghizadeh, M. R., Janossy, I., and Smith, S. D. (1985). "Optical bistability in bulk ZnSe due to increasing absorption and self-focusing," Appl. Phys. Lett. **46**, 331. (2.10, 3.5.1)

Tai, K. (1984). "Nonlinear optical transverse effects: cw on-resonance enhancement, cw off-resonance interference rings, crosstalk, intracavity phase switching, self-defocusing in GaAs bistable etalon, self-focusing and self-defocusing optical bistability, and instabilities," PhD dissertation, Optical Sciences Center, University of Arizona.

Tai, K., and Gibbs, H. M. (1984). "Self-defocusing and self-focusing optical bistability using thin-sample phase encoding," talk WM2, Optical Society of America 1984 Annual Meeting, Technical Program. (3.9.2)

Tai, K., Gibbs, H. M., Hopf, F. A., and Moloney, J. V. (1984). "Instabilities in self-focusing and self-defocusing thin-sample optical bistability," talk WM3, Optical Society of America 1984 Annual Meeting, Technical Program. (6.3)

Tai, K., Gibbs, H. M., Hopf, F. A., Moloney, J. V., LeBerre, M., Ressayre, E., and Tallet, A. (1984). "Instabilities in self-defocusing and self-focusing optical bistability," International Conference on Lasers '84, San Francisco. (3.9.2)

Tai, K., Gibbs, H. M., and Moloney, J. V. (1982). "Intracavity phase switching and phase-plane dynamics of a bistable optical device," Opt. Commun. **43**, 297. [5.7]

Tai, K., Gibbs, H. M., Moloney, J. V., Weinberger, D. A., Tarng, S. S., Jewell, J. L., Gossard, A. C., and Wiegmann, W. (1984). "Self-defocusing and optical crosstalk in a bistable optical etalon," p. 415 in Optical Bistability 2, C. M. Bowden, H. M. Gibbs, and S. L. McCall, eds. (Plenum Press, New York). [2.9]

Tai, K., Gibbs, H. M., Peyghambarian, N., and Mysyrowicz, A. (1985), "Mirrorless dispersive optical bistability due to cross trapping of counterpropagating beams in sodium vapor," Opt. Lett. **10**, 220. [3.9.1]

Tai, K., Gibbs, H. M., Rushford, M. C., Peyghambarian, N., Satchell, J. S., Boshier, M. G., Ballagh, R. J., Sandle, W. J., LeBerre, M., Ressayre, E., Tallet, A., Teichmann, J., Claude, Y., Mattar, F. P., and Drummond, P. D. (1984). "Observation of continuous-wave on-resonance 'self-focusing'," Opt. Lett. **9**, 243 (1984). (3.9.2)

Tai, K., Moloney, J. V., and Gibbs, H. M. (1982a). "Optical crosstalk between nearby optical bistable devices on the same etalon," Opt. Lett. **7**, 429. [2.9]

Tai, K., Moloney, J. V., and Gibbs, H. M. (1982b). "Phase-plane dynamics of optical bistability switching," J. Opt. Soc. Am. **72**, 1769. [5.7]

Talkner, P., and Hänggi, P. (1984). "Activation rates in dispersive optical bistability with amplitude and phase fluctuations: A case without detailed balance," Phys. Rev. A **29**, 768. [6.5]

Tarng, S. S. (1983). "External switching of a bistable GaAs etalon," PhD dissertation, Optical Sciences Center, University of Arizona.

Tarng, S. S., Gibbs, H. M., Jewell, J. L., Peyghambarian, N., Gossard, A. C., Venkatesan, T., and Wiegmann, W. (1983). "Optical bistability in a GaAs-AlGaAs etalon using a diode laser," J. Opt. Soc. Am. **73**, 1969. (1.2, 3.6)

Tarng, S. S., Jewell, J., Peyghambarian, N., Gossard, A. C., Venkatesan, T., and Wiegmann, W. (1984). "Use of a diode laser to observe room-temperature, low-power optical bistability in a GaAs-AlGaAs etalon," Appl. Phys. Lett. **44**, 360. [3.6]

Tarng, S. S., Tai, K., Jewell, J. L., Gibbs, H. M., Gossard, A. C., McCall, S. L., Passner, A., Venkatesan, T. N. C., and Wiegmann, W. (1982). "External off and on switching of a bistable optical device," Appl. Phys. Lett. **40**, 205. [5.5]

Tarng, S. S., Tai, K., Jewell, J. L., Gibbs, H. M., Gossard, A. C., and Wiegmann, W. (1981). "External switching of an optical bistable GaAs etalon," J. Opt. Soc. Am. **71**, 1639. [5.5]

Tarng, S. S., Tai, K., Moloney, J. V., Gibbs, H. M., Gossard, A. C., McCall, S. L., Passner, A., and Wiegmann, W. (1982). "Overshoot switching and cross talk in optical bistability," J. Opt. Soc. Am. **72**, 1770. [3.6, 5.6]

Tarucha, S., Minakata, M., and Noda, J. (1981). "Complementary optical bistable switching and triode operation using $LiNbO_3$ directional coupler," IEEE J. Quantum Electron. **QE-17**, 321. [4.4]

Taylor, H. F. (1978). "Guided wave electrooptic devices for logic and computation," Appl. Opt. **17**, 1493. [3.10.5]

Tewari, S. P. (1980). "The absorption spectra of optical bistability with dispersion," Opt. Commun. **34**, 273. [2.8]

Tewari, S. P., and Hassan, S. S. (1982). "Optical bistability and reversed switching effect in a three-photon resonant medium," Opt. Acta **29**, 1091. [3.10.2]

Tewari, S. P., and Tewari, S. P. (1979). "Dispersive bistability of a homogeneously broadened medium," Opt. Acta **26**, 145. [2.5.1]

Tewari, S. P., and Tewari, S. P. (1980). "Optical bistability of a homogeneously broadened medium with dispersion," Opt. Acta **27**, 129. (2.4.3.)

Tewari, S. P., Tewari, S. P., and Das, M. K. (1980). "Relaxation behaviour of dispersive bistability of homogeneously broadened medium," Indian J. Phys. **54B**, 151. [2.5]

Thibault, G., and Denariez-Roberge, M.-M. (1985). "Thermooptical effects observed in CdS_xSe_{1-x} semiconductor doped glasses," Canad. J. Phys. **63**, 198. [3.5.2]

Tomlinson, W. J., Gordon, J. P., Smith, P. W., and Kaplan, A. E. (1982). "Reflection of a Gaussian beam at a nonlinear interface," Appl. Opt. **21**, 2041. [3.10.4]

Tooley, F. A. P., Smith, S. D., and Seaton, C. T. (1983). "High-gain signal amplification in an InSb transphasor at 77K," Appl. Phys. Lett. **43**, 807. [5.4]

Toyozawa, Y. (1978). "Population instability and optical anomalies in high density excited system," Solid State Commun. **28**, 533. [2.10]

Toyozawa, Y. (1979). "Bistability and anomalies in absorption and resonance scattering of intense light," Solid State Commun. **32**, 13. [2.10]

Tsang, W. T., and Olsson, N. A. (1983). "Cleared coupled cavity (C^3) semiconductor lasers for optical logic and gateable mode-locked operation," post-deadline paper THU11, Conference on Lasers and Electro-Optics Technical Digest (IEEE, New York). [1.2]

Tsang, W. T., Olsson, N. A., and Logan, R. A. (1983). "Optoelectronic logic operations by cleaved-coupled-cavity semiconductor lasers," IEEE J. Quantum Electron. **QE-19**, 1621. [1.2,1.3]

Tsukada, N., and Nakayama, T. (1981). "Triode operation of an intrinsic optical bistable device," IEEE J. Quantum Electron. QE-17, 164. (5.3)

Tsukada, N., and Nakayama, T. (1982a). "Optical bistability from interference effect between one- and two-photon transition processes," Phys. Rev. A **25**, 947. [3.10.1, 5.3]

Tsukada, N., and Nakayama, T. (1982b). "Modulation of optical bistability by an additional laser beam," Phys. Rev. A **25**, 964. [5.3]

Uchinokura, K., Inushima, T., Matsura, E., and Okamoto, A. (1981). "Intrinsic optical bistability near the phase transition temperature in semiconductor ferroelectric SbSI," Ferroelectrics **38**, 901. [3.8.3]

Ulbrich, R. G. (1981). "Energy relaxation of optically excited electron-hole pairs: non-equilibrium electron and phonon distributions in semiconductors," p. 68 in Proceedings of the International Conference on Excited States and Multiresonant Nonlinear Optical Processes in Solids, Aussois, France, March 18-20 (Les Editions de Physique, Orsay, France). [3.6, 5.5, 7.3]

Umegaki, S., Inoue, H., and Yoshino, T. (1981). "Optical bistability using a magneto-optic modulator," Appl. Phys. Lett. **38**, 752. [4.5]

Umegaki, S., Tanaka, S., and Inoue, K. (1984). "Optical flip-flop with simple configuration of optical bistable devices," p. 446 in Optics in Modern Science and Technology, Conference Digest of the 13th Congress of the International Commission for Optics, Sapporo.

Vach, H., Seaton, C. T., Stegeman, G. I., and Khoo, I. C. (1984). "Observation of intensity-dependent guided waves," J. Opt. Soc. Am. B **1**, 509 [3.10.5] and Opt. Lett. **9**, 238 [3.10.5].

Valera, J. D., Seaton, C. T., Stegeman, G. I., Shoemaker, R. L., Mai, Xu, and Liao, C. (1984). "Demonstration of nonlinear prism coupling," talk FO3, Optical Society of America 1984 Annual Meeting, Technical Program; Appl. Phys. Lett. **45**, 1013. (3.10.5)

Vallée, R., Delisle, C., and Chrostowski, J. (1984). "Noise versus chaos in acousto-optic bistability," Phys. Rev. A **30**, 336. [6.3]

Van den Broeck, C., and Hänggi, P. (1984). "Activation rates for nonlinear stochastic flows driven by non-Gaussian noise," Phys. Rev. A **30**, 2730. (6.5)

van der Pol, B. (1922). "On oscillation hysteresis in a triode generator with two degrees of freedom," Phil. Mag. **43**, 700. [1.3] van der Pol, B. (1934). "The nonlinear theory of electric oscillations," Proc. Inst. Radio Engrs. **22**, 1051. [1.3]

van der Ziel, J. P., Tsang, W. T., Logan, R. A., Mikulyak, R. M., and Augustyniak, W. M. (1981). "Subpicosecond pulses from passively mode-locked GaAs buried optical guide semiconductor lasers," Appl. Phys. Lett. **39**, 525. [6.1]

Venkatesan, T. N. C. (1977). "The theory and experiment of an optical device exhibiting bistability and differential gain," PhD Dissertation, City University of New York. [3.2, 3.3, 5.1]

Venkatesan, T. N. C., Gibbs, H. M., McCall, S. L., Passner, A., Gossard, A. C., and Wiegmann, W. (1979). "Optical modulation by optical tuning of a Fabry-Perot cavity," IEEE J. Quantum Electron. **QE-15**, 8D. (5.2)

Venkatesan, T. N. C., Gibbs, H. M., McCall, S. L., Passner, A., Gossard, A. C., and Wiegmann, W. (1979). "Optical pulse tailoring and termination by self-sweeping of a Fabry-Perot cavity," Opt. Commun. **31**, 228. [5.2]

Venkatesan, T., Lemaire, P. J., Wilkens, B., Soto, L., Gossard, A. C., Jewell, J. L., Gibbs, H. M., and Tarng, S. S. (1983). "All-optical data switching in an optical fiber link using a GaAs bistable optical device," J. Opt. Soc. Am. **73**, 1969. [3.6] (1.2)

Venkatesan, T., Lemaire, P. J., Wilkens, B., Soto, L., Gossard, A. C., Weigmann, W., Jewell, J. L., Gibbs, H. M., and Tarng, S. S. (1984). "All-optical data switching in an all-optical fiber link using a GaAs optical bistable device," Opt. Lett. **9**, 297. [3.6]

Venkatesan, T. N. C., and McCall, S. L. (1977a). "Optical bistability and differential gain between 85 and 296°K in a Fabry-Perot containing ruby," Appl. Phys. Lett. **30**, 282. [1.2, 3.3]

Venkatesan, T. N. C., and McCall, S. L. (1977b). "Optical bistability and differential gain in a Fabry-Perot interferometer containing ruby," Bull. Am. Phys. Soc. **22**, 38. [3.3]

Vincent, D., and Otis, G. (1981). "Hybrid bistable devices at 10.6 μm," IEEE J. Quantum Electron. **QE-17**, 318. [4.5]

Vinokurov, G. N., and Zhulin, V. I. (1982). "The peculiarities of the hysteresis phenomena in the light reflection from the amplifying media," p. 184 in Proceedings of the XI National Conference on Coherent and Nonlinear Optics, Yerevan, USSR.

Vlad, V. I. (1981). "Opto-electronic bistable devices using twisted nematic liquid crystal cells for real-time optical information processing," Rev. Roum. Phys. **26**, 1097.

Vlad, V. I. (1982). "Liquid crystal light valve with bistability controlled by the optoelectronic feedback from a TV camera and image analyzer," Rev. Roum. Sci. Tech. Ser. Electrotech. Energ. **27**, 421.

Vlad, V. I. (1982). "Opto-electronic bistable devices in real-time optical information processing," Opt. Commun. **41**, 411. (4.5)

Vlad, V. I. (1984). "Bistable image preprocessor using a liquid crystal cell controlled by the optoelectronic feedback from a TV camera and image analyzer," p. 170 in Optics in Modern Science and Technology, Conference Digest of the 13th Congress of the International Commission for Optics, Sapporo. (7.4)

Waldrop, M. M. (1984). "Artificial intelligence in parallel," Science **225**, 608. [7.4]

Wallis, R. F., Maradudin, A. A., and Stegeman, G. I. (1983). "Surface polariton reflection and radiation at end faces," Appl. Phys. Lett. **42**, 764.

Walls, D. F. (1982). "Multistability in coherently driven nonlinear systems," Appl. Phys. B **28**, 101. [3.10.2]

Walls, D. F., Drummond, P. D., Hassan, S. S., and Carmichael, H. J. (1978). "Non-equilibrium phase transitions in cooperative atomic systems," Prog. Theor. Phys., Supplement **64**, 307. [2.8]

Walls, D. F., Drummond, P. D., and McNeil, K. J. (1981). "Bistable systems in nonlinear optics," p. 51 in Optical Bistability, C. M. Bowden, M. Ciftan, and H. R. Robl, eds. (Plenum Press, New York). [2.8, 6.5]

Walls, D. F., Kunasz, C. V., Drummond, P. D., and Zoller, P. (1981). "Bifurcations and multistability in two-photon processes," Phys. Rev. A **24**, 627. [3.10.1]

Walls, D. F., and Milburn, G. J. (1981). "Quantum fluctuations in nonlinear optics," NATO Summer School on Quantum Optics and Experimental General Relativity, Bad Windsheim, Germany. [6.5]

Walls, D. F., and Zoller, P. (1980). "A coherent nonlinear mechanism for optical bistability from three level atoms," Opt. Commun. **34**, 260. [3.10.2]

Walls, D. F., Zoller, P., and Steyn-Ross, M. L. (1981). "Optical bistability from three-level atoms," IEEE J. Quantum Electron. **QE-17**, 380. [3.10.2]

Watkins, D. E., Phipps, Jr., C. R., and Thomas, S. J. (1980). "Determination of the third-order nonlinear optical coefficients of germanium through ellipse rotation," Opt. Lett. **5**, 248. [7.3]

Watson, E. A. (1981). "Transverse effects in optical bistability and superfluorescence," M.S. dissertation, Optical Sciences Center, University of Arizona. [2.9]

Weaire, D., Wherrett, B. S., Miller, D. A. B., and Smith, S. D. (1979). "The effect of low power nonlinear refraction on laser beam propagation in InSb," Opt. Lett. **4**, 331. [3.7, 7.3]

Wedding, B., Gasch, A., and Jager, D. (1983). "Self-pulsing and transients of a Fabry-Perot interferometer with quadratic nonlinear medium," J. Appl. Phys. **54**, 4826.

Wedding, B., and Jäger, D. (1982). "Bistability observed in a Fabry-Perot interferometer with quadratic nonlinear medium," Appl. Phys. Lett. **41**, 1028. [3.10.7] (5.7, 6.1)

Wedding, D., and Jäger, D. (1984). "Optical bistability and instability in nonlinear silicon Fabry-Perot interferometers," talk M-1, European Conference on Optics, Optical Systems and Applications.

Wehner, M. F., Chrostowski, J., and Mielniczuk, W. J. (1984). "Acousto-optic bistability with fluctuations," Phys. Rev. A **29**, 3218. (6.5)

Weinberger, D. A. (1984). "Optical bistability in ZnS and ZnSe thin-film interference filters and in GaAs and CuCl etalons," Ph.D. dissertation, Optical Sciences Center, University of Arizona.

Weinberger, D. A., Gibbs, H. M., Li, C. F., and Rushford, M. C. (1982). "Room-temperature optical bistability in thin-film interference filters," J. Opt. Soc. Am. **72**, 1769. [3.5]

Weiner, J. S., Chemla, D. S., Miller, D. A. B., Pinczuk, A., Gossard, A. C., Wiegmann, W., and Burrus, C. A. (1984). "Highly anisotropic optical properties of single-quantum-well waveguides," talk FO1, Optical Society of America 1984 Annual Meeting, Technical Program. (3.10.5)

Weiss, C. O. (1982). "Optically bistable N_2O laser," Opt. Commun. **42**, 291. (1.3)

Weiss, C. O. (1983). "'Turbulence' (chaos) in a laser," p. 12 in Laser Spectroscopy VI, H. P. Weber and W. Lüthy, eds. (Springer-Verlag, Berlin). (6.3)

Weiss, C. O., and King, H. (1982). "Oscillation period doubling chaos in a laser," Opt. Commun. **44**, 59. (6.3)

West, L. C. (1983). "Construction of a general purpose digital computer based on optical logic elements," p. 44 in Conference on Lasers and Electro-Optics, Technical Digest (IEEE, New York). (7.4)

Weyer, K. G., Wiedenmann, H., Rateike, M., MacGillivray, W. R., Meystre, P., and Walther, H. (1981). "Observation of absorptive optical bistability in a Fabry-Perot cavity containing multiple atomic beams," Opt. Commun. **37**, 426. [3.2]

Wherrett, B. S. (1983a). "A comparison of theories of resonant nonlinear refraction in semiconductors," Proc. Roy. Soc. (London) A **390**, 373. [7.3.1]

Wherrett, B. S. (1983b). "Optical bistability - on reflection," paper 11 in Programme of Sixth National Quantum Electronics Conference, University of Sussex, 19-22 September.

Wherrett, B. S. (1984). "Fabry-Perot bistable cavity optimization on reflection," IEEE J. Quantum Electron. QE-20, 646. [3.5]

Wherrett, B. S., and Higgins, N. A. (1982a), "Theory of nonlinear refraction near the band edge of a semiconductor," Proc. Roy. Soc. **379**, 67. [7.3.2]

Wherrett, B. S., and Higgins, N. A. (1982b). "Landau level non-linear refraction in semiconductors," J. Phys. C: Solid State Phys. **15**, 1741. [7.3.2]

White, I. H., and Carroll, J. E. (1983a). "The bistable performance of closely coupled twin stripe lasers," in Colloquium on 'Non-linear Optical Waveguides' (IEE, London).

White, I. H., and Carroll, J. E. (1983b). "New mechanism for bistable operation of closely coupled twin stripe lasers," Electron. Lett. **19**, 337. (1.3)

White, I. H., Carroll, J. E., and Plumb, R. G. (1982). "Closely coupled twin-stripe lasers," IEE Proc. **129**, Part I, 291. (1.3)

White, I. H., Carroll, J. E., and Plumb, R. G. (1983). "Room-temperature optically triggered bistability in twin-stripe lasers," Electron. Lett. **19**, 558. (1.3)

Williams, W. E., Soileau, M. J., and Van Stryland, E. W. (1984). "Optical switching and n_2 measurements in CS_2," Opt. Commun. **50**, 256. [5.2]

Willis, C. R. (1977). "Cooperative effects and fluctuations in optical bistability," Opt. Commun. **23**, 151. [6.5]

Willis, C. R. (1978). "Derivation of a master equation for optical bistability," Opt. Commun. **26**, 62. [6.5]

Willis, C. R. (1981). "Complex order parameters in quantum optics first order phase transition analogies," p. 431 in Optical Bistability, C. M. Bowden, M. Ciftan, and H. R. Robl, eds. (Plenum Press, New York).

Willis, C. R. (1983). "Effect of laser-frequency fluctuations on optical bistability," Phys. Rev. A **27**, 375. (6.5)

Willis, C. R. (1984). "Effect of laser amplitude and phase fluctuations on optical bistability," Phys. Rev. A **29**, 774. (6.5)

Willis, C. R., and Day, J. (1979). "The effect of dispersion on optical bistability," Opt. Commun. **28**, 137. [6.5]

Winful, H. G. (1980). "Optical bistability in periodic structures and in four-wave mixing processes," PhD dissertation, University of Southern California. [3.10.3]

Winful, H. G. (1981). "Effect of end reflections on distributed-feedback bistable optical devices," IEEE J. Quantum Electron. **QE-17**, 164. (3.10.5)

Winful, H. G. (1982a). "Light-induced pitch dilation and bistable reflection in cholesteric liquid crystals," J. Opt. Soc. Am. **72**, 1720. [3.4]

Winful, H. G. (1982b). "Nonlinear reflection in cholesteric liquid crystals: mirrorless optical bistability," Phys. Rev. Lett. **49**, 1179. [3.10.6]

Winful, H. G. (1984). "New models for optical chaos," p. 89 in Conference Digest of the United States-Japan Seminar on Coherence, Incoherence, and Chaos in Quantum Electronics, Nara, M. D. Levenson, and T. Shimizu, coordinators. (2.10, 6.4)

Winful, H. G., and Cooperman, G. D. (1982). "Self-pulsing and chaos in distributed feedback bistable optical devices," Appl. Phys. Lett. **40**, 298. [6.4, 3.10.5]

Winful, H. G., and Marburger, J. H. (1980a). "Hysteresis and optical bistability in degenerate four-wave mixing," Appl. Phys. Lett. **36**, 613. [3.10.3]

Winful, H. G., and Marburger, J. H. (1980b). "Hysteresis and bistability in degenerate four-wave mixing," J. Opt. Soc. Am. **70**, 657. (3.10.3)

Winful, H. G., Marburger, J. H., and Garmire, E. (1979a). "Theory of bistability in nonlinear distributed feedback structures," Appl. Phys. Lett. **35**, 379. [3.10.5]

Winful, H. G., Marburger, J. H., and Garmire, E. (1979b). "Theory of bistability in nonlinear distributed feedback structures," J. Opt. Soc. Am. **69**, 1421. [3.10.5]

Wittke, K. J., Galanti, M., and Volk, R. (1980). "n_2 - measurements at 1.32 μm of some organic compounds usable as solvents in a saturable absorber for an atomic iodine laser," Opt. Commun. **34**, 278. [7.3]

Won, J. W. (1983). "Optical bistabilities, phase transitions, and Q-switch characteristics of an N_2O laser with a saturable NH_3 absorber," Opt. Lett. **8**, 79. (1.3)

Wright, E. M., Firth, W. J., and Galbraith, I. (1985). "Beam propagation in a medium with a diffusive Kerr-type nonlinearity," J. Opt. Soc. Am. B **2**, 383. (2.9)

Wu, F. Y., Ezekiel, S., Ducloy, M., and Mollow, B. R. (1977). "Observation of amplification in a strongly driven two-level atomic system at optical frequencies," Phys. Rev. Lett. **38**, 1077. [6.2, C]

Wynne, J. J. (1969). "Optical third-order mixing in GaAs, Ge, Si, and InAs," Phys. Rev. **178**, 1295. [7.3]

Wysin, G. M. (1980). "Optical bistability with surface plasmons," M.S. Thesis, University of Toledo, Toldeo, Ohio. [3.10.5]

Wysin, G. M., Simon, H. J., and Deck, R. T. (1981). "Optical bistability with surface plasmons," Opt. Lett. **6**, 30. [3.10.5]

Yabuzaki, T., Kitano, M., and Ogawa, T. (1984). "Optical bistability with symmetry breaking and related phenomena," p. 78 in Conference Digest of the United States-Japan Seminar on Coherence, Incoherence, and Chaos in Quantum Electronics, Nara, M. D. Levenson, and T. Shimizu, coordinators. (3.10.2)

Yabuzaki, T., Okamoto, T., Kitano, M., and Ogawa, T. (1984). "Optical bistability with symmetry breaking," Phys. Rev. A **29**, 1964. [3.10.2]

Yajima, H., Sudo, E., Yumoto, J., and Kashiwa, K. (1984). "Optical-optical guided-wave modulator," Appl. Phys. Lett. **45**, 214. [4.5]

Yao, S. S., Karaguleff, C., Gabel, A., Fortenberry, R., Seaton, C. T., and Stegeman, G. I. (1985). "Ultrafast carrier and grating lifetimes in semiconductor-doped glasses," Appl. Phys. Lett. **46**, 801. (3.10.5)

Yariv, A. (1975). Quantum Electronics, Second Edition (John Wiley, New York). [C]

Ye, Pei-Xian, and Zhang, Yi-Xiang (1982). "Theory of optical bistability in two-photon resonance media," Acta Phys. Sin. **31**, 779.

Yoshino, T., Umegaki, S., Inoue, H., and Kurosawa, K. (1982). "Light intensity stabilization using highly-efficient Faraday rotator," Jpn. J. Appl. Phys. **21**, 612. [4.5]

Yu, F.T.S. (1983). Optical Information Processing (John Wiley, New York). [7.4]

Yuan, J. M., Liu, E., and Tung, M. (1983). "Bistability and hysteresis in laser-driven polyatomic molecules," J. Chem. Phys. **79**, 5034.

Yuan, J.-M., Tung, M., Feng, D. H., and Narducci, L. M. (1983). "Instability and irregular behavior of coupled logistic equations," Phys. Rev. A **28**, 1662. (6.3)

Yuen, H. P. (1976). "Two-photon coherent states of the radiation field," Phys. Rev. A **13**, 2226. [6.5]

Yuen, S. Y. (1983a). "Third-order optical nonlinearity induced by effective mass gradient in heterostructures," Appl. Phys. Lett. **42**, 331. [7.3.1]

Yuen, S. Y. (1983b). "Degenerate four-wave mixing due to intervalence band transition in p-type mercury cadmium telluride," Appl. Phys. Lett. **43**, 479. (7.3.1)

Yuen, S. Y., and Becla, P. (1983). "Saturation of band-gap resonant optical phase conjugation in HgCdTe," Opt. Lett. **8**, 356. [7.3.5]

Yumoto, J., and Otsuka, K. (1985). "Frustrated optical instability: self-induced periodic and chaotic spatial distribution of polarization in nonlinear optical media," Phys. Rev. Lett. **54**, 1806. (6.4)

Yumoto, J., Yajima, H., Ishihara, S., Sekiguchi, Y., Yamaya, K., and Nakajima, M. (1983). "Blocking oscillation of optical bistable devices using dc drift phenomena of $LiNbO_3$ waveguide," Appl. Phys. Lett. **42**, 780. (4.5, 6.1)

Yumoto, J., Yajima, H., Ishihara, S., Shimada, J., and Nakajima, M. (1982). "A novel integrated optical bistable device," Jpn. J. Appl. Phys. Suppl., 275.

Zaidi, H. R. (1983). "Effect of dipole-dipole interaction on optical bistability," Phys. Rev. A **28**, 3119.

Zakharov, Y. P., Nikitin, V. V., Semenov, A. S., Uspenskii, A. V., and Sheheglov, V. A. (1967). "Concerning the theory of optically coupled GaAs p-n junction lasers," Sov. Phys. Solid State **8**, 1660. (1.3)

Zakman, Z. M., Collins, S. A., and Gerlach, U. H. (1979). "Multiple stable levels in a liquid crystal light valve using optical feedback," J. Opt. Soc. Am. **69**, 1422. (7.4)

Zardecki, A. (1980). "Time-dependent fluctuations in optical bistability," Phys. Rev. A **22**, 1664. [6.5]

Zardecki, A. (1981). "Fluctuations of optically bistable systems in the bad-cavity limit," Phys. Rev. A **23**, 1281. [6.5]

Zardecki, A. (1982). "Noisy Ikeda attractor," Phys. Lett. A **90A**, 274. [6.3]

Zhang, Hong-Jun, Dai, Jian-Hua, Yang, Jun-Hui, and Gao, Cun-Xiu (1981). "A bistable liquid crystal electro-optic modulator," Opt. Commun. **38**, 21.

Zheludev, N. I., Zadoyan, R. S., Kovrigin, A. I., Makarov, V. A., Pershin, S. M., and Podshivalov, A. A. (1983). "Instability of the amplitude and polarization of ultrashort light pulses exciting a semiconductor optical resonator," Sov. J. Quantum Electron. **13**, 843.

Zhong, B. A., and Sun, Y. N. (1984). "A study on optical bistable devices using Ti-diffused $LiNbO_3$ branched waveguides," p. 444 in Optics in Modern Science and Technology, Conference Digest of the 13th Congress of the International Commission for Optics, Sapporo. (4.4)

Zhu, Shi-Yao (1984). "Study of two-photon optical bistability," Acta Phys. Sin. **33**, 16.

Zhu, Z.-F., and Garmire, E. (1983a). "Optical bistability in BDN dye," p. 108 in Conference on Lasers and Electro-Optics, Technical Digest (IEEE, New York) [1.4, 3.5]

Zhu, Z.-F, and Garmire, E. (1983b). "Optical bistability in BDN dye," IEEE J. Quantum Electron. **QE-19**, 1495. [1.4,3.5]

Zhu, Z.-F., and Garmire, E. (1983c). "Optical bistability in four-level nonradiative dyes," Opt. Commun. **46**, 61. [1.4, 3.5]

GLOSSARY OF SYMBOLS

ROMAN SYMBOLS

A	Optical absorption: $P_I = AP_I + P_T$; p. 86.
a	HWHM radius of the intensity profile; p. 75.
B_{1S}	Binding energy of the ground-state exciton; Fig. 3.6-14.
C	$\equiv \dfrac{\alpha_0 L_{RC}}{4T} = \dfrac{\alpha_0 L_{FP}}{2T}$; Eq. (2.1-23) and Eq. (2.2-19).
c	Speed of light in vacuum.
c_V	Specific heat; p. 121.
D	Diffusion coefficient; p. 90.
D	$\equiv 1 + \Delta_0^2 + X$; p. 45.
$d(r)$	Excitation density; p. 85.
E	Intracavity electric field slowly varying complex amplitude; Eq. (2.1-1).
E_B	Backward electric field slowly varying complex amplitude; p. 24.
E_F	Forward electric field slowly varying complex amplitude; p. 24.
E_I	Incident electric field slowly varying complex amplitude; p. 24.
E_i	$=$ ImE. Imaginary part of complex electric field amplitude; Eq. (2.1-1).
E_R	Reflected electric field slowly varying complex amplitude; p. 24.
E_r	$=$ ReE. Real part of complex electric field amplitude; Eq. (2.1-1).
E_T	Transmitted electric field slowly varying complex amplitude; p. 24.
E_x	Resonance energy at zero wavevector for the transverse exciton; p. 155.

E_{XX}	Resonance energy at zero wavevector for the transverse biexciton; p. 155.
$\mathcal{E}$	Electric field (includes carrier wave oscillating at an optical frequency); p. 19.
F	$\equiv \kappa E(\gamma_T \gamma_L)^{-1/2}$; Eq. (2.1-5).
F	$\equiv n_0 a^2/\lambda L$. Fresnel number for a cylinder HWHM radius a and length L. This definition is 9.06 times smaller than $n_0 \pi w_0^2/\lambda L$, another often-used definition; p. 75.
$\mathscr{F}$	Finesse; $\pi\sqrt{R_\alpha}/(1 - R_\alpha)$, $R_\alpha = (R_F R_B)^{1/2}\exp(-\alpha_B L)$; [Below Eq. (D-4)].
G	$= \frac{dI_T}{dI_I}$. Differential gain; Eq. (2.2-20).
$g(\Delta\omega)$	Normalized inhomogeneous distribution; Eq. (2.1-9).
$\hbar$	Plank's constant divided by 2π.
$\mathscr{H}$	Hamiltonian (scalar, even though it appears bold).
I	$= I_C$. Intracavity intensity; p. 2.
I_I	Input intensity; p. 1.
I_s	Saturation intensity; p. 3.
I_T	Transmitted intensity; p. 1.
$I_\downarrow$	Input intensity for which I_T switches down; p. 1.
$I_\uparrow$	Input intensity for which I_T switches up; p. 1.
K	Thermal conductivity; p. 121.
k	Boltzmann's constant.
k	$n_0 2\pi/\lambda$; p. 19.
k	$\equiv R\alpha L/(1-R)$; Eq. (2.2-4).
L	Length of nonlinear medium in ring cavity; Fig. 2.1-1.
L_{FP}	Length of Fabry-Perot etalon filled with nonlinear medium; p. 27.
L_{RC}	Length of nonlinear medium in ring cavity; p. 22.

L_T	$= 2L + 2\ell$. Total length of ring cavity; p. 21.
L_W	GaAs well thickness; Fig. 3.6-14.
$2\ell + L$	Length of empty par of nonlinear ring cavity; see Fig. 2.1-1.
ℓ_D	Diffusion length; p. 85.
N	Excitation parameter, for example carrier density or temperature; p. 86.
N	Number density of atoms; p. 32.
N_a	Upper state atomic number density; p. 25.
N_b	Lower state atomic number density; p. 26.
N_c	Free-carrier density; p. 90.
N_T	Total number of atoms; p. 297.
n	Total index of refraction; p. 316.
n_0	Background (intensity-independent) refractive index; p. 30.
n_2	Nonlinear index of refraction; n_2I is the intensity-dependent change in refractive index; p. 30.
n_{EX}	Contribution of the exciton to the refractive index; p. 137.
P	Light-beam power.
$P_\uparrow$	Optical bistability minimum switch-on power; p. 90.
p	Electric dipole moment; p. 19.
$\mathcal{P}$	Atomic polarization oscillating at an optical frequency; Eq. (2.1-2).
Q	$\equiv \bar{v} + i\bar{u}$; Eq. (2.1-4).
R	$= 1 - T$. Mirror intensity reflection coefficient; p. 21; Eq. (3.8-4).
R_B	Intensity reflection coefficient of the output mirror; p. 353.
R_F	Intensity reflection coefficient of the input mirror; p. 353.
R_α	$= (R_F R_B)^{1/2}\exp(-\alpha_B L)$. Effective reflectivity in the presence of background absorption. [Below Eq. D-4; p. 353].
$S^\pm$	$\equiv \bar{u} \pm i\bar{v}$; p. 63.

Sup — Find largest value of subsequent quantity as X is varied over all positive values; p. 94.

T — Mirror intensity transmission; p. 2.

T_c — Thermal conductivity time; p. 121.

T_1 — $= 1/\gamma_L$. Homogeneous relaxation time; upper state lifetime; p. 25.

T_2 — $= 1/\gamma_T$. Transverse relaxation time.

T_2^* — $= 1/\gamma^{2*}$. Inhomogeneous broadening relaxation time.

t_R — Cavity round-trip time; p. 196.

$\mathcal{T}$ — $\equiv I_T/I_I$; Eq. (2.6-1).

u — Real (in phase) dispersive component of the polarization when the SVEA components of the electric field are a real amplitude ξ and phase ϕ; Eqs. (2.4-3), and (C-45).

$\bar{u}$ — Real part of complex polarization amplitude; Eq. (2.1-2) when the SVEA electric field is complex $E = E_r + iE_i$.

v — Imaginary (90° out-of-phase) absorptive component of the polarization when the SVEA components of the electric field are a real amplitude ξ and phase ϕ; Eqs. (2.4-3) and (C-46).

$\bar{v}$ — Imaginary complex polarization amplitude; eq. (21.-2) when the SVEA electric field is complex $E = E_r + iE_i$.

w — Atomic two-level inversion; p. 19 and Eq. (C-47).

w_0 — Waist of Gaussian electric field amplitude ; Ea. (2.9-14).

X — $\equiv |x|^2$. Normalized output intensity; Eq. (2.1-26).

X_e — Value of X for $\frac{d^2Y}{dX^2} = 0$; p. 23.

x — Fraction of some element in a semiconductor such as Al in $Al_xGa_{1-x}As$; Fig. 3.6-13.

x — $= F(L, t) = \kappa E_T(\gamma_T\gamma_L)^{-1/2}$; Eq. (2.1-18).

$x_{\downarrow L}$ — Normalized ouptut field amplitude in the lower branch just at switch-off, i.e., $y = y_\downarrow$.

$x_{\uparrow L}$ — Normalized output field amplitude in the lower branch just before switch-on; i.e., $y = y_\uparrow$; Fig. 2.4-1.

$x_{\downarrow U}$ Normalized output field amplitude in the upper branch just before switch-off; i.e., $y = y_{\downarrow}$; Fig. 2.4-1.

$x_{\uparrow U}$ Normalized output field amplitude in the upper branch just after switch-on, i.e., $y = y_{\uparrow}$; Fig. 2.4-1.

$\tilde{x}(\lambda)$ $\equiv \int_0^\infty dt\, e^{-\lambda t} x(t)$. Laplace transform; Eq. (2.8-22).

Y $\equiv |y|^2$. Normalized input intensity; Eq. (2.12-26).

y $\equiv \kappa E_I(\gamma_T \gamma_L T)^{-1/2}$; Eq. (2.1-17).

$y_{\downarrow}$ Normalized input field amplitude at which the output switches to a low-transmission state; Fig. 2.4-1.

$y_{\uparrow}$ Normalized input field amplitude at which the output switches to a high-transmission state; Fig. 2.4-1.

GREEK SYMBOLS

α $= \dfrac{\alpha_0}{1 + \Delta^2}$

α_B Background absorption (intensity-independent), but may be frequency dependent; p. 54.

α_{EX} Exciton peak absorption coefficient; p. 137.

α_0 $= 8\pi\omega N p^2/(n_0 \hbar c \gamma_T)$. Low-intensity peak absorption coefficient; Eq. (2.1-12).

β $= \beta_0 + \beta_2 I_T$. Eq. (2.3-6).

β_0 Cavity-laser phase detuning. Same as $\Delta\phi_0$; Eq. (2.3-6).

β_0' $= \beta_0 - 2\pi m$ Cavity-laser phase detuning from the closest peak; Eq. (2.1-16).

$\beta_2 I_T$ Intensity dependent phase shift; Eq. (2.3-6).

Γ_{xx} FWHM of the biexciton absorption curve; p. 155.

$\Gamma_{xx,0}$ FWHM of the biexciton absorption curve at low light intensities; p. 155.

γ $= \gamma_L = \gamma_T$ when $\gamma_L = \gamma_T$.

γ_c — $= cT/(2Ln_0)$. Cavity relaxation rate; p. 62.

γ_L — Relaxation rate of the longitudinal polarization w; Eq. (C-72).

γ_T — Relaxation rate of the transverse polarization (u and v); Eq. (C-73).

Δ — $\equiv \Delta\omega/\gamma_T$. Atom-laser detuning normalized to atomic HWHM linewidth; p. 20.

Δt — $= (L_T - L)/c$; Eq. (2.1-15).

$\Delta\beta_0$ — Change in intracavity phase shift induced by an external switching pulse; p. 234 and Figs. 5.7-5,6.

$\Delta\phi$ — Relative input phase between two input electric fields in crosstalk calculations; Fig. 2.9-8.

$\Delta\phi_0$ — Cavity-laser detuning. Same as β_0.

$\Delta\omega$ — $= \omega_a - \omega$. Atom-laser angular frequency detuning; p. 19.

δn — Light-induced change in refractive index.

$\delta(x)$ — Delta function.

$\delta\lambda_{EX}$ — HWHM width of the exciton resonance; p. 137.

ε — Complex dielectric function; p. 155.

ε_0 — Complex dielectric function at zero frequency; p. 155.

ε_∞ — Complex dielectric function at high frequency; p. 155.

ζ — Voltage increment per detected photon in the shot-noise hybrid experiment of Section 6.5.

θ — $\equiv R\beta_0'/T = 2\pi(\nu_c - \nu)L_T R/cT$; Eq. (2.1-28).

κ — $= 2p/\hbar$; p. 32.

λ — Laser wavelength in vacuum.

λ_c — Wavelength of control beam; p. 212.

λ_s — Wavelength of a signal beam whose transmission is affected by a control beam at wavelength λ_c; p. 212.

μ — Bifurcation parameter in Ikeda instability studies. For a hybrid bistable system it is proportional to the product of input intensity and feedback gain.

ν	Laser frequency; p. 2.
ν_a	Center frequency of atomic resonance; p. 19.
ν_c	Frequency of Fabry-Perot resonant cavity peak transmission; p. 21.
ν_{FP}	Frequency of Fabry-Perot cavity peak transmission; p. 2.
ξ	Real slowly varying amplitude of the electric field; Eq. (2.4-2).
ρ	Mass density; p. 121.
ρ	$= r/w_0$ or r/r_p. Normalized radial variable; p. 346.
$\tilde{\rho}$	Density matrix of a two-level atom in a frame rotating at ω; Eqs. (2.4-6).
$\sigma(X)$	$= \int_{-\infty}^{+\infty} g(\Delta) \frac{1 + i\Delta}{1 + \Delta^2 + X} d\Delta = \sigma_r + i\sigma_i$; Eq. (2.1-30).
τ	$= t - z/c$. Retarded time; p. 71.
τ_c	$= 1/\gamma_c$. Empty-cavity decay time.
τ_M	Medium response time; p. 198.
τ_p	Input intensity pulse duration.
τ_R	Response time in section 5.6; p. 272.
τ_R	$= \frac{8\pi\tau_0 n_0}{3N\lambda^2 L}$. Superfluorescence time; Eq. (2.4-18).
τ_0	Spontaneous emission time; Eq. (C-71).
$\tau_\downarrow$	Optical bistability switch-off time; p. 314.
$\tau_\uparrow$	Optical bistability switch-on time; p. 314.
ϕ	Real slowly varying phase of the electric field; Eq. (2.4-3).
$\chi^{(3)}$	Third-order nonlinear susceptibility; p. 316.
Ω_c	$= \kappa\lvert E_T\rvert/\sqrt{T} = \kappa\lvert E\rvert$. Intracavity Rabi frequency; pp. 67 and 253.
ω	Laser angular frequency; p. 19.
ω_a	Atomic resonance angular frequency; p. 19.

INDEX